101

WITHDRAWN

CAMBRIDGE PALEOBIOLOGY SERIES
2 *Patterns and Processes of Vertebrate Evolution*

Patterns and Processes of Vertebrate Evolution is a new text providing an integrated view of the forces that influence the patterns and rates of vertebrate evolution, from those at the level of living populations and species to those that resulted in the origin and radiation of the major vertebrate clades. The evolutionary roles of behavior, development, continental drift, and mass extinctions are compared with the importance of variation and natural selection that were emphasized by Darwin. The book is extensively illustrated, showing major transitions from fish to amphibians, dinosaurs to birds, and land mammals to whales. No work since Simpson's *Major Features of Evolution* has attempted such a broad study of the patterns and forces of evolutionary change.

Undergraduate students taking a general or advanced course on evolution, as well as graduate students and professionals in evolutionary biology and paleontology, will find this book of great interest.

CAMBRIDGE PALEOBIOLOGY SERIES

Series Editors:

D. E. G. BRIGGS, *University of Bristol*
P. DODSON, *University of Pennsylvania*
B. J. MACFADDEN, *University of Florida*
J. J. SEPKOSKI, *University of Chicago*
R. A. SPICER, *University of Oxford*

CAMBRIDGE PALEOBIOLOGY SERIES is a new collection of books in the multi-disciplinary area of modern paleobiology. The series provides accessible and readable reviews of exciting and topical aspects of paleobiology. The books are written to appeal to advanced students and to professional earth scientists, paleontologists, and biologists who wish to learn more about the developments in the subject.

Books in the Series:

1. *The Enigma of Angiosperm Origins*   Norman F. Hughes
2. *Patterns and Processes of Vertebrate Evolution*   Robert L. Carroll

Patterns and processes of vertebrate evolution

ROBERT L. CARROLL

McGill University

PUBLISHED BY THE PRESS SYNDICATE OF THE UNIVERSITY OF CAMBRIDGE
The Pitt Building, Trumpington Street, Cambridge CB2 1RP, United Kingdom

CAMBRIDGE UNIVERSITY PRESS
The Edinburgh Building, Cambridge CB2 2RU, UK   http: //www.cup.cam.ac.uk
40 West 20th Street, New York, NY 10011-4211, USA   http: //www.cup.org
10 Stamford Road, Oakleigh, Melbourne 3166, Australia

First published 1997
Reprinted 1998

Printed in the United States of America

Typeset in Garamond

A catalogue record for this book is available from the British Library

Library of Congress Cataloguing-in-Publication Data is available

ISBN 0-521-47232-6 hardback
ISBN 0-521-47809-X paperback

Dedicated to the work of
GEORGE GAYLORD SIMPSON,
who brought paleontology into
the evolutionary synthesis
and established the study
of large-scale evolution
on a strong scientific footing.

Contents

Preface   *page* xi
Acknowledgments   xv

1. **Current problems in evolutionary theory**   1
   Introduction   1
   Large-scale evolutionary phenomena   9
   Obstacles to uniting analysis of short- and long-term
   evolutionary processes   11
   Vertebrates as a model for the study of evolution   14

2. **Theories of evolution at the level of populations
   and species**   19
   Darwin's view of evolution at the population level   19
   Dobzhansky and Mayr and the modern understanding of
   the role of species in evolution   21
   The fossil record   24
   Stratigraphy   25
   Eldredge and Gould and the theory of punctuated equilibria   27

3. **Evolution in modern populations**   34
   Introduction   34
   Evolutionary change in immigrant populations   36
   Evolution within the Galápagos finches   38
   Significance of changes among Darwin's finches to longer-term
   evolutionary phenomena   50
   Other examples of evolutionary change on islands   52
   Clines   52
   Summary of evidence for significant change within species   54
   Speciation   54
   Summary   56

4. **Limits to knowledge of the fossil record and their
   influence on studies of evolution**   57
   Introduction   57
   Limitations of the fossil record   57
   Changes in the relative completeness of the fossil record
   over time   65
   Dating geological events and processes   68
   Rates of evolution   72
   Summary   81

5.   Patterns of evolution among late Cenozoic mammals   82
Introduction   82
The Plio–Pleistocene ice ages   84
Testing punctuated equilibria   85
Recognition of stasis   86
Directional evolution   90
Anagenetic origin of species   102
Morphological change at the time of speciation   104
Species selection   106
Summary of evolutionary patterns among late Cenozoic
mammals   110
Rates of evolution among late Cenozoic mammals   110

**6.   Patterns of evolution of nonmammalian vertebrates in
      the late Cenozoic**   115
Amphibians and reptiles   115
Birds   118
Stickleback fish   119
The cichlid fishes of the East African Great Lakes   123
Explanations for the rapid evolution of cichlids   132
Species-level evolution among the cichlids of the East African
Great Lakes   137
The cichlid radiation as a model to bridge the levels of
microevolution and macroevolution   139
Summary   144

**7.   The influence of systems of classification on concepts
      of evolutionary patterns**   145
Patterns and processes at the species level   145
Linnean classification   146
Phylogenetic systematics   150
Monophyly and paraphyly   151
Monophyletic groups of vertebrates   155
Naming and defining clades and included groups   160
The impact of phylogenetic systematics on the study
of evolution   165
Summary   166

8.   Evolutionary constraints   167
Introduction   167
Historical constraints   169
Chemical constraints   172
Material constraints   173
Summary   179

9.   Evolutionary genetics   180
Introduction   180
Basic models of genetics   180

The shifting balance theory of evolution   190
Polygenic or quantitative inheritance   192
Analysis of quantitative traits   193
The effect of selection on quantitative traits   196
Rates of accumulation of quantitative traits   201
Nature of genes for quantitative traits   202
The enigma of low selection coefficients for long-term
evolutionary change   202
Genetic constraints   208

10. Development and evolution   212
Genetics and development   212
Heterochrony   214
Homeobox genes   215
The phylotypic stage   220
Hox genes in chordates   221
The origin of craniates   222
Developmental processes and the evolution of the skull
and axial skeleton   225
The evolution of fins and limbs   227
The origin of tetrapod limbs   230
Developmental processes of tetrapod limbs   236
Morphogenesis and evolution of tetrapod limbs   240
Integration of developmental biology with the evolutionary
synthesis   258
Development and macroevolution   262
Summary   264

11. Physical constraints   266
Constraints on body form in fast-swimming vertebrates   266
Primitively aquatic vertebrates   268
Secondary aquatic adaptation among groups with terrestrial
ancestors   274
Flight   274
Terrestrial constraints   282
Discussion   286
Transfer of substances across membranes   286
Heat absorption and transfer   289
Miniaturization   290
Summary   295

12. Major evolutionary transitions   296
Introduction   296
The origin of terrestrial vertebrates   300
The origin of birds   306
The origin of mosasaurs   324
The origin of whales   329
General features of major transitions   336

13. Patterns of radiation   340
Introduction   340
The Cambrian explosion   341
Radiations among primitively aquatic vertebrates   349
Paleozoic and Mesozoic tetrapods   350
Early Cenozoic mammals   352
Later Cenozoic mammals   358
Birds   359
Discussion   359

14. Forces of evolution   362
Forces of evolution evident at the level of populations
and species   362
Additional factors of long-term evolution   365
Evolutionary trends   365
Continental drift   368
Mass extinctions   376
Summary   388

15. Conclusions and comparisons   389
General features of vertebrate evolution   389
Is a distinct theory of macroevolution necessary?   391
Agenda for the future   392
Comparisons   394
Final conclusions   403

Glossary   405
References   411
Index   439

Preface

The study of large-scale evolution is undergoing a period of explosive growth through the influx of new theories and information gained from many areas of research. In addition to paleontology, these include genetics and developmental biology, systematics, geology, and astronomy. The most evident contributions have come from an enormous increase in data from fossils that demonstrate the patterns and rates of evolution from the level of populations to those of the largest-scale radiations. It is only since the publication of George Gaylord Simpson's groundbreaking books on evolution that fossils have revealed the nature of transitions between major adaptive groups of vertebrates, and only in the past fifteen years that phenomena at the species level have been studied in detail from fossil evidence. Understanding of the evolutionary significance of fossil data has been greatly augmented by increased accuracy of geological dating and knowledge of sedimentary processes. In addition, the rise of phylogenetic systematics and molecular biology have provided means of determining relationships that enable the historical patterns of origins and radiations to be established with greatly increased reliability. Long-term field studies, especially those of the Grants on the Galápagos finches and of Greenwood and others on the East African cichlid fish, have provided an extensive basis for comparison between patterns and rates of evolution in modern populations and those that can be reconstructed from the fossil record.

Developmental biology is now providing the most exciting new contributions to understanding the mechanisms of large-scale structural change. Knowledge of regulatory systems revealed by the *Hox* genes demonstrates a hierarchical pattern of control that casts an entirely new light on the interrelationships between mutations and selection in governing the modification of major anatomical structures. While developmental biology opens a new avenue to the study of major evolutionary changes, the concept of evolutionary constraints provides a framework for investigating the long-term constancy of other aspects of body form and function.

Knowledge of plate tectonics gathered since the early 1960s demonstrates how changes in the position and configuration of the continents and oceans have governed global climates and patterns of dispersal throughout the history of life. These factors have had a major influence on the course of evolution that was unsuspected by Darwin and was not incorporated into the new evolutionary synthesis. Both geological and astronomical studies have provided evidence to test the importance of mass extinctions.

Another key impetus to the study of evolution over the past twenty-five years has been the writings of Stephen Jay Gould and his colleagues. Several features of evolution that they have emphasized are stressed in this study:

1. the significant differences between the patterns of evolution postulated by Darwin for long periods of time and those observable from the fossil record;
2. the difficulties of explaining the patterns of long-term evolution on the basis of processes whose studies are limited to living species; and
3. historical contingency and other evolutionary constraints.

Whatever the final evaluation of Eldredge and Gould's theory of punctuated equilibrium, their papers have spurred paleontologists to decades of zealous fieldwork to determine the empirical patterns of evolution at the species level.

Because studies in these many disciplines are dependent on different techniques and data sets, communication has become increasingly difficult, and widely differing concepts of evolution are now hypothesized to explain phenomena at various time scales and in different taxonomic groups. The purpose of this book is to provide an integrated view of the patterns and processes of evolution ranging from those studied primarily in living organisms to those expressed only over vast stretches of geological time. It is specifically intended to help bridge the gap or break down the barriers of subject matter that have long isolated studies of evolution by paleontologists and those working with modern populations. Attention is focused on a single large group, the vertebrates, whose study has long been devoted to investigation of evolutionary problems. It is hoped that information gained from vertebrates can provide a model for comparison with patterns of evolution observed in other major groups of organisms.

The general conclusions of this study differ from those of Darwin, the modern evolutionary synthesis, and the theory of punctuated equilibrium in being much more pluralistic. The rates and patterns of evolution as well as the forces of evolutionary change are observed to be extremely diverse and variable, differing significantly between modern populations and species and those illustrated by the fossil record over millions and hundreds of millions of years. The relative expression of stasis and directional change varies from character to character, within taxa over time, and especially between taxa. This variation is especially obvious if comparison is extended to all major groups of organisms, including prokaryotes and protists. Although Mendelian and population genetics are important for understanding the mechanics of evolutionary change, behavior and external factors of the physical and biological environment are more significant in determining the rate, direction, and nature of change over long periods of time.

This book grew from an attempt to establish the degree to which evolutionary processes studied among modern populations can explain large-scale changes in anatomy, behavior, and way of life that occurred over tens and hundreds of millions of years. The chapters are organized so that this problem can be investigated step by step. The text begins with studies of phenomena at the level of populations and species, based on fossil as well as living groups, and proceeds to the analysis of patterns and inferred processes of evolution over hundreds of millions of years, involving the highest taxonomic ranks. Intercalated are chapters on geological and biological subjects, ranging from sedimentation and continental drift to genetics, systematics, and development, that are necessary for an understanding of evolutionary studies over a broad range of time scales. These subjects are presented so

that readers with a background in either biology or geology will be able to understand the great range of factors that have influenced evolution. *Patterns and Processes of Vertebrate* was written with a broad readership in mind, from undergraduate and graduate students studying evolution or paleontology, to anyone who is
interested in the history of life.

Acknowledgments

Many colleagues, friends, and family members, spanning the years since my childhood, deserve thanks for contributing to my interest and continuing study of paleontology and evolution. My father introduced me to paleontology when I was five by showing me a collection of fossils he used in teaching, including trilobites and other Paleozoic invertebrates. He and my mother encouraged my interest through collecting trips in the Devonian and Carboniferous of Michigan and the Jurassic, Eocene, and Oligocene of the western United States. As a graduate student, my training was strongly influenced by an incomparable group of professors at Harvard: Ernst Mayr, Brian Patterson, Al Romer, George Gaylord Simpson, Harry Whittington, Ernest Williams, and E. O. Wilson.

I very much appreciate the suggestions provided by those who read some or all of the manuscript. Colleagues at Redpath Museum – Graham Bell and David Green – read drafts of the entire text, as did Hans-Dieter Sues of the Royal Ontario Museum, Toronto. The chapter on genetics was reviewed by Dan Schoen and Kurt Sittmann of the Department of Biology at McGill. Don Kramer provided useful references on the importance of behavior to large-scale evolution. Sean Carroll, Brian Hall, and Paul Lasko contributed many helpful suggestions that were incorporated in the final version of the chapter on development and evolution.

Preliminary versions of the manuscript served as a text for an undergraduate course in evolution given at McGill over the past five years. The students demonstrated just how difficult it is to convey concepts of evolution that span time scales ranging from days and years to millions and hundreds of millions of years. I thank my graduate students, postdoctoral fellows, and others working at the Redpath Museum for their contribution of papers and ideas that otherwise may have been missed, as well as for numerous informal discussions that gave a broader viewpoint to many of the subjects that were covered. John Alroy, Michael Caldwell, and Christine Janis allowed me to make use of their unpublished manuscripts so that this text could reflect the most recent research in their fields. I especially appreciate the assistance of Dan Riskin, a student at the University of Alberta, Edmonton, for his efforts to integrate new data bearing on the pattern and rate of evolution among early Paleocene placentals.

The greatest assistance in preparing this text has been provided by Elena Roman through computer drafting and labeling of nearly all of the illustrations, and ensuring consistency among the text, figures, and bibliography. My son David assisted by computing the changes in allelic frequency under different selection coefficients that are illustrated in Chapter 9. Access to the biological and paleontological literature was greatly facilitated by the staff of the Blacker–Wood Library: Eleanor Mac-

Lean, Ann Habbick, Jennifer Adams, and Yvonne Mattocks. The Natural Sciences and Engineering Research Council of Canada provided grant support for several of the research projects reported in this volume.

Production of this book benefited greatly from the efforts of the staff of Cambridge University Press. Robin Smith provided the initial stimulus to write a book on this subject and gave advice and encouragement during the preparation of the manuscript. His editorial assistant at the time, Susan Ghanbarpour, supplied extensive technical advice. The production editor, Michael Gnat, contributed greatly to the final appearance of the volume.

Finally, my heartfelt thanks to my wife, Anna Di Turi, who provided support and encouragement throughout the preparation of the book, in addition to reading the entire manuscript.

1 Current problems in evolutionary theory

> Thus, macroevolution and microevolution are decoupled in the important sense that macroevolutionary pattern cannot be deduced from microevolutionary principles.
> — Stebbins and Ayala (1981)

Introduction

Evolution is the greatest unifying principle in biology. It explains both the astounding diversity of life and the extraordinary constancy of the basic molecular constituents and chemical processes common to all organisms. E. O. Wilson, an evolutionary biologist and author of the widely respected book *The Diversity of Life* (1992), argues that there may be as many as 100 million living species, although only 2 million have yet been described and named. George Gaylord Simpson, who made the greatest contribution to integrating evolutionary theory with knowledge of the fossil record, estimated that as many as 50 billion species may have existed throughout the 3.5 billion years of life on Earth (Simpson 1952). (*Note:* American billions [10^9] are used throughout this volume.)

In size and complexity, life ranges from single-cell bacteria, as small as 10^{-13} g in mass, to gigantic sequoias and whales, exceeding 10^8 g and composed of billions of cells, controlled by hundreds of thousands of genes. Organisms occupy all habitats on the Earth's surface, from the deepest trenches of the ocean to the air above the highest mountains.

Underlying the diversity of form, physiology, and way of life, all living things are composed of nearly identical chemical constituents – specific sugars, amino acids, fats, lipids, and nucleic acids – and have basically similar metabolic processes. Most important, all organisms are linked to one another by a common genetic heritage. All of the millions of living species are the product of continuous evolutionary change since their origin from a common ancestry more than 3.5 billion years ago.

The publication in 1859 of Charles Darwin's *On the Origin of Species by Means of Natural Selection* marks the beginning of our understanding of how evolutionary change has molded the history of life on Earth. Darwin not only provided convincing evidence that all organisms, living and fossil, are related to one another and that all aspects of their biology are the result of millions of years of evolution, but also proposed a well-reasoned mechanism for change: natural selection. Soon after the publication of *The Origin of Species,* nearly all biologists acknowledged the basic facts of the evolutionary history of life, but the significance of natural selection and other evolutionary processes are still subject to serious scientific debate. In fact, there is currently more controversy regarding the mechanisms of evolution than at any time since the early years of the twentieth century. Alternative theories have been proposed to explain nearly all aspects of evolution, from the role of natural selection within populations to the largest-scale patterns and processes of evolution in plants and animals over the past 500 million years.

Why should such a widely accepted concept in an extensively studied discipline still face such fundamental problems? One of the major reasons is that studies of modern populations and evidence from the fossil record give the impression of very different patterns and rates of evolution.

Although Darwin's theory sought to deal with evolution over all time scales, almost all of his evidence was drawn from the modern biota. In the absence of adequate evidence from fossils, he simply extrapolated the patterns and processes that he could study in living organisms to the uncounted millions of years of the history of life. This is most clearly shown by the only illustration that appeared in the first edition of *The Origin of Species* (Fig. 1.1). Darwin used this figure twice in successive paragraphs, first to illustrate the pattern of evolution over tens to hundreds of thousands of generations within individual populations and species, and later to show the pattern of change over millions and hundreds of millions of generations. He argued that both the patterns and processes of evolution were essentially identical over these vastly different time scales. Although many biologists and popular textbooks since the time of Darwin have perpetuated this concept of the history of life, other scientists have argued that neither the patterns nor the processes of evolution that can be studied in living populations are adequate to explain the conspicuous differences in morphology, physiology, and way of life of the major groups of microorganisms, plants, and animals, or the major patterns and different rates of evolution observed in the fossil record.

The most graphic demonstration of the inadequacy of Darwin's hypothesis of the constancy of evolutionary patterns over all time scales can be seen by comparing his hypothetical representation of the patterns of evolution for both very short and very long periods of time with the patterns of evolution that have since been reconstructed on the basis of the fossil record of multicellular plants and animals over the past 500 million years (Figs. 1.2–1.4).

The diagram used by Darwin to illustrate evolution both at the level of populations and species and over the vast expanse of geological time is characterized by gradual and continuous change. Most populations within species, or families within orders, diverge progressively. Some lineages continue with little change, but most eventually become extinct. The entire adaptive space is occupied by the groups diagramed, and the rate of change, indicated by the slope of the lines, remains fairly constant.

The patterns established from the fossil record of the major groups of vascular plants, vertebrates, and nonvertebrate metazoans are conspicuously different. There are relatively few major lineages, all of which are very distinct from one another. Gaps between the lineages indicate that adaptive space is not fully occupied. Instead of showing gradual and continuous change through time, the major lineages appear suddenly in the fossil record, already exhibiting many of the features by which their modern representatives are recognized. It must be assumed that evolution occurs much more rapidly *between* groups than *within* groups. For

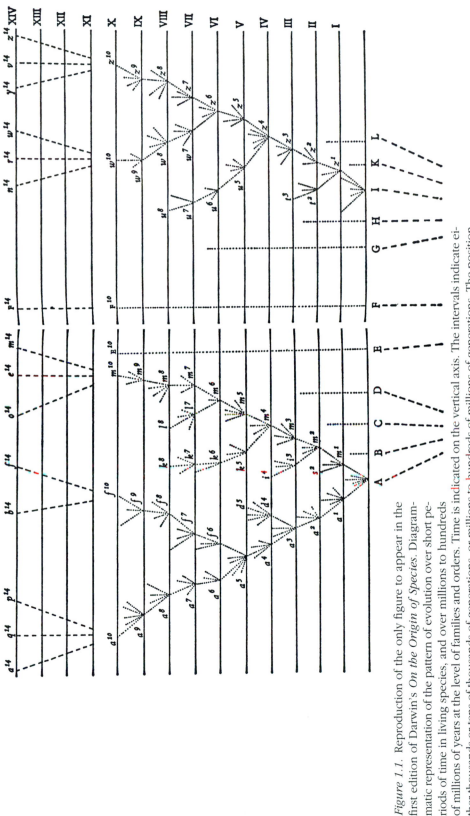

Figure 1.1. Reproduction of the only figure to appear in the first edition of Darwin's *On the Origin of Species*. Diagrammatic representation of the pattern of evolution over short periods of time in living species, and over millions to hundreds of millions of years at the level of families and orders. Time is indicated on the vertical axis. The intervals indicate either thousands or tens of thousands of generations, or millions to hundreds of millions of generations. The position along the horizontal axis indicates the degree of morphological difference. The angle of the lines represents the rate of evolution; vertical lines indicate evolutionary stasis.

most of their evolutionary history, fundamental aspects of the anatomy and way of life of these lineages do not change significantly. Very few intermediates between groups are known from the fossil record.

This pattern is most conspicuous for the major groups of nonvertebrate metazoans (Fig. 1.2). Multicellular animals are first known in the late Precambrian, some 580 million years ago (Lipps and Signor 1992), but the early portion of their fossil record is very poorly known because they lacked external skeletons and so were preserved only under unusual conditions. Evidence of the presence and early diversification of multicellular animals is provided primarily by the tracks, trails, and impressions they left in the sediments in which they lived. Then, over a very short time in the Early Cambrian, they underwent an explosive radiation. Over a period of approximately 5 million years, they gave rise to all the major groups alive today, as well as a smaller number of extinct groups (Bowring et al. 1993). By 525 million years ago, all of the living phyla and most of their constituent classes had diverged. Sponges, arthropods, primitive chordates, echinoderms, bryozoans, brachiopods, molluscs, annelids, and so on were all recognizable by this time. Among the molluscs, all the modern classes – scaphopods, gastropods, monoplacophorans, and pelecypods – were known by the end of the Lower Cambrian. Although the remains of most of the major groups are represented by animals with a calcareous or phosphatic exoskeleton, other lineages, including animals without heavily mineralized body parts, are known from the Burgess Shale and other localities with depositional environments favoring the preservation of soft-bodied organisms (Gould 1989).

In all the major lineages, the earliest known members had already achieved the basic body plan of their living descendants. They differed in details, but most can be readily allied with modern classes. Some major groups of Paleozoic organisms, such as the trilobites, and many of the strange animals described from the Burgess Shale are without living descendants, but they can be recognized as belonging to larger taxonomic groups common today. Although trilobites differ in the nature of their mouth structures and details of limb anatomy, they can be united with living arthropods on the basis of more general characters, such as the jointed articulation of their limbs and nature of segmentation. Few fossils are yet known of plausible intermediates between the invertebrate phyla, and there is no evidence for the gradual evolution of the major features by which the individual phyla or classes are characterized.

Among the vertebrates, the radiation of placental mammals shows a similar pattern (Fig. 1.3). Scattered, fragmentary remains of primitive placental mammals are known as early as the Lower Cretaceous, but their fossils only become common in the latest Cretaceous and earliest Cenozoic. By the beginning of the Eocene, approximately 8.5 million years later, all the major groups living today, as well as many extinct orders, had differentiated. They retained some primitive features, but critical characteristics by which we recognize such divergent groups as carnivores, horses, elephants, and primates were already evident. These remained essentially constant during the subsequent 57 million years separating them from the living fauna.

The pattern of large-scale evolution of vascular plants is similar in the presence of a relatively small number of major groups that appeared suddenly in the fossil

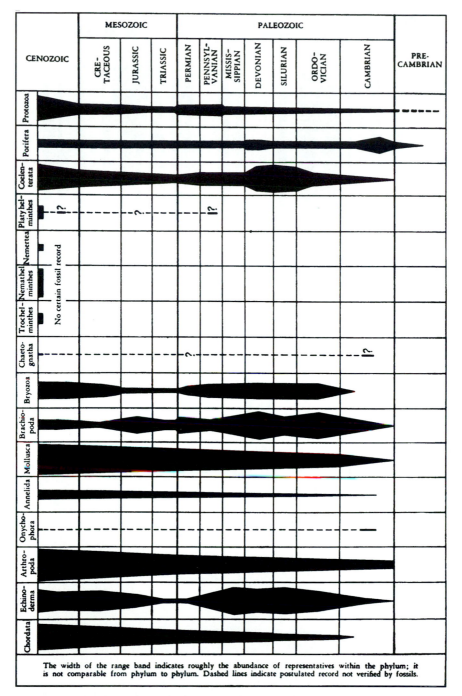

Figure 1.2. Geological ranges of the principal animal phyla based on the fossil record. The width of the range band indicates roughly the abundance of representatives within each phylum; it is not comparable from phylum to phylum. Dashed lines indicate postulated record not verified by fossils. Roughly horizontal orientation of each lineage indicates the constancy of the basic morphological pattern. (Other conventions that have been used in drawing up such phylogenies, which strongly influence the way in which evolution is visualized, are discussed in Chapter 7.) From Shrock and Twenhofel (1953), *The Principles of Invertebrate Paleontology.* Reproduced with the permission of McGraw–Hill, Inc.

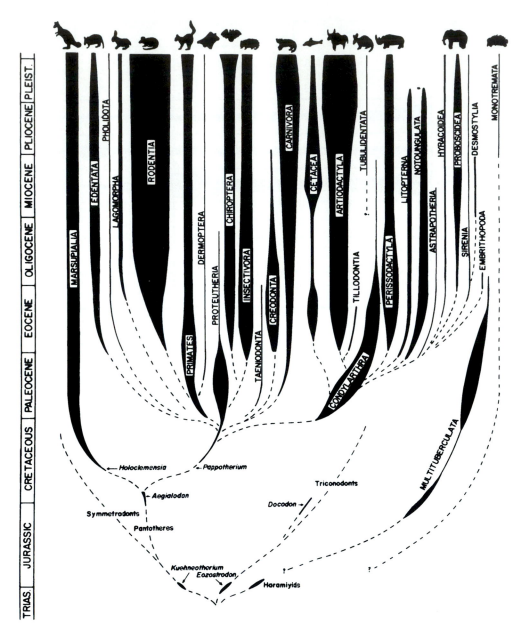

Figure 1.3. Radiation of placental mammals in the Cenozoic. Conventions as in Fig. 1.2. From Gingerich (1977).

record and then persisted with little fundamental change for hundreds of millions of years (Fig. 1.4). By the end of the Devonian, many of the major subdivisions of vascular plants living today can be recognized on the basis of distinctive reproductive structures. All primitive land plants reproduce via tiny spores contained in sporangia. The major taxonomic groups are distinguished by the position on the plant at which the sporangia are attached. The most primitive condition may be illustrated by the extinct Rhyniopsida, in which the sporangium is in a terminal position, at the tip of the plant. In the Lycopsida (represented in the modern flora by the

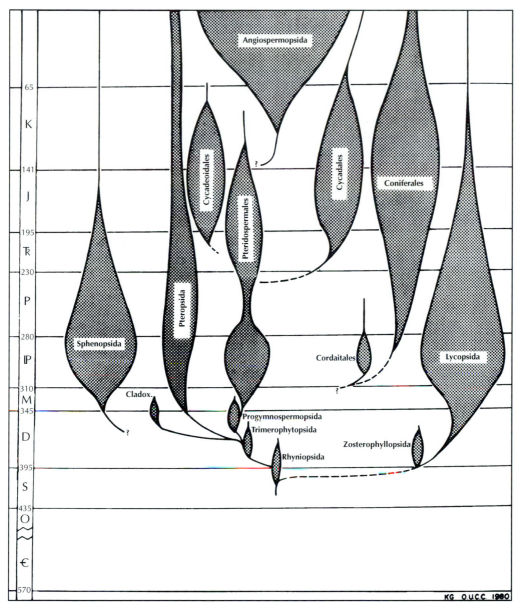

Figure 1.4. Generalized phylogenetic relationships of major vascular plant groups as understood in 1980. Conventions as in Figs. 1.2 and 1.3. From Knoll and Rothwell (1981).

ground pine *Lycopodium* and *Selaginella*) the sporangia are located at the base of the leaves. The ferns (Pteropsida) typically attach their sporangia to the lower surface of the leaves. The sporangia in the Sphenopsida (horsetails) are arranged in whorls at the top of the plant. These basic patterns have been maintained for more than 350 million years. Few, if any, intermediates are known between these patterns.

Several minor groups, which may represent stages linking rhyniopsids and more advanced plants, are recognized from the Devonian and early Carboniferous. In addition, there are two major groups of vascular plants that were common in the

Paleozoic and Mesozoic, the Cycadeoidales and Pteridospermales, that have no living descendants. The Cycadeoidales resemble living cycads but are not closely related. The Pteridospermales, or seed ferns, have been suggested as being related to flowering plants, but differ from them in most aspects of their anatomy and pattern of reproduction, rather than representing a transitional sequence linking them to more primitive vascular plants.

One of most important changes in reproduction that occurred among early land plants was the origin of the seed. This was achieved by the end of the Devonian in lineages recognized as ancestral to all more advanced vascular plants. By the Late Pennsylvanian, 300 million years ago, seed-bearing conifers had already achieved a general body form and physiology similar to that of their living descendants. The last major evolutionary advance among vascular plants was the emergence of flowering plants (the angiosperms) in the Early Cretaceous. As yet, there is no firm evidence of the lineage from which they evolved, nor known fossils that document the many changes in anatomy that separate them from more primitive vascular plants (Sanderson and Donoghue 1994; Crane, Friis, and Pedersen 1995). By the end of the Cretaceous, nearly all the living orders of angiosperms had become differentiated from one another, and most of the major characters by which their living representatives are recognized had evolved. In no cases can the gradual evolution of these characters or groups be documented.

Not all groups of unicellular plants and animals have an equally complete fossil record, but the record, to the extent that it is known, shows a pattern similar to that of multicellular eukaryotes. The same basic evolutionary pattern applies to all groups for which there is an adequate fossil record (Table 1.1).

Progressive increase in knowledge of the fossil record over the past hundred years emphasizes how wrong Darwin was in extrapolating the pattern of long-term evolution from that observed within populations and species. If the patterns of evolution over time scales of millions and hundreds of millions of years are so dif-

Table 1.1. *Comparison of patterns of evolution at micro- and macroevolutionary levels*

Microevolution (based on Darwin's observations of modern populations and species)	Macroevolution (based on the fossil record)
Most populations within species diverge progressively over time.	Major lineages appear suddenly in the fossil record and then persist for long periods of time within the same structural and adaptive framework.
Some lineages show stasis, but they have little potential for evolutionary survival.	
Evolutionary rates are relatively similar from lineage to lineage, and over time within a single lineage.	Evolutionary change appears to be very rapid during the origin of major groups and much slower within groups.
The entire range of possible morphologies is represented by some populations within the history of a species.	Significant gaps separate structural and adaptive patterns that are characteristic of major groups.

ferent from those that Darwin postulated for modern populations and species, can the process of natural selection that he established on the basis of living species adequately explain long-term evolutionary phenomena?

Large-scale evolutionary phenomena

Evolutionary processes such as variation and selection that are common to the level of populations and species must occur over all time scales; but there are many other aspects of evolution that are evident only over time scales much longer than those that can be studied among living organisms. The following phenomena are of particular concern to biologists:

1. *The origin of major new structures.* Biologists have long struggled with the conceptual gap between the small-scale modifications that can be seen over the short time scale of human study and major changes in structure and ways of life over millions and tens of millions of years. Paleontologists in particular have found it difficult to accept that the slow, continuous, and progressive changes postulated by Darwin can adequately explain the major reorganizations that have occurred between dominant groups of plants and animals. Can changes in individual characters, such as the relative frequency of genes for light and dark wing color in moths adapting to industrial pollution, simply be multiplied over time to account for the origin of moths and butterflies within insects, the origin of insects from primitive arthropods, or the origin of arthropods from among primitive multicellular organisms? How can we explain the gradual evolution of entirely new structures, like the wings of bats, birds, and butterflies, when the function of a partially evolved wing is almost impossible to conceive?

2. *The extremely irregular occupation of adaptive space as opposed to the nearly continuous spectrum of evolutionary change postulated by Darwin.* Although an almost incomprehensible number of species inhabit Earth today, they do not form a continuous spectrum of barely distinguishable intermediates. Instead, nearly all species can be recognized as belonging to a relatively limited number of clearly distinct major groups, with very few illustrating intermediate structures or ways of life. All of us can immediately recognize animals as being birds, turtles, insects, or jellyfish, and plants as conifers, ferns, or orchids. Even with millions of living species, there are only a very few that do not fit into readily recognizable taxonomic categories. Of all living mammals, only the tree shrews are difficult to classify. Are they primitive relatives of primates (our own distant relatives) or closer to the true shrews and moles among the insectivores? Even among the hundreds of thousands of recognized insect species, nearly all can be placed in one or another of the approximately thirty well-characterized orders.

One might hypothesize a very different pattern among extinct plants and animals: Fossils would be expected to show a continuous progression of slightly different forms linking all species and all major groups with one another in a nearly unbroken spectrum. In fact, most well-preserved fossils are as readily classified in a relatively small number of major groups as are living species. Nearly all mammals that lived in North America and Europe during the past 50 million years can be classified among the seventeen living orders. Most, although certainly not all, of these species can be readily related to living carnivores, rodents, bats, whales, elephants, horses, cows, and so on.

Compared with the millions of specimens of trilobites that have been collected, there are very few that might be thought to bridge the gap between trilobites and any other group of extinct arthropods. The number of species that bridge the gaps between dinosaurs and more primitive reptiles and between dinosaurs and birds is very small compared with the number that everyone recognizes as dinosaurs. How do we account for the extremely irregular distribution of basic body plans in space and time under a theory of evolution based on gradual and continuous change?

3. *The apparently much greater rates of evolution during the origin of groups than during their subsequent duration.* This and other questions of the rates of evolution were the major focus of Simpson's book *Tempo and Mode in Evolution* (1944), as well as of the more recently published *Rates of Evolution* (Campbell and Day 1987). It is difficult to apply the near constancy of natural selection advocated by Darwin to the differences between very rapid and very slow rates of evolution seen over long time scales of the fossil record. If natural selection is continuously acting on all organisms within modern populations, why have some animals and plants, such as cockroaches, horsetails, and the horseshoe crab *Limulus,* remained almost unchanged for the past 325 million years, whereas in the same period of time, small, scaly, cold-blooded animals resembling primitive living lizards gave rise to the greatly altered dinosaurs, birds, and mammals?

4. *The cause and nature of major radiations.* One of the major features of large-scale evolution is the sudden appearance and rapid radiation of major groups that dominated the Earth's biota for tens or hundreds of millions of years. Can such explosive radiations be explained on the basis of the apparently slow and small-scale changes that can be seen in modern species?

5. *The cause and significance of mass extinction.* Despite our current concern over extinction caused by human activities, the modern biota gives us no evidence of the catastrophic extinctions that marked the close of the Paleozoic and Mesozoic eras, nor of how important mass extinction may have been to the course of evolution.

These problems emphasize the difficulties of direct comparison between evolutionary patterns observed in living populations and those revealed by the fossil record. This discrepancy has led to the use of two contrasting terms to apply to evolution at different scales of time and different degrees of change: **microevolution** (involving phenomena at the level of populations and species) and **macroevolution** (evolutionary patterns expressed over millions and hundreds of millions of years). Because these patterns of evolution appear so different, it is natural to suggest that different processes may also be necessary to explain evolution within groups over short periods of time, and both within and between groups over long periods of time. One of the major problems to be considered in this study is the degree to which the processes of evolution hypothesized by Darwin on the basis of modern populations and species can serve to explain the major patterns of evolution revealed by the fossil record.

Biology is not alone in having to deal with phenomena involving very different scales of magnitude. In the late nineteenth century, all the problems of physics concerning the movement and attraction of matter appeared to be solvable on the basis of Newtonian mechanics. Following the discovery of radioactivity by Antoine-Henri Becquerel in the 1890s, it became necessary to establish an entirely new set of concepts to deal with the mechanics of matter at the subatomic level. Knowledge of the gravitational attraction of planetary bodies provided very little insight into the manner of interactions of subatomic particles.

A closer analogy is provided by changes in concepts of large-scale geological processes within the past forty years. Until about 1960, geology was dominated by the uniformatarian concepts developed by James Hutton and Charles Lyell during the late eighteenth and early nineteenth centuries. They argued that nearly all geological processes since the consolidation of the Earth's crust could be explained on the basis of phenomona that can be directly observed today. There is no reason to think that the processes of physical and chemical erosion, nor the gradual accumulation of sediments in the oceans are any different now than they were in the past. Earthquakes and volcanic eruptions are associated with mountain building today and presumably contributed in a similar way to changes in the configuration of the Earth's surface throughout its history. Based on these concepts, it was assumed that the continents had long been in their present positions, although large areas may have been covered by extensive inland seas, and the configuration of the margins differed as a result of changes in sea level and the degree of uplift or erosion of mountain chains.

All our ideas regarding the permanency of continents changed in the late 1950s as a result of increased knowledge of the nature of the rocks beneath the oceans. For the first time, it was recognized that the configuration of all the ocean basins is continually changing, and has been doing so since the solidification of the Earth's crust. Large-scale changes in the configuration of ocean basins, continents, and major mountain chains are now explained through the theory of **continental drift**, or **plate tectonics**. None of these processes was recognized nor could have been explained on the basis of the assumptions of the uniformitarian theory of geology. No geological processes that we can observe on a day-to-day basis could predict the large-scale, long-term movements of the continents. The evolutionary patterns and processes that can be studied on a daily basis also failed to predict the *patterns* of evolution over hundreds of millions of years. Can they explain the *processes* of large-scale, long-term evolution?

Obstacles to uniting analysis of short- and long-term evolutionary processes

It remains difficult to develop an integrated model of evolution that incorporates evidence from studies of modern populations and the fossil record. Although organisms living today are but a continuation of an evolutionary sequence begun 3.5 billion years ago, the information that can be gained from the study of living forms differs radically from that which can be learned by studying their fossil remains.

Thousands of scientists throughout the world make use of living organisms to study short-term evolutionary phenomena at the genetic, organismal, and population level. It is at least potentially possible to establish the nature and rate of mutational change for every gene in every organism. Work now under way on the human genome, and that of fruit flies, nematodes, and yeast, will provide a general model for the DNA content and function of all genes. Changes in gene frequency in populations from generation to generation can be used to establish the relative importance of mutation rate, population size, and natural selection in controlling the rate and direction of evolutionary change.

This data base is almost unlimited in terms of the number of organisms that may be studied and the detail of study possible at the biochemical and genetic level. Most hypotheses can be tested in a diversity of groups, and experimental studies can be carried out both in the laboratory and in the field. Attention has been focused on finer and finer details of population structure and changes at the molecular and genetic levels.

On the other hand, studies of living populations have inherent temporal limitations. The process of speciation may occasionally occur over less than a thousand years, but no speciation events among vertebrates have been documented during the whole of human history. The longevity of most species exceeds a million years, so we have no possibility of directly viewing the complete time span of more than a tiny number of exceptionally short-lived species. Many of the species that make up the living biota have arisen within the past 10 million years, but we have no way of studying the many billions of extinct species except from their fossil record. No changes that have been observed in modern populations involve more than a few genes or exceed minor modifications in structure, physiology, and behavior. The modern fauna provides no evidence of the prior existence of such groups as the dinosaurs and flying reptiles that dominated Earth for tens of millions of years, nor of the catastrophic extinction that they faced 65 million years ago. Only the fossil record provides direct evidence of the patterns and processes of evolution over the 3.5 billion years that preceded the appearance of the modern biota.

Unfortunately, while fossils provide a broad view of long-term evolutionary phenomena, they provide very little evidence of the processes that can be studied in living organisms. Fossils provide no direct evidence of the nature of mutations or mutation rates involved in the evolution of novel structures or physiological patterns, nor means to measure selection coefficients. Moreover, only a fraction of the characters that can be studied in living populations are preserved in fossils, which are almost always limited to bony or calcareous skeletons. A great number of organisms, from bacteria and nematodes to cartilaginous lampreys and hagfish, have no hard parts and only very rarely leave any fossils at all. Soft tissues, metabolic processes, and behavior of extinct organisms can be reconstructed only on the basis of their homologues in living species. Fossils only rarely show the outline of their soft anatomy and, even more uncommonly, internal organs. Very unusual conditions of preservation retain details at the cellular level, but these are too rare for systematic study of evolutionary change. Tracks and trails of soft-bodied organisms, footprints, and nests of dinosaur eggs provide occasional glimpses of the behavior of extinct organisms, but these can only rarely be associated with particular species or provide an overview of changing patterns within major groups.

The most spectacular recent advance in the study of extinct organisms has been the capacity to recover DNA from fossils, enabling sequencing of particular genes in individual species at a level of accuracy nearly equivalent to that available to scientists studying living organisms. Unfortunately, the conditions necessary for the preservation of DNA appear so limiting that one can expect to be able to sample only a very few organisms within the vastness of geological time. Because of the degree of degradation of the DNA in even the best-preserved fossils, it is unlikely that more than a very small portion of the genes involved in basic metabolic processes can be recovered. On the basis of present techniques, knowledge of the organism as a whole remains even more restricted from DNA studies than from other fossil remains.

The inherent limitations of fossil remains, the incompleteness of sedimentary sequences containing fossils, and the limited precision of dating impose inevitable restrictions on what comparisons are possible between the evolutionary patterns and processes in living species and those that are known only from fossils. The most detailed comparisons are restricted to fossils from the most recent deposits, covering only the past 1–5 million years, and a few older examples in which conditions of preservation and deposition enable us to view change at a scale approaching that of modern populations.

Because of the different subject matter studied, research on micro- and macro-evolutionary phenomena are typically published in different scientific journals. Studies of modern populations are frequently process-oriented and appear in such journals as *Evolution, Genetics, Genetical Research,* and *Heredity.* Those on fossils are typically oriented toward anatomy and stratigraphy and are published in *Journal of Paleontology, Palaeontology, Palaeontographica,* and *Journal of Vertebrate Paleontology.* More recently founded journals, including *Historical Biology, Evolutionary Biology,* and particularly *Paleobiology,* attempt to bridge the gap but typically emphasize more long-term evolutionary phenomena.

Another serious barrier to developing a unified theory of evolution encompassing data from both living and fossil populations is the fact that very few scientists have devoted equal attention to problems over these different time spans. Of the thousands of persons studying evolution in living populations, very few are also concerned with evolution over longer periods of time. Paleontologists must be cognizant of the basic publications dealing with the evolution in living species, but very few have actually worked with variability in modern populations. Only a handful of individuals in past and present generations have made important contributions to the study of evolution at both the population level and over the long periods of time represented by the fossil record. These included Simpson, who brought paleontology into the modern evolutionary synthesis in the late 1940s and 1950s, and Bjorn Kurtén, who pioneered efforts to study rates of evolution in extinct populations. Others, notably Michael Bell (1988) and his colleagues (Bell, Baumgartner, and Olson 1985) and a host of paleontologists studying evolution in Quaternary mammals (whose work is summarized by Martin and Barnosky [1993]) have contributed greatly through studies of horizons in which the fossil record provides data that can be studied in nearly as much detail as the bones of animals in modern populations. This work provides the vital factual link between patterns and processes that can be studied directly in living species but can only be assumed to have occurred in most fossil sequences.

Despite these particular studies, there remains a very wide conceptual gap between phenomena at these two time scales. Paleontology and the study of the modern biota have had an irregular history of contribution to evolutionary theory. Paleontology has contributed more than any other discipline to demonstrating the historical fact of evolutionary change, but vertebrate paleontologists did not generally accept the theory of natural selection until well into the twentieth century. Between 1930 and 1970, evolutionary thought was united into a single "new synthesis" combining contributions from classical and population genetics, study of the nature and significance of speciation, and information from the fossil record (Mayr and Provine 1980). Although patterns of evolution illustrated by the fossil record made a strong contribution to the modern evolutionary synthesis, hypotheses involving processes of change were largely based on studies of modern populations, strongly influenced by Darwin's theory of natural selection. During the past twenty years, in contrast, new hypotheses of evolutionary processes have been proposed primarily on the basis of reevaluation of the fossil record.

The most conspicuous challenge to Darwinian selection theory in the past forty years is the theory of **punctuated equilibria**, first proposed by Eldredge and Gould in 1972 and since elaborated in numerous other papers (Stanley 1975, 1979; Gould and Eldredge 1977, 1993; Vrba 1980, 1984; Gould 1982, 1995a; Eldredge 1985a,b). This theory deals specifically with evolution at the level of species but has consequences that extend to long-term and large-scale evolutionary patterns and processes. The most striking tenets of punctuated equilibrium are that natural selection has little influence in changing the structure, physiology, or behavior of species subsequent to their origin. Most changes are associated with the origin of species through the process of speciation. Evolutionary trends are generated not within species, but by differential survival within groups of species.

These ideas have prompted a renewed debate extending over a broad range of evolutionary topics, from the level of change within populations to long-term evolutionary patterns. Most important, they have forced evolutionary biologists and paleontologists to examine currently available data more objectively and to search for more information, especially the careful collection and detailed study of fossils from relatively recent strata, in an effort to bridge the gap between micro- and macroevolutionary phenomena.

Several books have dealt with the problems of macroevolution and the integration of evolutionary processes over all time scales, but they leave many questions unresolved (Stanley 1979; Eldredge 1985a,b; Levinton 1988; Hoffman 1989). These publications attempt to address the problems very broadly, using examples from all groups of eukaryotes. One major way in which the present investigation differs from most studies of large-scale evolution is that it is focused on a single major taxonomic group: the vertebrates.

Vertebrates as a model for the study of evolution

Although all of life is unquestionably interrelated and shares the same basic molecules and metabolic processes, there are significant differences between major taxonomic groups that would be expected to affect patterns and processes of evolution. Eukaryotes clearly have different evolutionary potential than prokaryotes because they have two sets of genetic information rather than a single one and regularly undergo recombination and sexual reproduction. Among eukaryotes, vascular plants exhibit hybridization and polyploidy much more commonly than do multicellular animals, resulting in different mechanisms for speciation and subsequent evolutionary change. Multicellular organisms have much more potential for anatomical differentiation and much different capacity for morphological and behavioral change than do unicellular forms.

The many phyla of multicellular animals have fundamentally different body plans and ways of life that make it very difficult to compare specific patterns of evolutionary change between them. Beginning with Simpson (1944) comparisons have repeatedly been made between the rates of evolution in clams and mammals, but it is difficult to establish quantitative measures that could be applied to both groups (Stanley 1975). Detailed evolutionary comparison between the different metazoan phyla is also difficult because their interrelationships are still subject to much controversy. Of even greater importance, the most significant periods in their evolution – their origins and early stages of differentiation – are extremely poorly documented in the fossil record (Lipps and Signor 1992).

Vertebrates have several advantages as a model for the study of long-term evolutionary processes. They unquestionably comprise a single **monophyletic group**; that is, they share a common ancestral species. This permits comparison of homologous structures and physiological processes throughout all the included lineages. Nearly all vertebrate species exhibit a single pattern of reproduction, involving recombination of gametes from two sexes. The only examples of single-sex, parthenogenetic populations that have been recorded include members of eight genera of bony fish, two amphibian genera, and fourteen lizard genera. There is no evidence that this pattern was ever widespread among vertebrates or that it is an important factor in large-scale patterns of evolution (Dawley and Bogart 1989). Nearly all vertebrates are free living. They are never colonial in a physical sense, although this is approximated in a social sense by naked mole rats (Jarvis et al. 1994). Vertebrates are rarely if ever truly parasitic on other organisms, although physical attachment of males and females does occur among a small number of deep-sea fish, and lampreys attach themselves to their prey for long periods of time.

The basic consistency of the anatomy, developmental processes, and reproductive mode of vertebrates acts as a control in what can be considered an experimental system. Within this system one may compare the effects of variables such as changing environments, ways of life, and different taxonomic affinities on the rates and patterns of evolution over different time scales.

Although they share a fundamentally similar body plan and reproductive mode, vertebrates are an extremely diverse assemblage in terms of their structure and way of life. More than 45,000 species have been named, and many more species, especially among oceanic fish, remain to be described (Minelli 1993). Among adequately known multicellular animals, only arthropods, with countless millions of insects, are known to be more specious. Vertebrates range in size from some fish and amphibians weighing only a few grams to whales weighing hundreds of tons. As a group, vertebrates are the most studied and best known of all eukaryotes. In addition, our own membership in this assemblage makes it easier for us to understand and take interest in their structure, behavior, and way of life.

The factor that contributes most significantly to the study of long-term evolution among vertebrates is their excellent fossil record (Carroll 1987). In contrast with other metazoan groups, the origin of a fossilizable skeleton occurred early in their radiation, so that most major events in their history are well documented. Not only are most vertebrate groups known from many fossils, but the part of the body that fossilizes, the skeletal system, provides evidence of many of the most important functions of the body. The skeleton directly reflects general body form, the mode of locomotion, many aspects of feeding, and the size of and, in some groups, details of the external form of the brain. The general position and mechanical function of individual muscles can be reconstructed in terrestrial vertebrates from the configuration of the bones and from tendon scars on their surface. In several groups, elements of the reproductive and respiratory systems are directly preserved, and one occasionally finds calcified traces of major blood vessels. Even muscle fibers and kidney tubules are sometimes preserved, although they provide little information other than documenting the basic constancy of these tissues among vertebrates. Even behavior can be studied via fossil footprints, burrows, fossilized excrement, nests and eggs, and so on (Thulborn 1990; Carpenter, Hirsch, and Horner 1994).

With the exception of the very earliest vertebrates, which probably possessed neither bone nor calcified cartilage, and some other groups that lost or greatly reduced their bony skeleton (the modern orders of jawless fish), there is a good fossil record of most major taxa throughout their history. There are still significant gaps in our knowledge of the origin of some groups (notably the turtles, salamanders, bats, and ichthyosaurs), and some periods of geological time and particular environments that are much less well known than others; however, the fossil record from the Middle Ordovician to the present is sufficiently well known that new discoveries are unlikely to change our general understanding of large-scale patterns of evolution (Fig. 1.5).

Other eukaryotic groups have contributed significantly to the study of long-term evolution. The exoskeletons of fossil arthropods are informative in reconstructing many aspects of the living animal, but the early stages in the evolution of most classes remain poorly known. The most important radiation occurred in the late Precambrian and Early Cambrian and was initiated among soft-bodied forms, for which the fossil record will always be deficient.

Tiny arthropods, such as ostracods, and unicellular organisms, such as foraminifera and diatoms, have the advantage of being preserved in great number in extremely long sequences of marine sediments. This enables their use to chart evolution at the level of populations, species, and genera for millions of years. Unfortunately their fossilizable remains reflect little more than general body shape and would not necessarily reflect important changes in their soft anatomy, physiology, behavior, or way of life.

Fossils of vascular plants provide an excellent record of the early stages of their radiation in the late Paleozoic and early Mesozoic, but the extremely important early radiation of angiosperms (flowering plants) has been very difficult to analyze because of the rare preservation of flowers and herbaceous (nonwoody) species (Crane et al. 1995).

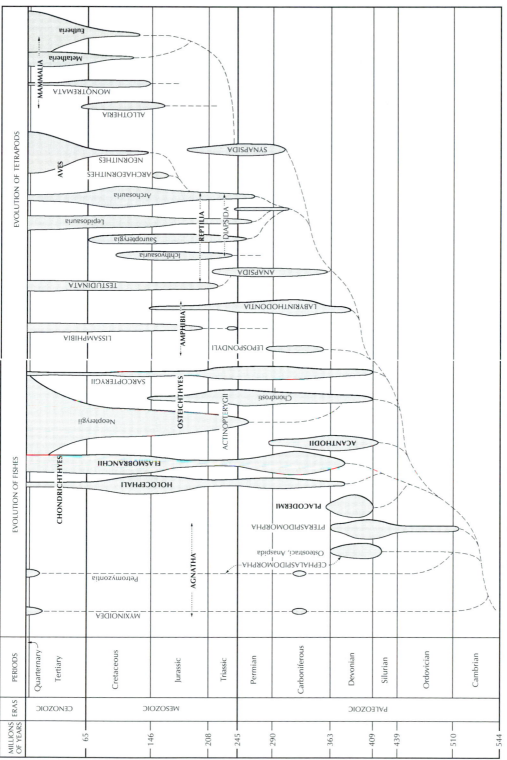

Figure 1.5. General pattern of vertebrate evolution. From *Analysis of Vertebrate Structure* by Hildebrand, 4th ed. Copyright © 1995. Reprinted by permission of John Wiley & Sons, Inc.

In short, no group shows a greater potential for the study of its evolutionary history than does vertebrates. What can be learned about the patterns and processes of vertebrate evolution can then be used as a basis for comparison of evolutionary phenomena in other groups, such as other metazoan phyla and vascular plants that have different reproductive modes, ways of speciation, and anatomical features.

The last comprehensive macroevolutionary analysis with an emphasis on vertebrates was undertaken by Simpson (1953). Since that time, there have been a number of advances in our knowledge of both fossil and living organisms that provide copious new data to evaluate the theories put forward by Darwin, the authors of the evolutionary synthesis, and more recent hypotheses.

Within the past ten years, advances in molecular biology have enabled study of developmental processes that hold the promise for explaining how basic body plans are produced and altered (Hall 1992; S. Carroll 1995). This information is already contributing to analysis of the significance of developmental processes in controlling the rate and extent of evolutionary change in vertebrates (Caldwell 1995a,b, 1996). Advances in molecular biology have also provided a whole new suite of genetic and biochemical characters that can be used to classify living organisms, and are now being extended to study of DNA in long-extinct plants and animals (Avise 1994, pp. 356–9).

The past twenty-five years have also witnessed an enormous increase in knowledge of the fossil record, with a near doubling of recognized species and growing attention to anatomical details and morphometric change that assist in evolutionary analysis. Accompanying the growth of knowledge of extinct organisms has been a revolution in the manner in which their relationships are established. This results from an ever-increasing use of phylogenetic systematics (Hennig 1966; Wiley 1981; Ax 1987). Willi Hennig's efforts to develop a system of classification that directly reflects evolutionary relationships is the most significant change in the approach to taxonomy since the work of Linnaeus.

During this same period, both the accuracy and the precision of radiometric dating have greatly increased the capacity to date the horizons in which fossils are found and to determine rates of evolution. The entirely new field of magneto-stratigraphy enables paleontologists to correlate the age of fossiliferous deposits worldwide on the basis of periodic reversals of the Earth's magnetic field (Mac-Fadden 1992).

All of these advances contribute to our understanding of evolutionary phenomena. This book serves primarily as a progress report in interpreting the patterns of vertebrate evolution from the fossil record. The amount of information that has accumulated since the time of Simpson's major works is sufficient for a major shift in emphasis. Instead of a largely theoretical consideration of large-scale evolution with a few examples from the fossil record, this book is an attempt to use the fossil record as a test for a variety of current evolutionary theories. At the species level, particular attention is paid to the many distinct hypotheses that have been elaborated under the general heading of punctuated equilibria. Subsequent chapters shift to larger-scale evolutionary phenomena to examine the concepts of evolutionary constraints and other ideas bearing on long-term conservation of basic body plans, major radiations and extinction, and the appearance of major evolutionary novelties.

2 Theories of evolution at the level of populations and species

Before we consider problems of large-scale and long-term evolution, it is necessary to begin, as did Darwin, with evolutionary phenomena that can be studied directly in living populations.

Darwin's views of evolution at the population level

Darwinian selection theory remains the primary model for the process of evolutionary change at the level of populations. Most of Darwin's arguments presented in *The Origin of Species* were based on phenomena that could be studied over short time scales. Although the full significance of evolution can only be seen at time scales covering millions or even billions of years, very little evidence of large-scale patterns was available during Darwin's time. The fossil record was still extremely poorly known and could not contribute to understanding of evolutionary processes at any level.

Darwin used the chart illustrated in Chapter 1 (see Fig. 1.1) primarily as a basis for discussing evolution at the level of populations within species. He considered several hypothetical species designated *A, B, C, . . . , L,* suggesting that over tens of thousands of generations the descendants of species *A* and *I* might become so different that they would be recognized as new species. Darwin argued that natural selection acted continuously on all characters of all species to bring about the optimal adaptation to their physical and biological environment. This does not necessarily result in constant or continuous change, however. Rather, Darwin stressed the irregular and intermittent action of natural selection. In discussing this chart, he stated that the actual pattern of evolution was probably much more irregular than was illustrated. Darwin also recognized that some features within variable species will remain fixed as a result of stabilizing selection. He further noted that some structures might change in ways that were not obviously adaptive as a result of correlation with other, more obviously adaptive changes, because of laws of correlation of growth.

Darwin (1859, p. 153) specifically stated that he saw no limit to the amount of change that is possible through the process of selection, but admitted that all change is dependent on the occurrence of new variations. He argued that the capacity for variation was the primary factor distinguishing species that remained conservative in their structure from those that were capable of diverging and forming new species. Darwin repeatedly emphasized the almost imperceptibly slow and gradual nature of change, requiring very great numbers of generations or long periods of time. This great emphasis on the gradual nature of change between species was certainly in response to the then-prevailing assumption of large, abrupt differences between essentially static species resulting from their individual and separate creation.

Darwin's view of the mode of evolution is typically expressed in terms of progressive transformation within a particular lineage, but he in fact outlined a wide variety of possible patterns and rates that would be expected at the level of populations and species. Darwin placed a great deal of emphasis on the divergence of populations. He considered that small, isolated populations might have the greatest potential for rapid change because natural selection could act most effectively in a small area with limited environmental variability, but he noted that the potential for long-term adaptation and survival might be greater in larger populations with a greater amount of total variability and greater challenges from the more varied environment. He suggested that the optimal model for rapid and long-term evolution would be a large area subject to repeated breakup and coalescence of environments and populations: The entire species could contribute to the variability available in the individual populations, and yet repeated isolation would contribute to locally more effective natural selection. He specifically noted that partially isolated populations might evolve more rapidly because of limited interbreeding with other populations that were subject to different selective pressures.

Darwin argued that many populations might last for brief periods of time but that only the most divergent were likely to survive, since natural selection would tend to eliminate intermediate forms. Although not stated in that form, one might suggest an analogy between Darwin's "population selection" and the hypothesis of species selection proposed by Stanley (1975, 1979). Darwin was not dogmatic on this point, however, but noted that the survival of particular lineages would ultimately depend on the nature of the places that were either unoccupied or imperfectly occupied by other species.

It is of particular interest that Darwin's chart did imply the frequency of stasis. Of eleven species illustrated, only two show progressive change; the other nine do not change perceivably over the thousands or tens of thousands of generations indicated on the chart. Darwin clearly indicated that such unvarying species, despite their numerical predominance in his chart, were extremely vulnerable to extinction.

Darwin envisaged the origin of new species as resulting both from the divergence of populations within species and from progressive change within a particular lineage. "In some cases I do not doubt that the process of modification will be confined to a single line of descent, and the number of the descendants will not be increased; although the amount of divergent modification may have been increased in the successive generations" (1859, pp. 163–4). Darwin did not recognize the distinction between speciation and phyletic evolution that was later emphasized by Dobzhansky (1937, 1951), Mayr (1942, 1963), and Simpson (1944, 1953).

The overall pattern suggested by Darwin was one of progressive radiation of populations and species with greater than average potential for variation, and the successive extinction of species with less capacity to respond to natural selection. At the level of populations and species, the number of lineages does not necessarily increase, and the adaptive space does not change appreciably.

See G. A. C. Bell (1996) for a recent analysis of the power of selection.

Dobzhansky and Mayr and the modern understanding of the role of species in evolution

For Darwin, acceptance of evolution depended on the capacity for lineages to change over many generations to the extent of giving rise to populations so different that they would be recognized as belonging to distinct species. Although partial isolation was recognized as contributing to the effectiveness of natural selection in producing change, Darwin did not differentiate the formation of species as a result of the division of lineages from the formation of species within a particular lineage. He showed continual divergence of lineages in his diagram, but this led to new species only gradually, over a great many generations, rather than as a direct result of their initial divergence.

An objective definition of individual species was essentially impossible within the context of Darwin's view of evolutionary change through time. Only arbitrary divisions were possible within a single evolving lineage. For this reason, there was much controversy regarding the objective nature of species during the early years following acceptance of evolutionary theory (Mayr 1982). The taxonomic differentiation of species within evolving lineages remains a very difficult problem for paleontologists. On the other hand, biologists working with modern species have relatively little difficulty differentiating one from another. Nearly all plant and animal species that are observed in the wild can be readily assigned to a particular species; only rarely are hybrids or other intermediates recognized (Diamond 1992). On this basis, Dobzhansky (1937) and Mayr (1942) formulated a species concept that pertained specifically to living organisms. According to Mayr (1963, p. 19): "Species are groups of actually or potentially interbreeding natural populations that are reproductively isolated from other such groups." This is referred to as the **biological species concept**, since it relies on the biological nature of species, rather than on their morphology, to differentiate them from one another.

The differentiation of species present in the living fauna may have resulted primarily from morphological or physiological change within lineages, as was emphasized by Darwin, or primarily by achievement of reproductive isolation. As a result, two patterns of evolution are recognized at the species level: **Phyletic evolution** is the progressive accumulation of change within a single lineage; **speciation** results from the splitting of lineages so that each is reproductively isolated from the other (Fig. 2.1). More generally, any evolutionary change within a lineage is referred to as **anagenesis**, and the multiplication of lineages is termed **cladogenesis.** According to Mayr (1963) and Simpson (1953), speciation per se is not necessarily accompanied by significant morphological change.

Whereas phyletic evolution results from natural selection and is directly associated with adaptation within a lineage, the initial stage of speciation is typically imposed on the species by external forces that may not reflect specific aspects of the adaptation of the parental species. Mayr (1963) argued that speciation among multicellular animals nearly always depends on the physical separation of populations or individuals of a species that originally occupied a single, geographically contiguous area. One might imagine a single land mass being dissected by rivers or glaciers, or portions of the ocean being separated by land, such as the elevation of the Isthmus of Panama during the late Cenozoic. Isolation may also occur as a result of the dispersal of organisms to places from which it is very unlikely that they or their descendants will ever return to reproduce with the parent populations. This commonly occurs in the colonization of oceanic islands and other isolated environments. If these barriers are large enough and sufficiently long lasting, phyletic evolution will proceed in each separate portion of the species, leading to different patterns of adaptation in the separate environments. Eventually, random changes in the genetic material will accumulate to the degree that reproduction between the two incipient species becomes difficult or impossible, even if they again inhabit the same area.

When species occupy the same area they are referred to as **sympatric**; when they occupy different areas, they are termed **allopatric**. As long as populations are allopatric, it is difficult to apply the biological species concept to determine whether or not they belong to the same species. Once they become secondarily sympatric (i.e., sympatric a second time), their capacity to produce viable and fertile hybrids determines whether or not they are accepted as separate species.

If incipient species are separated for a significant period of time in areas where different physical and biological conditions result in significant adaptive change, they may have only limited interactions if they become secondarily sympatric. In this case both may survive, with limited subsequent change, within the same general area. At the other extreme, if the populations have been separated only for a short period of time, they may be fully interfertile and hybridize completely. If separation occurs for an intermediate length of time and is followed by sympatry, the two incipient species may have not diverged greatly from one another in their adaptational specializations. If they then compete strongly for the same resources, it is probable that one or the other will rapidly become extinct. If extinction does not occur, there may be time for selection to act on both populations to amplify any structural and/or behavioral traits that enable them to reduce competition for the same space or resources – for example, feeding on different prey, or nesting in different areas. This is referred to as **character displacement**. If the incipient species have adapted to quite different ways of life while allopatric, they may, when they become sympatric, avoid competition yet still have the capacity to interbreed. In this case, the hybrids, if intermediate in structural and behavioral characteristics, may be at a disadvantage relative to both parents and so be selected against. Since formation of hybrids with a lower potential for survival is a waste of the parents' gametes and other aspects of their reproductive efforts, selection will act to reinforce any aspects of behavior that would reduce the potential for interbreeding. This is likely to accentuate differences in physical features as well, further distinguishing recently sympatric species.

This model of speciation, which appears nearly universal among vertebrates, is referred to as **allopatric speciation** because it depends on physical separation of the parental species into two reproductively isolated populations. Some agents of isolation, such as separation of lithosphere plates associated with continental drift, may divide species into large, subequal portions, but speciation is more likely to affect peripheral populations with much smaller numbers of individuals than the main population. Such peripheral isolates may have already adapted to somewhat different environmental conditions than the remainder of the species. For this reason, one may think of the process of speciation as beginning even when the species as a whole retains some degree of genetic continuity. Speciation as envisaged

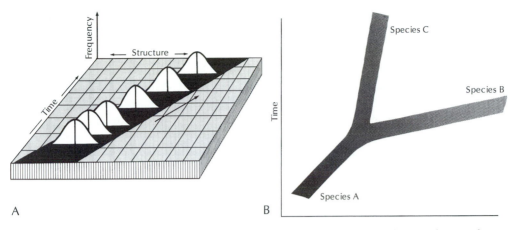

Figure 2.1. Patterns of evolution at the level of species. **A,** Phyletic evolution: change over time in a given lineage, as emphasized by Darwin. Division between successive species is arbitrary (from Moore, Lalicker, and Fischer [1952], *Invertebrate Fossils.* Reproduced with the permission of McGraw–Hill, Inc.). **B,** Speciation through reproductive isolation of lineages. Phyletic evolution occurs within each lineage.

by Dobzhansky and Mayr is hence not fully distinct from the pattern of species formation hypothesized by Darwin, for both may begin at the level of partially isolated populations. Mayr referred to speciation via marginal, partially isolated populations as **peripatric speciation**.

In the speciation of plants, as opposed to that of most vertebrates, polyploidy and other changes in the number of chromosomes are common factors. This may occur spontaneously in a particular lineage or arise as a result of hybridization. Because many plants are self-fertile and/or capable of vegetative reproduction, a single modified individual may give rise to a new species that is instantly reproductively isolated from other members of the population in which it lives. This is an example of **sympatric speciation**. Some biologists suggest that sympatric speciation may be possible among animals in which partially isolated populations within a widely ranging species may be subject to markedly different selective pressures and depart significantly from other populations in their morphology and behavior. For speciation to occur under these conditions, there would have to be preferential mating within rather than between populations. Even a very low percentage of interbreeding between populations (as little as 0.01 percent per year) is sufficient to maintain a degree of genetic homogeneity within an entire species such as that of the lizard *Lacerta agilis,* which ranges across most of temperate Eurasia (Yablokov, cited in Ayala and Valentine 1979). Sympatric speciation may be possible among vertebrates, but few if any specific examples have been demonstrated, even among the extremely speciose cichlid fishes in East African lakes (Echelle and Kornfield 1984). Peter and Rosemary Grant (1989) described the conditions that would be necessary for the occurrence of sympatric speciation among vertebrates, and concluded that it is a very rare phenomenon.

Simpson (1944, 1953) followed the work of Dobzhansky and Mayr in emphasizing phyletic evolution rather than speciation as the primary focus for evolutionary change. One would expect to see morphological change immediately following renewed sympatry in incipient species, but this would not occupy a significant portion of the duration of a species.

The nature of species and speciation continues to be a central concern of evolutionary biologists. Numerous, divergent views of the definition and nature of species continue to be elaborated (e.g., collected papers edited by Otte and Endler [1989]). They range from strong support of sympatric speciation (at least in groups such as insects, in which parasites and herbivores may have extremely restricted, but changeable, host species [Diehl and Bush 1989]) to a new definition of species that extends Dobzhansky's concentration on sexually reproducing organisms to the vast number of asexual forms (Templeton 1989). The most extreme viewpoint is taken by Nelson (1989), who argues that species may not have a unique role in evolution.

Cracraft's (1989) viewpoint is important for its emphasis on the evolutionary role of the major subdivisions of species (subspecies, for most authors), arguing that they, not the larger assemblage of potentially interbreeding individuals of the biological species concept of Mayr, are the actual units of continuing genetic interchange that have the potential for adaptation to different habitats and geographical regions. This view, in fact, continues that of Darwin in emphasizing the importance of subdivisions of species diverging toward the status of species on their own.

Despite the apparent controversy exemplified by the papers in the Otte and Endler volume, most current research that deals with evolution at the level of the species demonstrates that it is practical and informative to work within the concept of the biological species elaborated by Dobzhansky and Mayr (Futuyma 1986).

The fossil record

By the time Darwin wrote *The Origin of Species,* many fossils had been discovered that demonstrated the prior existence of a diversity of plants and animals completely different from any extant organisms. Entire groups, such as dinosaurs, the flying pterosaurs, and several types of marine reptiles were common in the Mesozoic but have no living representatives. Other groups, such as fossil elephants, could be related to modern families, but these particular genera had not survived into the modern world.

Fossils discovered in Darwin's time clearly showed major changes in Earth's biota during geological history, but the fossil record was not sufficiently well known to trace the evolution of any of the living species. The first human fossils, those of Neanderthals, were found just three years before publication of *The Origin of Species,* but their status as members of a population significantly more primitive than modern humans was not established until many years later. The specific relationship of Neanderthals to modern humans has still not been satisfactorily resolved (Trinkaus and Shipman 1993). The single most significant fossil to be discovered in the nineteenth century was the skeleton of the ancestral bird, *Archaeopteryx.* It was collected in Germany in 1861 and was hailed as a missing link uniting birds and reptiles. This specimen provided the first convincing evidence of a link between two major groups of vertebrates. On the other hand, it contributed no information as to the patterns and processes of evolution at the species level, for it was isolated by tens of millions of years and enormous gaps in morphology from its nearest possible relatives.

Even among apparently well-known sequences, such as the lineages of Cenozoic mammals found in the western United States in the late nineteenth century, most fossil-bearing beds were separated from one another by long periods of time. The horse genera known at the time of Darwin were separated from one another by tens of millions of years.

What Darwin eventually hoped to find were "infinitely numerous transitional links" (1859, p. 310) joining all forms of life. In his chapters on the inadequacies of the fossil record he asked, "Why then is not every geological formation and every stratum full of such intermediate links? Geology assuredly does not reveal any such finely graduated organic chain; and this, perhaps, is the gravest objection which can be urged against my theory" (p. 280). He answered this in terms of the nature of the fossil record: "The geological record is extremely imperfect and this fact will to a large extent explain why we do not find interminable varieties, connecting together all the extinct and existing forms of life by the finest graduated steps. He who rejects these views on the nature of the geological record, will rightly reject my whole theory" (p. 342).

Despite more than a hundred years of intense collecting efforts since the time of Darwin's death, the fossil record still does not yield the picture of infinitely numerous transitional links that he expected. In contrast, a very different pattern of the distribution of fossil organisms has been established by paleontologists.

Stratigraphy

It should be emphasized that the study of fossils and their practical use in stratigraphy long predated the writings of Darwin. By early in the nineteenth century, Alexandre Brongniart, Georges Cuvier, and William Smith independently recognized that most sedimentary deposits were characterized by particular suites of fossils. These fossil assemblages could be used to identify comparable strata over hundreds and even thousands of miles. Fossils helped to demonstrate that the relative sequence of different strata was also constant over large areas, providing the basis for a geological time scale that was eventually extended over the Earth's entire surface. The study of fossils became extremely important in the tracing of deposits of commercial value such as coal, limestone, and shale, and later in the discovery of oil and gas deposits, as well as the engineering of canals, railroads, and other construction projects. At the same time, the establishment of the relative sequence of beds and their correlation from one area to another were contributing to an understanding of the geological history of Earth. Paleontology developed initially as the handmaiden of stratigraphy rather than as a means to understand evolution. Although fossil vertebrates and remains of the leaves and woody structures of plants have been studied primarily in reference to their biological nature, most paleontologists who study the remains of microorganisms, nonvertebrate metazoans, and plant spores and pollen are mainly concerned with the use of fossils for characterizing particular geological horizons.

Most papers in publications such as *Journal of Paleontology* and *Palaeontology* that deal with descriptions of stratigraphic sequences and particular taxonomic groups include stratigraphic ranges of all species that are discussed. These charts are designed for use in correlation and relative dating (Fig. 2.2). They rarely indicate relationships. Such charts give the impression that all members of each species retain a very uniform morphology throughout their range, at least to the extent that most specimens can be readily recognized and distinguished from those in higher or lower beds. This is certainly true in a practical sense. Stratigraphers who are not themselves paleontologists depend on these data for establishing correlations throughout the world. Correlations based on the ranges of fossils are commonly accurate within a range of half a million years, and may be accurate to levels of approximately a hundred thousand years.

It is the very constancy of their morphology that makes fossil species so useful in characterizing particular stratigraphic horizons. The great attention given to using fossils as stratigraphic indicators must somewhat bias collection and study toward common fossils that do retain a nearly constant morphology through an extensive section of strata. Nonetheless, detailed faunal studies have not documented the many cases of progressively changing species that would be expected if this were a common phenomenon. For most of the fossil record few, if any, well-documented cases have been presented that show progressive change within species. The inadequacies of several putative examples were discussed by Gould and Eldredge (1977).

Extensive use of fossils as stratigraphic indicators in the nearly 150 years since the publication of *The Origin of Species* showed that a great many fossil species either have long time spans (hundreds of feet of strata, or tens of millions of years) or are restricted to a single horizon. Careful reading of the text of these papers reveals numerous cases in which transitions between species are recognized, but most of the range of most species appears to be without conspicuous change. This gives an overall picture that is very different from that expected by Darwin when discussing the inadequacies of the fossil record. Darwin recognized that stasis was a common feature of the history of species, but he argued that species with little variability would not long survive in competition with other species that were more responsive to natural selection. The fossil record suggests just the reverse: Species that show little variability were obviously both common and extremely long lived. Fossils showing transition between species are rare and confined to short stratigraphic intervals. Few well-studied examples from the record of fossil invertebrates show a pattern of evolution such as that predicted by Darwin, with gradual and progressive change within and between species over long periods of time. Most paleontologists and evolutionary biologists were not concerned by these observations, however, since it was assumed that the fossil record was still too poorly known to show what was happening at the level of populations. Long-term stasis was not considered contrary to the writings of Darwin, even if it appeared to be expressed by most species. Anything might be happening in the intervals between known species, but there was no reason to think that change resulted from any process other than those postulated by Darwin.

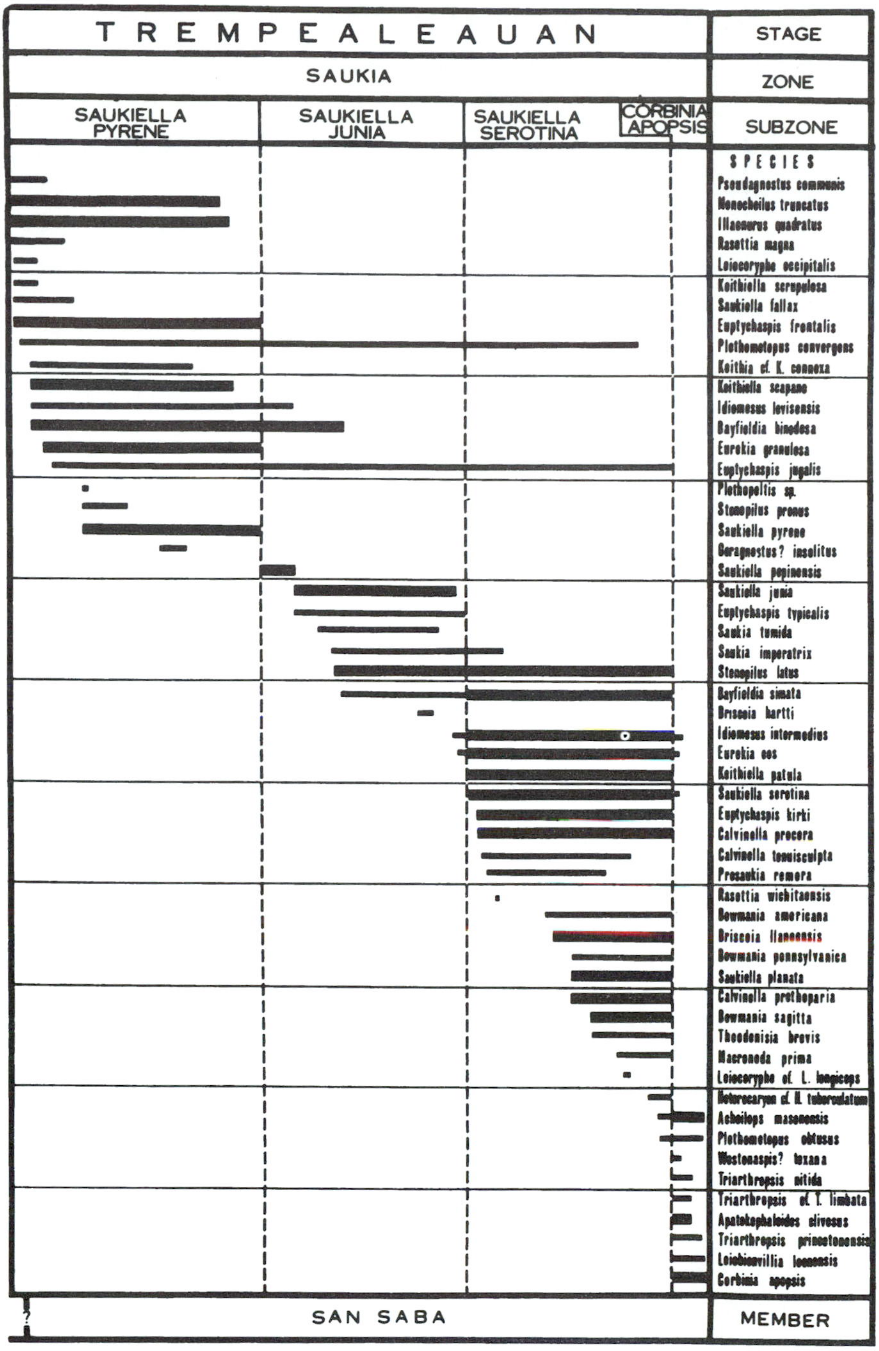

Figure 2.2. Stratigraphic ranges of Upper Cambrian trilobites. From Longacre (1970).

Eldredge and Gould and the theory of punctuated equilibria

This view was suddenly challenged in a paper by Eldredge and Gould (1972), "Punctuated equilibria: An alternative to phyletic gradualism." They suggested that what appeared to be the picture from the fossil record – conspicuous morpholog-

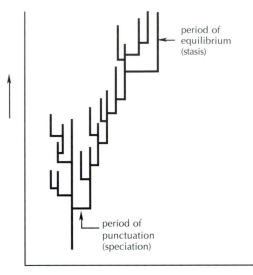

period of
equilibrium
(stasis)

period of
punctuation
(speciation)

degree of difference from original form

Figure 2.3. The pattern of evolution termed "punctuated equilibrium" by Eldredge and Gould, in which most morphological change occurs during the process of speciation and most of the duration of a species is characterized by stasis. From Gould (1982).

ical gaps between species and long periods during which particular anatomical patterns are maintained – was the literal truth. The long periods of little change or stasis were considered to be periods of equilibrium between the species and its environment. The gaps were thought to represent short periods of rapid change, or punctuation, brought about by speciation. They illustrated punctuated equilibrium as a series of straight vertical lines, connected by short horizontal segments (Fig. 2.3). This is clearly distinct from the pattern illustrated by Darwin and implies very different underlying processes. Darwin did not feel that all evolutionary change occurred at the same rate; some of the lines in his diagram are vertical, implying stasis, but his writings indicate that he felt that many, or perhaps most, species changed continually and steadily. Eldgredge and Gould, in contrast, felt that most change within a species occurred at the time of speciation, and that most of the history of species was represented by stasis. Gould later (1982, p. 84) specifically suggested that approximately 99 percent of the history of a species was passed in stasis.

Gould (1993) was careful to emphasize that gaps between species were not the result of saltation, in the form of large instantaneous changes from generation to generation, as had been proposed by the geneticists De Vries (1906) and Goldschmidt (1940) and the paleontologist Schindewolf (1950), but changes over periods from 5,000 to as much as a hundred thousand years that frequently pass unnoticed in the fossil record because of the incomplete nature of sedimentary sequences. Gould referred to such changes as being instantaneous in terms of the geological record, although they are still very long in human and ecological time scales.

Gould and Eldredge cited the allopatric and peripatric models of speciation established by Mayr (1954, 1963) as the logical cause for the pattern seen in the fossil record. If sympatric speciation is extremely rare among metazoans, one would not expect to see examples of speciation within a particular geographic area; despite expectations by paleontologists, one should not be able to document divergent lineages within a local stratigraphic sequence. If physical division of the gene pool were necessary for initiating speciation, one would expect that new species would arise elsewhere, following separation, and would only be able to cohabit with the parental lineage following sufficient morphological change to avoid close competition. In the well-known mammalian faunas from the Tertiary of North America, the appearance of new species in a particular area is frequently attributed to migration from elsewhere. Speciation and return to sympatry might well be expected to occur during the relatively short intervals suggested by Gould and Eldredge. A single lineage represented by different but closely related species in successive horizons can be attributed to speciation followed by replacement of the native species by its newly sympatric descendant, as opposed to anagenesis within a single lineage.

Emphasis on factors associated with speciation to explain gaps in the fossil record is not in itself controversial. What is much more difficult to accept is the assertion by Eldredge and Gould that nearly *all* differences that characterize species are a result of speciation, to the exclusion of phyletic evolution. Whereas Darwin, Mayr, and Simpson emphasized phyletic evolution to the near exclusion of speciation as a factor in morphological change, Gould and Eldredge took the opposite view. In their first paper on this subject, Eldredge and Gould stated: "Tracing a fossil species through any local rock column, so long as no drastic changes occur in the physical environment, should produce *no* pattern of constant change, but one of oscillation in mean values" (p. 95).

If nearly all species are essentially static throughout their duration, change must be explained by mechanisms other than phyletic evolution. Eldredge and Gould argued that most change results from speciation. Change during speciation may be attributed initially to the small population size that may accompany this process. As suggested by Darwin, small isolated populations may evolve especially rapidly in response to the specific selective pressures of a local environment, in contrast to populations that range over a larger area and may be subject to many conflicting selective pressures. Change during allopatry can be explained through adaptation to different environments, but this does not differentiate change following speciation from change that may occur within the species in its original environment if that environment itself changes. Change can be also explained by selection for character displacement and elaboration of premating isolating mechanisms in incipient species that have recently returned to a sympatric distribution.

Even more contentious than the association of major change with speciation are the reasons given to explain why species do not show significant change during roughly 99 percent of their duration following origin by speciation. Gould and Eldredge specifically reject the notions that either interbreeding between populations or stabilizing selection are powerful enough to maintain the perceived stability of species. Following Mayr, they argued for some sort of genetic cohesiveness in the nature of species. Eldredge and Gould suggest: "The answer probably lies in a view of species and individuals as homeostatic systems – as amazingly well-buffered to resist change and maintain stability in the face of disturbing influences"

(1972, p. 114). This lacks any specific mechanism. According to Futuyma (1989, p. 558): "From a genetic point of view, the hypothesis that Eldredge and Gould (1972) proposed to explain the patterns claimed for the fossil record appears untenable."

In the absence of any significant change within species, and only random changes at the time of speciation, Eldredge and Gould suggest that long-term evolutionary trends are produced by differential survival of species, a phenomenon elaborated by Stanley (1975, 1979) under the name **species selection**. This is discussed in more detail in Chapter 5.

Eldredge and Gould (1972) and Gould and Eldredge (1977) not only emphasized the very different appearance of the fossil record compared with that postulated by Darwin, but also discussed at length the tendency for paleontologists to support the gradualist views of Darwin even when they are at odds with the overwhelming evidence to the contrary. They cited many papers in which paleontologists described change between species as being gradual when the only real evidence was a gap in morphology between two distinct but presumably closely related lineages. In other cases, "gradual transitions" were shown to comprise several successive species, each of which exhibits no significant change during its duration. Eldredge (1985b) provided a detailed example drawn from the study of Paleozoic trilobites, a group with which he is particularly familiar. Campbell's (1977)

Figure 2.4. Phylogeny of the trilobite group Phacopinae. From Campbell (1977).

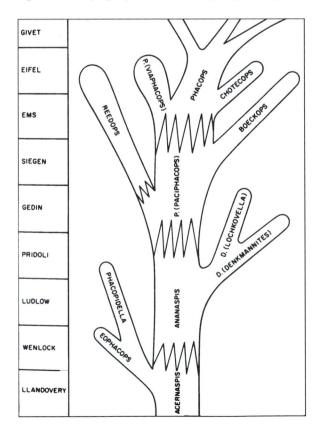

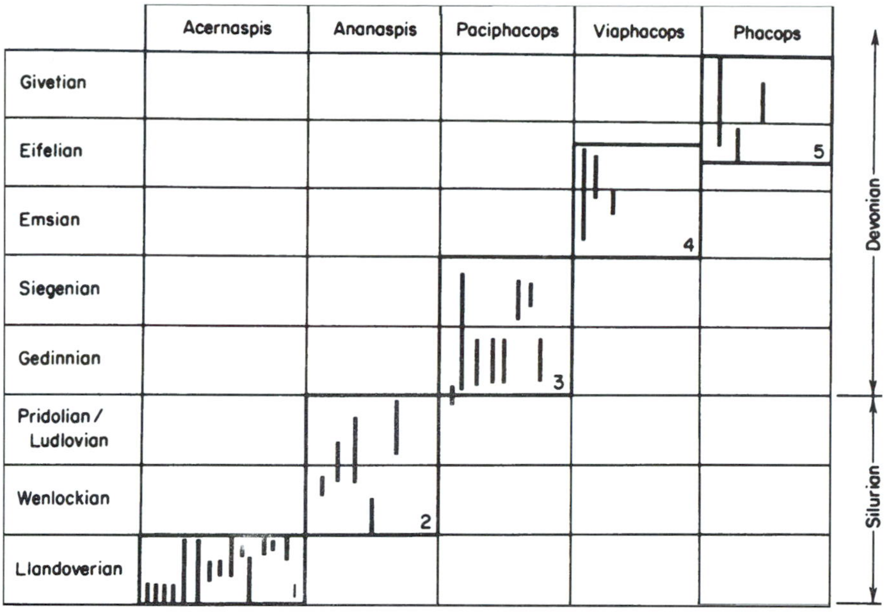

Figure 2.5. Eldredge's illustration of the stratigraphic ranges of better-known species of five genera of the Phacopinae. From Eldredge (1985b), *Unfinished Synthesis,* Oxford University Press.

phylogenetic scheme for the Phacopinae through 60 million years of Lower Silurian through Middle Devonian strata (Fig. 2.4) gives the impression of continual progressive change; but the data consist of a series of almost unchanging species, in clearly defined genera, most of which are restricted to particular geological horizons (Fig. 2.5). No real transformations can be demonstrated between any of the species. Eldredge argued that significant environmental changes reflected in the subdivisions of the major stratigraphic units were the cause of abrupt changes in the trilobite lineages, which remained essentially static within each unit.

On a personal note, I must confess to having observed the predominance of stasis in the fossil record when I was 17, but failed to grasp its significance. For a science project, I had expected to be able to trace evolutionary change among the very rich assemblage of brachiopods that I was collecting in the Middle Devonian Alpena Formation in Northern Michigan. After a brief effort to find an appropriate succession of fossiliferous beds in one locality, I realized that no matter where I collected in this formation, I found the same species. I could recognize a great variety of species, but each was clearly distinct from the others, with no evidence of progressive change. Approximately ten years later, Eldredge searched these same deposits for progressive evolution within the trilobite genus *Phacops* and incorporated his observations in the theory of punctuated equilibria (as reported in Eldredge [1985a]).

In order to make their interpretations appear as different as possible from those of other paleontologists, Eldredge and Gould coined contrasting terms to characterize the alternative hypotheses. They designated theirs *punctuated equilibria,* and provided a new phrase for the ideas they attributed to other paleontologists, **phyletic gradualism**. Phyletic gradualism should not be confused with phyletic evolution, coined previously to distinguish evolution within a given lineage from speciation. The term *phyletic gradualism* was coined to emphasize a constant, slow rate of evolution.

It should be noted that the word *gradual* in this context is used by Eldredge and Gould in a different way than that emphasized by Darwin. Darwin used the word *gradual* primarily in the sense of continuous (without gaps) to distinguish it from saltatory change associated with instantaneous creation. This usage is in accord with the primary definition in dictionaries of his day, such as the *Oxford English Dictionary* and Johnson's *Dictionary of the English Language* (1832 edition): "Of or pertaining to degree; only in gradual difference" and "proceeding by degrees; advancing step by step; from one stage to another." In contrast, Gould and Eldredge have used *gradual* as meaning at a regular and slow rate, which is clearly distinct from the extremely variable *rates* of evolution accepted by Darwin and Simpson. This distinction is quite evident in the view of evolution attributed to Darwin and paleontologists in general by Eldredge and Gould:

> In this Darwinian perspective, paleontology formulated its picture for the origin of new taxa. This picture, though rarely articulated, is familiar to all of us. We . . . identify the following as its tenets:
>
> (1) New species arise by the transformation of an ancestral population into its modified descendants.
> (2) The transformation is even and slow.
> (3) The transformation involves large numbers, usually the entire ancestral population.
> (4) The transformation occurs over all or a large part of the ancestral species' geographic range. (1972, p. 89)

It is difficult to argue with viewpoints attributed to an entire discipline, but one can compare these statements with the writings of Simpson, who was the primary spokesman for paleontologically oriented evolutionary theory.

Tenet 1 does accord with the writings of both Darwin and Simpson, and is probably the feeling of most paleontologists. The emphasis on transformation clearly distinguishes this interpretation from the nearly instantaneous change resulting from speciation advocated by Eldredge and Gould.

Tenets 2–4, on the other hand, do not follow from the writing of either Darwin or Simpson, who clearly argued for a much broader interpretation of the patterns and rates of evolution. Simpson in particular stressed that taxa of any level might undergo especially rapid rates of evolution at the time of their origin. One may question the meaning of the word *entire* in the third tenet. Darwin, Simpson, and Mayr all wrote of the importance of small, isolated populations in evolutionary change. If Eldredge and Gould equate *entire* with large populations occupying much of the range of a common species, tenet 3 does not accord with the writings of earlier authors, nor necessarily with those of the majority of practicing paleontologists. If *entire* is applied to all members of a small, isolated population, it would be accepted by many paleontologists but would not differ from the ideas of Eldredge and Gould.

Eldredge and Gould argue that these statements imply several consequences, two of which seem especially important to paleontologists:

1. Ideally, the fossil record for the origin of a new species should consist of a long sequence of continuous, insensibly graded intermediate forms linking ancestor and descendant.
2. Morphological breaks in a postulated phyletic sequence are due to imperfections in the geological record.

Most paleontologists would accept the first of these consequences, although the word *long* would be subject to argument. Eldredge and Gould's hypothesis differs only in the relative time required for the change. They do not advocate saltation. Simpson and many other evolutionary biologists admit the possibility of relatively rapid short-term evolution during transitions between taxa at all taxonomic levels. Paleontologists frequently cite migration, not gaps in the record, as explaining replacement of one lineage by a closely related form in a stratigraphic sequence.

All of Gould and Eldredge's points are succinctly phrased in the abstract of their 1977 paper:

> We believe that punctuational change dominates the history of life: evolution is concentrated in very rapid events of speciation (geologically instantaneous, even if tolerably continuous in ecological time). Most species, during their geological history, either do not change in any appreciable way, or else they fluctuate mildly in morphology, with no apparent direction. Phyletic gradualism is very rare and too slow, in any case, to produce the major events of evolution. Evolutionary trends are not the product of slow, directional transformation within lineages; they represent the differential success of certain species within a clade – speciation may be random with respect to the direction of a trend. (p. 115)

These same beliefs are again expressed in their most recent paper on this subject (Gould and Eldredge 1993).

The most striking claim made by Gould and Eldredge is that natural selection is not a significant force in controlling the pattern or rate of evolution at the level of species. Thus they deny the primary tenet of evolutionary theory as proposed by Darwin.

Can these points be maintained in view of the evidence provided by modern studies of living populations?

3 Evolution in modern populations

Gould and I claimed that stasis – nonchange – is the dominant evolutionary theme in the fossil record. It is characteristic of most species that have ever lived. – Eldredge (1985a, p. 128)

Clines are widespread and occur in the majority of, if not all, continental species. – Mayr (1963, p. 362)

Introduction

Darwin and the authors of the punctuated equilibrium model saw evolution from two contrasting points of view. Although Darwin used natural selection to explain evolution throughout the history of life, the evidence he used was drawn almost entirely from modern populations. Eldredge and Gould, in contrast, elaborated their hypothesis primarily on the basis of the known fossil record, but argued that stasis and punctuation were dominant forces throughout evolution and hence should be recognizable in living species.

Eldredge and Gould claimed that the theory of punctuated equilibria grew out of consideration of the nature of speciation elaborated by Dobzhansky and Mayr on the basis of living populations. In contrast, their assertion that natural selection is not significant in producing long-term evolutionary trends is completely contradictory to the concepts of both Darwin and the authors of the modern synthesis.

Although the pattern of evolution throughout the entire history of species can only be established on the basis of data from the fossil record, knowledge of the living fauna provides important evidence of the potential for long-term evolutionary change within species, and can be used in evaluating the degree of correlation between speciation processes and morphological change.

Darwinian selection theory was based on four major premises:

1. All populations exhibit variability.
2. A significant portion of that variability is inherited.
3. Only a small proportion of the individuals in each generation survive to produce the next generation.
4. Traits that favor the survival of individuals in one generation will be preferentially perpetuated in the next.

These first propositions can be readily demonstrated on the basis of the anatomy, protein polymorphism, genetics, and demography of living populations. There was also a fifth proposition:

5.   Progressive changes in traits from generation to generation will result in long-term evolutionary change.

This was accepted by the authors of the modern synthesis but not by Eldredge and Gould, who argued that very little change occurs during the longevity of species. They viewed changes within populations as being essentially random in direction. Evolutionary change resulting from natural selection was considered capable of tracking minor, short-term environmental variations, frequently in an oscillatory manner, but of having little long-term significance. These conclusions appear to contradict most of the studies of evolution within living populations that have been carried out during the past 150 years. The problem is this: Can observations made over a few years, or even the lifetime of a scientist, be extrapolated to the hundreds of thousands or millions of years of most species' longevity?

Darwin's best examples of large-scale morphological change within known species were primarily limited to domestic plants and animals. In these cases, historical and archaeological records going back several thousand years documented morphological changes as great as those that separate well-established species, but the changes were related to much more intense selection than would normally be observed in nature. No examples are known of changes of this magnitude in natural populations during such a limited period of time. Although the manner of selection is clearly very different from that encountered in nature, these examples do demonstrate the genetic, developmental, and physiological capacity for extremely rapid change of considerable magnitude in a diversity of different vertebrate groups. The magnitude of change is equivalent to that expected by Darwin to lead from one species to another.

Natural populations do show very rapid changes in the frequency of individual genetic traits. Many studies have documented radical changes in allele frequencies on a seasonal basis in natural populations such as the land snail *Cepaea* (Cameron 1992) and over decades and centuries in the peppered moth *Biston betularia* (Kettlewell 1973). The best-documented examples are from invertebrate groups in which the agents of selection are simple and well demonstrated (in both cases, avian predators on different color morphs), and the genetic background well established. However, these changes are minor in terms of the organism as a whole and are not progressive over longer time scales: The proportion of moths exhibiting industrial melanism has decreased dramatically as air pollution has been reduced, and fossils of *Cepeae* that preserve color banding show that color polymorphism has been retained for thousands of years. The expression of particular alleles is unquestionably oscillatory and has led to no permanent changes over this period of time. Endler (1986) has tabulated a great many recent studies of natural selection in modern populations, but most are of very short-term duration, and longer-term oscillation or random change around a common mean cannot be precluded.

Few changes within nondomesticated species over the span of human history have been documented as being primarily unidirectional or of the magnitude that would lead to the recognition of a new species through phyletic evolution. However, such a limited period of time samples only a very small percentage of the duration of species, which averages well over a million years. Less than 1 percent of the amount of change necessary for the recognition of distinct species would be expected to occur during the period of written history. Thus even the most representative of the early cave paintings would not be expected to document differences within living species, although they do illustrate woolly mammoths, mastodons, and rhinoceroses that have since become extinct.

To demonstrate significant, progressive phyletic evolution in living species, we must look for special cases in which populations have evolved much more rapidly than average, or seek examples where large-scale change is indirectly reflected in modern populations. Among the most striking examples of rapid evolutionary change are those that have resulted from immigration of organisms from one continent to another. Although many of these cases are the result of human agency, comparable invasions have occurred frequently under natural conditions, as in the colonization of oceanic islands.

Evolutionary change in immigrant populations

One of the most thoroughly studied cases of evolutionary change among immigrant populations is that of the house sparrow *Passer domesticus,* which was introduced into eastern North America between 1852 and 1860. Descendants of a small number of birds from central England rapidly spread throughout North America from southern Mexico into southern Canada. Johnston and Selander (1971) measured sixteen characters from the bony skeleton of 1,752 birds from thirty-three localities distributed throughout the range of the species. All showed statistically significant interlocality variation.

Regression analyses were performed relative to fifteen independent environmental variables. Significant correlations were found between many variables; the most conspicuous were between overall body size and geographic distribution. The largest birds occur in central and eastern Canada, the Great Plains, and the Rocky Mountains. Those from the southwestern United States and Texas are intermediate in size, and those from the West Coast, the Gulf Coast, and Mexico are the smallest. Correlation with winter and summer temperatures in these areas indicates that body size has adapted to climatic conditions as predicted by Bergmann's (1847) and Allen's (1877) ecogeographic rules. Sparrows with small body cores and relatively long limbs can thermoregulate most efficiently at high ambient temperatures, and those with large body cores and relatively short limbs can thermoregulate most efficiently at low ambient temperatures.

Despite the relatively low degree of variability expected in a small founder population from a single geographical area, their descendants managed to adapt to an extremely wide range of local climatic conditions in only 50–115 generations.

In order to establish the relative amount of evolutionary change that these sparrows had undergone, Johnston and Selander compared both intra- and interpopulational variation in North America and Europe. They found no statistical difference between populations collected in Germany during 1815–54 and those of 1963–5 – approximately the same period of time during which extensive variability had evolved in North America. Intrapopulation variability is limited in both North America and Europe, and does not differ significantly in the two continents.

Table 3.1. *Interlocality variability in skeletal characters of house sparrows of Europe and North America*

Character	Range in Means[a] Males Europe	Eng.–Ger.	N.A.	Females Europe	Eng.–Ger.	N.A.
Skull length	1.50	0.40	1.20	1.40	0.80	0.70
Premaxilla length	0.63	0.31	0.35	0.58	0.33	0.33
Dentary length	0.51	0.24	0.30	0.44	0.30	0.31
Mandible length	1.00	0.78	0.80	1.30	0.80	0.50
Narial width	0.13	0.05	0.15	0.11	0.06	0.16
Sternum length	1.40	0.07	1.20	1.72	0.90	0.80
Sternum depth	1.03	0.23	0.70	1.31	0.57	0.65
Humerus length	0.90	0.50	0.90	1.10	0.80	0.80
Ulna length	1.10	0.70	1.20	1.07	1.00	0.70
Femur length	0.60	0.50	0.90	0.90	0.60	0.76
Tibiotarsus length	1.40	0.80	1.70	2.20	1.20	1.10
Tarsometatarsus length	1.10	0.50	0.90	1.60	1.10	0.80

[a]For any character, the maximum character-state value minus the minimum character-state value equals the range in means ($Y_{max} - Y_{min}$ in mm).
Abbreviations: Eng., England; Ger., Germany; N.A., North America.
Source: Johnston and Selander (1971).

Interpopulational differences in male sparrows from North America are equal to or greater for all character measures than for individual English or German populations (Table 3.1). Five characters show equal or greater variability than that exhibited in all European samples. Female sparrows in North America show less striking change: Only six characters show equal or greater variability than a single population, and only one shows equal or greater variability than all European populations.

The greater degree of interpopulational variability in Europe can be attributed to the much longer period of time during which the European populations have adapted to particular environments. Presumably the European populations began to adapt to local conditions soon after the end of the last ice age, approximately 10,000 years ago. Although North American sparrows have not yet achieved the same degree of local specialization as their European relatives, the basic pattern is clearly established in 1 percent of the time. Even the period of time available to the European sparrow is short in geological terms and would be difficult to quantify in strata earlier than the Quaternary (the past 1.8 million years of geological time); but it does show the capacity for significant evolutionary change within a species under conditions of natural selection.

Despite their extensive range and clearly established specialization to local conditions, all of the house sparrows in Europe or in North America can be placed in a single species. The changes observed are consistent, directional, and occur within a particular species, just as argued by Darwin. The strong correlation with climatic conditions indicates that changes in body proportions are significant to the survival of the individual populations. Although one might consider the introduction of sparrows into North America as equivalent to a speciation event, the differentiation that has occurred within each continent is certainly confined to a single species.

Gradually slowing rates of evolutionary change would be expected to continue in North America, at least until achieving the degree of difference seen in European populations. The stasis observed over the past 150 years in European populations suggests that these characters may have achieved an optimal size for the present conditions. Relative to the probable life span of this species, the changes in both European and North American populations are of very short duration as viewed by the geological time scale. Their pattern of evolution, if seen in the fossil record, would appear punctuated, as suggested by Eldredge and Gould; yet it was the result of natural selection, acting in a consistent and directional manner, as argued by Darwin. Although change in these characters may have reached a point of equilibrium in the European populations, it is impossible to know how much more change may occur in North America, where gradual and progressive accommodation to even more extreme environments may continue for thousands or millions of years. At this time scale, we lose the capacity for direct measurement in living populations, but there is no obvious reason why continued evolution within species would not eventually lead to the more extreme changes that we see in the birds of the Galápagos and Hawaiian islands.

Evolution within the Galápagos finches

Darwin, the authors of the modern synthesis, and Eldredge and Gould all accepted that the rate of evolutionary change may be most rapid in small isolated populations. Small islands hence have a strong potential as models for the study of evolution at a human time scale. Oceanic archipelagos – in which colonization from the mainland is an extremely rare event, and movement from island to island is possible but not common – provide ideal conditions for speciation and interaction between closely related species. Evolutionary biologists have taken advantage of such opportunities in a number of island groups, including the Hawaiian Islands (Raikow 1977; Otte 1989), the Canary Islands (Thorpe, McGregor, and Cumming 1993), and the Galápagos (Bowman, Berson, and Leviton 1983; Steadman and Zousmer 1988). Particularly informative studies have been carried out on the birds of the Galápagos Islands by Lack (1947) and over the past twenty-five years by Peter and Rosemary Grant, their daughters, and colleagues (P. R. Grant et al. 1985; P. R. Grant 1986, 1994; B. R. Grant and P. R. Grant 1989, 1993).

The Galápagos archipelago, some thousand kilometers west of the coast of Equador, provides a nearly ideal locality for the study of evolution (Fig. 3.1). It was here that Darwin first began to consider the possibility of the transmutation of species. The local diversity of species whose closest relatives lived on the mainland of South America argued for migration and subsequent evolutionary change, rather than individual creation.

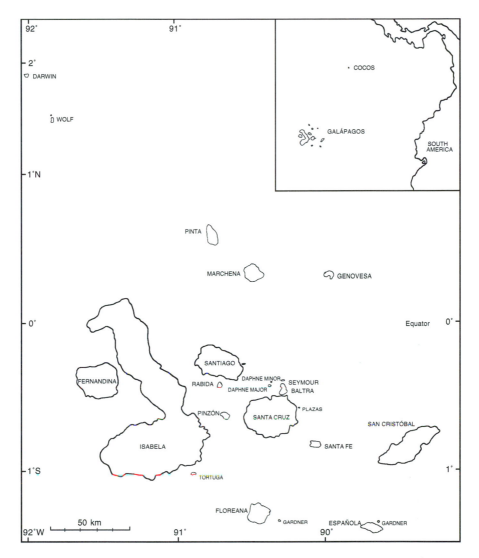

Figure 3.1. Map of the Galápagos Islands. Redrawn from Peter R. Grant (1986), *Ecology and Evolution of Darwin's Finches.* Reproduced by permission of Princeton University Press.

The Galápagos Islands have the special advantage for evolutionary studies of being subject to sharply changing climatic conditions. Within most years there are marked seasonal differences between warm, wet periods from January to May and cooler, dry periods in the remaining months. In addition, extreme differences in the total annual rainfall may occur at intervals of approximately ten years. During the period of the Grants' study, annual rainfall ranged from 0 mm to more than 2,400 mm within three years, resulting in radical changes in the vegetation and hence the availability of food for the birds being studied (Fig. 3.2). The changing climatic conditions place extreme selective pressures on the bird populations over short periods of time, leading to high mortality and differential survival of different phenotypes.

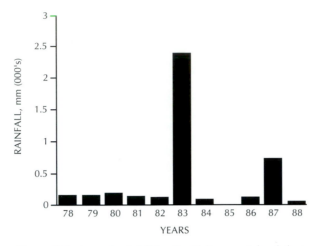

Figure 3.2. Annual rainfall for the Galápagos Islands between 1978 and 1988. Records for 1985 are not missing; no rain fell that year. From Grant and Grant (1989), *Evolutionary Dynamics of a Natural Population,* University of Chicago Press.

Figure 3.3. Skulls and jaw muscles of a variety of finches from the Galápagos Islands. Redrawn from Bowman, "Morphological differentiation and adaptation in the Galápagos finches." Copyright © 1961, University of California Press.

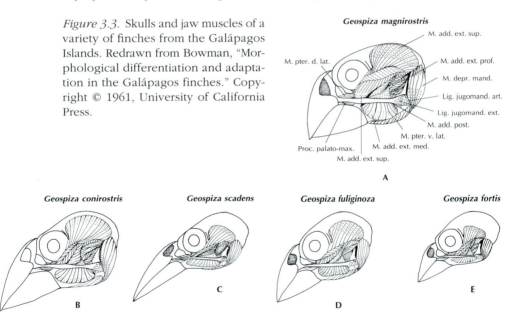

Darwin collected a few representatives of the birds that were later to be termed *Darwin's finches.* He recognized that they could be interpreted as showing progressive evolutionary change and speciation from birds now common on the mainland of South America, but did not appreciate the full diversity of species in the Galápagos or the significance of their distribution on the individual islands. So different were some species from finches elsewhere that Darwin identified them as warblers and tanagers.

Fourteen finch species, placed in four genera, are currently recognized on the Galápagos Islands and Cocos, 600 km to the northeast. They can be divided into four major adaptive types: six species of ground finches, six species of tree finches, a warbler finch, and the Cocos finch. They vary in size from less than 10 g to more than 40 g. All the species can be distinguished from one another by the shape and

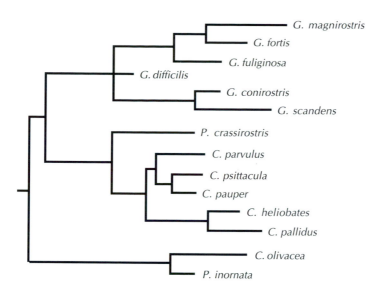

Figure 3.4. Phylogeny of the Galápagos finches. From Peter R. Grant (1986), *Ecology and Evolution of Darwin's Finches* (after Lack 1947). Reproduced by permission of Princeton University Press. See also Stern and Grant (1996).

size of their beaks, which are specialized for a variety of foods ranging from the eggs of seabirds, to arthropods, flowers and nectar of the common cactus *Opuntia,* and many sizes of seeds (Fig. 3.3). The ways of acquiring food also differ markedly from species to species and from population to population. The woodpecker finch *Cactospiza pallida* and the mangrove finch *Cactospiza helibates* feed on insect larvae and termites that they pry from holes in dead branches with twigs or cactus spines held in their beak. Populations of the sharp-beaked ground finch *Geospiza difficilis* on the isolated islands of Wolf and Darwin peck the developing feathers of the tail and wings of seabirds and drink their blood.

The anatomy and phylogeny of the Galápagos finches are well known from studies by Bowman (1961), Sarich (1977), and Yang and Patton (1981), providing a firm basis for evolutionary studies (Fig. 3.4). Based on the amount of molecular difference from species to species, Sarich (1977) established that the thirteen finch species on the Galápagos and the one on Cocos Island diverged from one another 1–2 million years ago. Within the Galápagos Islands proper, the greatest degree of phylogenetic separation is between the warbler finch *Certhidea* and all the other species. Based on protein polymorphisms, Yang and Patton (1981) suggest that they have been separated for 0.5–1 million years.

The islands are far enough from the mainland that no more than one or two colonization events have been hypothesized (Grant 1986, p. 254). There are seventeen major islands and many of smaller size. Individual birds are capable of flying from one island to another throughout the archipelago, but very few birds fly to distant islands in any one breeding season; thus the populations are genetically isolated. Hybridization between species is possible (both within and between islands) and in some cases is significant in maintaining genetic variability in small populations, but each species has maintained its integrity for as long as the island fauna has been studied. Most of the major islands are large enough and varied enough in their flora and topography to support many species (Table 3.2). Competition is avoided by specialization for different sources of food, reflected by the fact that species living on the same island always differ more from one another in the proportions of their beaks than do populations of the same species that live on separate islands (Fig. 3.5).

The large number of related species, the number of islands of different size, differing habitats, and differing distances from each other provide the possibility for hundreds of separate studies of environmental interactions between local populations and evolutionary response to environmental change. The Grants have concentrated on particular examples that they have studied over long periods of time.

One of their most detailed studies involved the large cactus finch *Geospiza conirostris* on the small, isolated island of Genovesa (Grant and Grant 1989). The small size of this island and the limited extent of particular habitats enabled the Grants to study a large percentage of the total number of individuals within particular populations. The degree of isolation made it possible to establish the amount of outbreeding with migrants from other populations of the same or closely related species from other islands. Over nearly a decade of study, the Grants came to know individually all the members of particular populations. They banded, measured, and weighed approximately 2,500 adult and young birds. They recorded the songs of all male birds and mapped their territories. They developed genealogies within family groups as a basis for establishing the heritability of a suite of traits governing body size, limb size, and the size and shape of the beak. Of particular importance, they documented in a quantitative manner the pattern of feeding throughout the life span of as many birds as possible, from season to season and from year to year, to establish its relationship to beak size and its importance to survival, reproduction, and evolutionary change. They demonstrated that the size and proportions of the beak were highly heritable and directly correlated with

Figure 3.5. Lateral profiles of the bills of six species of Geospizinae from Indefatigable Island. Redrawn from Bowman, "Morphological differentiation and adaptation in the Galápagos finches." Copyright © 1961, University of California Press.

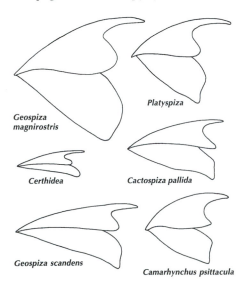

Table 3.2. *Distribution of the species of Darwin's finches on the seventeen major islands (>1.5 km²) of the Galápagos*

Island	*Geospiza* magnirostris	fortis	fuliginosa	difficilis	scandens	conirostris	psittacula	pauper	parvulus	*Platyspiza* crassirostris	*Cactospiza* pallida	heliobates	olivacea
Seymour		B	B		B								B
Baltra		B	B		B								B
Isabela	B	B	B	(E)	B		B		B	B	B	B	B
Fernandina	B	B	B	B			B		B	B	(B)	B	B
Santiago	B	B	B	B	B		B		B	B	B		B
Rábida	B	B	B		B		B		B	B			B
Pinzón	B	B	B		(B)		E		B	E	B		B
Santa Cruz	B	B	B	E	B		B		B	B	B		B
Santa Fe	B	B	B		B		B		B	(E)			B
San Cristóbal	E	B	B	(E)	B				B		B		B
Española			B			B							B
Floreana	E	B	B	E	B		B	B	B	B			B
Genovesa	B			B		B							B
Marchena	B	B	B		B		B			B			B
Pinta	B	B	B	B	B		B		(B)	B			B
Darwin	B			B									B
Wolf	B			B									B

Abbreviations: B = breeding; (B) = probably breeding; E = extinct; (E) = probably present as a breeding population formerly and now extinct.

Source: From Peter R. Grant (1986), *Ecology and Evolution of Darwin's Finches*. Reproduced by permission of Princeton University Press.

the choice and capacity to feed upon particular foods. Slender beaks were more effective in penetrating soft fruit and feeding on nectar, whereas the most massive bills were associated with cracking open the cactus seeds and stripping bark from the trunks in search for insect prey.

In this and other species, birds bred during the wet season, when there was an abundance of food for the young. All birds feed on exposed insects, fruit and flowers of the *Opuntia* cactus, and small seeds. There is substantial mortality among birds during their first year, but it does not seem to be correlated with any of the morphological features that were measured. Mortality among fledglings appears to be largely random but may be related to their general capacity to feed themselves and avoid predators. Under normal climatic conditions, mortality among adults is very low, and individual birds may live for up to ten years.

Between 1978 and 1982, the amount of rainfall was within the normal range, and the wet and dry seasons were of average duration. During the dry seasons, adult survival was generally unrelated to phenotype. The extremely long-lasting El Niño conditions in 1983 led to eight months of heavy rain, totaling more than 2,400 mm. This extraordinarily wet year was followed by drought conditions that continued until 1986: In 1984 there were only 59 mm of rain, and in 1985 there was no rain at all. These unusual environmental conditions provided an excellent natural experiment to study the effects of selection. During the extended rainy season, when there was an abundance of food, the population grew very rapidly and selection appeared to have little effect on the measured features of body size and beak configuration. The only exception was a slightly higher survivorship of female birds with long, narrow bills, but this was not statistically significant.

At the beginning of the drought, selection was first evident in relationship to behavior. In 1984, forty-five of eighty-seven birds that were studied continued to search out food that was common during more moist conditions, picking arthropods from vegetation and eating small seeds. As these food items became progressively less available, these birds all died. The other forty-two birds shifted to other food sources that required additional effort. Their behavior changed to tearing up rotting cactus pads in search of larval arthropods and stripping the bark off trees to extract insects. Even among this group, only fourteen survived.

At the beginning of the drought, there was no apparent difference in the physical attributes of the birds that survived, but storm conditions during El Niño led to serious damage to the *Opuntia* cactus, whose flowers and soft fruit had served as a common food source, especially favored by birds with long narrow bills. Selection initially acted against this trait (Table 3.3) while favoring birds with shorter but still slender bills, since these were more effective in penetrating soft cactus pads filled with arthropods. As the pads became harder with progressive drying, selection next favored the birds with the deepest bills. This character was strongly linked to overall size. Between 1983 and 1985, the number of birds declined from 103 to 75 to 46. Among the surviving population, the relative length of the bill was greater by an average of 2.5 percent.

Although only about 20 percent of the original population survived to the end of the drought and there was strong selection on bill size and shape during this period, there was no overall change, since selection acted successively in different directions. At the beginning of the study, there appeared to be a slight selective ad-

Table 3.3. *Change in character dimensions in the large cactus finch* Geospiza conirostris *during the dry period 1983–5*

Period	Character	Coefficients		
		$\beta \pm$ S.E.	s	N
1983–4	Bill length	-0.22 ± 0.08**	-0.24**	137
	Bill depth	-0.05 ± 0.08	-0.07	(0.54)
	Tarsus length	-0.02 ± 0.07	-0.10	
1984–5	Bill length	-0.05 ± 0.18	0.14	74
	Bill depth	0.37 ± 0.19*	0.31**	(0.39)
	Tarsus length	0.14 ± 0.19	0.22**	
1983–5	Bill length	-0.28 ± 0.22	-0.01	137
	Bill depth	0.27 ± 0.21	0.15	(0.19)
	Tarsus length	0.24 ± 0.17	0.17	

Note: Coefficients are standardized directional selection gradients ($\beta \pm$ standard error) and differentials (s), and their statistical significance is indicated by *($P < 0.05$) or **($P < 0.01$). The number of individuals present at the beginning of each interval is given under N, and the proportion surviving follows in parentheses.

Source: From Grant and Grant (1989), *Evolutionary Dynamics of a Natural Population,* University of Chicago Press.

vantage among females for long narrow beaks. This was followed by selection for short beaks in both sexes, and then for deep beaks (closely correlated with long beaks) in the final period of the drought.

The strongest selection at the height of the drought so reduced the population and the survivorship of the progeny that it would soon have led to extinction of the population without further anatomical change. On the other hand, if drought conditions had lessened but had not been completely relieved, it might have led to progressive increase and eventual fixation of alleles producing the largest possible sized beaks.

In another study, of *Geospiza fortis* on Daphne Major (a tiny island near the center of the archipelago), a severe drought in 1977 led to strong directional selection. From mid-1976 to the end of 1977, population size dropped from 1,200 individuals to 180. During this period the availability of seeds dropped from 10 to 3 g/m^2, and the remaining seeds were of larger size and harder to crack. Mortality affected primarily small birds, with beaks smaller than that necessary for cracking large seeds. The progeny born in 1978 were 0.31 standard deviations larger, on average, than the 1976 populations. The average beak depth increased by about 4 percent. Males are normally about 5 percent larger than females, which placed them at a strong selective advantage, resulting in them outnumbering females in 1978 by 5 or 6 to 1. The amount of change during this period was augmented by sexual selection, for larger males with larger bills were preferentially selected by females. Further study of this species during the drought following the 1982–3

El Niño event demonstrated similar directional change in beak proportions (Grant and Grant 1993).

In these cases, the finch populations were able to respond quickly to selection because of the large amount of genetic variability that was available, despite recurrent episodes of reduced population size and strong directional selection. In the case of *Geospiza conirostris,* the Grants attribute the continuing high variability to a low percentage of hybridization (1–4 percent annually) with other species living on the same island.

Clearly, such intense selection cannot continue for long; nor will it result in much more extensive change, since the genetic variation in the population will eventually be exhausted, but more important, an optimal size for dealing with the largest seeds available will soon be reached. Grant concluded:

> Although I have no reason to believe the change was an initial step on a long-term phyletic trajectory, it can be extrapolated to show the forces of selection necessary to propel a species along such a trajectory. Selection thus fluctuates in direction within a generation, and possibly between generations as well, driven by the unpredictable and large fluctuations in rainfall. (1986, p. 375)

This is exactly the pattern of change within species postulated by Eldredge and Gould to explain the long-term stasis of species. Although there are minor fluctuations in the availability of food on a seasonal basis and drastic changes approximately once in a decade, the morphology of the populations on a particular island may remain essentially constant over long periods of time.

The multiplicity of different islands within the Galápagos archipelago also provides data for study of longer-term evolutionary processes. Bowman (1961) and Grant (1986) demonstrated that each island has a distinctive flora, which strongly influences the types of food available to the finches. This is specifically expressed in the proportion of small, relatively soft seeds to large seeds that can be cracked only by birds with the deepest bills. All but the smallest islands have numerous finch species as permanent, breeding populations, but the particular species differ from island to island. Without more complete knowledge of their fossil record, it is impossible to determine the history of invasion of the individual islands, but the current pattern of species distribution has been nearly constant for the past 150 years.

Since the first collections by Darwin, naturalists have had difficulty assigning populations living on different islands to one species or another. Repeated studies by a series of taxonomists using many different methods – most recently, protein polymorphisms (Yang and Patton 1981; Polans 1983) – now provide a reliable means of identifying the species to which most of the island populations belong. This work shows that the morphology (specifically, the proportions of the beak) may differ in a statistically significant manner between populations of a particular species that are found on different islands. Grant (1986) has shown that beak shape is correlated with two factors: the nature of the seeds on the different islands, and the nature of the other finch species.

Among the nine species belonging to the genera *Geospiza* and *Camarhynchus,* all are more different from one another when they occur together on the same island than when they occur separately on different islands. When one of the species is the only member of a particular genus on an island, it is morphologically inter-

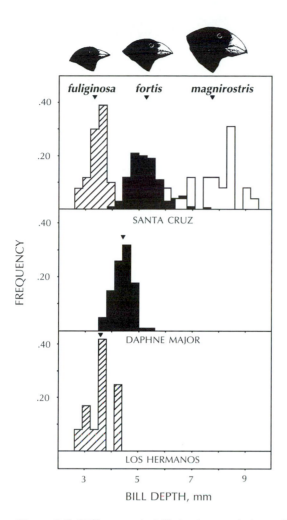

Figure 3.6. Differences in bill size in populations of the same finch species living on different islands. Frequency distribution of beak depths of adult males in populations of ground finches (*Geospiza*). From Peter R. Grant (1986), *Ecology and Evolution of Darwin's Finches.* Reproduced by permission of Princeton University Press.

mediate between the size it has when coexisting with other species and the size of the most similar member of the same genus (Fig. 3.6). This degree of differentiation is correlated with feeding on different sizes of seeds, which enables more than one species to inhabit the same area with a minimum of competition for food.

Particular pairs of finches are rarely if ever found together on the same island; this presumably results from their having such a similar morphology that they would be unable to avoid competition. Food resources on the islands of Genovesa and Española are similar, but they have different species of *Geospiza*. On Genovesa there are three species, *G. magnirostris, G. conirostris,* and *G. difficilis.* On Española there are only two, *G. fuliginosa* and *G. conirostris. G. fuliginosa* fills the role played by *G. difficilis* on Genovesa and has a comparably sized beak. *G. conirostris* replaces *G. magnirostris.* Two species of this genus that are common on other islands, *G. fortis* and *G. scandens,* are absent on Genovesa and Española. Their feeding role appears to be taken by *G. conirostris,* which does not occur on the other large islands.

The greatest environmental difference shown by populations of the same species occurs in *Geospiza difficilis,* which occupies humid forests on the islands of Pinta, Santiago, and Santa Cruz, but on Genovesa and Wolf lives in the arid zone. On Pinta, Santiago, and Santa Cruz, the arid zone is occupied by *G. fuliginosa.*

There is no evidence as to how the combination of species on particular islands was established, but the consequences are quite clear. The configuration of the beak depends on the available food and the potential for competition with other finches. Finches, especially when young, readily fly from island to island, but whether they remain and establish new breeding populations depends on the availability of a distinct niche that is not occupied by resident finches. If different species with similar feeding behavior are already present, selection will act to alter the feeding behavior and ultimately the beak shape both of the newcomers and of the resident species (i.e., character displacement) so they will not be in direct competition with each another. If this does not occur, it is unlikely that both can persist in a single habitat.

None of the studies of the Galápagos finches is of long enough duration to document the course of a successful colonization of a new island. Presumably the period of character displacement is relatively short, and so the fossil record would give the *pattern* of a punctuated event. On the other hand, the *process* of punctuation is not that hypothesized by Eldredge and Gould, in that it does not require speciation. Although movement from island to island has a partially isolating effect, as in the case of allopatric speciation, it can occur without any diminution in the potential for interbreeding by birds that fly from island to island. Although differences in proportions of the bills are sufficient to distinguish most members of particular island populations from one another, it is still possible to recognize them as members of the same species. Speciation may eventually occur between populations on distant islands, but this may not occur until long after they have achieved status as distinct populations. Clearly, the changes in behavior and anatomy are driven by selection for effective adaptation to differences in the environment, rather than by the fact of separation per se. In contrast with the fluctuating evolutionary changes observed on a single island, changes that occur from island to island result in differences that may last as long as do the floral and faunal differences between the islands.

The Cocos finch, *Pinaroloxias inornata,* provides an example of a very different evolutionary phenomenon than the species on the Galápagos. Cocos is a small island, 47 km² in area, 630 km northeast of the Galápagos, with only a single finch species. This island differs in showing little seasonality. The average rainfall is 8 m a year, and the entire island has a single vegetative zone. The one finch species shows very little morphological variability, but the individual birds are extremely variable in their feeding habits. The species feeds from ground to canopy, but individual birds concentrate on leaves, flowers, small seeds, arthropods, nectar, or fruit. This species runs almost the entire gamut of food sources divided among thirteen species on the Galápagos (Werner and Sherry 1989). This one species has a much wider niche than did its mainland predecessors, faced with competition from numerous other small passerines. Presumably the small size and environmental homogeneity of Cocos has precluded speciation or even population-level specialization on one food type or another. Competition between individuals is reduced by behavioral specialization. It is not yet known whether this behavioral specialization has a genetic component, but it persists throughout the life of an individual. The Cocos finch is as distinct from mainland finches as are those on the Galápagos, suggesting an equivalent period of isolation, but there is very little likelihood of either speciation or further morphological change on the single island.

Fossils of at least nine of the living species from the island of Floreana have been described by Steadman and Zousmer (1988), but radiometric dating does not document an age much greater than 2,400 years. Other sites mentioned by Steadman et al. (1991) have the potential for extending this span to approximately 4,500 years. No fossils have been described that differ anatomically from the living species, although they do document the presence of *Geospiza magnirostris* and *G. difficilis* on the island of Floreana, where they no longer live. Unfortunately, nothing is known of their earlier fossil record, precluding establishment of large-scale patterns and rates of evolution during that period.

Plant remains from cores taken from sediments in El Junco Lake on San Cristóbal show a series of climatic changes during the past 50,000 years. Conditions have alternated from similar to the present to dryer and warmer. For the past 3,000 years, as well as 6,200–8,000 and 34,000–48,000 years ago, the conditions were similar to the present. During the intervening periods, it has been warmer and dryer. This was particularly marked 10,000–34,000 years ago, when sea levels were lower during the last major glacial advance. Climatic conditions prior to 48,000 years ago are not known.

Periods of climatic consistency may have resulted in relative evolutionary stasis, but changes between these periods may have led to rapid changes in behavior and structure. Persistently different conditions would probably lead to rapid fixation of particular genes, giving a punctuated appearance if it were recorded in the fossil record, but within species, rather than being associated with speciation events.

Everyone who has studied the Galápagos archipelago emphasizes that it exhibits a fauna caught in the midst of evolutionary change. This is most clearly manifest among Darwin's finches, which have been studied in such admirable detail by the Grant family. The endemic species are all clearly distinct from their mainland progenitors, reflecting approximately a million years of separate evolution, but the process of speciation continues among the island forms. Fourteen species are currently recognized on the basis of consistent morphological differences and limited interbreeding; but Grant (1986) admits that separate species status could easily be argued for at least two populations that are morphologically similar but genetically isolated from populations on widely separated islands, or that one or two morphologically distinct species that interbreed freely when in contact could be combined.

If rainfall increased and became more regular from season to season and from year to year, species that had specialized to avoid competition for reduced resources during the dry season and dry years might hybridize more extensively and eventually reduce the number of distinct species. On the other hand, geographically isolated populations of the same species might become so specialized that hybrids would be selected against if the parental populations returned to sympatry, leading to clearly defined species status.

The phenomena of local extinction and renewal of populations on particular islands, documented historically and from the fossil record, show that combination and isolation of species are occurring continually. This too contributes to coalescence or further separation of incipient species. This process has presumably been going on for at least a million years, since the first colonization of the archipelago, and will presumably continue as long as the islands are inhabited by the finches.

Archipelagos may represent extreme examples of evolutionary processes, but the island phenomenon is paralleled by many other examples of shifting geographical isolates, such as scattered lakes, hills or mountains dotting a lowland, or discrete shallow bays on the edge of a large lake, or along the margin of an island. Under these conditions there may be no clear separation between speciation events and other periods in the duration of a species. Most vertebrate species comprise numerous populations that are recognizably distinct at any one instant of geological time. Even at a human time scale, it is clear that populations shift in space and may coalesce or become more separate in the manner of an ever-changing mosaic. Under most conditions, the assemblage of populations will retain the integrity of a single species. Individual populations are far more likely to become extinct rather than to found new species, but if the areas of isolation are sufficiently large and not already occupied by similar species, there is a greater potential for the success of a new lineage. In the case of the Galápagos finches, this potential was realized in two stages: first when the archipelago was initially colonized from the mainland, and subsequently when each of the separate islands, or discrete habitats on the larger islands, was colonized. The initial colonizations may have occurred more than a million years ago, but the full differentiation of the species into ecologically and reproductively isolated units has not yet been fully achieved. This extremely long process of speciation may be attributed to the particular degree of isolation possible on these islands and to the radical and extremely irregular changes in seasonal and annual rainfall that may have persisted throughout much of this time.

Grant (1986, chap. 16) discussed the limitations of further evolution of the finches on the Galápagos Islands. Because of their inherent structural and developmental constraints, as well as competition from seabirds and large predators, the finches are extremely unlikely to radiate beyond the finch–warbler adaptive zone. More specifically, the largest living species are unlikely to become significantly larger because they are already capable of dealing with the largest and hardest seeds available. The number of species is unlikely to increase significantly since this is governed by the number of islands and the degree of distinction between the available habitats. Species number might be reduced, however, if the rainfall increases and becomes more regular, leading to a reduction in the number of distinct habitats, as on the island of Cocos. We may expect fluctuational changes to continue within individual populations and species as long as present climate and geological conditions prevail; but major and long-term changes in weather conditions, or major new volcanic eruptions, might so change the archipelago that significant evolutionary changes would occur.

Significance of changes among Darwin's finches to longer-term evolutionary phenomena

Because the small size of the populations and the extreme climatic conditions result in very high selection coefficients and rapid but short-term changes in the phenotype, evolution of the Galápagos finches provides an excellent model for evolutionary change within living species. These examples show that no significant change occurs under favorable conditions, when food is readily available; only during prolonged droughts is survivorship closely correlated with physical attributes. These observations suggest that evolutionary change is correlated with severe and relatively long-lasting changes in environmental factors. This corresponds with Gould and Eldredge's view of the predominance of stasis during the period when a species is at equilibrium with its environment. The absence of long-term evolutionary trends among species appears to be supported for the time period of the Grants' study, and (with much less documentation) for the 4,500 years of the fossil record of the islands. On the other hand, the nature and degree of change observed within species is such as would result in significant differences if continued for a sufficiently long period of time under directional selection. Since the finches arrived in the Galápagos a million years ago, the local climate has undergone several long-term oscillations, which may have led to long-term evolution among species that have survived. Unfortunately, this cannot be demonstrated from the short span of the fossil record on these islands.

Morphological changes that are significant to the survival of the finches do occur within particular populations of single species. There is no evidence from the study of the Galápagos finches that speciation, or even splitting of populations, is necessary for the accumulation of significant anatomical changes. This too requires documentation over longer time periods.

These examples illustrate why we can see rapid and apparently directional changes in modern populations over short periods of time whereas the more coarse-grained data available from the fossil record will show what appears as stasis. There is, however, nothing in the observations of the Galápagos finches that contradicts either the writings of Darwin or the conclusions of the evolutionary synthesis. Even up to the time of the writings of Simpson (1944, 1953) there was very little detailed evidence of the patterns of evolution among natural populations. Diagrams of the rate and pattern of change during the duration of species were entirely hypothetical. On the other hand, both Darwin and Simpson repeatedly emphasized the likelihood that the rate of evolution was extremely variable and irregular, encompassing both stasis and episodes of rapid change. Eldredge and Gould were correct in emphasizing this phenomenon at the species level, but these detailed examples demonstrate that it can be fully explained according to the tenets of Darwinian selection theory.

The assumption that changes observed in populations over short periods of time typically result in long-term trends and culminate in large-scale, long-term modifications of the morphology is certainly incorrect, although it is frequently implied if not explicitly stated in both biology and paleontology textbooks. The Galápagos examples demonstrate that most changes observed at the population level are short-term responses to changes in the environment that are frequently reversed or redirected within a brief period of time. On the other hand, these studies do demonstrate the capacity within populations for rapid, measurable change that may, under unusual circumstances, lead to species-level differentiation.

The extent of most changes at the population level are constrained by factors of the physical and biological environment, and rarely transgress the average variability of the species. However, these observations demonstrate not the near impossibility of significant change within species postulated by Gould and Eldredge, but merely its improbability over short periods of time. As long as competition, resources, predation, and the physical environment remain relatively static, there is little likelihood that selection will result in significant changes in the mean of morphological or physiological traits of a population. However, when the environment changes or populations migrate to another habitat, there is no evidence that inherent constraints of the genetics or developmental biology of the individuals preclude rapid and far-reaching changes that enable them to accommodate to the altered conditions.

It should be mentioned that these particular examples, drawn from small, isolated populations in a widely fluctuating environment, are very different from those emphasized by Gould and Eldredge, based on large, generally widespread populations preserved in the marine fossil record.

Other examples of evolutionary change on islands

Other examples of evolutionary change, showing a greater time span and greater magnitude of change, but with much less information regarding the pattern of change during the process, have been documented over the past 10,000–12,000 years from other island populations (Sondaar 1977). Among large mammals, dwarfing is a common response to island isolation. A particularly interesting example is provided by a population of the woolly mammoth, *Mammuthus primigenius,* on Wrangel Island in the Siberian Arctic (Vartanyan, Garutt, and Sher 1993). This island became isolated from the mainland between 13,000 and 12,000 years B.P. (before present) and is now 200 km off the coast. Mammoth bones of this age are comparable in size to the mainland specimen of the species, but teeth dated from 3,730 to 7,620 years B.P. were 20–25 percent smaller. Other dwarfed elephants show much greater reduction of the limb bones than the teeth, indicating that this population was of small dimensions overall. Lister (1993) suggests that they weighed only one-third as much as normal members of this species. Dwarfing is attributed to the effect of insularity, combined with response to the warming climate and changing vegetation of the postglacial world. Such a great degree of reduction presumably required genetic change and developmental adjustment, rather than being simply an effect of ecophenotypic response to inadequate diet. Other dwarf elephant species are known from Malta, Sardinia, and islands off the California coast.

Clines

Even 10,000–12,000 years are too short a time to observe more than a tiny fraction of the average period of species longevity. It is impossible to examine the extent of change that occurs within the full duration of a species without a thorough knowledge of its fossil record over a period of one to several million years. On the other hand, we can compare, by analogy, the patterns of character variation shown in the distribution of living plants and animals in space with the probable pattern of their distribution in time.

Nearly all adequately studied plant and animal species that occupy extensive ranges show some characters that vary according to their geographical distribution. The most systematic differences are those expressed in **clines**: gradients of phenotypic or genotypic change in a population or species correlated with the orientation of features of the environment such as altitude, latitude, temperature, and dryness. According to Mayr (1963), clines are expressed in nearly all terrestrial species. The simplest to understand are those involving body proportions and temperature control in warm-blooded mammals and birds. The general pattern is characterized as Bergmann's rule: Body size in geographically variable species averages larger in the cooler parts of the range of a species (Bergmann 1847). This may involve as much as a twofold difference in total body mass. Although it is not recognizable in every species, Rensch found relatively few exceptions in 1936 (summarized by Mayr [1963]). The rule is followed in 92 percent of Palaearctic birds, 87.5 percent of Malay birds with Palaearctic relatives, 74 percent of North American birds, 81 percent of North American mammals, and 60 percent of Western European mammals.

If we assume that most species originate from small, isolated populations, the geographical variation that is exhibited by wide-ranging species must have evolved within those species as they multiplied and extended their range. Clinal variation, which is known for a great number of well-studied species, shows that species do manifest large-scale, long-term progressive change within the period of their duration, contrary to the assumptions of Eldredge and Gould. In addition to body size, many other aspects of anatomy and physiology are recognized in living populations that would not be reflected in the fossil record. Geographical differences that can be attributed to climate have been postulated for numerous Quaternary mammals (Martin and Barnosky 1993).

More general comparison between geographic variation in fossil and living species is complicated by the nature of the organisms and the characters being studied. Among living organisms, study has concentrated on terrestrial forms, which are subject to marked difference in climate relative to latitude. This is particularly true of birds and mammals. In contrast, the vast majority of fossils that have been described are marine invertebrates, living in habitats that show relatively less systematic latitudinal change. In addition, these invertebrates would not be expected to show the type of change observed in endothermic vertebrates.

Gould and Eldredge make little reference to clines in their publications. In their 1977 paper they suggest that examples of gradual and progressive change described by Gingerich may be the result of migration of species exhibiting clinal variation rather than change over time within a local species. It would seem that intraspecific clinal differences that evolved were evidence of significant change within a species just as much as was change within a local population over time.

The extensive nature of phyletic or anagenetic change within species is also exemplified by *ring species,* in which a series of local populations or subspecies are distributed around a physical barrier such as the Arctic Ocean or the Himalayan plateau. Numerous examples were cited by Mayr (1963). Wake, Yaney, and Frelow (1989) analyzed a case in the plethodontid salamander species *Ensatina eschscholtzii* within which seven subspecies are arrayed in a ring around the Central Valley of California. Interbreeding is common among most of the adjacent populations but not those at one point at the southern extremity of the valley, which is assumed to be where the most distantly related populations meet one another. Separation to the extent of nearly complete reproductive isolation has occurred without genetic barriers between any of the other subspecies that succeed one another on the east and west of the valley. The two southern subspecies are sympatric, but there is little if any evidence of hybridization. In addition to genetic isolation, they are distinguishable by morphological and ecological differences of a degree expected in full species. This phenomenon almost certainly resulted from progressive change over space and time within a succession of populations that extended their range from north to south, spreading out on either side of the valley, and finally meeting in the south after a sufficient period of time that differences in their morphology, behavior, and pattern of reproduction had reached the level of distinct species. This process presumably began no more than 10,000 years ago, following the end of Pleistocene glaciation.

Summary of evidence for significant change within species

At the time scale that can be studied in living populations, morphological change that is significant to survival has been documented in many taxa. Particularly good examples are provided by recent immigrant populations and populations on islands. Evidence from clines and ring species among living taxa show morphological trends in space that are analogous with long-term trends in time.

Progressive changes within populations are frequently associated with temporal or geographic differences in the environment. The amount of change is constrained by the length of time involved and the approach to optimal conditions. The amount of variability expressed in most natural populations is sufficient to permit change at a much greater rate than is usually recorded in the fossil record. Natural selection rarely acts in the same direction long enough to deplete the variability present at any one time. The time period of human study is not sufficiently long to establish the point at which further change would be precluded because of the absence of new mutations.

Speciation

The writings of Dobzhansky (1951) and Mayr (1963) have emphasized the distinction between speciation and phyletic evolution and the importance of speciation (or cladogenesis) to the generation of diversity. Eldredge and Gould argue further that speciation is also necessary for the generation of significant morphological change.

In order to evaluate this hypothesis, it is necessary to specify what aspect of speciation is to be emphasized. According to Gould (1982, p. 87): "Reproductive isolation and the morphological gaps that define species for paleontologists are not equivalent. Punctuated equilibrium requires either that most morphological change arise in coincidence with speciation itself, or that the morphological adaptations made possible by reproductive isolation arise rapidly thereafter." Although these factors may coincide fairly closely in time, they result from very different underlying processes. Since reproductive isolation among vertebrates is typically associated with geographic separation, it is common that newly isolated species will be exposed to a different environment than the parental species and respond rapidly to different selection pressures. In this matter, there is no dispute between Eldredge and Gould and the authors of the evolutionary synthesis. The question is whether reproductive isolation per se is necessarily correlated with morphological change.

Larson (1989) emphasized the need to differentiate the reproductive aspects of speciation from the morphological changes that are typically used to differentiate species. This can now be achieved in modern populations through electrophoretic analysis of protein polymorphisms. The restriction of different alleles to particular populations previously included in a single species shows that many are genetically isolated and should be recognized as distinct species, although they show little if any morphological differences. A great many cases of morphologically similar but reproductively isolated species, termed **sibling species**, are recognized (Mayr 1963). Among vertebrates they are particularly common in fish, amphibians, and rodents. Not enough work has yet been undertaken to provide an overall percentage for vertebrates as a whole. Among birds, which are the most completely known taxonomically of any vertebrate class, less than 5 percent of the species are considered sibling species, but samples from other classes suggest a higher percentage. Larson cited numerous studies among plethodontid salamanders, including one in which he identified a minimum of fifteen speciation events, only three of which were associated with substantial morphological change. The other species could not be distinguished morphologically on the basis of living specimens, and fossil remains would certainly have been included in a single species. Speaking in reference to both living and fossil species, R. A. Martin (1993, p. 261) stated: "There are no examples of speciation in rodent evolution of which I am aware in which there was rapid and distinctive change in form to the extent that we would recognize the descendant species as something drastically different from its ancestor."

Extremely extensive speciation has occurred recently among the cichlid fish of the East African Rift Valley lakes (Echelle and Kornberg 1984). The species are morphologically distinct, but only in a few characters associated with feeding and reproduction. Where speciation has occurred within the past few hundred years, the few differences (such as color) can be associated primarily with sexual recognition. The evolution of these fish is discussed extensively in Chapter 6.

The many cases of sibling species demonstrate that speciation per se is not necessarily associated with substantial morphological change. Barton and Charlesworth (1984) provided evidence that very small population size associated with speciation is probably not conducive to significant morphological change.

Futuyma (1989) pointed out that the genetic isolation that is key to speciation may facilitate rapid morphological change in other ways. In the absence of gene flow from the genetically and adaptationally more diverse parental species, the new species, if initially low in numbers and geographically isolated, can respond much more rapidly to differing forces of selection in the new environment. In summary he states: "Thus, reproductive isolation provides evolutionary changes with a degree of permanence that may cause a pattern of punctuation and stasis in the fossil record" (p. 574). However, the genetic fixation of some recently derived features at the time of speciation does not preclude the continuation of previously established evolutionary trends following speciation, as is claimed by Eldredge and Gould. Rapid change immediately following speciation is not sufficient on its own to produce a punctuated pattern, which requires a cessation of change during the remainder of a species' longevity. As we have seen in the case of the house sparrows, change may continue to occur at a rapid pace, even within populous, widespread species. The amount of change is influenced by the magnitude of differences between the local environments, rather than by the prior occurrence of geographic isolation. The phenomenon is strictly Darwinian.

Not only may morphological changes continue to accumulate after the initiation of reproductive isolation, but differentiation may also have begun well before. Most species are normally divided into subspecies and populations, any one of which has the potential for giving rise to a separate species if it is isolated from the rest of the parental taxon. The structural and behavioral differentiation of a population or subspecies that will become a separate species may frequently begin while there is still a possibility of genetic interchange with the rest of the species. This is clearly demonstrated by clines and ring species but is presumably the case for any partially isolated population on the margin of a species' range.

Since many species consist of two or more subspecies, and most comprise many recognizable populations, it is clear that there is only a very small possibility that any particular subspecies or population will give rise to a distinct species. Assuming that speciation among vertebrates nearly always requires the presence of a physical barrier precluding gene flow, only those populations that are so isolated will have the opportunity of giving rise to a new species. Unless there is some correlation between the nature and/or amount of morphological change that occurs in populations and subspecies and the possibility of their achieving genetic isolation, there is no reason to think that the amount of change that occurs in populations that will ultimately give rise to new species will be any greater than that which occurs in the many other populations that will not.

Summary

Study of modern populations casts doubt on two of the tenets of punctuated equilibrium:

1. Significant, directional morphological change does occur within species.
2. There is little evidence that significant change is correlated with speciation, if that term is restricted to factors directly associated with the initial isolation of a new species.

For longer-term consideration of both of these factors, we must turn to the fossil record.

4 Limits to knowledge of the fossil record and their influence on studies of evolution

Introduction

Modern studies of evolution at the population level support most of the conclusions reached by Darwin. In particular, natural selection has been shown to be an effective force in altering the phenotype and the underlying allele frequency, at least over time scales up to hundreds of years. Plant and animal breeding shows the great potential for response to selection at a much higher rate than is known to occur in the fossil history of major groups. There is no evidence that inherent factors of development or genetic homeostasis limit the amount of change within species. However, even at this scale, the nature and extent of evolutionary change can be seen to be restricted by factors of the physical and biological environment. The dimensions of the beaks of the Galápagos finches are ultimately limited by the size of the seeds on which they feed. Once an optimal size is reached, change ceases, unless the size or hardness of the seeds change or the available food is replaced by other sources. The size of the birds and their beaks is also constrained by the presence of other potentially competing finch species.

The primary factor limiting the study of evolution in modern populations is one of time. Even the fastest-evolving forms do not depart from the general pattern of their ancestors over the entire history of humans. The most disparate dog or cattle breeds are members of the same species, and intermediate forms have the potential to interbreed. The complete process of speciation cannot be observed within modern populations, even over the entire span of human history: Although some modern species may be less than a thousand years old, there are no cases in which the entire life span of a species can be traced. The origin of major new structures, body plans, and ways of life have few if any counterparts in changes observed in modern populations; these can only be studied through the fossil record.

Limitations of the fossil record

To a biologist, it may appear a simple matter to establish the patterns of evolution within the history of a species through the careful collection and dating of fossils. In fact, the nature of geological processes places inherent limits on both the initial preservation of fossils and their potential for accurate dating that greatly restrict our capacity to study evolutionary patterns at the level of populations and species. Only a small fraction of the number of species that now live are known from the fossil record, and very few species are known from sequences of fossils demonstrating their morphology during successive periods of time. Before presenting specific data documenting patterns of evolution, it is necessary to discuss the limits of resolution throughout the fossil record and how they have influenced evolutionary thought.

Growth of knowledge of the fossil record

Awareness that fossils are the remains of once-living organisms goes back at least as far as the writings of Zanthos of Sardis, who lived 500 years before Christ; but it was not until the late eighteenth century that fossils began to contribute to evolutionary thought. The discovery in the eighteenth and early nineteenth century of dinosaurs, gigantic marine reptiles, and distinctive mammals that did not belong to living species showed that many animals that had once lived on Earth had become extinct. These discoveries raised the possibility that there had been a succession of different types of organisms during the history of life. Jean-Baptiste de Lamarck (1744–1829) demonstrated that invertebrate fossils discovered in the Paris Basin showed a progressively higher percentage of modern forms in beds that accumulated from the Eocene to the Pliocene. He concluded that this was the result of progressive evolutionary change. On the other hand, the early paleontologists Georges Cuvier (1769–1832), Richard Owen (1804–92), and Louis Agassiz (1807–73) were not convinced that the fossil record supported the idea of evolutionary change. The Canadian paleontologist John William Dawson, in his *Modern Ideas of Evolution* (1890), attributed the succession of different dominant life-forms during the Earth's history to a series of creations and extinctions.

By the time Darwin wrote *The Origin of Species,* a considerable variety of fossils were known, representing many major groups of extinct plants and animals; but they did not document progressive change within particular lineages or transitions between any of the major kinds of organisms. Darwin lamented the inadequate knowledge of the fossil record in two chapters but assumed that fossils would eventually confirm his hypothesis. His diagrammatic representation of evolutionary patterns over short and long time spans was strictly hypothetical.

The late nineteenth century was a time of extensive fossil collecting, especially in the American West, leading to the discovery of a great variety of dinosaurs and Cenozoic mammals. Lineages such as those of fossil horses and elephants appeared to show progressive change over long periods of time, with extensive modification of the skeleton leading, as Darwin postulated, to very different body forms. Thomas Huxley used these discoveries to argue forcibly for the acceptance of evolutionary theory. By the early twentieth century, nearly all biologists and paleontologists accepted that the diversity of life on Earth was the result of evolutionary change over the millions of years of its history. In contrast, few influential scientists accepted Darwin's theory of natural selection as an explanation for the process of evolutionary change.

In fact, knowledge of the fossil record was still far too incomplete to determine the pattern or rate of change within individual species that would be necessary to evaluate Darwin's theory of natural selection. The horse fossils that so effectively documented large-scale evolutionary change were separated from one another by tens of millions of years (Fig. 4.1). Forms such as *Hyracotherium [Eohippus], Meso-*

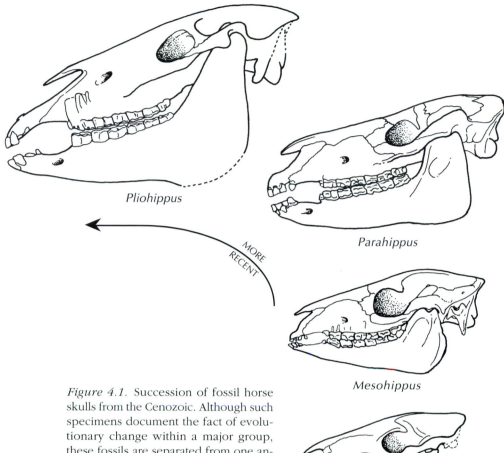

Figure 4.1. Succession of fossil horse skulls from the Cenozoic. Although such specimens document the fact of evolutionary change within a major group, these fossils are separated from one another by tens of millions of years and provide no evidence of the patterns or rates of change within species. Reprinted from Gregory (1957), *Evolution Emerging,* Macmillan Magazines Limited.

Pliohippus

MORE RECENT

Parahippus

Mesohippus

Hyracotherium (Eohippus)

hippus, Parahippus, and *Pliohippus* were each very distinct from one another, with few, if any, intermediates. Any pattern of evolution could have been postulated for the intervening lineages. It might have been a slow and steady progression, which seemed to be implied from a superficial interpretation of Darwin's writings, or highly irregular in pattern and rate. Darwin's only limiting condition was that evolution acted gradually, in the sense of showing no marked changes between generations: "*Natura non facit saltum.*" Even the presence of very large gaps between particular morphological patterns, resulting from saltational change, could not be precluded by what was known of the fossil record in the nineteenth century – nor, for that matter, in many lineages as they are known today. Some paleontologists emphasized apparently sudden anatomical changes within fossil lineages and elaborated saltationist theories of evolution (Schindewolf 1950), whereas others were impressed with long-term, apparently unidirectional evolution and postulated inherently driven orthogenetic forces that would control the direction of evolution in a teleological manner (Cope 1896; Osborn 1934).

Increasing knowledge of classical and population genetics during the twentieth century was shown by Dobzhansky, Mayr, and Simpson to support the major precepts of Darwinian selection theory. Neither orthogenesis nor saltational changes of the nature proposed by late-nineteenth- and early-twentieth-century paleontologists is consistent with what is now known of the way in which change occurs in the genetic material at either the molecular or chromosomal level.

Natural selection had become accepted by most paleontologists by the 1940s (Simpson 1944; Jepsen, Mayr, and Simpson 1949). Unfortunately, the fossil record was still too incompletely known to demonstrate the actual pattern of evolutionary change within species that would be necessary to determine the nature of the response of populations to natural selection.

Two very influential books by Simpson, *Tempo and Mode in Evolution* (1944) and *The Major Features of Evolution* (1953), established the contributions of paleontology to the evolutionary synthesis. They also illustrated limits to the study of evolution resulting from inadequacies of our knowledge of the fossil record at that time. In *Tempo and Mode in Evolution,* Simpson was forced to deal with many major problems in a largely theoretical and qualitative rather than quantitative manner. Attention was focused on the potential for the discoveries of Mendelian and population genetics to explain the patterns of evolution at all levels, but detailed cases from the fossil record were not available. Simpson was unable to cite any examples to illustrate evolutionary patterns within species. Rates and patterns of evolution were discussed at the level of genera, but even at this level the fossil record was still so incompletely known that the genus appeared as an essentially typological entity. The lack of evidence for the origin and specific relationships of groups at all taxonomic levels was discussed at length. Simpson introduced the concept of *quantum evolution* in an effort to explain why there was so little evidence of intermediate forms linking species or higher taxa that occupied different adaptive zones. He felt that such evolutionary shifts must have occurred rapidly, within small populations that had little chance of being preserved as fossils. *The Major Features of Evolution* was the latest attempt by a vertebrate paleontologist to provide a unified picture of evolutionary theory. Simpson took advantage of increased knowledge of the fossil record since the writing of *Tempo and Mode,* but the degree of temporal and phylogenetic resolution was not much greater.

Numerous publications supporting and elaborating the conclusions of the evolutionary synthesis have been published subsequently (e.g., Tax 1960; Simpson 1961; Mayr 1963; Dobzhansky et al. 1977), but no major changes in the general concept of evolutionary patterns or processes are evident from the early 1950s to the publication of Eldredge and Gould's paper on punctuated equilibria in 1972. In contrast with the assumption of Simpson and most other paleontologists that the fossil record was still far too incompletely known to provide direct evidence of evolutionary patterns at the level of species, they argued that what was already known of the remains of countless species accurately reflected their actual pattern of evolution, and that this pattern was entirely different from that postulated by Darwin and accepted by most modern biologists (see Chapter 2). Eldredge and Gould declared that the morphological gaps between species seen in the fossil record were not simply a result of gaps in sedimentary sequences but an accurate reflection of rapid change taking place in small populations that were geographically separated from the ancestral lineage. They emphasized the great longevity of individual species and the continuing absence of any well-documented fossil evidence of changes within species to argue that natural selection was incapable of producing progressive evolutionary modification within species.

Both the sudden appearance of new species and the long-term stasis within species emphasized by Eldredge and Gould on the basis of the fossil record appeared sharply at odds with the observations of modern species. Their ideas were so controversial that an entire generation of paleontologists have devoted their energies to either substantiate or refute them. Whole issues of the journals *Special Papers in Palaeontology* (1985) and *Paleobiology* (1983, vol. 3[4]) were devoted to data bearing on these topics. The most important aspect of this work was a change in emphasis in paleontology to studies of evolution at the level of species. This work is now beginning to provide the link between microevolutionary patterns and processes in living species with the larger-scale evolutionary patterns that can be studied only from the fossil record. Detailed information from that record should serve to test whether patterns and rates of evolution over tens of thousands and millions of years differ significantly from those observed in modern populations.

The capacity of the fossil record to test any theory regarding the patterns and processes of evolution depends on the accuracy of measuring the rates of evolution. This in turn depends on both the completeness of the fossil record and the ability to date the fossils accurately and over short time intervals.

The probability of fossilization and the irregularity of sedimentation

Many scientific endeavors have practical and/or theoretical limitations. In paleontology, there are limits to what can be learned from the fossil record that are fundamental to the nature of geological processes. Knowledge of these limitations is necessary to understand what one can expect to learn of evolutionary patterns both at the level of species and for larger-scale evolutionary phenomena.

As a result of predation, scavenging, and decay, only a tiny fraction of the organisms that once lived are fossilized. Simpson (1944) suggested that perhaps as many as 1 in 5,000 members of a particular vertebrate species might be fossilized in especially productive horizons. In some marine deposits, there may be millions of specimens of common invertebrates, such as brachiopods, ammonites, or bivalves, but other species are represented by only one or two specimens. An informative example involving Lower Triassic ammonoids is provided by Smith and Patterson (1989). Of 1,194 species named from this interval, 46 percent were known from only a single specimen and 20 percent from two specimens. We have no way of knowing how many ammonoid species may have existed during this time for which no specimens have yet been found. Compilation of a variety of marine taxa by Raup and Sepkoski (1986), ranging in age from mid-Permian to Pleistocene, showed that 38 percent of 9,250 genera were restricted to a single horizon. In a sample including echinoderms and fish from the same time interval, Smith and Patterson found that 68 percent of the species came from a single horizon. Nothing can be said about the pattern of evolution in species known from a single horizon, but such organisms do give an indication of the probable absence of other species, and the likely incomplete nature of the fossil record of other taxa.

For the entire geological time scale, Raup (1987) estimated that approximately a quarter of a million species are currently known as fossils. He compared this with Simpson's (1952) estimate of the total number of species that have lived as 50 billion!

The fossilization and long-term preservation of organic remains require that they be buried by sediments that prevent further physical, chemical, and biological breakdown. Water within sediments typically carries soluble minerals such as calcium carbonate and silica that infiltrate the openings in bones, shells, and woody material and enable them to be preserved with little subsequent alteration in structure or histology (Behrensmeyer and Hill 1980). The rate and degree of continuity of sedimentation are hence of extreme importance in determining whether or not organisms will be preserved as fossils and how accurately the fossils represent the pattern and rate of evolution within a species. The absence of fossils may reflect the absence not of particular groups of animals or plants but simply of the appropriate conditions for their preservation (Figs. 4.2–4.4).

Rates of sedimentation are extremely irregular in all depositional environments. Schindel (1982) compiled data on rates of sedimentation from 25,000 deposits ranging from recent sediments to some that have been accumulating over more than 100 million years (Fig. 4.5). These examples cover eight major environments: rivers, deltas, coastal wetlands, reefs and shallow-water carbonates, bays and lagoons, lakes, inland seas, and the bathyal–abyssal zones. The most general feature in all these environments is that the rate of sedimentation is inversely proportional to the time over which it occurs. Sedimentation may be very rapid over short intervals, but over long ones it is reduced or interrupted by periods of nondeposition of varying length. The problems of determining rates of deposition and degrees of completeness of sedimentary sequences over different time intervals are further discussed by Anders, Krueger, and Sadler (1987).

One may think of the oceans as being areas of long-term, continuous deposition, but even in deep basins the rate of sedimentation is governed by the relief of the nearest land and the amount of exposure of coastal areas. These differences are of sufficient magnitude to alter the apparent pattern of evolution between gradual and punctuational (MacLeod 1991). In depositional environments that would be expected to preserve terrestrial vertebrates, such as overbank and floodplain deposits or deltas, deposition may be initiated and terminated on a seasonal or sporadic basis, depending on climate and weather. Much more time may be represented by gaps between periods of deposition than by sediments.

At a larger scale, marine transgressions and regressions lasting hundreds of thousands or millions of years may leave large gaps in the record of both shallow marine and near-shore terrestrial communities. Deposition may be continuous for short periods, but the longer the time involved, the more numerous are larger-scale gaps; thus the average rate of sedimentation appears progressively lower.

Stasis, phyletic gradualism, and punctuated change differ in the degree to which they can be established from stratigraphic records of fossils. Stasis is potentially the simplest phenomenon to demonstrate, since two fairly complete fossils separated by a significant stratigraphic interval are sufficient to establish that continuous directional change had not occurred. Even if morphological change were discovered in fossils of the same species preserved within the interval between them, it would not falsify this conclusion if stasis is defined as oscillation about a mean.

Figure 4.2 (right). Evolutionary history of a taxon undergoing progressive evolutionary change while giving rise to three diverging lineages. Black areas indicate time of sedimentation allowing preservation of limited segments of the history of this group. Gaps between black areas are times without a sedimentary record. Redrawn from Skelton (1993), *Evolution: A Biological and Palaeontological Approach,* Addison–Wesley–Longman (publ.), copyright holder Open University.

Figure 4.3 (below). History of sedimentation during the duration of the taxon shown in Fig. 4.2. Redrawn from Skelton (1993), *Evolution: A Biological and Palaeontological Approach,* Addison–Wesley–Longman (publ.), copyright holder Open University.

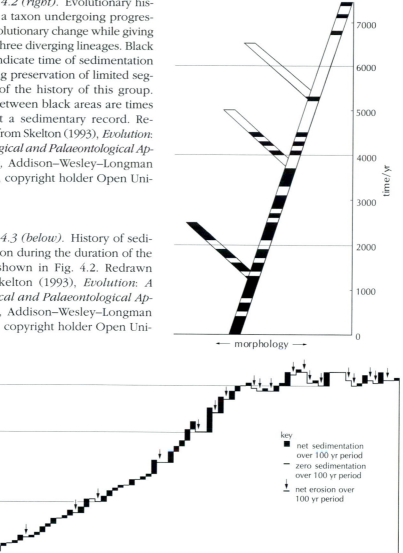

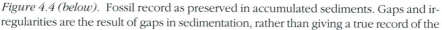

Figure 4.4 (below). Fossil record as preserved in accumulated sediments. Gaps and irregularities are the result of gaps in sedimentation, rather than giving a true record of the

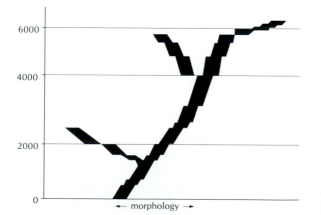

pattern and rate of evolution. One derived lineage is completely missing, and a second appears suddenly, without intermediate forms linking it with its immediate ancestor. The rates of evolution appear to change within the main lineage, although this did not occur in the actual history of the group. Most fossil sequences are much less completely known than this example indicates. Redrawn from Skelton (1993), *Evolution: A Biological and Palaeontological Approach,* Addison–Wesley–Longman (publ.), copyright holder Open University.

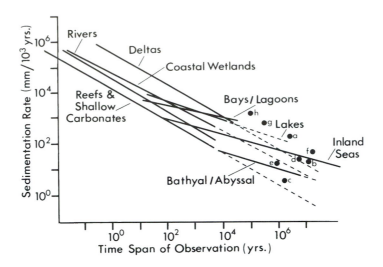

Figure 4.5. Log plot of sedimentation rates vs. time spans over which each rate is calculated. Solid lines are major axes of variation for 250 sedimentation rates compiled for eight depositional environments. Dashed lines are extrapolations of trends to time spans longer than cases compiled. No corrections for compaction have been made. From Schindel (1982).

Long-term directional change can also be readily demonstrated for organisms preserved in deposits resulting from continuous, relatively rapid sedimentation. In contrast, it is very difficult to disprove that an apparently punctuated pattern was not actually the result of an interrupted stratigraphic record of phyletic gradualism. One can imagine a situation in which short periods of rapid and continuous deposition are separated by long stretches without deposition. The periods of rapid deposition might have a rich assemblage of fossils but from such a short period of time that little evolutionary change is evident. This case would give the impression of a strongly punctuated pattern even if evolution had actually occurred at a slow and constant rate for the entire sequence. Since all sedimentation is to some degree irregular and episodic, there is an inherent bias toward a punctuational pattern for any mode other than stasis. On the other hand, it is unlikely that alternating periods of deposition and nondeposition would be such as to produce the impression of gradualistic change from a truly punctuated pattern. The extreme rarity of gradualistic change cited by Eldredge and Gould argues strongly against a bias in this direction.

There are several ways in which gaps in a sedimentary sequence can be detected, but it is much more difficult to estimate their duration. It is frequently possible to recognize surfaces that have been exposed to erosion, desiccation, or soil formation. Marked changes in lithology may also suggest intervening periods of nondeposition and changes in deposition rate. Deposits of the same total thickness may result from different length periods of deposition or from gaps of different duration and would give the appearance of different patterns of evolution. From knowledge of a number of sedimentary sequences within the same formation, it may be possible to establish the most probable pattern for the evolving species. This appears most effective in deep-sea deposits that can be studied from many drill cores (MacLeod 1991).

Schindel's data (Fig. 4.5) show that different depositional environments differ consistently in their completeness over a wide span of time intervals. Varved lake deposits show particularly regular depositional sequences that may represent yearly increments. This would seem an ideal circumstance to study evolutionary change; unfortunately, lakes commonly last relatively short periods of time. For example, the large Rift Valley lakes of East Africa (Echelle and Kornfield 1984) and comparable lakes from the Triassic and Early Jurassic of North America (McCune, Thomson, and Olsen 1984) are subject to periodic drying up, with cycles of approximately 20,000 or 100,000 years associated with regional climatic changes. The climatic changes are linked to the *Milankovitch cycles* – changes in the orientation of the Earth's axis of rotation and the eccentricity of its orbit (Olsen 1986). The period during which these lakes could be inhabited is long enough for the occurrence of speciation and adaptation to different feeding modes, and some species show considerable structural change, but there is only limited evidence of the lineages surviving beyond the lakes in which they evolved (McCune 1996). In the case of such lakes, the end of the depositional cycle not only interrupts the fossil record but results in the extinction of nearly all the contained lineages.

Changes in the relative completeness of the fossil record over time

The problem of incomplete sedimentary sections is common to all geological periods. The longer the section studied, the greater the missing portion, but this problem may be no greater in the Paleozoic than in the Tertiary since sedimentary processes within a particular environment were essentially the same. On the other hand, the chance of erosion or tectonic activity resulting in the loss or inaccessibility of sedimentary sequences increases progressively with age. Whole continents may be lost by subduction under other lithosphere plates (see Chapter 14). Very few areas of the Earth's crust that formed more than 3.5 billion years ago are still present and accessible to geologists. The older the fossils, the less likely they are to survive and be collected, no matter how numerous or widespread they may have been in life.

Even though the fossil record of vertebrates began only about 500 million years ago, particular periods of time and their faunas remain extremely poorly known. Ordovician vertebrates, limited primarily to near-shore marine deposits, are known from only three major sites, one each in Australia and North and South America. There are few if any deposits of Paleozoic age from which it is possible to study evolutionary patterns at the level of species, at least among vertebrates. There are many rich localities with a diversity of species, but most have very limited time spans; only very general patterns of the radiation of major groups can be established. During the Carboniferous, when the first major radiation of amphibians and reptiles occurred, only about twenty important fossil localities have yielded a fauna of terrestrial vertebrates (Smithson 1985; Milner 1987) (Fig. 4.6A). Most are restricted to less than a square mile in extent and are located almost exclusively in North America and Europe, limited to a narrow band only a few degrees from the paleoequator. Nearly all species are known from only a single horizon. The temporal extent of most of the horizons has not been estimated, but oxbow lakes – which constitute the depositional environment at Linton, Ohio; Jarrow, Ireland;

and Nyrany, Czech Republic – normally last 1,000–10,000 years before filling with sediments (DiMichele and Hook 1992). The lake system at East Kirkton, Scotland, is thought to have existed for no more than about 40,000 years (Clarkson, Milner, and Coates 1993). Other horizons may be similarly limited in time. Omitting the Stephanian, at the very end of the Pennsylvanian, Carboniferous deposits bearing terrestrial vertebrates may represent less than 1 percent of the total duration of this period; that is, more than 99 percent of the elapsed time is not represented by any fossils of terrestrial vertebrates. Because of large gaps in the fossil record in the Carboniferous, it is still impossible to determine the interrelationships of most lineages of amphibians that have been described (Carroll 1992) (Fig. 4.6B).

For the Mesozoic, the interrelationships of the major groups are better established, but very few species are sufficiently well known for detailed comparison to be made between specimens from successive horizons or different localities. Only in exceptional circumstances, such as varved lake deposits, is it possible to study populations representing a great many short time intervals within the duration of particular species (McCune 1996). Only from these localities is it possible to describe the pattern of evolution within particular lineages over much of the duration of the species. Although dinosaurs appear to be a well-known assemblage, there are very few cases in which remains of particular species are known from deposits that succeed one another in time, or in which evolution at the species level has been studied. Work by Horner, Varricchio, and Goodwin (1992) is unique in showing what may be interpreted as anagenetic transformations between genera. Deposits from the Upper Cretaceous Two Medicine Formation covering less than half a million years show four examples of transitions between previously recognized genera from underlying and overlying strata. The long interval of time during which erosion and destructive tectonic processes could have acted since deposition will always limit the number of localities for which such studies are possible within the Mesozoic.

Within the Cenozoic, we have a much more complete knowledge of the overall stratigraphic record, especially in North America. This is summarized in a comprehensive volume edited by Woodburne (1987), *Cenozoic Mammals of North America: Geochronology and Biostratigraphy* (see also Woodburne 1996). According to Lindsay and Tedford (1990, p. 609): "North American land mammal ages now come very close to representing all of Cenozoic time." This degree of completeness has enabled many studies of evolution at the level of species and genera (Gingerich 1976, 1982, 1987; Rose and Bown 1986; MacFadden 1992). Nevertheless, detailed studies of sedimentary processes indicate that even in the Tertiary there is a limit to the level of dependable resolution. For the fluvial deposits in which many Tertiary mammals are found, no succession of fossils can be dated within less than approximately 10,000 years, and few can be dated consistently within intervals of less than a hundred thousand years. Processes of original deposition, reworking of sediments, and collection on surfaces subject to erosion result in time averaging of individual fossils over spans up to tens of thousands of years (Behrensmeyer and Hook 1992, tab. p. 80).

Gingerich (1987 and references cited therein) has studied patterns of evolution in a great number of mammalian lineages preserved in Palaeocene and Eocene deposits. For example, in 1976 he described the phylogenetic patterns of three genera of primitive placental mammals from the Early Eocene Willwood Formation in

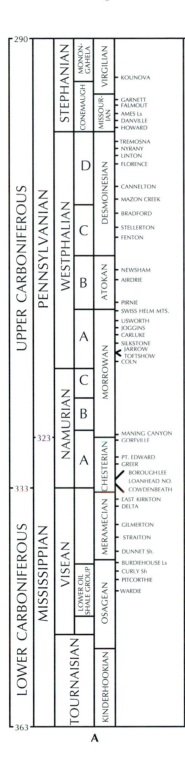

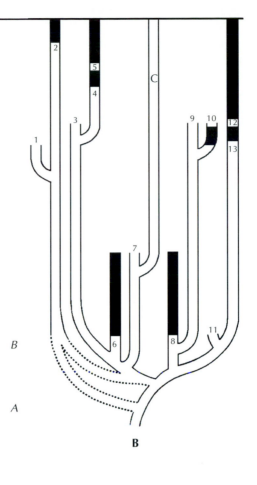

Figure 4.6. **A,** Stratigraphic horizons within the Carboniferous from which terrestrial amphibians and reptiles have been found. No fossils are known from the intervening horizons. Based on data from Milner (1987) and Smithson (1985). **B,** Fossil record of nectridean amphibians from the Upper Carboniferous. Solid lines indicate horizons from which fossils have been found. Gaps indicate absence of fossil remains. Key to genera and species of nectrideans: 1, *Scincosaurus;* 2, *Sauravus;* 3, *Diceratosaurus;* 4, *Diploceraspis conemaughensis;* 5, *Diploceraspis burkei;* 6, *Keraterpeton;* 7, *Batrachiderpeton;* 8, *Urocordylus;* 9, *Ctenerpeton;* 10, *Ptyonius;* 11, *Lepterpeton;* 12, *Sauropleura scalaris;* 13, *Sauropleura pectinata;* C, lineages leading to the Lower Permian genus *Diplocaulus.* There is no way of estimating the time between points *A* and *B,* during which the three families of nectrideans diverged. From Carroll (1984).

Wyoming, based on samples from fifty-eight relatively evenly spaced horizons within a 550-m section. He concentrated on changes in dimension of the first lower molar tooth, which showed progressive increase or decrease over time both within and between successive species. Cases of branching speciation were also postulated. Subsequent studies of the conditions of sedimentation in these localities by Dingus and Sadler (1982) indicated that one could depend on some sedimentation occurring within each period of a million years, a 4 in 7 chance of it oc-curring within each hundred-thousand-year interval, and a 1 in 2 chance within 10,000-year intervals. Extensive collection may result in the discovery of addition-al fossils, but it would not result in finer temporal resolution.

At this level of completeness, Dingus and Sadler conclude that we would not expect to resolve evolutionary processes lasting less than hundreds of thousands of years. Gaps within fossil lineages of shorter duration might result from lack of deposition and have no evolutionary significance.

Gingerich (1982, 1987) subsequently worked in other areas within the Eocene in which the rates of deposition were up to twice as rapid as in the Willwood For-mation. Nevertheless, he concluded that biological processes that occurred on a temporal scale of less than 5,000–10,000 years were effectively beyond resolution in the fluvial deposits in which most Tertiary mammals were found.

With an exceptionally long and continuous stratigraphic record, it should be possible to date fossils from several intervals within the length of species duration; yet in practice this is rarely the case, even in the Tertiary. The vagaries of preserva-tion, the irregularities of deposition, and the infrequent availability of beds that can be used for radiometric dating still make it difficult to establish whether the rarity or absence of fossils linking species is a consequence of a punctuated evolution-ary pattern or a result of geological factors, even in the mid-Cenozoic.

Even at the level of relationships between species and genera, the fossil record of Cenozoic mammals shows many conspicuous gaps. This is clearly demonstrat-ed by a recent phylogeny of middle Miocene horses illustrated by MacFadden (1992) (Fig. 4.7). Although these horses are an extensively studied group with a rich fossil record in North America, many well-documented lineages remain sepa-rated from one another by gaps of millions of years, well beyond the probable du-ration of speciation events. For example, there is a gap of more than 3 million years at the base of lineage 4, and another of 2.5 million years in the lineage postulated as the common ancestor of lineages 4 and 5. There is no fossil record at all of the lineages involved in five successive speciation events leading to lineages 6 and 17, covering a time span of more than half a million years. These breaks in the fossil record may result from gaps in sedimentary sequences or reflect the absence of horses in particular geographical areas or environments. They provide no informa-tion at all regarding evolutionary processes.

Dating geological events and processes

As knowledge of the fossil record has improved, so have methods to date events in the Earth's history and within the evolution of fossil groups. Despite the limits of precision in establishing the age of fossils and the length of time between fossilif-erous horizons on the basis of the thickness of sediments (emphasized earlier in this chapter), this method remains the one most widely used for dating individual fossils in the field. It also continues to form the primary basis for estimating the rate of evolutionary processes within species.

Fossils themselves have long been used as a means of dating stratigraphic hori-zons. By the early nineteenth century it was recognized that different fossil assem-blages typically occurred in a particular sequence within geological strata, which

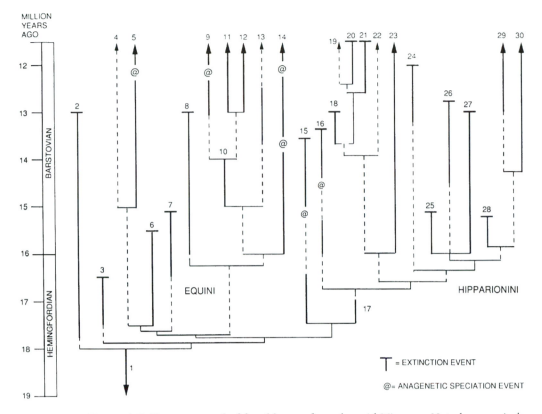

Figure 4.7. Known record of fossil horses from the mid-Miocene. Note long periods for which fossils of particular lineages are not known (dotted lines), and absence of transitional forms between almost all lineages (thin horizontal lines). This is more likely to result from lack of deposition of sediments in areas where horses were living than from aspects of evolutionary change within this group. From MacFadden (1992).

served as a basis for establishing their relative age. The major units of the currently accepted geological time scale were established initially on the basis of a succession of fossil-bearing strata. Even without knowledge of their age in years, the presence or absence of particular species allowed geologists to establish the relative sequence of a great number of subdivisons within each larger unit. Some of these subunits are now known to last no more than a hundred thousand years, providing a very effective means of dating over very large areas of the planet. Dating of fossil vertebrates frequently depends on their association with fossils of invertebrates or plants (especially pollen and spores) that can be dated relative to other geological horizons. Fossil vertebrates themselves are used to establish the major divisions of terrestrial Cenozoic deposits in North America (Woodburne 1987) and Europe (Savage and Russell 1983). Although fossils may serve as the basis for recognizing particular stratigraphic horizons, independent means of dating in years must be established in order to determine rates and patterns of evolution.

Radiometric dating

The first widely accepted effort to apply laws of physics to dating Earth was undertaken late in the nineteenth century by Lord Kelvin (William Thomson), who calculated the probable age of the world on the assumption that it had cooled progressively from the molten state. His date of approximately 12 million years was a serious blow to evolutionary theory, being far shorter than Darwin and others thought necessary for the gradual unfolding of change and diversification. Ironically, the reason that Lord Kelvin's suggestion for a very short history of Earth was incorrect – the fact that its heat has been maintained by the decay of radioactive elements – was also the basis for establishing a correct age. Radioactive decay has generated sufficient heat to maintain the Earth's temperature at a nearly constant level for more than 4 billion years. Soon after the discovery of radioactivity by Becquerel in the 1890s, it was recognized that radioactive decay occurred at a nearly constant rate. By 1913, when his *The Age of Earth* was published, Arthur Holmes had used the decay of uranium to lead to provide the first estimates of the planet's age based on an absolute time scale. This was to provide a dependable framework for the timing of all geological events and the history of life on Earth.

Except for ^{14}C, whose rate of decay is so rapid that it can be used for dating events only over the past 50,000–70,000 years, isotopes used in radiometric dates are associated primarily with volcanic and other igneous rocks. Most fossils are dated indirectly by their presence in beds immediately above or below beds that have few or no fossils of their own but contain isotopes that can be dated.

A number of different elements have been used for radiometric dating (Odin 1982; Geyh and Schleicher 1990; Harland et al. 1990), but most currently accepted dates were established from ^{40}K–^{40}Ar decay. These isotopes can be studied effectively in both lava flows and ash falls. The latter provide an especially useful means of dating fossil in the Cenozoic of North America: They were frequent and widespread, and may be closely associated with beds containing vertebrates.

As with our knowledge of the fossil record, the accuracy of radiometric dating progressively lessens as one goes back into the geological record, and fewer and fewer dates have been established. Use of ^{40}K–^{40}Ar decay can date with an uncertainty of only 1–2 percent, and the more recently developed $^{40}Ar/^{39}Ar$ decay has an uncertainty as small as 0.1 percent. The latter method is very accurate for events over periods of tens of thousands to a million years. For events occurring 10 million years ago, the expected error is approximately 10,000 years, which is equivalent to the highest expected resolution from most stratigraphic sequences. For the early Cenozoic, the best possible dating yields an expected error of approximately 65,000 years (or a current practical limit, applied to the Cretaceous–Tertiary boundary, of 100,000 years [Dingus 1984]). Although this is within the range of most species' longevity, it would not suffice to track differences at the level of populations, and may be shorter than the period during which speciation occurs. For short time intervals in the early part of the Tertiary, radiometric dating is less accurate than estimates based on sedimentary processes.

Magnetostratigraphy

Another method of dating, which does not itself become progressively less accurate with time, is magnetostratigraphy (MacFadden 1992, chap. 6). The polarity of the Earth's magnetic field has reversed many times in the planet's history, over very irregular intervals of time (Fig. 4.8). The timing of reversals is dependent on radio-

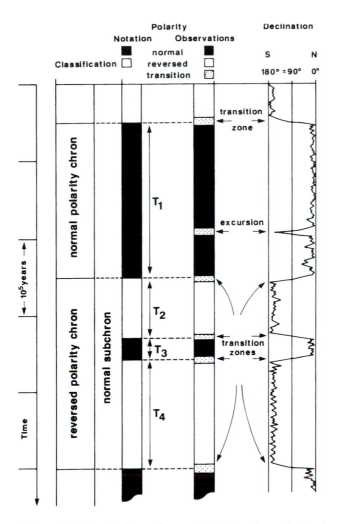

Figure 4.8. Terminology of magnetostratigraphy: polarity chrons, polarity subchrons, transition zones, and excursions. From Harland et al. (1990).

metric dating of associated strata, but determination of the time in one place is sufficient for correlation with other beds showing the event everywhere on Earth, since the shift in polarity affects the entire planet simultaneously.

Polarity may be either normal, as at present, or reversed, with the needle of the compass pointing south rather than north. The relative duration of the episodes is sufficiently irregular that their sequence can be correlated in widely separate areas, just as one can date events through comparison of the irregular spacing of tree rings. Changes in magnetic polarity have the advantage of being detectable in any sediments containing ferromagnetic minerals, whereas rocks appropriate for radiometric dating are only rarely deposited in close association with beds containing fossils. The average duration of each period of normal or reversed polarity during the Cenozoic is approximately 330,000 years. For this time interval, $^{40}K-^{40}Ar$ dates currently average only 200,000 years apart, but they are directly applicable to only a small geographical area.

A small number of more accurate dates may be achieved by concentrating on the time of reversal, which takes only 5,000–10,000 years, so the age of the small percentage of fossils that overlap a change in polarity can be dated very accurately no matter how long ago the reversal occurred. However, much less than 1 percent of fossils can be expected to be datable with this level of accuracy; moreover, this method provides only one date rather than the minimum of two to bracket the duration of biological processes such as speciation or anagenetic change.

Paleomagnetic reversals have been extensively studied as far back as the early Jurassic, but most of the Cretaceous is dominated by a single polarity and so provides no significant date lines (Fig. 4.9). For the Paleozoic, little information is yet available except for the dominantly reverse polarity recognized for the Upper Carboniferous and Permian (Harland et al. 1990).

Rates of evolution

Radiometric and paleomagnetic dating provides an effective basis for determining the age of fossils and hence the rates of evolution of the species to which they belong. Quantitative study of evolutionary rates began with Simpson (1944, 1953) and Haldane (1949). Simpson discussed three major ways to measure the rate of evolution: genetic, morphological, and taxonomic.

Morphological change

The most obvious measure of evolution is in terms of quantitative difference in morphological features seen in sequences of fossils. This was formalized by Haldane (1949) in terms of the equation,

$$r = [\ln(x_2) - \ln(x_1)]/(t_1 - t_2)$$

where r is the rate of change, x_1 is the initial dimension and x_2 the final dimension of the character, and $(t_1 - t_2)$ is the amount of time between the ages of x_1 and x_2. Haldane expressed the rate in **darwins** (d), the amount of change per million years. Linear measures are transformed to natural logarithms to minimize the effect of absolute size, so that valid comparisons can be made between organisms of very different size.

Measurements in darwins have been made in hundreds of different taxa, fossil and living, and show a great range of different rates both within and between groups. It was originally assumed that the rate of evolution might be similar within particular groups (e.g., faster in mammals than clams) but might differ during different periods of evolution (e.g., faster during the origin of a group and slower during subsequent periods), thus furnishing a comparable means of measuring rates over all spans of geological time. Work by Kurtén suggested that for mammals, rates were higher during the Pleistocene than in the Tertiary, and higher yet in the Holocene (Recent). He attributed this difference to more powerful selection in the Pleistocene and Holocene in relationship to periodic glaciation and the radical changes since the retreat of the last continental ice sheets. A radically different interpretation was given by Gingerich (1983) on the basis of a compilation of 521

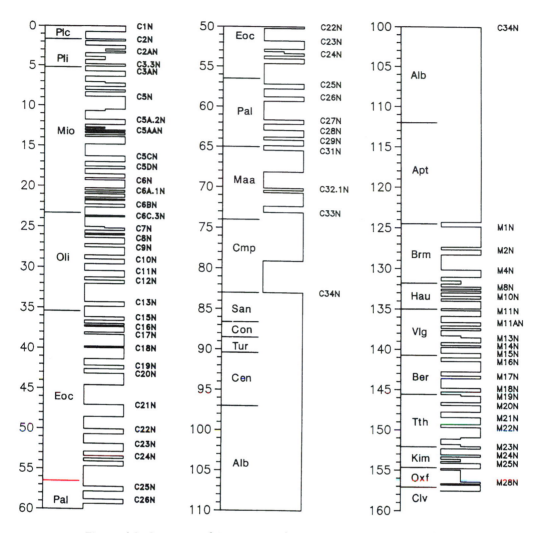

Figure 4.9. Summary of Cenozoic and Mesozoic magnetic-reversal time scale. Normally magnetized intervals lie to the right of the polarity graph, reversely magnetized intervals to the left of it, and mixed intervals in the central portions. From Harland et al. (1990).

measurements made over a great range of different time spans, from 1.5 years to 350 million years. He discovered a strongly negative correlation between the span of time over which change was measured and the apparent rate of evolution – from a typical rate of 0.1 d in fossils to an average of 60,000 d in laboratory selection experiments. He argued that the apparently very slow rates in fossil groups could be attributed to averaging out changes that might be very rapid for short periods of time, but were of short duration and so would appear much slower if measured over long periods (Fig. 4.10).

He illustrated this graphically through an example of changes in the Eocene primate *Cantius* over different intervals of time (Gingerich 1987). Fossils were collected from fifteen horizons, separated by 40,000–45,000-year intervals. From this sequence, he measured fourteen rates of change at 40,000–45,000-year intervals, thirteen rates of change between successive 80,000–90,000-year intervals, twelve

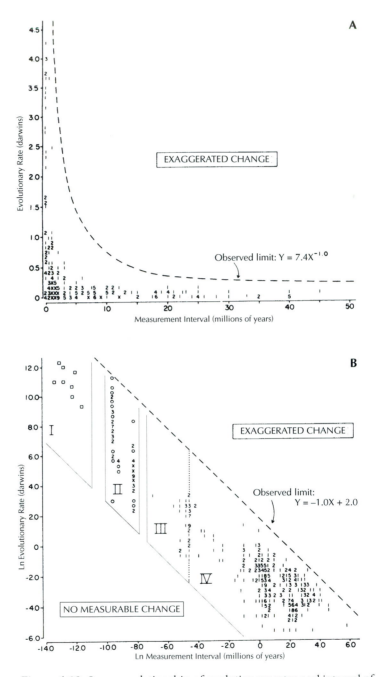

Figure 4.10. Inverse relationship of evolutionary rates and interval of time over which rates were measured. **A,** Central portion of distribution of 521 morphological rates measured in darwins over intervals ranging from 1.5 years to 350 million years. Observed limit (dashed line) is derived from that in B. **B,** Logarithmic transformation of entire distribution shown in A. The observed limit and the distribution as a whole have a slope approximately −1.0. Domains I–IV correspond to rates from laboratory selection experiments (open squares), historical colonization events (open circles), post-Pleistocene faunal recovery from glaciation, and fossil invertebrates and vertebrates. Reprinted with permission from *Science*, vol. 222, Gingerich. Copyright © 1983. American Association for the Advancement of Science.

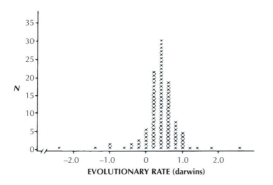

Figure 4.11. Histogram of 105 evolutionary rates representing all combinations of comparisons between fifteen samples of three or more specimens, taken from successive stratigraphic intervals totaling approximately 0.6 million years of geological time. From Gingerich (1987).

rates of change between 120,000–135,000-year intervals, and so on. The resulting rates showed a normal distribution, ranging from >2.0 d negative change (reduction in size) to >2.0 d positive change (Fig. 4.11). The mode is +0.4 d, reflecting the overall tendency toward successively larger specimens in higher horizons.

Gingerich emphasized that biologically significant comparisons of evolutionary rates can only be made over equivalent periods of time. The rapid rates of evolution observed in modern populations almost certainly could not be sustained over long periods of geological time. Under unidirectional selection pressure, the trait might quickly reach an optimal or structurally limited size and cease to evolve, or under changing selection pressure might undergo reversal or long-term oscillation, thus greatly reducing the net long-term rate of evolution.

The use of Haldane's equation remains a practical measure of change in linear dimensions when comparing similar intervals of time; MacFadden (1992) used it extensively in his recent study of horse evolution. It must be recognized, however, that nearly all measures over short periods of time will result in high rates, whereas measures over long periods of time will imply much slower rates. As a result, evolutionary rates will superficially appear to be faster during the Quaternary and Late Tertiary, where temporal resolution is far closer than in earlier geological periods.

Because of the problem of measuring rates of evolution over different time intervals, Haldane (1949) suggested that the only valid measure of evolutionary rate would be from generation to generation. This was followed up by Gingerich (1993a,b) in proposing a new measure of evolutionary rate, the **haldane** (h). This is described by the following equation:

$$\text{rate (h)} = \frac{(\ln x_2 / s_{\ln x}) - (\ln x_1 / s_{\ln x})}{t_2 - t_1} = \frac{\bar{z}_2 - \bar{z}_1}{t_2 - t_1}$$

where $\ln x_1$ and $\ln x_2$ are natural logs of measurements at times t_1 and t_2, $s_{\ln x}$ is the pooled standard deviation of $\ln x_1$ and $\ln x_2$. One haldane is defined as change by a factor of σ (i.e., one standard deviation) per generation.

The haldane has a theoretical advantage over the darwin in measuring evolutionary rates because it is based on generation time rather than years. Gingerich refers to the average change for a single generation, measured in standard deviations, as the **intrinsic evolutionary rate**. In common with the darwin, the intrinsic evolutionary rate must be inversely proportional to the number of generations over which it is measured. It can only be calculated in living populations, or for very limited cases in the fossil record, such as varved lakes, from which yearly intervals can be determined and factored into probable generation times. For most examples from the fossil record, the actual intervals that can be sampled cover a great many generations. Gingerich applied this procedure to length and width measurements of the lower first molar tooth in successive samples of the primitive Eocene horse *Hyracotherium grangeri,* for which he established that the minimum number of generations between samples was approximately 6,750. He calculated the average intrinsic rate as 0.225 h, with a 95 percent confidence interval in the range 0.056–0.870 h.

This is a very broad confidence interval. In addition, the large number of generations that were averaged would give a figure far below the expected rate for some generations that exhibited high rates of change. One may compare, for example, the extremely variable intergenerational changes in metric characters among Darwin's finches in normal and drought years. This particular example shows both the difficulty of applying this measure to the fossil record and, on the other hand, the potential for the application of a very important measure of clear evolutionary significance across the gap between modern and fossil populations.

Gingerich calculated the intrinsic rate of change in two modern populations: one involving laboratory selection in the mouse, the other natural selection in the marine snail *Littorina*. For the mouse, a line selected for increased body size (with approximately 80 percent of each litter eliminated in each generation) increased at an average rate of 0.180 h per generation over a period of twenty-three generations. Study of spire reduction, attributed to selection by a predatory crab, in *Littorina* covered three intervals of seventeen, sixty-nine, and eighty-six generations, with a median rate of change per generation of 0.148 h.

So far, attempts to determine the intrinsic evolutionary rate have been conducted in only an extremely small number of species. Nevertheless, the preliminary results are potentially of great significance: They indicate that the amount of change from generation to generation, even when averaged over many generations, is large relative to the amount of difference that is commonly used to distinguish subspecies and species. Lande (1986), following Simpson (1961) and Mayr (1969), used two phenotypic standard deviations to characterize the limit of differences recognized between populations within a subspecies. Subspecies would be expected to exhibit differences greater than two phenotypic standard deviations, which is only twenty times that difference exhibited between individual generations in the examples studied by Gingerich.

Two species of *Hyracotherium, H. grangeri* and *H. aemulor,* which follow one another in stratigraphic sequence in the section studied by Gingerich, differ from one another in the size of the lower first molar by 4.15 σ. This is only forty times the amount of change postulated for an average intrinsic evolutionary rate, or eighteen or nineteen times the intrinsic evolutionary rate calculated for this character in *Hyracotherium grangeri*. At least 13,500 generations separate these two species, but there is no evidence of the pattern of evolution between them. Gingerich refers to this as an *interval of punctuation*. If the change in dimensions of the molar occurred unidirectionally, as if under selection, at a constant rate over 13,500 generations (the minimum duration of their stratigraphic separation), the average rate would be much slower (0.0002 h) than the average rate measured over intervals of 6,750 generations in the ancestral species.

An extremely important observation resulting from this study is that the intrinsic rate of evolution (the amount of change per generation) may be similar for all characters, whether their pattern of evolution over time scales of thousands or more generations shows stasis or clearly directional change, and whether they are measured within or between species. At the level of morphological features within species, the haldane provides a limiting measure of the rate of change but no information regarding the direction or rate of evolutionary change at longer time scales, even within short-lived populations.

Although the intrinsic rate of evolution is the fastest possible rate of change, in all cases the actual rate of evolution, termed **net rate of evolution** by Gingerich, is much slower and may appear as stasis or random walk. From these observations, Gingerich concludes that control of evolutionary rates is not inherent, since rates of change from generation to generation are normally orders of magnitude faster than are necessary to account for the most rapid changes recognized over longer periods of time, even in the most stringent selection experiments. Control of the rate and direction of evolution must be extrinsic to the organism and population, residing in physical and biological aspects of the environment.

Simpson and Gingerich concentrated on linear dimensions as a fundamental measure of anatomical change. Changing ratios of body proportions, such as limb length versus snout–vent length, or the width and breadth of the skull, are also frequently used to describe evolutionary change, but they are not amenable to expression in darwins or haldanes. Bookstein et al. (1985) have developed procedures that can be used to describe change in shape in two dimensions. Many taxa are distinguished from one another by differences in the number of particular skeletal elements, such as vertebrae, fin rays, digits, or phalanges. These are referred to as **meristic changes**. Such changes are used frequently in taxonomic definition of groups, but not commonly as a measure of evolutionary rate among vertebrates, because of the rarity of well-dated sequences. Debraga and Carroll (1993) used the total number of anatomical changes that occurred over a particular time span as a measure of the relative rates of evolution in the transition *between* two major groups and the rates of evolution *within* the ancestral and descendant groups.

Changes in physiology and behavior are even more difficult to compare and nearly impossible to measure in a quantitative manner. Nevertheless, changes in behavior may have been the primary factor in such major transitions as the emergence of land vertebrates from the water, and in the origin of the many lineages of secondarily aquatic vertebrates. Increase in metabolic rate was certainly key to the origin of both ornithurine birds and mammals.

Taxonomic rates

Another way of measuring evolution is in terms of the longevity of taxa and the rate of their origination and extinction. The rapid appearance and persistence of numerous distinct lineages is a conspicuous aspect of the early evolution of major taxonomic groups. Species, genera, families, orders, and even phyla have been used for analysis, depending on the scale of the radiation. Simpson (1944, 1953) made extensive use of genera rather than species in evaluating evolutionary rates because of the practical limitations of the fossil record as it was known at that time. Although genera might comprise several species, more data was then available on the longevity of genera than of species, which were frequently known from single horizons. Subsequently, studies showed broadly similar patterns of evolution for genera and species, indicating that this procedure is appropriate for comparative studies (Novacek and Norell 1982) (Fig. 4.12). Closer approximation to estimating the duration of taxa can theoretically be made by using species longevity, since the species has the potential for more objective definition. (See Chapter 7 for discussion of the relative objectivity of taxa at different levels.)

Most studies of the longevity of taxa have been based directly on the known temporal range of fossil remains (Fig. 4.13). This certainly underestimates their total range. Unless the fossil record reveals a speciation event that marks the end of a particular species, or that species continues to the present, it is impossible to ascertain when a species actually begins and ends. The time of its origin can be estimated through analysis of relationship with its most closely related sister-group. This will nearly always indicate a significantly longer range than indicated by the fossil record directly, as demonstrated by Novacek and Norell (1982). There is no way to be certain of the actual time of extinction, since even today taxa that were

Figure 4.12. Plot of late Cretaceous and early Tertiary primates showing similar diversity patterns at both species and generic levels. From MacFadden (1992).

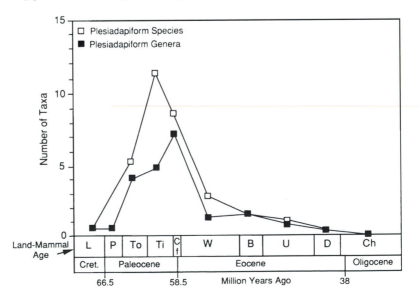

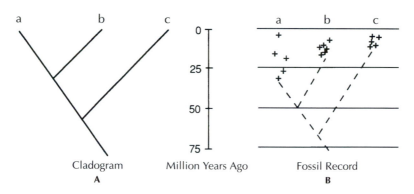

Figure 4.13. Divergence times based on cladistic interrelationships (**A**) of three hypothetical taxa a, b, and c showing conflict between cladistic hypothesis and known fossil occurrence (**B**). Because of an incomplete fossil record, the biostratigraphic origination times are typically later than those predicted from the cladistic branching patterns. From MacFadden (1992), modified from Novacek and Norell (1982).

thought to be long extinct are occasionally recognized in the modern biota. The discovery of a coelacanth in 1938, after this group was thought to have died out in the Cretaceous, is a familiar modern example, but many less well-known cases are reported from the fossil record, such as the extension of the range of large labyrinthodont amphibians from the Triassic into the Jurassic and Cretaceous (Milner 1989).

Although the rate of taxonomic evolution can be only loosely quantified, it is a useful way to compare rates of evolution at different stages in the longevity of a major group and between related groups; for example, Hulbert and MacFadden's (1991) study of horse evolution in the middle Miocene showed that it began with very rapid speciation, followed by a steady state between origination and extinction.

Molecular evolution

Simpson suggested that it might also be possible to determine the rate of evolution by measuring the total amount of genetic change between two generations or within a succession of generations. Efforts to do so were made by Dobzhansky (1947) on the basis of *Drosophila*. Simpson (1953, p. 4) concluded his discussion by stating, "There is little hope that fossils can ever be used for any extensive study of genetic evolution directly as such." Unfortunately, Simpson did not foresee the revolution brought about by the development of techniques that made it possible to retrieve and sequence DNA from fossils. This has made it practical to demonstrate base-pair differences in small segments of particular genes in many extinct organisms, including horses (Higuchi et al. 1987), insects (Cano et al. 1993), and plants (Soltis, Soltis, and Smiley 1992). Although it is potentially possible to use the amount of change between fossil and living taxa to determine the rate of change in a particular gene within the group, the rarity of fossil DNA makes this impractical as a general means of comparison, at least for the foreseeable future (Smith and Littlewood 1994).

The ease with which DNA and RNA can now be sequenced in living organisms does provide a means for establishing relationships and measuring relative rates of evolution in all taxonomic groups. This is particularly important in groups for which the fossil record is incompletely known and for study of the sequence of divergence of populations and subspecies living today. In common with changes in morphology, changes in DNA do not occur at the same rates in all taxa, or even in the same taxa at different periods of its evolution. Mitochondrial DNA changes much more rapidly than nuclear DNA, and different portions of the same gene evolve more rapidly than others. It is thus possible to select particular types of DNA whose rates of change are most appropriate for the time span of the problem being studied. Since there is an almost infinite amount of DNA, it is not surprising that details of phylogenies will differ depending on what genes are used. Several journals are devoted specifically to molecular studies of evolution (e.g., *Journal of Molecular Evolution, Molecular Biology and Evolution,* and *Molecular Phylogenetics and Evolution*).

Molecular evidence is used primarily for establishing the nature of interrelationships of different lineages and the relative time of their divergence. By itself, it cannot establish the actual time when lineages originated. Like other aspects of evolution, change in DNA is broadly correlated with elapsed time, but not directly so. For any particular problem, the molecular clock must be "set" on the basis of some other evidence, such as the fossil record or a geological event. If we are fairly certain of the timing of a geological event, such as the beginning of separation of South America and Africa as a result of continental drift, we can use this information to give a minimum time for the separation of populations that inhabited those continents. The time at which their genome was identical must have been prior to that split. The total amount of change can then be divided by the elapsed time to give an approximate rate of change. If this rate is fairly constant, it can be used to establish the sequence of divergence of individual lineages within the groups that subsequently evolved separately in South America and Africa. This provides information as to the rate of divergence and the relative longevity of the living members of these lineages. If the evidence provided by several different genes in all the taxa being studied is consistent, it supports the initial assumption of the near constancy of rate of genetic change.

So far, none of the genes that have been used in taxonomic studies provides direct evidence of the evolution of morphological structures. Although phylogenies based on molecular evidence may accurately reflect the relative time of divergence of vertebrate groups, they cannot be used to establish how or when specific characters by which we recognized members of these groups evolved. Taxonomic groups are recognized by the presence of one or more unique morphological traits by which all known members can be identified. Although the divergence of a group may be accompanied by the emergence of a clearly defined new character, it is very unlikely that more than a single character will arise in synchrony with the divergence, or that even a single character will be fully developed at this moment. No matter how accurate the phylogenetic resolution, the fossil record remains the primary basis for establishment of the patterns and rates of morphological change.

Summary

It is clear from the nature of sedimentary processes and larger-scale geological phenomena that the preservation of the remains of organisms is both extremely rare and extremely irregular. In contrast with studies of evolutionary patterns and processes based on living populations, the paleontologist is very limited in the evidence that is available. Particular problems, such as the nature of the transition between major groups, or the response at the species level to changes in the environment, must be studied on the basis of particular examples of specific taxa over limited time intervals.

The capacity to determine the patterns and rates of evolution is limited by the fact that both the number of fossils that can be recovered and our ability to date events decline progressively in older strata. Prior to the Mesozoic, wide morphological gaps separate nearly all orders of fish and amphibians. Very few adequately known species are known with certainty from more than one locality or geological horizon. In the Mesozoic, relationships between the major groups can be more accurately established, but evolution at the level of genera and species is still largely unknown. At least among terrestrial mammals, the record is far more completely known in the Tertiary, with many relationships well established at the generic and specific level. Patterns of evolution within vertebrate species, in contrast, are typically beyond the resolution of sedimentary records and radiometric dating. Prior to the late Cenozoic, only a few cases can be confidently established as illustrating punctuated equilibria or phyletic gradualism.

Gould (e.g., 1982, 1990) has argued that the acceptance of the theories of punctuated equilibria and species selection depends on their quantitative dominance. In fact, this is extremely difficult, if not impossible, to document from the fossil record. Among vertebrates, it is difficult to determine the relative importance of stasis and phyletic evolution even for particular characters within a single species over a restricted period of time. Establishing the relative frequency of punctuated versus phyletic change for life as a whole, or even for mammals throughout the Cenozoic is beyond the capacity of our current knowledge of the fossil record, and this, in turn, is but an unknown fraction of the actual history of the species that lived within this time period.

The remainder of this book deals with specific examples of patterns of evolution studied through the fossil record. These examples were selected because they are well documented and demonstrate particular features that are important for understanding how evolution operates in different vertebrate groups and over different time scales. They also illustrate some of the difficulties common to the study of fossils. I have attempted to approach this study in a systematic manner, proceeding from the level of species to progressively higher taxonomic groups and larger-scale evolutionary phenomena, but this has not always been possible: Even the largest-scale evolutionary events must have begun at the level of species.

5 Patterns of evolution among late Cenozoic mammals

Introduction

Detailed knowledge of the patterns and rates of evolution throughout the duration of species is necessary to establish a link between observed patterns of change over short periods of time in living populations and the larger-scale phenomena that can only be studied over the long periods of geological history. For the Paleozoic and most of the Mesozoic, very few fossils of any vertebrate species are known from more than a single stratigraphic horizon, making it impossible to establish their actual longevity or any aspects of rates or patterns of evolution during their duration. It is only during the latter part of the Cenozoic that the fossil record is sufficiently well known that systematic study of change within species is possible.

In this chapter, we concentrate on the last 5 million years of the Cenozoic, which provide the most complete record of evolution at the species level and the most accurately dated. This period includes three geological epochs: the Holocene, which covers the 10,000 years since the retreat of the last great continental glaciation; the Pleistocene, which began approximately 1.64 million years ago (mya); and the Pliocene, from 5.2 million to 1.64 mya (Fig. 5.1). The term Quaternary is widely used to include both the Holocene and the Pleistocene. These dates follow the most recent geological time scale (Harland et al. 1990). Other publications that were used extensively in compiling this chapter give somewhat different dates for the beginning of the Pleistocene: Savage and Russell (1983) and Williams et al. (1993) place the beginning of the Pleistocene at 1.8 mya, which is near the beginning of the Olduvai magnetic reversal event, and Martin and Barnosky (1993) suggest 1.7 mya. The Pliocene and Quaternary encompass the entire duration of most living mammalian species. Approximately 90 percent of living European mammals are known from the Pleistocene (Kurtén 1968).

Figure 5.1 (facing). Time scales for the Cenozoic. Separation into a series of epochs *(column 4)* was originally based on the increasing percentage of living species within a succession of molluscan faunas. The land-mammal ages are based on a succession of mammalian faunas in North America (Woodburne 1987). The ages in years are based on radiometric dates, correlated with a rich marine record of invertebrates and protistans. Large-scale correlation of both marine and terrestrial sediments is possible through comparison with irregular reversals of magnetic polarity. Both the marine and terrestrial biotas are influenced by gradual but irregular cooling throughout the era. During the past 2 million years, changes in temperature became much more rapid and extreme.

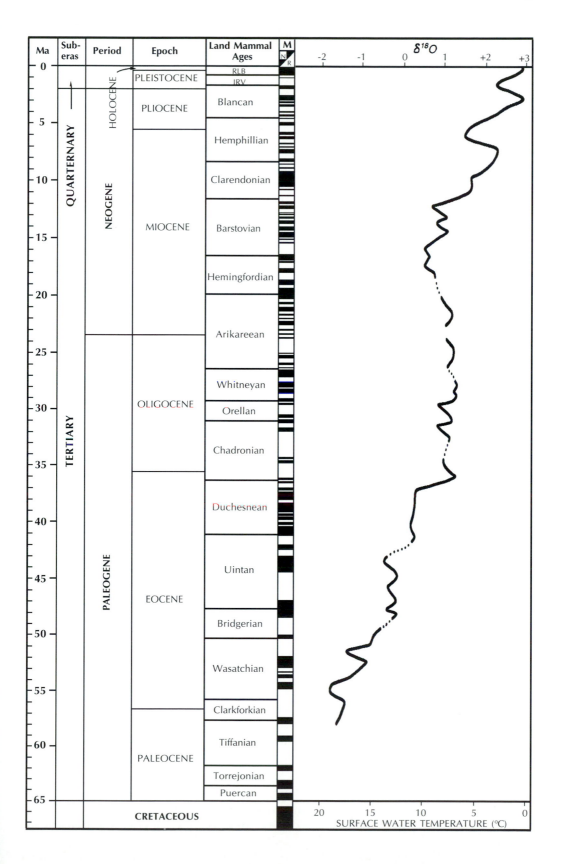

Especially precise dating of Holocene and late Pleistocene fossils is possible through measure of ^{14}C decay, which can be applied directly to bone and wood with a range of error that may be as low as 2–4 percent. Unfortunately it is effective over a range of only 50,000–70,000 years, but many dates established through the use of ^{40}K–^{40}Ar and ^{40}Ar/^{39}Ar decay are available for earlier in the Cenozoic. Reversals of magnetic polarity occurred only three times during the Pleistocene, but seventeen reversals can be used for dating Pliocene sediments. During the late Pliocene and Pleistocene, these means of dating are augmented by repeated episodes of dramatic climatic change, continental glaciation, and extensive regression and transgression of the sea; all provide additional means of gauging the time and rate of biological changes.

These changing conditions, which were among the most stressful in the history of the Earth's biota, would be expected to force rapid response on the part of plants and animals in the form of dispersal, physiological accommodation, or evolutionary change. A further advantage of research in this time period is that most Pleistocene species are still living today, so that aspects of their descendants' physiology and behavior can be used to judge the environments in which they lived earlier in the Quaternary. Modern animals with a good fossil record also provide a means for establishing the degree of concordance between species established on the basis of the differences in the skeleton (the only way in which fossils can be assigned to particular species) and species recognized on the basis of the biological species concept, which can be applied only to living organisms.

The Plio–Pleistocene ice ages

It was long thought that there were four major episodes of glacial advance and retreat in both North America and Europe that were broadly synchronous. It is now realized that the previously recognized glacial episodes of the later Pleistocene comprise two or more periods of glacial advance and retreat, and that many other periods of continental glaciation preceded them, back into the late Pliocene.

Work by Shackleton and Opdyke (1977) on the relative concentration of oxygen isotopes O^{16} and O^{18} in the oceans (preserved within the shells of calcareous foraminifera and coccoliths and siliceous diatoms) provides a reliable basis for establishing the pattern of climatic change throughout the Cenozoic. Water formed from the lighter isotope evaporates more rapidly; hence, glacial ice, which forms from rainwater, has a lower proportion of O^{18} than does seawater. As more and more water is withdrawn from the sea to form the ice sheets, the seawater accumulates a higher proportion of O^{18}. The temperature at which the snow in ice sheets condensed can be established directly on the basis of the isotope ratio: A change of 0.6– 0.8 parts per thousand results from a change of 1 °C at the site of condensation of water vapor. Thus the temperatures over the land and in the sea, as well as the times of glacial advance and retreat, can all be calculated.

The oceanic temperatures declined at a slow and irregular rate throughout the Cenozoic but more rapidly since the late Pliocene, indicating the onset of continental glaciation (see Fig. 5.1). Major ice sheets built up rapidly in the North Polar regions approximately 2.4 mya. The first continental glaciation began in the northern continents roughly 1.8 mya, just before the beginning of the Pleistocene. From that time until about 0.9 mya, the marine isotope records suggest that periods of ice advance followed one another at approximately 41,000-year intervals. From 0.9 mya through the last major continental glaciation, the interval increased to roughly a hundred thousand years. The maximum extent of the last glacial advance (termed the Wisconsin in North America and the Würm in Europe) occurred about 17,000 years ago, with a rapid retreat beginning 10,000 years ago. This marks the end of the Pleistocene and the beginning of the Holocene.

The intervals of glacial advance are comparable with different periods of the Milankovick cycles (Crowley and Kim 1994). Forty-one thousand years is the period over which the obliquity of the Earth's axis relative to the plane of the ecliptic changes from 21.8° to 24.4°. One hundred thousand years is the period of change in the eccentricity of the Earth's orbit around the Sun, which varies from nearly circular to somewhat elliptical. Since these cycles control both the distance and the orientation of the Earth's axis of rotation relative to the Sun, they presumably played a part in the cyclic nature of the climatic change of the Pleistocene; but many other factors of the Earth's geography, including extent of plant cover, circulation of the oceanic and atmospheric currents, and changing albedo, must have influenced the duration of these cycles as well (Williams et al. 1993). There is no obvious reason why the period of the cycles would have changed 0.9 million years ago. In general, the periods of cooling have been much longer than the periods of warming and melting following the peak of glaciation, but brief periods of warming (*interstadials*) are also recorded within the periods of major continental glaciation, and short periods of dramatic cooling have occurred during interglacials. There is no firm evidence as to whether the Holocene marks the end of the Pleistocene glacial epoch or is merely an interglacial that will eventually be followed by continuing episodes of extensive continental glaciation.

Although it may eventually be possible to correlate the many episodes of glacial advance and retreat with changes in the vertebrate fauna, this is currently possible only for the later part of the Pleistocene, and even here it is frequently difficult to establish which glacial or interglacial within a long series is represented by particular fossil horizons.

Testing punctuated equilibria

The importance of Pliocene and Quaternary mammals for the study of evolutionary patterns was first emphasized by Kurtén (1960). Hibbard (1949) demonstrated the potential for extensive collection of mammalian remains through sieving and washing of vast quantities of fossiliferous sediments that formed the data base for most subsequent studies. Barnosky (1987) emphasized the unique importance of the Quaternary in testing the hypothesis of punctuated equilibria and provided numerous examples of recent studies of evolutionary rates and patterns. Many additional studies on Quaternary mammals were presented in the volumes edited by Fejfar and Heinrich (1990a) and Martin and Barnosky (1993).

Recognition of stasis

> Most species, during their geological history, either do not change in any appreciable way, or else they fluctuate mildly in morphology, with no apparent
> direction.   - Gould and Eldredge (1993, p. 115)

Before examining the evidence from the fossil record, it is necessary to quantify the differences between the patterns of evolution expected on the basis of the alternative theories of punctuated equilibrium and Darwinian phyletic evolution. Gould (1982) specified that species might be expected to exhibit a period of rapid change equivalent to approximately 1 percent of their longevity associated with their origin through speciation, but that the remaining 99 percent would typically be spent in stasis. Accepting Stanley's (1979) estimate of 2 million years for the average duration of mammalian species, the period of rapid change should occupy about 20,000 years. The remaining 1,980,000 years would be expected to be characterized by stasis.

Gould and Eldredge have defined stasis only loosely. For example, Gould (1982) described stasis as mild fluctuation in morphology. Gould and Eldredge (1993, p. 223) used the phrases ". . . stability (often for millions of years) of paleontological 'morphospecies'" and "absence of change in lineages." Although *stability, equilibrium,* and *absence of change* are used interchangeably by Gould and Eldredge, these terms may be interpreted as expressing a spectrum of permissiveness. Complete absence of change within a lineage is almost inconceivable in view of the observed variability of modern populations and the inevitable change in genotype from generation to generation in sexually reproducing organisms. Stability and equilibrium, on the other hand, imply some degree of latitude. *Equilibrium* suggests movement about a mean, with little or no net change over significant time spans. The context of their use of the term *stability* implies retention of sufficient similarity that all members of a lineage would be placed in the same species. Martin and Barnosky (1993, p. 6) define stasis as "the meandering of measured character or shape variables around grand means (the equilibrium)." Lande (1986, p. 351) defined the duration of stasis for a particular character in a fossil sequence as "the maximum period in which the mean phenotype remains within subspecific bounds, i.e., x changes by less than two phenotypic standard deviations measured within populations." These definitions imply that change *within* a species is less than that which commonly occurs *between* species. Levinton (1988) defines stasis more stringently as a statistically lower deviation from a starting condition than that expected by chance. This follows the view of Mayr as well as Eldredge and Gould that intraspecies stasis is not a random phenomenon involving passive drift of neutral characters but a positive attribute of species at equilibrium, restricted by some sort of inherent homeostasis. As demonstrated by Bookstein (1987), chance phenomena, such as random walk and genetic drift, have no constraint to remain close to a mean value. They are more likely to mimic directional change than stasis. Gingerich (1993b) provided a graphical means to distinguish stasis from the greater scope of change expected from random processes.

In contrast, Darwin placed no limits on the amount of change that might occur within a lineage, but argued that it could lead to differences great enough to justify recognition of a distinct species. Simpson (1961) suggested that populations succeeding one another in time within an evolving lineage could be distinguished as distinct species when the ranges of observed changing characters did not overlap. Recognition of change of this magnitude within a particular lineage, without evidence of any speciation events, would be well beyond the bounds expected from stasis.

Eldredge and Gould's original concept of stasis clearly applied to the entirety of species characteristics. The similarity of species over millions of years involves all characters that can be studied in the fossil record. In contrast, most studies or tests of stasis are restricted to individual characters, as in Eldredge's (1985b) example of the number of facets in the eye of a trilobite. It is natural for this to be the case, as it is practical to measure the change (or lack thereof) of single characters but impractical or impossible to judge the amount of change in enough characters to describe species fully.

For fossils, we are automatically limited to the study of hard parts of the skeleton that are commonly preserved. No aspects of soft anatomy, physiology, or molecular constituents can be taken into consideration in evaluating the importance of stasis in the fossil record. In the case of pelycopods and many single-celled organisms, the only character that can be conveniently studied is the external covering. For the fossil record, stasis in this structure is equivalent to stasis of the entire organism. In most vertebrates, the skeletal remains include many more elements, all of which should be studied in order to demonstrate or refute the occurrence of stasis at the species level. This is rarely, if ever, practical: For most groups, complete skeletons are only seldom found, and the chance of finding a geological succession of complete skeletons is even smaller. Even if such a succession were found, in most cases it would be both extremely expensive and time consuming to prepare and study large numbers of complete skeletons, although exceptions are exemplified by studies of fossil fish (McCune et al. 1984; Bell 1988).

In practice, most elements of the vertebrate skeleton are relatively conservative, and attention can be focused on the smaller number that do change within a particular species. Among mammals, teeth are the single most diagnostic elements: They are the most durable portion of the skeleton and exhibit a great number of distinct anatomical features. Nearly all mammalian species can be identified by the configuration of the occlusal surface of the molar teeth. There is typically some degree of variability of this and other tooth characters within species, so they serve as an ideal means of evaluating evolutionary change. The configuration of the teeth is closely tied to diet, as were the bills of the Galápagos finches discussed in Chapter 3, which were the focus of attention even where all other body proportions could be studied. Mammalian teeth are also indicative of total body size within a particular taxonomic group, so that constant tooth dimensions may imply relative stasis in other aspects of the skeleton. For these reasons, many studies of evolutionary rate among mammals have concentrated on dental characteristics (e.g., Rose and Bown 1986; Gingerich 1987).

Examples of stasis. Several examples of stasis were summarized by Barnosky (1987):

1. The shrew *Blarina brevicauda* (Guilday et al. 1978): minor fluctuations in length of mandibular molar row over a period of at least 8,600 years.

2, 3. The hamsters *Cricetulus bursae* and *Rhinocricetus ehiki* (Kurtén 1968): minor fluctuations in length of tooth row over a period of approximately 300,000 years.

4. The ground squirrel *Spermophilus townsendii* (Rensberger, Barnosky, and Spencer 1984): minor fluctuations of width-to-length ratios of third upper molar over a period of at least 33,000 years, and possibly for as long as 320,000 years.

Lich (1990) and Anderson (1993) provided a particularly thorough analysis of stasis in the rodent *Cosomys primus,* whose fossils were preserved in ten successive horizons within 300 ft of strata in the Pliocene of Idaho (Fig. 5.2). Radiometric ages are provided by two volcanic ashes within this sequence, but these can be used only indirectly for establishing the total time span. They indicated minimum and maximum rates of sedimentation of 1.8 and 6.6 ft per thousand years, representing 45,000–165,000 years. The average interval between the horizons represents 3,000–10,000 years, but they are quite irregular in length. The study concentrated on the first lower molar, for which the following traits were measured: mean molar length, width, and posterior loop width. Differences in the mean phenotype for all three characters between successive populations were less than two phenotypic standard deviations and were not correlated with stratigraphic succession. A one-way ANOVA test revealed no significant differences among any of the populations. The use of a nonparametric runs test revealed that variation in the metric means from one population to the next was due to random fluctuation ($P > 0.05$) rather than the result of an evolutionary trend. For the entire sequence, the null hypothesis of random walk was rejected through use of a random walk test (Bookstein 1987) supporting the hypothesis of stasis.

Stasis was unquestionably demonstrated in these characters for a substantial period of time, but two limitations of this study should be noted.

1. *Limited temporal span.* According to Gould, stasis would be expected to occur for approximately 99 percent of the duration of a species. The longevity of this species is approximately 1.2 million years. Hence, stasis has been demonstrated for only 4–14 percent of this range, depending on the actual time span of this sedimentary sequence. Stasis could be refuted by the demonstration of greater changes in these particular characters over any of the remaining 86–96 percent of the longevity of the species.

Stasis during some period of species duration is documented in many of the studies reported in the book edited by Martin and Barnosky (1993). For example, Barnosky (chap. 3) described stasis during some portion of the time span of *Microtus pennsylvanicus* for all characters studied, but most of these showed significant change during other periods.

2. *Limited number of characters.* More important, only a single character complex in one tooth has been studied. Stasis within this species could be refuted by the discovery of any character that does show significant, directional change during any of the 99 percent of its duration following origin through speciation. For example, Zakrzewiski (1969) reported the presence of four molar triangles rather

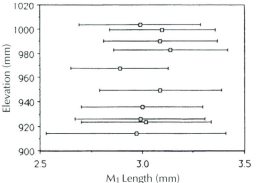

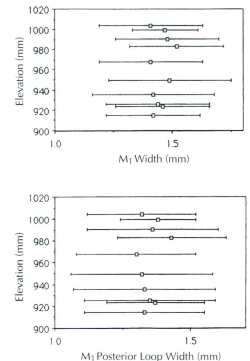

Figure 5.2. Stasis in the rodent *Co-somys primus*. Measurements of dimensions of the first lower molar in specimens collected from a series of strata in Idaho. The samples are from ten horizons, representing 45,000–165,000 years. A nonparametric runs test revealed no significant pattern (at the 0.05 significance level) in change in the mean molar dimensions over time. From Anderson (1993).

than three in specimens of *Cosomys primus* in some horizons. He suggested that these represented a different ecological setting, but this implies a biologically significant, if short-term, evolutionary change.

As quantified by Gould (1982), punctuated equilibrium is a highly asymmetrical theory in its requirement for demonstration versus its vulnerability to contradiction. Full support requires knowledge of the pattern of evolution for 99 percent of the longevity of any species; yet the theory can be refuted by knowledge of any interval of the species greater than 1 percent of its duration. In fact, the evidence so far available indicates that long-term stasis of all readily measured characters is limited to relatively few lineages of late Cenozoic mammals. A quantitative measure of the proportion of lineages of mammals in which stasis is described in dental characters is provided by Chaline (1987). He recognized a total of 140 lineages of microtine rodents (lemmings, voles, and muskrats) in the late Cenozoic of North America and Eurasia, of which 98 are living today, but considered that only 38 are well documented in the fossil record; within these, he recognized only three cases of stasis in dental characters. The *Pliolemmus antiguus* lineage is static for nearly 1.5 million years. *Pliomys episcopalis* shows no morphological changes from its appearance 2.5 million years ago until it disappeared, 2 million years later. The *Ungaromys weileri-nanus* lineage exhibits only a small tendency of triangle closure over 2 million years. Although not analyzed in a quantitative manner, these appear to be among the best examples of stasis from the late Cenozoic. Changes in characters other than dentition might eventually be demonstrated, but teeth seem to be the most consistent source of variability in most rodents that have been studied.

Within this particular group of rodents, stasis is thus demonstrated in 8 percent of the adequately studied lineages, whereas 92 percent show significant morphological change within an appreciable portion of their longevity. (The extent of change exhibited by this group of rodents is described below in the section "Longer-term directional change.")

Summary of stasis. Stasis has been shown to occur in individual osteological characters in numerous mammalian species for periods of tens or even hundreds of thousands of years of the late Cenozoic. In a few species, most elements of the dentition remain essentially unchanged for all of their known duration. This supports the contention of Eldredge and Gould that stasis may be an important feature of evolution at the species level, but it does not support the further claim that phyletic evolution within species is so rare that it does not normally lead to significant change within species. Most of the species of late Cenozoic mammals that have been studied in detail exhibit some characters that change significantly during some periods of their duration, although many of these lineages also show periods in which there is little measurable change. Clearly, Eldredge and Gould's claim that most species do not change in any appreciable way during their geological history is not supported by data from late Cenozoic mammals.

Directional evolution

> Phyletic gradualism is very rare and too slow, in any case, to produce the major events of evolution.   – Gould and Eldredge (1977, p. 115)

Gould's definition of stasis is so limiting in its scope for variation and demanding in its extent throughout the duration of a species that almost any unidirectional change that can be measured, at any time except for the first few thousand years of a species' longevity, would be sufficient to refute its suggested occurrence. The fact that very few late Cenozoic mammals have been shown to exhibit stasis throughout most of their duration is not in itself sufficient to demonstrate that significant change, such as that which differentiates contemporary species, does occur within species. There is, however, abundant evidence that this is the case. Several examples were cited by Barnosky (1987):

1. The sagebrush vole, *Lagurus curtatus,* showed marked change in the pattern of its molars over a period that lasted 33,000–320,000 years. Early in the sequence, a new variant in the upper third molar tooth appeared in which a deep reentrant made the anterior and middle vertical ridges more prominent than the posterior one. The frequency of the new morph increased progressively from 0 to 100 percent within the succeeding horizons. In the same sequence, the common pattern of the first lower molar changed progressively from having four triangles to having five (Rensberger et al. 1984).

2. Between more than 350,000 years ago and about 50,000 years ago, the molars of the water vole *Arvicola cantiana* exhibited progressive

change in size, the ontogenetic age at which roots grow, and the thickness of enamel (Stuart 1982).

3, 4. Other examples of directional change in teeth were reported in the pigs *Mesochoerus limnetes* and *Metridiochoerus andrewsi* (Harris and White 1979).

A particularly good example is provided by Hulbert and Morgan's (1993) study of the giant armadillo *Holmesina,* which inhabited Florida continuously from approximately 2.3 million to 12,000 years ago. Only a single lineage is recognized, which was not affected by immigration from other areas. Specimens are known from more than twenty localities, distributed temporally in a more or less uniform way throughout the range of the lineage. Although up to three sequential species have been recognized by some workers, there is no evidence of cladogenesis within this assemblage, nor any points of subdivision that are not obviously arbitrary. The amount of change throughout the sequence is substantial, with increase in the length of the limb bones of up to 150 percent and in the area of the osteoderms of 225 percent (Fig. 5.3). The area of tooth occlusion increased less rapidly than total body size. Many of the changes were episodic, or variable-rate phyletic change, with rates differing by a factor of 3–10 during some intervals. There is no overall evidence of stasis, or association of change with speciation events.

These examples demonstrate readily analyzed single characters or character complexes that change within a species in a unidirectional manner. The duration of change is much longer than the period suggested by Eldredge and Gould for association with speciation events, and may occur at various times within the longevity of the species. The origin and spread within the population of new elements, such as additional areas of tooth occlusion, general increase in size, or readily recognizable differences in proportions result in the descendant forms being clearly distinct from early members of the same species. They certainly meet the criteria for "significant" change expressed by R. A. Martin (1993, p. 232): "Morphological change in a phyletic sequence can be considered significant if variation in the sequence is considerably greater than the expression of intraspecific morphological variation and/or at least equal to interspecific variation as demonstrated by the closest living relatives of the taxon under investigation."

Mosaic evolution. Other studies show much more complex situations in which different characters, or the same character at different periods of time or from different stratigraphic sequences, show a variety of different patterns of evolution, varying from stasis to various rates of directional change. This is referred to as **mosaic evolution**.

An informative example is provided in papers by Barnosky (1990, 1993), in which he evaluates both geographical and temporal change in a single living species. *Microtus pennsylvanicus* is a meadow vole that has been common throughout northern North America for the past 450,000 years. Modern populations range from isolated enclaves in Mexico and Florida to the northern edge of the continental margin in Canada and Alaska. Barnosky studied individual populations from the extremes of the range in Alaska, Hudson Bay, and Colorado, as well as both modern and Pleistocene populations in Virginia. He concentrated on the following aspects of the third upper molar tooth: length, width, amount of closure between triangles 1 and 2, perimeter and area of the entire occlusal surface, perimeter and area of the posterior loop, and shape factors of the entire occlusal surface and the posterior loop (Fig. 5.4). The pattern of change varied considerably among these characters, both geographically and in time.

Concentrating on the fossil and living populations in Virginia, there is measurable increase in the width of the teeth relative to their length, as well as in the degree of confluence between triangles 1 and 2, between samples from the interval 30,000–18,000 years ago and the present. In contrast, four characters showed stasis: simplicity of posterior loop, tooth length, and shape factors of the entire tooth and the posterior loop.

The modern Colorado population differs from both the fossils and modern populations in Virginia in both shape factors and in the ratio of the perimeter of the entire occlusal surface to the perimeter of the posterior loop. On the other hand, the modern population in Colorado resembles the fossil population in Virginia in the ratio of tooth width to length.

Decrease in tooth length can be described as a south-to-north cline, with statistically significant differences between Virginia and Alaska and between Colorado and Hudson Bay, but lesser difference between the other localities. Eastern populations are characterized by a greater width/length ratio than teeth from western populations. Triangles 1 and 2 are more confluent in both fossil and modern samples from Virginia than are those from Alaska, Colorado, and Hudson Bay, and the posterior loop shows much less frequent presence of a second lingual reentrant angle. The latter character is present in 0–5 percent of the Virginia populations and 20–60 percent in the other localities. The lesser development of reentrants in both the fossil and living populations in Virginia is probably a primitive feature, judging by the elaboration of additional triangular occlusal surfaces in many other lineages of microtine rodents. The Colorado population is geographically the most isolated, and is the most derived in the greatest relative perimeter of the posterior loop, the total shape factor, and the posterior loop shape.

Among these particular traits, some show regional specialization, other demonstrate clinal variation of continental scope, and some of those studied from fossil and living populations in Virginia show temporal variation.

More recently, Barnosky has concentrated on temporal variation in fossils from a number of localities in the Appalachians (Fig. 5.5). This study encompassed approximately the last 53,000 years of the Quaternary, including modern populations. The fossils came from localities in Pennsylvania, Virginia, and Tennessee in which bones and teeth had come from fissure fillings and cave deposits. In each case, some but not all of the horizons were dated directly by radiocarbon techniques. The average interval between horizons in the three fossil localities ranged from 1,625 to 8,800 years, including the period between the last fossil horizon and the present. This research concentrated on tooth characters similar to those studied previously in modern populations.

In none of the three deposits is there close correlation among any of the four characters that were measured. Each appears to have its own pattern and rate of evolution. Only a single character (posterior loop shape) in a single deposit (Baker Bluff Cave), showed clearly directional change over the interval studied: 19,100 ± 850 years. At the New Paris locality length, width/length, and posterior loop

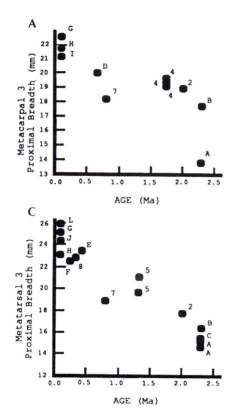

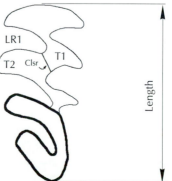

Figure 5.3. Directional change in limb size over 2.5 million years in the giant armadillo *Holmesina*. From Hulbert and Morgan (1993).

Figure 5.4. Occlusal view of the third upper molar of the microtine rodent *Microtus pennsylvanicus* showing dimensions that were measured in a study of mosaic evolution. Abbreviations: Clsr, closure; LR1, lingual reentrant; T1, triangle 1; T2, triangle 2. Modified from Barnosky (1990).

shape all showed stasis over a period of 1,300 years, as judged by a one-way ANOVA and Tukey's HSD test. Other characters at other localities showed a mixture of stasis and rapid change over varying intervals of time. Little if any correlation could be established between any morphological changes and changes in climate.

The results of these studies demonstrate a very complex pattern of evolution of these traits in relationship to geography and particular periods of time. The total range of this species is 450,000 years, so that the segment studied here amounts to approximately 12 percent. Neither stasis nor unidirectional change is dominant. One might argue that most characters showed considerable fluctuation but without clearly demonstrating directional change. The variable length of time represented by the different fossil deposits, uncertainties regarding the time between sampled horizons, and differing sample size make statistical analysis difficult.

The course of dental change during this period in the evolution of *Microtus pennsylvanicus* somewhat resembles the pattern of evolution that might be expected from a fossil record of the Galápagos finches or other extensively studied modern populations. Neither displays the simple pattern of continuing unidirectional change that might naïvely be expected from models of phyletic evolution. Rather, they show that the complex, ever-changing patterns of evolution that can be seen in modern populations can also be demonstrated by detailed studies of the fossil record over periods of time intermediate between those of modern populations and the millions of years covered by most studies of large-scale change.

Longer-term directional change. The most basic differences between the model of evolution proposed by Darwin and elaborated in the evolutionary synthesis, and that formulated by Eldredge, Gould, and Stanley, is whether or not evolution within a particular lineage, unmarked by speciation events, could be so extensive as to lead to forms that would be recognized as distinct species. The examples just discussed are limited to change within individual lineages that are typically considered as belonging to single species. Numerous other examples of evolutionary change have been described that transcend the range of variability generally accepted as characterizing individual species. Most such sequences extend over several million years, but all lack the detailed resolution that exemplify the cases just studied.

Because of the irregularities of depositional processes and local loss of sediments through erosion, one can never expect to discover very long sequences of continuous deposition within a particular geographical area. The maximum local sequence, not including living forms, that was documented by Barnosky (1993) was only 34,000 years in duration, which is well less than 10 percent of the usual species longevity. One must accept a trade-off between fine-scale sampling of relatively short sequences, as in these examples from the Late Quaternary, or less close sampling for periods approximating that of species longevity.

The sampling problem is well illustrated by examples of larger mammals discussed by Lister (1993). The moose, *Alces,* can be divided into three successive species on the basis of major changes in the proportions of the antlers over a period of 1.8 million years (Figs. 5.6, 5.7); but since its remains come from only three or four adequately dated horizons, the actual pattern of evolutionary change within the

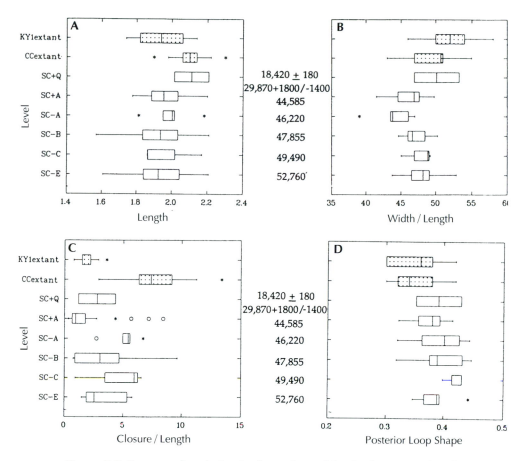

Figure 5.5. Patterns of evolution in dimensions of the third upper molar of *Microtus pennsylvanicus* over 52,760 years in the late Pleistocene and Holocene. Each dimension follows a different pattern of evolution. None is clearly directional. From Barnosky (1993).

individual species cannot be determined. The fossil record of the proboscidean genus *Mammuthus* shows progressive increase in the frequency of the lamellae that make up the tooth crown. This is documented from ten successive horizons, spanning a period of approximately 2.25 million years. There is overall progressive change, but with more irregular fluctuation among specimens from the latest horizons that are separated by the shortest time intervals (Figs. 5.8, 5.9).

In studies of a million years or more, the overall patterns of evolution become easier to follow than in studies covering shorter periods of time. There are two reasons for this difference:

1.   The longer intervals between horizons result in the elimination of minor fluctuations resulting from short-term differences in the physical and biological environment.
2.   The late Cenozoic lineages cover a time of long-term, basically unidirectional climatic change.

Although the entire Earth has undergone progressive cooling from the Eocene to the present, this was strongly accentuated during the past five million years.

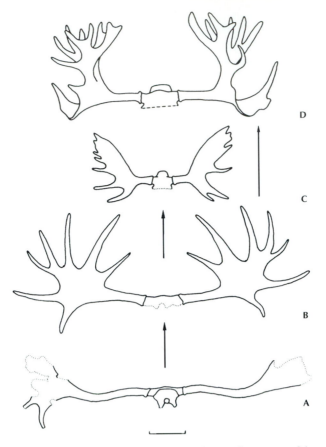

Figure 5.6. Conspicuous change in the configuration of the antlers of the moose, *Alces,* in a sequence of four species over a period of 1.6 million years. **A,** *A. gallicus;* **B,** *A. latifrons;* **C,** *A. alces,* **D,** *A. scotti.* From Lister (1993).

Figure 5.7. Change of proportions in antlers beam of Eurasian moose through the Quaternary in a series of three species over 1.6 million years. The pattern of evolution could be interpreted as gradual and directional, but the fossil record is so incomplete that could equally well be interpreted as punctuational. From Lister (1993).

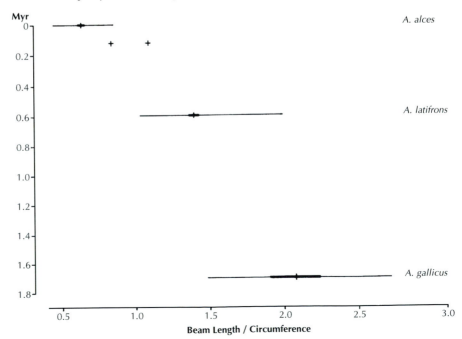

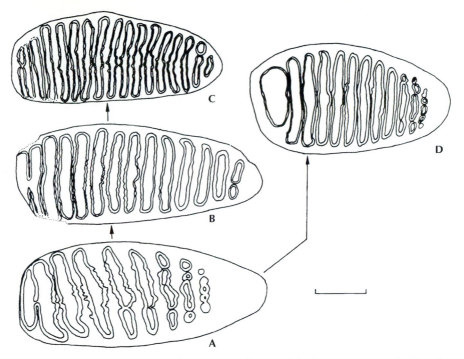

Figure 5.8 Change in lamellar frequency and enamel thickness in upper third molar of the Quaternary genus *Mammuthus* over 2.25 million years, through a succession of four species. **A,** *M. merdionalis;* **B,** *M. trogontherii;* **C,** *M. primigenius;* **D,** *M. columbi.* From Lister (1993).

Figure 5.9. Dimensions of lamellar frequency of European mammoth molars through the Quaternary. Samples 1–3, *M. merdionalis;* sample 4, *M. trogontherii;* samples 5–10, *M. primigenius.* For each sample, the mean, range, and standard error of the mean are shown. From Lister (1993).

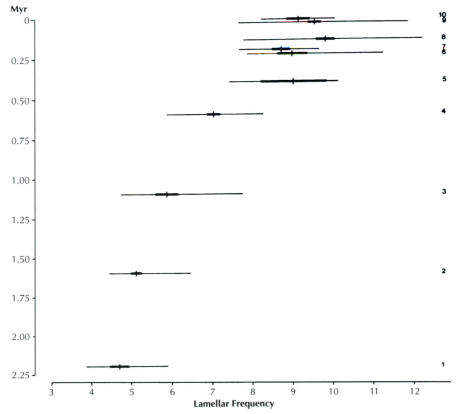

Both cooling and drying during this period led to significant changes in the vegetation, with a great increase in the area of grasslands in the south and steppe and tundra in the north. A great many mammalian lineages evolved to take better advantage of these newly available or expanded environments. This is demonstrated most clearly by marked, progressive changes in the nature of their dentition (Janis and Fortelius 1988; R. A. Martin 1993). Change in the teeth was necessary because grass is naturally more abrasive than other vegetation as a result of the accumulation of opaline silica in the tissues during growth. In addition, grass grows close to the ground and is commonly covered with dust and sand that are ingested with the food. This was accentuated in the dry and windy conditions of the late Cenozoic savannah.

The teeth of herbivorous mammals were modified in many ways to facilitate feeding on grass. The occlusal surface is generally flattened and expanded in area. The surface becomes more complex with the addition of many enamel lamellae that extend across the entire tooth (in the case of elephants) or by the formation of many sharp reentrant angles of the superficial enamel (rodents). The enamel is harder than the dentine that makes up the rest of the tooth's surface, and so stands out as a series of cutting edges. The histological structure of the enamel may change so as to resist the pressure along these cutting edges more effectively (Fig. 5.10). In addition, the height of the tooth increases so that it will last longer despite the severe wear at the occlusal surface. This results in very tall, prismatic teeth. The tall, narrow columns of enamel are strengthened by the elaboration of a new tissue, cementum, that covers the lateral surface of the tooth; in advanced rodents, it fills the deep reentrants of the enamel in the molar teeth. As the height of the tooth crown increases, the size of the tooth root is both relatively and absolutely reduced. To compensate for the loss of attachment area provided by the roots, specialized areas of the lateral surface of the crown (termed *dentine tracts*) take over as areas of attachment for the periodontal ligament (Fig. 5.11).

Growth of mammalian teeth occurs at the base of the roots as long as they are open. Once the roots are closed (primitively, at the time the teeth erupt), growth ceases. In many lineages of mammals that feed on abrasive food such as grass, progressive ontogenetic delay in the formation of the roots led to the condition in which they never closed, enabling the teeth to continue growing throughout the life of the animal. This is the ultimate solution to the problem of tooth wear.

Changes in these characters are evident in many lineages of mammals over the last 5 million years of the Cenozoic. They are graphically illustrated among artiodactyls, horses, and elephants, but the most thoroughly studied evidence of change within species is provided by rodents, whose fossils are extremely numerous. Among the rodents, most attention has been focused on an assemblage within the family Cricetidae termed the microtines, or arvicolids, which includes lemmings, voles, and muskrats (Chaline and Laurin 1986; Repenning 1987; Fejfar and Heinrich 1990b; L. D. Martin 1993; R. A. Martin 1993). Chaline (1987) recognized 140 separate lineages of microtines from Eurasia and North America, ranging from the late Miocene to the present. The microtines, like the large mammals of the steppe and tundra, specialized toward feeding on grass that had become increasingly common as a result of the progressively cooler and dryer climate of the late Cenozoic.

One of the best-documented sequences is that of the lineage leading to the liv-

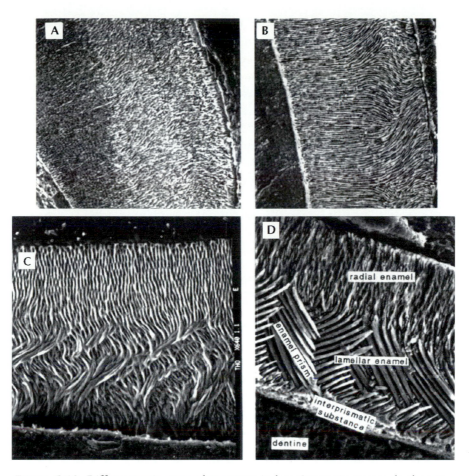

Figure 5.10. Differences in enamel structure in late Cenozoic mammals showing changes to accommodate greater stress. A–C, pocket gophers: **A,** *Pliogeomys* (5 million years B.P.) showing thick radial enamel; **B,** *Geomys* (2 million years B.P.) showing enamel thinning and the beginning of enamel structure; **C,** modern *Thomomys* showing advanced enamel structure. **D,** *Microtus paroperarius* (0.6 million years B.P.) showing radial and lamellar enamel. C and D at twice the magnification of A and B. From L. D. Martin (1993).

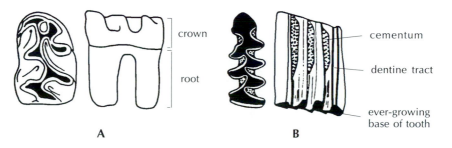

Figure 5.11. Primitive and advanced cricetid rodent molars to show differences associated with feeding on grass. **A,** The primitive cricetid *Rotundomys,* upper Miocene. **B,** The advanced microtine *Arvicola terrestris,* Late Quaternary. Both in occlusal and lateral views. Redrawn from Fejfar and Heinrich (1990b).

ing muskrat *Ondatra zibethicus* (L. D. Martin 1984; R. A. Martin 1993, 1996). It is usually described as extending through a succession of six species, belonging to two genera. The lineage is first recognized in the Upper Pliocene, approximately 3.75 million years ago. The most conspicuous change is in size, calculated from tooth dimensions as 103 g in the earliest species to 1,600 g at the height of the Wisconsinian glaciation (16,000 years ago), and then declining to 840 g in the Holocene. Such a change in size suggests major differences in physiology and way of life. Size change appears essentially continuous for the entire range of the lineage but differs in both rate and direction. The average rate of change over 3.75 million years is calculated as 0.73 d_t. For the past 600,000 years, the rate is actually slower (0.40 d_t), although the *amount* of change was much greater since the initial size was greater. Dwarfing over the past 16,000 years proceeded at a rate of 0.21 d_t. All these rates in darwins have been adjusted to million-year intervals.

Increase in size was accompanied by progressive change in numerous dental characters:

1. increase in the number of triangles on M_1 (first lower molar) from five to seven;
2. increase in height of crown and dentine tract;
3. loss of enamel atoll;
4. addition and spread of cementum in reentrant angles; and
5. decrease in number of roots from three to two, and great delay in their formation, so that they are much smaller.

These changes were initiated at different times and proceeded at different rates. The height of the crown and the height of the dentine tract appear to increase progressively through the sequence, but the rates cannot be precisely quantified because of the extensive gaps between horizons (Fig. 5.12). Although relatively few horizons have been sampled within the duration of each nominal species, the length and width of the occlusal surface of the teeth show continuous increase over more than 3 million years, with no evidence of punctuation (Fig. 5.13).

Although muskrats are usually divided into six successional species, R. A. Martin (1993) recognizes only a single species, since there is no evidence of branching within this assemblage. He attributes the absence of speciation to their aquatic way of life, which enables them to be distributed widely in interconnected water courses without greatly restricting the amount of gene flow among distant populations.

The European water rats, *Mimomys* from the Pliocene and the Pleistocene genus *Arvicola,* show a similar pattern of dental evolution over a succession of eight nominal species extending over more than 3 million years. Size remained small in this sequence, which has been sampled from thirty-five horizons (Chaline 1987) (Fig. 5.14). The European water rats spanned a succession of four climatic phases. Change appears to be continuous, within the limitation of sampling, with no periods of stasis.

The pattern of evolution within the muskrat and water rat lineages contradicts two of the major tenets of punctuated equilibrium: A great deal of change occurred in single, very long-lived lineages, and these species exhibited large population size and great geographical extent throughout their evolution.

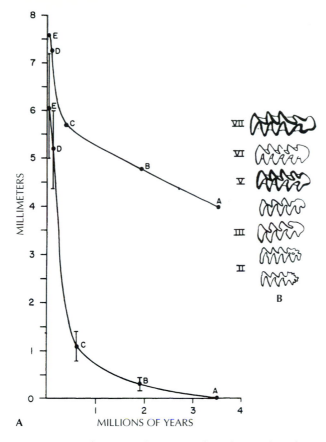

Figure 5.12. Change in characters of muskrat molars during the late Cenozoic. Graph **A** shows change in dentine tract heights on lower first molar (lower curve) and tooth lengths (upper curve). A, Hagerman local fauna; B, Borches and Grandview local faunas; C, Cudahy local fauna; D, Wisconsinan fauna; and E, modern muskrats. Part **B** shows lower first molars arranged in stratigraphic sequence, with oldest at the bottom. From L. D. Martin (1993).

Figure 5.13. Measurements of lower first molars of muskrats showing increase in size from older (Arvicoline Zone II) to younger (Zone VII) arvicoline zones in North America. Note that the size increases are overlapping, except for that area where no samples were available (Zone IV). From L. D. Martin (1993).

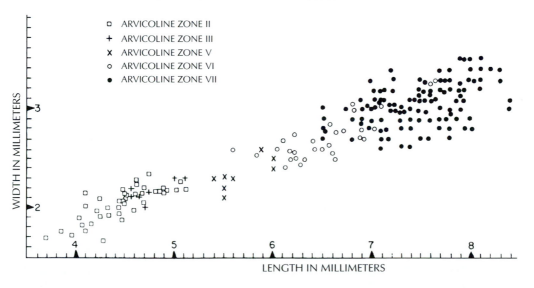

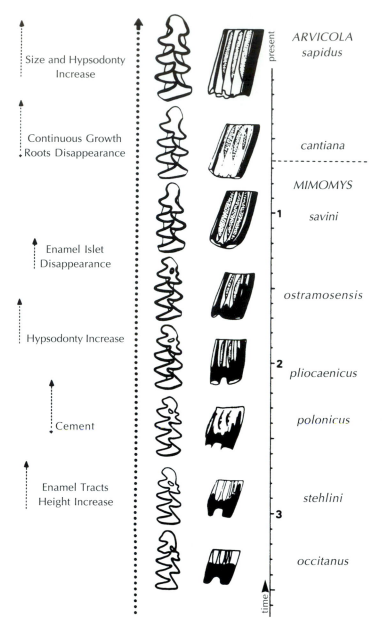

Figure 5.14. Directional change in *Mimomys (Hintonia) occitanus–Arvicola sapidus* lineage, showing the major stages in hypsodonty increase, increase of dentine tract height, appearance of cement, disappearance of enamel islet, disappearance of root and acquisition of continuous growth, size increase. This lineage has been sampled from thirty-five successive horizons. From Chaline (1987).

Anagenetic origin of species

Based on the great extent of morphological change, it has been customary to separate the lineage leading to modern muskrats into six species, belonging to two genera. The water rats are divided into eight species, also in two genera. This clearly exceeds the requirements for the evolution of one species from another in the Darwinian sense of phyletic evolution. Maglio (1973) provided additional examples of sequential species among elephants. Seven of the seventeen Quaternary elephant species can be documented as having arisen through anagenesis. Nine species are thought to have arisen by cladogenesis, but this is not unexpected according to Darwinian models of evolution.

In a broader study, Kurtén (1968) stated that 25 of the 111 modern species of mammals that first appeared during the Pleistocene in Europe were linked to ancestral species by transitional forms. Barnosky (1987) noted that this does not necessarily indicate that the remaining 86 species arose suddenly as a result of speciation events, but that the majority of species in question have still not been studied in detail. Based on the conventional definition of species, it is clear that many lineages of late Cenozoic mammals do show phyletic evolution from one species to another.

The recognition of a multitude of fossil sequences spanning the range of morphological change attributed to distinct species provides convincing evidence for the importance of phyletic evolution. On the other hand, it raises both theoretical and practical problems regarding the definition of species. In long, highly fossiliferous stratigraphic sequences, it becomes very difficult to define the points of separation between a succession of species on objective morphological criteria. Note, for example, the overlap of tooth size among nominal muskrat species (see Fig. 5.13). The problem is further complicated if there are a number of different traits evolving at different rates.

As a way around this problem, R. A. Martin (1993) suggested that only a single specific name be applied to a succession of populations within which there is no evidence of speciation events, no matter how much evolutionary change has occurred. In the case of the muskrats, he classified all in the living species *Ondatra zibethicus*. Extension of this practice would lead to a substantially lower number of fossil species than are currently recognized among well-known groups. He suggested a convention whereby previously accepted species' names for segments of such an evolving lineage would be added to the binomial so as to indicate that they are members of a continually changing morphological series. In the case of the muskrats, the name of the oldest recognized "species" would be written *Ondatra zibethicus / minor*. He refers to forms originally designated as successional species as **chronomorphs**.

At least in the late Cenozoic, the mammalian fossil record is sufficiently complete that one can expect to find at least a few fossils of nearly all the terrestrial species that existed. This makes it possible to recognize most speciation events, and in many cases to establish whether a particular lineage may be divided into separate species on the basis of speciation events.

If divisions between species were based entirely on the occurrence of speciation, it would be impossible, by definition, for new species to arise by anagenesis within a single lineage. One can, nevertheless, measure the amount of anagenetic change within a lineage and compare it with the amount of difference there is between two closely related, contemporary species. If the amount of change within a particular lineage over time is equal to or greater than the amount of difference between contemporary species, anagenetic change is significant, in contrast with the tenets of punctuated equilibrium.

Morphological change at the time of speciation

> [E]volution is concentrated in very rapid events of speciation (geologically instantaneous, even if tolerably continuous in ecological time).
> – Gould and Eldredge (1977, p. 115)

So far, we have concentrated on change within species. This accounts for most of the differences that are seen between well-studied species of late Cenozoic mammals, but we should also test Eldredge and Gould's hypothesis that major morphological changes occur at the time of speciation.

The importance of sibling species in modern vertebrate taxa, especially among small mammals, is sufficient to show that speciation is not necessarily coupled with significant morphological change (Mayr 1963). R. A. Martin (1993) noted that in the modern rodent genera *Microtus, Reithrodontomys,* and *Peromyscus,* all have a great many species that are very similar to one another.

It remains difficult to demonstrate the degree to which change is concentrated at the time of speciation in fossil forms. The fact that speciation in vertebrates is typically associated with geographic isolation makes it unlikely that a continuous sequence of fossils will be found in a particular area representing both ancestral and descendant species. On the other hand, rapid migration from one continent to another provides an opportunity to study the pattern of evolutionary change in relationship to speciation events. Repenning, Fejfar, and Heinrich (1990) described the case in a group of cricetid rodents:

> Thus, in intermittently connected faunal provinces and with as rapidly evolving mammals as the arvicolids of the Northern Hemisphere, it frequently has happened that newly arrived immigrants are inseparable from forms that remained in their territory of emigration but, because of isolation following the dispersal event, the immigrants rapidly evolved into forms recognizably distinct and requiring different names. (pp. 389–90)

Rapid, morphological change, however, clearly takes longer than the initial speciation event and can be associated with adaptation to a new environment, rather than being an immediate genetic response to separation from the parental lineage.

Many examples have been cited among late Cenozoic mammals in which changes within well-known species are greater than those that occur at the time of speciation. R. A. Martin (1993) stated generally that he knew of no examples in rodent evolution in which there was rapid and distinctive change accompanying speciation. For example, the cotton rat *Sigmodon* has persisted for over 4 million years in North America and given rise to a host of species that differ slightly from one another in size and the nature of their dentition, but the genus as a whole shows almost no change other than a slight increase in size throughout its duration. The great amount of speciation has certainly not led to any significant changes in morphology or way of life, although *Sigmodon* has as long a history as our own genus, *Homo.*

Goodwin (1993) did recognize one example of significant evolutionary change likely associated with speciation among twelve species of the prairie dog *Cynomys,* but he concluded by stating, "[I]t is clear that significant evolutionary change in prairie dogs has not been dependent on speciation" (p. 128).

Jones, Choate, and Genoways (1984) pointed to what must have been rapid

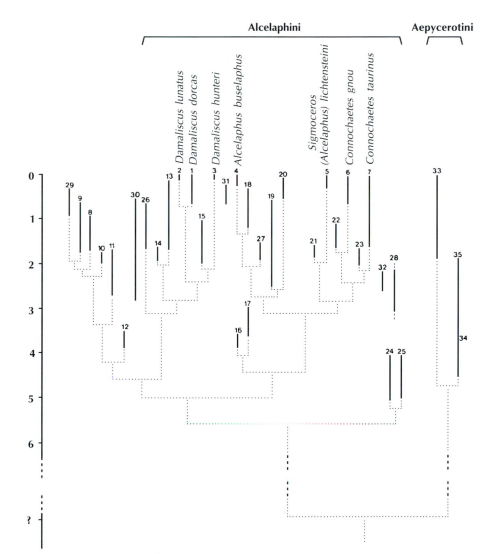

Figure 5.15. Phylogeny of the antelope tribes Alcelaphini and Aepycerotini. Vrba (1984) suggested that this group might provide an example of species selection, and that the rate of evolution between species might be significantly more rapid than the rate within species. Gaps in the fossil record appear far too long for this assemblage to provide evidence of any evolutionary processes. From Vrba (1984), "Evolutionary pattern and process in the sister-group Alcelaphini–Aepycerotini (Mammalia: Bovidae)" in *Living Fossils,* Springer-Verlag, Inc.

change at the time of speciation between the shrews *Blarina brevicauda* and *B. carolinensis* from the *B. brevicauda talpoides* lineage in the late Irvingtonian, followed by half a million years of stasis. No transitional morphologies can be detected. The actual duration of this speciation event cannot be established, but it was certainly much shorter than the subsequent evolution of the derivative species.

Vrba (1984) suggested that the time span between particular primitive and derived species within the antelope tribe Alcelaphini was sufficiently short that evolution must have been much more rapid than it was within the derived species. Unfortunately, the fossil record is too incomplete to establish quantitatively the amount of change in any particular lineage at the time of speciation (Fig. 5.15). The problem of long gaps in the fossil record will continue to raise the specter of special aspects of speciation and the origin of higher taxa as long as many stratigraphic sequences remain poorly known. On the other hand, adequately known fossil sequences support the views of Darwin and the authors of the evolutionary synthesis, who argued that morphological change is associated primarily with phyletic evolution within species, rather than being the result of speciation events.

Species selection

> Evolutionary trends are not the product of slow, directional transformation within lineages; they represent the differential success of certain species within a clade. . . .   – Gould and Eldredge (1977, p. 115)

Documentation of many examples in which significant, unidirectional change has occurred within species indicates that long-term evolutionary trends can be generated at the level of populations and species. Although a few examples have been cited in which changes in morphology are closely correlated with speciation, it has not been proven that speciation is necessary for the generation of evolutionary trends, nor that morphological change is predominantly concentrated at or near the time of speciation events. In contrast with the arguments of Stanley (1975, 1979) or Gould and Eldredge (1993), it is not necessary to postulate that evolutionary trends depend on sorting at the species level (species selection). Nevertheless, evidence from the late Cenozoic may be examined to determine the degree to which species selection might influence the course of evolution.

As elaborated by Stanley, species selection is visualized as the differential survival of lineages within a **clade** (a group whose origin can be traced to a single ancestral species) as a result of differential rates of speciation and/or extinction (Fig. 5.16). Such differentials may direct the course of evolution of particular traits in the absence of significant phyletic evolution, or even in the face of phyletic evolution in the opposite direction.

Barnosky (1987) accepted Vrba's (1984) suggestion that the phylogeny of alcelaphine bovids might exemplify trends produced by species sorting (see Fig. 5.15). There are general trends within the group for increased size and reduction of premolars, but both have exceptions among late-appearing species. On the other hand, the most dramatic evolutionary change evident in this assemblage is the elaboration of a great variety of shapes and sizes of the horns, with every species having a unique pattern. As elements of species recognition, the distinctive form of the horns must be unique to each species. This is an unlikely element to evolve by the accumulation of progressively more distinct structures as a result of the extinction of a succession of more primitive species. Rather, the presence of intermediate forms would force character displacement within species (an anagenetic process) in order to ensure reproductive isolation between species.

There is little evidence among late Cenozoic mammals for progressive change of characters within clades as a result of successive elimination of species exhibiting more primitive and intermediate character states. Among the best documented cases of major, long-term anatomical change are those illustrated by muskrats and water rats, in which there is no evidence for speciation within lineages that have

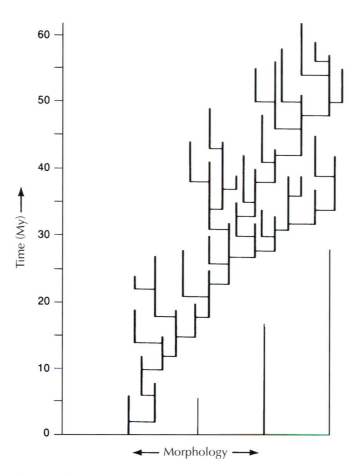

Figure 5.16. Hypothetical phylogenetic trend produced by species selection. Average rate of speciation increases toward the right, and average rate of extinction, which is inversely proportional to average species duration, decreases in the same direction. Speciation events moving to the left are equal in number to those moving to the right and contribute the same total amount of morphological change, so that direction of speciation plays no role in the formation of the trend. From Stanley (1979).

lasted more than 3 million years. In contrast, cotton rats (R. A. Martin 1993) have speciated extensively over a similar length of time. The cotton rat clade shows minor increase in size, but no other aspects of their anatomy show progressive change.

   To evaluate the significance of species selection or sorting as these ideas have been proposed by Stanley and Gould, it is not sufficient to demonstrate cases of differential survivorship among similar, closely related species; instead, it must be shown that evolutionary trends among species are produced by selection between species rather than by selection within species.

   The problem with all proposed cases of species selection is that it is very difficult to envisage situations in which selection among individuals *within* a particular species is not sufficient to explain evolutionary trends that are attributed to selection *between* species. The latter could only be demonstrated conclusively in cases where selection was operating in different directions at the two levels. Robert

Martin (1993, pp. 265–7) presented a possible example to explain the 3-million-year trend of increasing size among fifteen species of the cotton rat *Sigmodon*. He suggested that if the trend toward large size in the cotton rat clade were mediated not by intraspecific competition among individuals but by a heritable species-level property such as the tendency toward greater patchiness in distribution, then species selection could be said to be operating. If *inter*specific competition were acting to select for larger size in cotton rats, it is difficult to argue that *intra*specific competition would not have the same effect. It seems nearly impossible to differentiate these alternatives on the basis of the fossil record. All putative examples of species selection or species sorting must ultimately rest on the relative force of interspecific and intraspecific selection. Selection resulting from competition, which must be key to species selection, has been very difficult to establish even among living species because selection typically acts to reduce competition between sympatric species. Excellent examples are provided in work by Rosemary and Peter Grant (1989) on the Galápagos finches, and by Schluter (1994). Study of competition in modern populations depends on thorough knowledge of individual interactions between organisms within particular environments. This is clearly impossible in species known only from fossils. It seems unlikely that the ultimate tests to determine the occurrence of species selection can ever be conducted on the basis of fossil evidence.

Since the amount of morphological change produced by species selection must ultimately depend on the rate of speciation, a more general test of its effectivity would be to compare the rate of morphological change with the number of species in different taxa. R. A. Martin (1992) found a strong inverse relationship between body mass and speciation rate among North American terrestrial mammals (Fig. 5.17). From this correlation one would also expect an inverse relationship between size and rate of morphological change if morphological change were tied to the rate and amount of speciation.

Kurtén (1968) listed forty-three examples of species transformation among European mammals during the Pleistocene, based on morphological criteria. These are distributed taxonomically as follows: three insectivores, one bat, one cercopithecoid, one hominid, two felids, four canids, seven mustelids, one hyaenid, six ursids, four elephants, one rhino, four equids, four cervoids, and four rodents. Eight of these transitions involve small mammals, twenty involve large forms, and the remainder are of intermediate size. This is directly opposite of what would be expected if the high rate of speciation in small mammals drove morphological change.

It may be argued that Pleistocene mammals are an extreme case, since large body size, associated with high capacity for retaining body heat, and large geographical range would inherently favor large animals in adapting to life in the steppes and tundra at the margins of continental glaciers. Nevertheless, these examples demonstrate that relatively slowly speciating forms have as much or even greater capacity for significant morphological change as small, rapidly speciating lineages.

Looking beyond late Cenozoic mammals, we should consider the potential importance of species selection in producing evolutionary trends from a more general perspective. It may be possible to attribute changes in single characters or closely linked character complexes to species selection, but it clearly would be

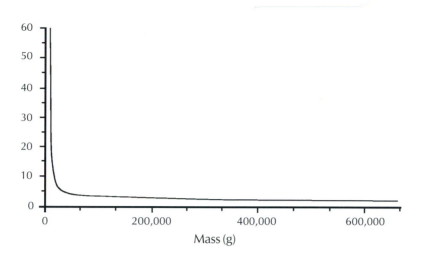

Figure 5.17. Inverse relationship between body mass and number of species per genus among North American terrestrial mammals. Simplified from Martin, R.A., "Generic species richness and body mass in North American mammals: Support for the inverse relationship of body size and speciation rate," *Historical Biology*, vol. 6, pp. 73–90. Copyright © 1992.

much less effective in explaining changes in a host of unrelated characters as, for example, those that distinguish hominid, rodent, or elephant species. The difference lies in the total amount of variability that is involved, the number of units that are subject to selection, and the frequency of selection.

In selection among individuals within a species, change occurs in every generation, with continuing modification of allele frequencies among variable traits, supplemented by the gradual appearance of new mutations. The total amount of variability is maintained by recombination arising from sexual reproduction, with every individual expressing a unique combination of characters on which selection can act. Although a large percentage of the population may fail to reproduce each generation, the remaining individuals retain a reservoir of alternative alleles for many loci that can serve as a basis for continuing selection. In species selection, it is assumed that little variability is expressed within each species, but rather resides among a number of related species within a clade. With selective elimination of species, a large proportion of the total variability within a clade is lost with every extinction.

The greatest strength of Darwin's theory of natural selection among individuals within a species is that the entire clade is involved in continuous change: The advantageous genes can spread throughout the entire species and be built upon with every generation. The weakness of species selection is that the total amount of genetic variability of the evolving clade is divided into a number of reproductively isolated units, so that many potentially advantageous attributes are lost with every event of species selection and so are unavailable for further change in the remaining members of the clade. In both of Stanley's publications on this subject (1975, p. 648; 1979, pp. 187–8) he noted that species selection is actually analogous to natural selection within asexual, rather than sexual, taxa. Within sexually reproducing species, the units of selection (individuals) interbreed, whereas, by the very definition of species, the units of species selection do not exchange genetic components from generation to generation.

A further but also very significant difference between individual and species selection is that the former occurs in every successive generation, whereas the latter can occur no more frequently than the time required for the production of new species, which may happen only a few times within 100,000 years. It is difficult to estimate the total number of speciation events that may occur in a clade, or how rapidly speciation may be initiated, but it certainly takes thousands or tens of thousands of years after the initial event of allopatric speciation for the new species to diverge sufficiently that they are reproductively isolated when (and if) they return to a condition of sympatry where they may be subject to species selection. Taking all these factors into consideration, the potential for change within species is many orders of magnitude greater than that which might result from differential selection *among* species.

Summary of evolutionary patterns among late Cenozoic mammals

Late Cenozoic mammals provide a wealth of information for testing the major tenets of Eldredge and Gould's theory of punctuated equilibrium. Although stasis of particular characters and character complexes has been demonstrated for significant periods of time in many lineages, only a very small number of species have been identified in which stasis is exhibited in all characters studied for most of the duration of the species.

Many lineages show phyletic evolution, with sufficient change over millions of years for a series of successional species to be recognized on morphological grounds, although no speciation events are documented.

Only a few examples of speciation events are correlated in time with significant morphological change. Lineages that dispersed from one continent to another during periods of lowered sea level are recognized as separate species, but most morphological change occurred subsequent to the actual speciation event, in relationship to the new environment. No major trends involving a complex of character changes can be demonstrated as having resulted from species selection.

Rates of evolution among late Cenozoic mammals

The pattern of evolution that can be seen among late Cenozoic mammals corresponds closely to that deduced by Darwin from modern populations. The accuracy and relatively close intervals of dating during this period should also provide a strong basis for establishing rates of evolution. What is immediately apparent is that evolution over this time scale appears to have occurred at a much slower rate than is documented in lab or field studies of modern populations, or has been recorded throughout the length of human history. We don't see changes occurring from season to season as in studies of the snail *Cepaea* (Cameron 1992), over years and decades as in the Galápagos finches (Grant 1986), or even over hundreds of years as in the peppered moth (Kettlewell 1973). This is largely a sampling problem, involving traits, time, space, and environments.

Little more than the skeleton is preserved in the fossil record; color (as in the peppered moth), soft anatomy, physiology, biochemistry, and behavior generally leave no record. These features change rapidly in the modern population, and presumably did so in extinct species, but simply cannot be observed in fossils.

Studies of time, space, and environments suffer from a related problem, that of "averaging." Radiocarbon and other methods of dating can potentially be accurate at the level of hundreds or thousands of years, but the processes of sedimentation and reworking of sediments, as well as the necessity of collecting fairly large samples, result in their representing fairly long spans of time. One might imagine collecting individual specimens from a series of exactly dated successional beds, but populations are sufficiently variable that you must have large numbers of specimens to record the range of variability. Paleontologists collect all they can from a few centimeters or a few feet of sediments and call this a single sample, although it clearly represents a substantial range of time. This is *temporal averaging*.

Most collections of late Cenozoic mammals have occurred within continental areas, where dispersal of populations occurs readily. It is difficult, if not impossible, to differentiate between changes over time in a particular resident population and differences between a succession of transient populations. This is in strong contrast with the degree of geographical control possible in studies of island populations, such as that of the Galápagos finches. Hence, most fossil collections average anatomical traits over space (*geographical averaging*) as well as time.

In addition to temporal and geographical averaging of fossil samples, the populations themselves must have been responding to continuously changing selective forces. In order to survive for millions of years, populations and species must be adapted to a wide range of differences in both the biological and physical aspects of the environment, such as weather, diet, and a range of predators. No species can afford to adapt to only a single, ideal environment. This we may term *environmental averaging*. Take, for example, the case of the snail *Cepaea* (Cain and Sheppard 1954), in which the frequencies of different color morphs change dramatically on an environmental basis. Despite the change in the selected phenotypes (dark color on dark soil, banded in the grass, and light in the sand), the range of variability of the underlying genotype remains stable for thousands or even millions of years, as revealed by fossils showing color banding. In the modern fauna, such rapid changes can be attributed to oscillating selection of qualitatively different alternative alleles or the accumulation of different additive genes for quantitatively different traits. The fossil record is largely blind to changes at this level.

Let us now turn from what we cannot see to what we can see. Over the 10,000 years since the retreat of the last continental glaciation, very little skeletal change can be recognized among the well-studied small to medium-sized mammals of the Holarctic. What change occurred was generally limited in scope to that observed in geographic variation of living members of these species (Graham, Semken, and Graham 1987). The fossil record documents repeated changes in the range of species in relationship to fluctuations in temperature and rainfall, but no consistent directional changes in morphology. Even in rapidly evolving groups such as rodents, elephants, and humans, 10,000 years is generally too short a time to record continuous directional changes, except for body size.

Within the longer span of the late Pleistocene, Kurtén (1968) reported cycles of size change in several species correlated with advance and retreat of continental ice sheets. Over a period of 35,000 years the woolly mammoth, *Mammuthus primigenius,* showed increase and decrease of limb length of approximately 25 percent, which suggests doubling and halving of body mass (Fig. 5.18). On the other hand, even the 50,000-year history of the vole *Microtus pennsylvanicus* studied by Barnosky (1993) is too short to show anything more than shifting, mosaic evolution in tooth dimensions. Davis (1987) reported a period of approximately 250,000 years for the gradual replacement of a molar pattern with six rather than five triangles in this species. In the muskrat (see Fig. 5.13) there is extensive overlap in tooth dimensions between species, lasting about 300,000 years, which closely matches the model of gradual change of species mean proposed by Darwin and others for phyletic evolution. Hominid species show a similar pattern of overlapping size range in cranial capacity over the past 3 million years (Fig. 5.19). Rightmire (1990) has emphasized that evolutionary change in cranial capacity within one species, *Homo erectus,* is difficult to differentiate statistically from stasis. However, when seen within the context of other hominid species over 2–3 million years, cranial capacity was clearly increasing, although the wide range of variability makes the rate of change difficult to quantify within each species.

We can now turn from the rate of change in individual characters to the rate of species replacement. The length of species duration during the Pleistocene was in the range of approximately 1.6–3 million years (Kurtén 1968). The successional species of muskrats and water rats persisted for an average of about half a million years, and elephants 500,000–1,250,000 years. It is interesting to compare these rates to those hypothesized by Darwin. In describing his illustration of the pattern of evolution, he suggested that phyletic transformation of species might take 14,000–140,000 generations. Accepting a generation time of about fifteen years for elephants, it would take 33,000–83,000 generations to evolve from one species to another. For small rodents, with a generation time of only seven weeks, it would require some 3.5 million generations. Therefore, mice take much longer to evolve than their generation time might predict, and elephants evolve much more rapidly; yet the time span suggested for the evolution of both groups is in the general range predicted by Darwin.

Above the level of species, an entire new fauna appeared in the late Cenozoic that exploited the newly developing arctic and tundra environment. This environment expanded rapidly from 2.5 million years ago, when severe cooling followed by extensive glaciation began in the Northern Hemisphere. At that time, there were no species specifically adapted for life in this environment. By 730,000 years ago, seven tundra and arctic species had evolved, including the reindeer, musk ox, tundra vole, Norway lemming, Arctic lemming, steppe mammoth, and the snow vole. They were joined by eight more species that appeared in succession throughout the later Pleistocene (Kurtén 1968).

Other aspects of evolution illustrated by late Cenozoic mammals involved large-scale dispersals and mass extinction. The many episodes of continental glaciation covering large areas of the Holarctic were accompanied by changes in sea level that resulted in repeated formation and severing of land links between Siberia and Alaska. These changing connections permitted intermittent dispersal of cold-adapted mammals throughout the northern continents. There was also extensive

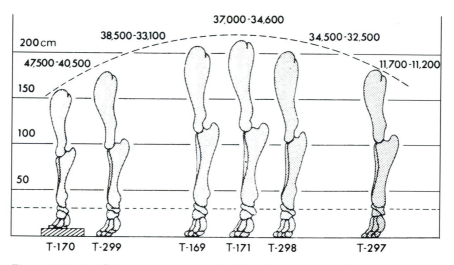

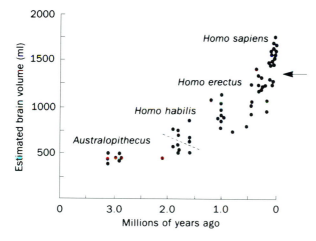

Figure 5.18. Oscillatory change in length of forelimb in the woolly mammoth, *Mammuthus primigenius*, 47,500–11,200 years ago. From Bjorn Kurtén (1968), *Pleistocene Mammals of Europe*, Weidenfeld & Nicolson (publ.).

Figure 5.19. Progressive increase in estimated brain volume in hominids over the past 3 million years. Arrow indicates modern mean value. Because of the broad range of values, it is difficult to demonstrate directional change within a particular species, such as *Homo erectus*, over time, but data from all species show a clearly directional trend. From Jones, Martin, and Pilbeam (1992).

dispersal between North and South America 2–3 million years ago as a result of the tectonic emergence of the isthmus of Panama (Webb 1978; Webb and Barnosky 1989).

In addition to dispersal, one of the most conspicuous aspects of faunal change during the late Cenozoic was extinction. Between 9,000 and 11,000 years ago, approximately 43 genera of mammals, most of large size, became extinct. This was particularly rapid and extensive in North America, with the extinction of camels, horses, elephants, saber-toothed tigers, cheetah, ground sloths, glyptodonts, and giant beavers. The cause of this mass extinction has long been debated because there were two major changes in the environment: one physical (the end of the ice age with the sudden drastic reduction in the extent of the arctic, tundra, and steppe environments) and one biological (the appearance of modern hunting techniques by an ever-expanding human population). What factors were most important in this extinction still cannot be established (Martin and Klein 1984). Among large mammals that did survive, many show significant size reduction since the end of the Pleistocene (Kurtén 1968; Sondaar 1977; Vartanyan et al. 1993).

6 Patterns of evolution of nonmammalian vertebrates in the late Cenozoic

> In contrast with physics or chemistry, in which particular atoms or molecules may exemplify patterns and processes common to all matter, biological examples are but samples of an extremely varied and never fully duplicated diversity of life.
>
> – Anonymous observation

The fossil record of late Cenozoic mammals provides the most comprehensive basis for establishing the patterns and rates of vertebrate evolution at the level of the species and genus. Study of many groups from the north Temperate Zone, including rodents, elephants, and artiodactyls, demonstrates progressive changes in dentition and ways of life in response to the replacement of forests by widespread grasslands. During the Quaternary, an entirely new fauna adapted to the tundra and arctic conditions of the far north.

Amphibians and reptiles

Study of north temperate Quaternary amphibians and reptiles (Holman 1991, 1993, 1995) shows a much different pattern. There is only limited evidence of either the origin or extinction of amphibian and reptile species during the Pleistocene compared with the appearance of more than a hundred new mammalian species during this period in Europe alone. Little evidence of skeletal change at the species level is apparent during the entire Quaternary. The current herpetofauna appears to have achieved most of its modern aspects by the late Pliocene, 3.5–2 mya.

Two factors contribute to what appear as very different evolutionary patterns exhibited by Quaternary amphibians and reptiles compared with mammals:

1. the relative capacity of the fossil record to document evolutionary change, and
2. differences in the biology of the two groups.

Study of evolution at the species level in fossil amphibians and reptiles is inherently much more difficult than in mammals. Whereas nearly 100 percent of modern mammalian species are recognized in the Pleistocene, less than 5 percent of the 4,014 living amphibian species have been identified from the fossil record, and no more than about 11 percent of living genera of amphibians and 13 percent of reptilian genera are recognized from fossil remains (Carroll 1977, 1984). Although the total number of modern genera and species is much lower within North America, a much higher percentage is known from fossils. Nearly all of the genera and approximately 60 percent of the 300–400 living species of amphibians and reptiles are recognized from Pleistocene deposits (Holman 1995). However, most of these identifications are based on a few isolated bones, and subtle differences in size and body proportions that are important in distinguishing living species cannot be documented from current knowledge of the fossil record. The complex and extremely durable teeth of mammals allow all species to be readily identified, and small but consistent differences can be easily recognized in evolving sequences. No elements of the skeleton are equally diagnostic in amphibians or reptiles.

The extremely conservative appearance of the evolutionary pattern may also be accounted for by physiological capacities and limitations of amphibians and reptiles that are different, at least in degree, from those of mammals and birds. As obligatory ectotherms, modern amphibians and reptiles lack the capacity to maintain a body temperature consistently above that of the environment, although some sea turtles and brooding pythons are exceptional in this regard. It would be impossible for any of the modern groups to be active in areas where the ambient temperature is continually near or below freezing. A few living genera manage to survive in areas of seasonal cold by hibernation or dormancy during the winter, but none could be said to have taken advantage of particular habitats or ways of life that became available because of the spread of continental ice sheets during the Pleistocene. On the other hand, a low metabolic rate confers an advantage on modern terrestrial amphibians and reptiles in their capacity to survive long periods with little food and/or oxygen, which enables them to survive the cold winters of the temperate zone (Holman and Andrews 1994).

Amphibians and reptiles appear to have accommodated to the conditions of the Quaternary rather than taking advantage of altered environments. Evidence from both North America (Holman 1991, 1995) and northern Europe (Holman 1993) shows that a major aspect of amphibian and reptilian accommodation was through rapid dispersal out of and back into areas subject to continental glaciation in response to the successive advances and retreats of the glaciers. Beyond the direct influence of the ice sheets, amphibian and reptile populations showed remarkable stability in the late Pleistocene and Holocene, despite significant local changes in temperature and rainfall that resulted in major geographical shifts in bird and mammal populations (Fay 1988; Van Devender and Mead 1978; Mead, Thompson, and Van Devender 1982).

In contrast with the large-scale view of amphibian and reptile evolution in the Quaternary provided by Holman, Pregill (1986) discussed shorter-term changes known to have occurred within the Holocene. He summarized data on twenty-eight occurrences of lizards known from fossil populations on the islands in many parts of the world, including the Canary Islands, Mascarenes, Galápagos, West Indies, and others. These illustrate what had been considered island gigantism, but Pregill concluded that they probably represent populations of normally large individuals that had inhabited these islands early in the Holocene and subsequently undergone dwarfism as a consequence of human disruption of the natural habitat and food resources. Although few if any of these cases were in response to natural change, the great decrease in size, up to a 50 percent reduction in body length, indicates the substantial capacity of these lizards for evolutionary change within a relatively short time frame (less than 10,000 years). Auffenberg (1958) reported significant size change in the box turtle, leading to very large size in the Pleistocene, but with subsequent reduction in most populations.

Blair (1965) described numerous cases of speciation or incipient speciation among North American amphibians during the Quaternary as a result of isolation following climatic change associated with glacial advances and retreats. Few are accompanied by significant morphological change. Modern techniques of genetic and protein analysis provide many additional examples of this phenomenon (e.g., Green et al. 1996).

One of the most important evolutionary events among amphibians in the late Cenozoic is the radiation of members of the tribe Bolitoglossini, within the salamander family Plethodontidae, in South America (Wake and Lynch 1976; Wake and Elias 1983). At least twenty-five species of the genus *Bolitoglossa* are endemic to South America, and *B. altamazonica* is widespread in the Amazon basin as far as 17° S. Although the more primitive members of this genus are restricted to montane environments in Central America and northern South America, the more southern species have adapted to jungle habitats, below 500 m. The species *B. digitigrada* is highly derived structurally in the reduction in the size and number of phalanges in both the front and hind limbs (Wake, Brame, and Thomas 1982). Unfortunately, this group has no fossil record, and the time over which this radiation has occurred can be dated only indirectly based on the degree of morphological and genetic differentiation and on the time at which they were able to reach South America. On the basis of the degree of differentiation, Vanzolini and Heyer (1985) suggest that radiation probably began prior to the formation of a permanent isthmian link, which occurred approximately 3 mya. On the other hand, the intolerance of salamanders to salt water would have made it difficult for them to have entered South America prior to about 9 mya, when tectonic activity resulted in a series of variably connected islands between Central and South America (Stehli and Webb 1985b).

Most osteological evidence suggests that the amount and rate of evolution among amphibians and reptiles during the Quaternary is much less than that of coexistent mammals; but do reptiles and amphibians support Gould's claim that significantly unidirectional change within species is an extremely rare phenomena? Although it is generally impossible to study long-term evolutionary change from the fossil record of living species of amphibians and reptiles, it can be judged indirectly through study of differences between geographically distinct contemporary populations. If extant populations and subspecies within a species differ consistently, and if those differences can be attributed to adaptation to different environments within the range of the species, they provide strong evidence of significant, directional evolutionary change within the species.

Many studies of variation among modern amphibian and reptilian species have been conducted, of which research by Thorpe and his colleagues is particularly informative as it involves island species for which the minimum time available for evolutionary change can be established by geological means. These researchers have documented cases of geographical variation among seven modern squamate species, belonging to four families in two orders, on the Canary Islands, the West Indies, and Taiwan (Thorpe 1987, 1991; Thorpe and Brown 1989, 1991; Brown and Thorpe 1991; Brown, Thorpe, and Baez 1991; Malhotra and Thorpe 1991; Thorpe and Baez 1993; Castellano, Malhotra, and Thorpe 1994). These studies all show significant variation in morphological features that can be correlated in a consistent manner with environmental differences from locality to locality within a single species. The following characters were studied: shade and patterns of coloration on different areas of the body; number and configuration of scales on the collar, trunk, and elsewhere; number of femoral pores; body dimensions; and body proportions. In nearly all cases, the characters show clinal variation in relationship to topographic features on the islands and resulting difference in climate and vegetation. There is no way to establish the rate of adaptation to the different environments, but it must have occurred within different populations of a single species, subsequent to the original colonization of each island. Most of these studies were conducted on islands where the earliest possible time of colonization can be dated as 2 million years ago or less. Most changes may have occurred rapidly, soon after the islands were populated, or evolution may have proceeded gradually throughout the time available. Alternatively, change may have been sporadic, or have tracked environmental changes in an erratic manner throughout the history of the island populations. Similar studies have been carried out on *Anolis* lizards in the Caribbean by Losos (1994) and his colleagues.

Although these studies provide excellent examples of evolutionary analysis of living populations, their results cannot be directly compared with studies of evolution over thousands or millions of years because none of the characters can be documented in fossils. Patterns of coloration are fossilized in some invertebrate groups – for example, on the wings of Carboniferous insects and the shells of Pleistocene snails and occasionally in fish – but not in lizards or snakes. Scales or their impressions are preserved in fossil squamates occasionally but not commonly enough for interpopulational comparison as in these examples. Body size and proportion may be known from individual fossils, but articulated specimens of amphibians and reptiles are not sufficiently common that statistical comparisons could be made between members of different populations of a particular species.

These characters may have varied significantly among ancient populations of amphibians and reptiles, but such variation would almost never be detectable from the fossil record. The rarity of fossil evidence for any particular pattern of evolution among individual species of Pleistocene amphibians and reptiles does not, in itself, indicate that phyletic evolution did not occur in these groups. Rather, since we know that changes in the soft anatomy and coloration may be conspicuous and relatively rapid among living species of amphibians and reptiles, we may logically conclude that similar changes also occurred earlier in the Quaternary, although this would be nearly impossible either to prove or disprove based on the fossil record.

It is almost certain that most amphibian and reptile species did exhibit relatively little change in their skeletal anatomy during the late Cenozoic. This, however, can be seen as but one end of a spectrum of rates of changes, with certain mammalian groups on the other. It in no way supports Gould and Eldredge's assumption that most species are incapable of phyletic evolution.

Birds

Studies by the Grants on the Galápagos finches (see Chapters 3 and 9) show how rapid change can be over short periods of time in bird populations. Unfortunately, limitations of the fossil record make it extremely difficult to establish evolutionary patterns over longer periods of time in the history of individual species (Alan Feduccia pers. commun.). Modern genera of all orders other than the passerines are known from the Miocene, indicating that birds as a group achieved relative modernity much earlier than mammals, with near stasis in skeletal features at the generic level over the past 15 million years (Fedducia 1996).

Stickleback fish

Observations of most living fish groups demonstrate only limited evolutionary change during the late Cenozoic, as is the case among amphibians, reptiles, and birds; but there have been few efforts to determine rates or patterns of change within individual species or genera.

One situation in which evolutionary time scales for both living and extinct fish species can be accurately established is in lakes. Most lakes are ephemeral at the geological time scale. Many are so shallow that they are subject to recurrent filling and desiccation at the scales of the Milankovich cycles (20,000, 41,000, or 100,000 years). Even in large lakes within tectonic basins, periodic changes in water level may be as great as several hundred feet, thus greatly changing the area of the lake, which may significantly affect the contained fauna. Periods of desiccation can be identified and dated from the nature of the sediments in the lake basin. In some lakes, the nature of sedimentation varies during the year, their beds can thus be dated at annual intervals.

Some lakes have the further advantage of being isolated from other bodies of water, so that biologists can be reasonably certain that most species living there have evolved within the lake rather than entering it from elsewhere. In such cases, if species are truly endemic to the lake, their evolutionary history can be no longer than the history of the current body of water in the lake basin.

Although the patterns of fish evolution during the late Cenozoic have not been analyzed in such a systematic manner as have those of amphibians, reptiles, and mammals, two groups provide insights into particular evolutionary phenomena: the stickleback fish of the north Temperate Zone and the cichlid fishes of the East African Great Lakes.

The threespine stickleback *Gasterosteus* (Fig. 6.1) provides an exceptional example in which patterns and rates of evolution can be studied in both modern and fossil populations (Bell 1988, 1994; Bell et al. 1985; Bell and Foster 1994a). Sticklebacks are members of the teleost order Gasterosteiformes, primitive percomorphs known as early as the Paleocene. *G. aculeatus* is widely distributed in coastal marine, brackish, and freshwater habitats around the northern margins of the Atlantic and Pacific Oceans. In addition to the three dorsal spines that give the group its name, most sticklebacks also have large spines extending ventrolaterally from the pelvic girdle, and possess a series of lateral plates that provide support and rigidity to the spines. Studies of modern fish show that the spines serve to discourage predation by larger fish.

Sticklebacks apparently originated in the marine environment but have repeatedly invaded freshwater habitats. More than a thousand populations of the threespine stickleback occupy small, isolated bodies of water adjacent to the sea coasts around the northern margins of the Atlantic and Pacific oceans. In more than twenty of these localities, the sticklebacks differ conspicuously from their marine ancestors in the extensive reduction or complete loss of the lateral plates, the pelvic girdle, and the accompanying spines. The loss of these protective structures is closely correlated with the absence of predatory fish in these particular bodies of water. The absence of predacious fish, in turn, is associated with larger populations of predacious insects, which may make use of the pelvic appendages in the capture of the sticklebacks. Thus, structures that have a high selective advantage in waters with fish predators appear to be strongly selected against in waters that lack larger fish. These bodies of water are isolated from one another, indicating that loss of the pelvic spines and girdles must have occurred separately in each of the populations. Although there is no fossil record of any of the living populations with reduced spines, the time available for their reduction can be established from geo-

Figure 6.1. Variation of body form and external features among North American populations of the stickleback fish, *Gasterosteus aculeatus*. Forms around the periphery have probably been derived independently from a marine ancestor, represented by the specimen shown in the middle. The conspicuous dorsal and pelvic spines and the lateral plates that are present in the marine form are variably reduced and lost in each of the freshwater populations. These changes have occurred within the past 10,000 years. The scale bars are all 1 cm. From Bell and Foster (1994b), *The Evolutionary Biology of the Threespine Stickleback*. Reproduced by permission of Oxford University Press.

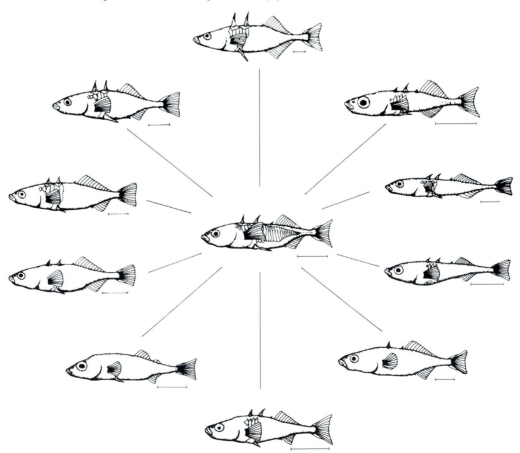

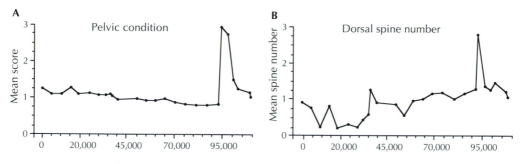

Figure 6.2. Change through time of the pelvic girdle complex and the dorsal spines in the stickleback fish *Gasterosteus doryssus* from the Miocene. **A,** Score of mean phenotype of pelvic girdle complex. **B,** Mean phenotype, summing the changes in the number of dorsal fin spines. Change in these characters occurred slowly but continuously through the first 92,000 years of the lake's history. The sudden peak at approximately 92,000 years represents the invasion of a different population, or a different species of this genus, which possessed the primitive pattern of the pelvic complex and spines. These structures were rapidly reduced during the final 15,000 years of the lake's history. From Bell (1988).

logical factors. In northern Europe and North America, the earliest time at which they could have entered freshwater streams and lakes is approximately 11,000 years ago, with the melting of the Pleistocene ice sheets. Although the exact pattern of evolution cannot be established, it is certain that these major changes in stickleback anatomy occurred very rapidly and independently in the different populations.

The speed of evolutionary change is remarkably confirmed by fossil evidence of a closely related stickleback, *Gasterosteus doryssus* from the late middle Miocene of Nevada (Bell 1988, 1994). More than 8,000 specimens have been collected from approximately 55 m of diatomaceous sediments from a shallow but extensive saltwater lake. The sediments were deposited in a pattern of alternating light and dark layers, comparable to the annual varves seen in modern lakes. Approximately 110,000 years of extremely regular deposition is recorded. The fauna of this lake is almost devoid of larger vertebrate predators, and for most of its history the sticklebacks have very reduced body protection. There is some evidence of predation by birds but no fossil evidence of the insect fauna.

Bell and his coworkers sampled the fauna at twenty-six intervals, separated by an average of approximately 5,000 years; individual samples covered about 525 years. Figure 6.2 illustrates changes in aspects of the pelvic girdle and dorsal spines. Except for scattered remains of large-spined fish in the earliest horizons, most of the fish from the first 92,000 years of the lake's history showed extensive reduction of the pelvic girdle and spines. Further reduction of the pelvic girdle continued at a very slow pace over the first twenty sampling intervals, while the number of dorsal spines increased in a very irregular manner.

Between samples from intervals 20 and 21, a distance of 1,086 varves, there is a dramatic change: the sudden appearance of fish with fully formed pelvic girdles and three dorsal spines. This is followed by a rapid drop in the frequency of these traits to a level approaching that of the previous samples. Additional sampling was done to gain more details of the changes between intervals 20 and 21 (Bell 1994).

Only low-spined fish (about fifty specimens) were found in the first 646 varves. The next interval, an approximately 122-year period, included seven spiny and twelve low-spined specimens. Forty-eight spiny specimens and no low-spined fish were present in the final 318-year sample. During this entire sequence, there are no specimens that show a morphological transition between the two distinct patterns. This is not an evolutionary event but rather a replacement of one population, or one species, by another. There is no evidence to explain how or why the spiny population suddenly entered the lake and replaced the low-spined forms.

Once the new population became dominant, evolutionary change occurred rapidly, with the average number of dorsal spines being reduced from 3 to 1.5 in 3,200 years; the mean pelvic girdle score was reduced by 90 percent in 7,800 years. Reduction proceeded at a slower pace to the end of the fossil record in these deposits. The period of rapid reduction of the spines and pelvic girdle in this Miocene lake is about one-third the time available for their reduction in post-Pleistocene lakes.

On the other hand, changes in the spines and pelvic girdle during the first 92,000 years occurred at a pace so slow that Bookstein (1988) calculated that it could be attributed to random walk rather than to selection. Bell (1994) argued that evidence from other stages in the evolution of sticklebacks in this lake demonstrated that changes were driven by selection, although the force of selection differed greatly at different stages in spine reduction.

Bell and Foster (1994b) pointed out that the threespine stickleback shows an unexpected pattern of evolution. If one considers the modern freshwater populations and those from the isolated Miocene lake in which pelvic girdle and dorsal spine reduction have taken place, very rapid morphological change has occurred as a result of strong selection over very short periods of time (Fig. 6.3). On the other hand, the parental marine species shows almost no morphological change over a period of 10–16 million years. Fossils of *Gasterosteus aculeatus* from marine deposits of middle Miocene age are essentially identical to living members of this species. Following Williams (1992), Bell and Foster (1994b) referred to this as **racemic evolution**, by analogy with a pattern of flower structure exemplified by the delphinium, in which there is a single, central stem, along which individual flowers grow at intervals on small stems. The term **iterative** might also be applied to this pattern of evolution.

Only a single other marine species of *Gasterosteus, G. wheatlandi,* is known from the modern fauna, but more than a thousand isolated populations of brackish and freshwater descendants are known at present, many of which are reproductively isolated and might be considered separate species. Probably none will survive more than a few tens or hundreds of thousands of years before being eliminated by changes in climate or topography. The same pattern may have been going on for at least 10 million years, without the survival of any of the protospecies that had diverged prior to the Quaternary. Racemic or iterative evolution within the stickleback fishes has resulted in the appearance of a large number of very similar derivative lineages, each restricted to a narrow adaptive zone. The cichlid fishes of the East African Great Lakes provide an example of a very different pattern of evolution.

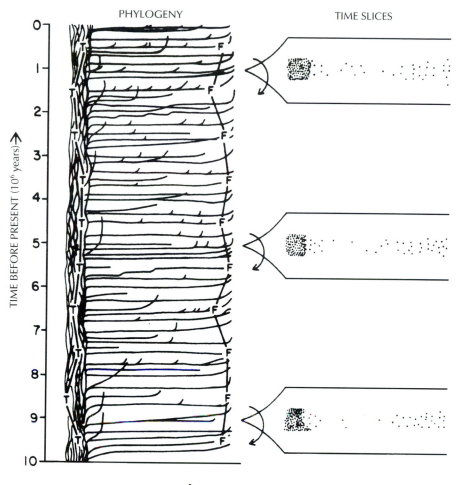

Figure 6.3. Phylogeny of pelvic reduction in the *Gasterosteus aculeatus* species complex since the Miocene. Interconnected marine lineages trend upward through time relatively unchanged, retaining the primitive fully formed pelvic phenotype. Most freshwater lineages (not shown) also retain the primitive pelvic phenotype. Lines diverging to the right represent freshwater lineages that evolved pelvic reduction and subsequently became extinct. Analysis of a series of fossils from marine deposits would reveal stasis. A series of fossils from lakes with reduced pelvic structures would give a false impression of stasis (F). Although rapid evolution occurs in each of the divergent populations, the global pattern of evolution would be one of stasis, within which there is extensive variability at the population level. From Bell (1988).

The cichlid fishes of the East African Great Lakes

While long-term climatic changes in the Northern Hemisphere set the stage for the origin of the tundra and arctic faunas, major tectonic events in the late Cenozoic of East Africa provided a comparable opportunity for the radiation of a particular group of teleost fish, the cichlids. Cichlids are relatively small, primarily freshwater fish that are widespread throughout the streams, rivers, and small lakes of South and Central America, Africa, and southern Asia (Keenleyside 1991a).

The fossil record of cichlids goes back to the Eocene of South America and Oligocene of Africa. The skeletal morphology of these early fossils is essentially similar to conservative living counterparts. Cichlids belong to a larger assemblage of teleosts, the Suborder Labroidei, which may have diverged from other percomorph fish by the Early Cretaceous (Stiassny 1991).

Until about 5 million years ago, cichlids remained a relatively conservative group in terms of habitat and phylogenetic diversity, although they had achieved a level of structural and behavioral specialization that set them apart from other teleosts. Parental care is highly developed throughout the group, and in many species, either the males or, much more commonly, the females brood the eggs and/or the larvae in the mouth. The feeding apparatus is distinctive in the high degree of elaboration of pharyngeal jaws, behind the normal jaw elements, which allows for great flexibility in the choice and manipulation of prey (Liem 1991).

About 300 cichlid species are distributed in the South and Central American river systems, with one genus having approximately 50 species; some 127 species are known from the river systems in Africa, outside the East African Great Lakes. In the number of species and their wide geographic distribution, cichlids living in the river systems of the Southern Hemisphere are relatively successful; yet in no areas are they the dominant fish group, nor is there other evidence to suggest their potential for a major adaptive radiation. In contrast, the cichlids that entered the newly formed East African lakes underwent an explosive radiation. They are now the most diverse of all vertebrate families in terms of species number and range of trophic specializations. In these attributes, they might be compared with much of the spectrum of radiation among placental mammals.

The formation of the East African Great Lakes can be attributed to rifting and crustal spreading that began approximately 12 million years ago. The greatest radiation of cichlids occurred in the three largest lakes, Malawi, Tanganyika, and Victoria (Fryer and Iles 1972) (Fig. 6.4). Lake Victoria is the second-largest freshwater lake in the world in terms of surface area, covering approximately 69,000 km²; Tanganyika has the sixth-largest surface area and Malawi the eighth. Lake Tanganyika, at 1,470 m, is second only to Lake Baikal in its depth, and Lake Malawi reaches depths of more than 700 m. In contrast, Lake Victoria is only 95 m at its deepest point, and most of the lake is less than 20 m deep.

These lakes have had a complex geological history, with each basin being occupied by bodies of water for varying periods of time. Malawi and Tanganyika were the result of gradual lengthening and deepening of the East Africa rift. Later, to the north, tilting of the Tanzanian shield formed a barrier that blocked major rivers that had flowed west from the drainage area now occupied by Lake Victoria. Tanganyika is by far the oldest of the cichlid-dominated lakes. The lake basin, according to geological evidence, is 9–12 million years old (Cohen, Soreghan, and Scholz 1993), although the current fish fauna appears to be no more than about 5 million years old (Meyer 1993). The basin of Lake Malawi may have formed as long as 2 million years ago, but the age of its fish fauna is estimated at 700,000 years. The Lake Victoria basin may be 250,000–750,000 years old, but its fish fauna is at most only 200,000 years old.

It has long been suspected that drought conditions in the late Pleistocene may have led to the drying up of Lake Victoria and the great reduction or complete elim-

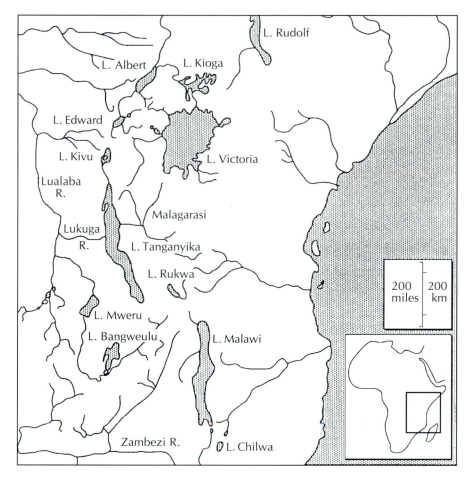

Figure 6.4. Geography of East Africa showing the location of the three Great Lakes and the major river systems. Redrawn from Fryer and Iles (1972).

ination of the fish fauna. This has recently been confirmed by Johnson et al. (1996), who dated the onset of renewed flooding of the lake basin at only 12,400 years B.P. The prior formation of a soil horizon over the entire lake basin precludes the possibility that any of the modern endemic fish had found refuge during the drought, and so the entire evolutionary history of this fauna spans less than 13,000 years.

Each lake has representatives of eight to nineteen families of noncichlid fish, but in biomass, species number, and trophic diversity, the cichlids are dominant in all lakes. There are 171 cichlid species that have been identified (in 49 endemic genera, grouped in 12 tribes) in Lake Tanganyika (Coulter 1991), more than 300 species in Lake Victoria (Barel et al. 1991; Meyer 1993), and at least 500 species in Lake Malawi (Ribbink et al. 1983; Owen et al. 1990). No lakes anywhere else in the world have as large a number of fish species.

The different ages of these lakes serve as a natural experiment to show the varying degrees of trophic differentiation, morphological change, and amount of speciation over different periods of time within a single family of vertebrates.

In order to establish the patterns of evolution within these lakes, it is first necessary to determine how the fish in each lake are related to one another. Because of the geographical separation of the lakes and the different times of their origin, the fauna in each lake might have evolved largely independently. On the other hand, some very similar, highly specialized cichlids occur in both Lake Tanganyika and Malawi, and other similar species pairs occur in both Lake Victoria and either Tanganyika or Malawi, leading Greenwood (1983, 1984) to suggest that the fauna of the three lakes shared a complex, interrelated history. This is difficult either to prove or disprove using anatomical information, since fish within a single family that have adapted to similar modes of feeding and locomotion are likely to resemble one another anatomically, whether or not they share an immediate common ancestry.

Within the past decade, study of restriction-fragment-length polymorphism and the DNA sequence of the mitochondrial genome have provided means of determining relationships that are thought to be largely independent of environmental and adaptational factors. Analysis of molecular data, summarized by Meyer (1993), indicates that the evolution of the cichlid faunas of the three major lakes occurred almost entirely independently (Fig. 6.5). The cichlids of Lakes Victoria and Malawi can each be traced to a single common ancestor: The genus *Astatotilapia,* which now lives in the rivers of East Africa, appears to be the closest relative of the Lake Victoria cichlids; the riverine genus *Rhamphochromis* is thought to be the closest relative of the Malawi cichlids. *Astatotilapia* and *Rhamphochromis* are believed to have shared a common ancestry with a particular tribe of cichlids from Lake Tanganyika, the Tropheini. Other cichlids spread from Lake Tanganyika to adjacent rivers but did not influence the evolution of cichlids in the other Great Lakes. These molecular studies demonstrate that the close morphological similarities between fish in different lakes are the result of convergent evolution, rather than being indicative of close relationships (Kocher et al. 1993).

The term **species flock** is often applied to groups of species, such as the cichlids in the East African Great Lakes, that share an immediate common ancestry and are restricted to a single geographical area (Echelle and Kornfield 1984).

Unfortunately, almost nothing is known of the fossil history of the fish species that are endemic to these lakes, so that the rate and pattern of evolution within each lake cannot be established directly. However, knowledge of living species from outside the lakes provides a good model for the nature of the colonizing species, and the geological history of the area provides strong evidence for the time at which the lakes were colonized, and so the maximum time available for the evolution of the fish lineages.

All of the East African Great Lakes were originally populated by species that had previously lived in the rivers of eastern Africa. The most primitive cichlids in the three major lakes are very similar to one another, and their ancestors presumably had similar feeding habits to primitive riverine forms. The range of structure and behavior of modern riverine cichlids living in South America was summarized by Lowe-McConnell (1991) who recognized the following adaptive types: those specialized to life in fast-running water, piscivores, bottom feeders, those dwelling in leaf litter, deep-bodied species, one possible zooplanktivore, and omnivores. Only

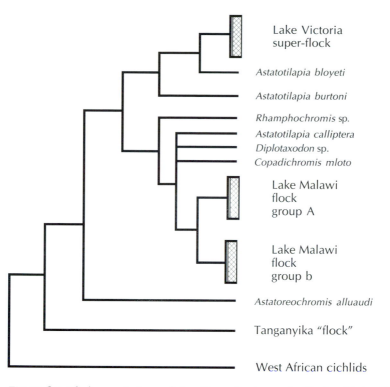

Figure 6.5. Phylogenetic tree relating the endemic species flocks of Lakes Victoria, Malawi, and Tanganyika. Other lineages represent species currently living in the river systems of East Africa. *Astatotilapia bloyeti* is a generalized haplochromine that is found throughout much of East Africa and probably resembles the ancestral species of the Lake Victoria flock. From Meyer (1993).

the species that adapted to life in very fast waters are highly specialized morphologically. Lowe-McConnell noted that the environments inhabited by all riverine cichlids fluctuate greatly in both physical and biological aspects, including water depth, current, temperature, predators, and prey. He argued that such continuing fluctuations made specialization to particular habitats very difficult. Instead, most riverine cichlids are very flexible in the nature of their diet and way of life in order to survive under constantly changing conditions. Speciation is common, but trophic differentiation is not conspicuous. This description also applies to the modern riverine species of Africa and presumably to the immediate ancestors of the species that entered Lakes Victoria, Malawi, and Tanganyika.

As these lakes formed, the cichlids found themselves in a physical environment very different from that to which they had previously been adapted. The geological basins were huge, so that as they filled with water they took on physical characteristics comparable with those of the ocean. These included their great size, with a prodigious diversity of physically distinct habitats, but also a much greater measure of stability in the properties of the water in any one place. Whereas adaptation to streams and rivers had depended on a high degree of flexibility and accommodation to change, the ocean-sized lakes provided opportunities for behavioral and structural specialization to many different habitats and ways of life. This may be judged by the number of trophic levels, the amount of morphological divergence, and the number of species.

Geological evidence demonstrates that Lake Victoria was the most recently occupied of the East African Great Lakes, with the present fauna having a duration of less than 13,000 years. The general appearance of the Lake Victoria cichlids is sufficiently similar that until recently nearly all were placed in the single genus *Haplochromis*. Most are of modest size, ranging in length from 70 to 220 mm. In 1974, Greenwood recognized 150–170 species within this genus, all but three considered endemic. He placed four additional haplochromine species in monotypic genera, and two other endemic species were assigned to *Tilapia*. Subsequent studies, summarized by Meyer (1993), have increased the estimated number to over 300 species divided into more than twenty genera.

Morphological differences that allow the recognition of such a large number of genera and species are largely concentrated in the head region and correlate closely with specialization for particular types of prey and modes of feeding (Greenwood 1981). These differences include skull proportions and size and configuration of the teeth of both the cranial and pharyngeal jaws. What appears as the most primitive dentition and cranial configuration is seen in primarily insectivorous fish, which feed chiefly on the benthonic larvae of Diptera. In both structural and behavioral features, they most closely resemble the primitive haplochromid species inhabiting rivers elsewhere in East Africa. From this adaptive type has emerged a complex radiation into a series of trophic levels, culminating in at least twenty distinct patterns of structural and behavioral specialization (Fig. 6.6). All of these adaptive types have been described by Greenwood (1974, 1981, 1984). The diet and mode of feeding have been determined by behavioral studies, as well as through stomach contents, and correlate closely with consistent but sometimes subtle differences in the cranial and dental anatomy.

A shift to feeding on phytoplankton apparently involved only minor cranial changes from the pattern of the primitive insectivores: an increase in the number but decrease in size of the teeth, and a two- or threefold increase in the length of the intestine. Considerably greater change is seen in fish that feed on algae grazed from rocks and rooted plants: The snout region is curved downward, the lower jaw is shorter and stout, and the teeth are longer, finer, and movably attached to the jaw bones.

From fish with a dentition adapted to scraping algae have apparently evolved species with a very different prey choice: the scales of other cichlids. Both patterns of feeding involve rasping, but other aspects of their behavior must have changed radically.

Two divergent types of mollusc feeders are recognized: those that crush the shells with the pharyngeal dentition and those that extract the prey from intact shells. The pharyngeal teeth in the former group are massive and blunt, but the oral dentition remains primitive. In contrast, fish that extract snails from their shells, or break the shells with the oral dentition, show much more specialization in the anterior portion of the skull and the oral teeth, yet retain a primitive pharyngeal dentition.

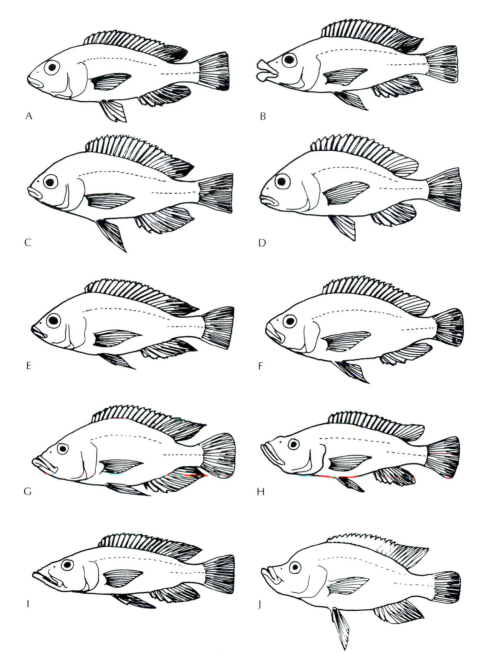

Figure 6.6. Body form in various Lake Victoria haplochromines showing the entire range of head and body shape represented among the endemic species. **A,** *Astatotilapia pallida* (generalized bottom feeder, close to the ancestral pattern); **B,** *Paralabidochromis chilotes* (specialized insectivore); **C,** *Macropleurodus bicolor* (specialized oral shelling mollusc eater); **D,** *Gaurochromis empodisma* (detritus eater); **E,** *Labrochromis ishmaeli* (specialized mollusc eater, pharyngeal crusher); **F,** *Limochromis obesus* (paedophage); **G,** *Harpagochromis skekii* (piscivore); **H,** *Harpagochromis cavifrons* (piscivore); **I,** *Prognathochromis macrognathus* (piscivore); **J,** *Pyxichromis parorthostoma* (diet unknown). This degree of diversity evolved within 12,400 years. Piscivores range from 150 to 250 cm; most other cichlids range from 100 to 130 cm. From Greenwood (1984).

Species feeding on insect larvae that burrow into rocks and wood have a shortened, strengthened lower jaw and an upper jaw capable of protruding downward. The teeth have sharply pointed crowns that protrude forward.

Although 30–40 percent of the cichlids feed on other fish, they are further subdivided within the general role of predators. Two distinct groups feed primarily on the embryos and larvae of other, mouth-brooding cichlids. Their mouths are widely distensible so that they can engulf the snout of the brooding female and force her to jettison her progeny into the mouth of the predator. Other species prey preferentially on eggs or young at particular growth stages.

Predators on adult fish are typically twice the length of other cichlids. The body is more streamlined and the jaws longer and more protrusile. The teeth are sharply pointed and unicuspid rather than bicuspid. The skull length may be increased relative to its depth.

In his summary of trophic diversification, Greenwood (1981) argued that most changes could be attributed to alterations in proportions of anatomical features common to all species. Nearly all could be produced by slight changes in differential growth patterns of the various skull regions, coupled with changes in the jaws and suspensorium. These changes are of the nature that might be expected to result from selectional sorting of quantitative genetic traits (see Chapter 9). Greenwood (1984, p. 149) stated that these fish provide "evidence of an ecological revolution brought about without a corresponding morphological one."

The extremely low degree of morphological differentiation of the many species of Lake Victoria cichlids is matched by the conservative nature of their mitochondrial DNA (mtDNA). In a study of fourteen species of nine representative genera in Lake Victoria, Meyer (1993) found no variation in 363 base pairs of the cytochrome b gene. Only two to three substitutions differentiate mitochondrial haplotypes in 440 base pairs of the control region of the mitochondrial genome. More variation has been documented in the homologous portion of mtDNA in living humans, belonging to a single species, than among nine genera of Lake Victoria cichlids, suggesting a very short time for their divergence and a very close common ancestry.

The radiation of the Lake Victoria cichlids provides one of the most informative examples of evolutionary change among living species that has been available for study. Unfortunately, the fauna has experienced massive extinction during the past twenty years (Barel et al. 1991). This has resulted from the introduction of the Nile perch, *Lates niloticus,* in the early 1960s as a more salable food fish than the smaller cichlids. The perch population expanded slowly until the end of the 1970s; then it suddenly increased at an explosive rate, so that this single species now dominates nearly the entire lake. The perch feed in most adaptive zones, except for the shallow margins of the lakes. In contrast with the cichlids, the Nile perch is a nearly indiscriminate feeder. By 1993, about 200 endemic cichlid species had become extinct (Meyer 1993). With the marked reduction in the cichlids, the population size of the Nile perch then declined, and they have shifted to feeding on shrimp, cyprinids, and juvenile members of their own species. Although the Nile perch entered Lake Victoria as a result of human activity, several other species of the same genus are endemic to Lake Tanganyika, where they are among the most abundant and ecologically important fishes, some reaching a length of over a meter (Coulter 1991). They feed heavily on cichlids, but that family is not in danger of extinction as is the case in Lake Victoria. Nile perch and cichlids also coexist in Lakes Albert and Turkana, but these lakes have a much smaller cichlid fauna relative to the three largest lakes.

Whether or not a diversity of cichlid species in Lake Victoria manages to recover from the invasion of the Nile perch, their prior history provides a good model for the initial stage in the evolution of the older faunas of Lakes Malawi and Tanganyika.

The entire trophic diversity seen in the cichlid fauna of Lake Victoria was apparently attained within less than 13,000 years. The only time when mammals underwent such a spectrum of trophic differentiation was at the very end of the Cretaceous and the beginning of the Cenozoic, a period of 5–10 million years. According to Feduccia (1995), the basic differentiation of the modern avian orders may also have occurred during this time span. The fossil record does not show any period in the evolution of other vertebrates during which there was such a high rate of speciation, with approximately a thousand new cichlid species in the East African lakes compared with no more than 500 previously. However, the fine anatomical distinctions between species that are recognized in living cichlids would probably not be visible in most fossil groups.

Although it is difficult to compare directly the nature of skeletal changes that occurred in fish and mammals, the appearance of twenty new genera of cichlids in Lake Victoria is somewhat comparable to the number of new arctic and tundra species that appeared during the Plio–Pleistocene. Cichlid evolution has proceeded for much longer periods in Lakes Malawi and Tanganyika than it has in Lake Victoria. There have been major changes in water level, but the total depth of these lakes is so great that large bodies of water must have persisted for at least 0.7 million years in the case of Lake Malawi, and for more than 5 million years for Lake Tanganyika. Both basins are older than this, and there may have been earlier cichlid radiations, but all of the currently recognized lineages appear to have diverged during these time intervals. There are seven major lineages in Lake Tanganyika, but as in the other lakes, molecular evidence indicates that they had a single common ancestor within the lake basin.

Despite the difference in the age of these faunas and the much greater depth and stability of the older lakes, the number of distinct trophic zones into which the cichlids of Lake Victoria evolved over less than 13,000 years is not exceeded in the much older lakes. In the case of Tanganyika, cichlid radiation may have been somewhat restricted by the nature of other fish families inhabiting the lake. Lake Tanganyika differs from Victoria and Malawi in having a large population of non-cichlid pelagic fish. These are dominated by four endemic species related to the Nile perch that are the primary predators in the lake. The absence of oxygen in the deeper portions of the lakes precluded the evolution of an abyssal fauna in either Lake Tanganyika or Malawi.

More generally, the basic homogeneity of the cichlid body plan, despite a myriad of small changes, suggests that there may be other constraints that restrict members of this family to relatively small body size and limit the degree of pelagic adaptation. Stiassny (1991) argued that the functional and behavioral versatility of the cichlid *Bauplan* facilitated trophic diversification without major structural modification. On the other hand, the very flexibility of this system may act as a constraint restricting change within particular limits. Work by Barel and his colleagues (Barel 1984; Barel et al. 1989, 1991) demonstrated that cichlids manifest a complex inter-relationship of functional systems, including feeding, locomotion, respiration, and reproduction, that may preclude the evolution of anatomical patterns that exceed the range of variation among the thousand or more known species. This may explain why the cichlids in Lakes Malawi and Tanganyika have not overshot the spectrum of adaptive zones exhibited in Lake Victoria. According to Liem (1991), size, in particular, may be limited by the specialized hydrodynamic "tongue" (described in the following section).

The number of cichlid species is substantially greater in Lake Malawi than in Victoria; it is smaller in Lake Tanganyika, but the morphological differences between members of the different trophic guilds are far greater (Fig. 6.7). This can be most simply interpreted as progressive evolution within each guild over time, accompanied by selection against intermediate forms. Generic distinction has been reached by more than twice the number of lineages recognized in Lake Victoria, and groups of genera are recognized as a higher taxonomic level, the tribe. However, even after at least 5 million years of evolutionary divergence, none of the genera have evolved to the extent that they have ever been considered members of a family distinct from the Cichlidae.

Explanations for the rapid evolution of cichlids

The rapid trophic differentiation among the cichlids of the East African Great Lakes and the proliferation of hundreds of endemic species may be unique events in vertebrate history. How can they be explained? Although other families of fish within the Great Lakes of East Africa also speciated and underwent considerable structural and behavioral specialization, there are more endemic species of cichlids known in these lakes than there are of all the other nineteen fish families.

Within these lakes, cichlids are unique in both the structure of their feeding apparatus and their reproductive habits. Liem (1991) and Meyer (1993) have emphasized the importance of the pharyngeal jaw apparatus, which greatly increases the flexibility of diets and feeding habits, whereas Dominey (1984) and McKaye (1991) have stressed the importance of their reproductive behavior.

Cichlids, in common with most labroid fish, are distinguished by the elaboration of the pharyngeal jaws, which have a well-developed articulation with the back of the braincase and can function independently from the normal, more anterior jaws (Fig. 6.8). The pharyngeal jaws, especially in mollusc crushers and the vast array of specialized predators, take over much of the food processing, allowing the oral jaws to evolve a variety of manipulative functions. Cichlids also use a complex system of water currents generated in the oropharyngeal cavity to manipulate the prey. Bemis and Lauder (1986) called this complex a *hydrodynamic tongue*.

Liem (1991) showed how changes in skull proportions enable cichlids to feed in fundamentally different ways without significantly altering the basic structure of the skull. The divergent head morphologies of three genera from Lake Tanganyika show different proportions of the buccal cavity that are associated with ram feeding (in which the predator opens its mouth and simply overtakes its prey), com-

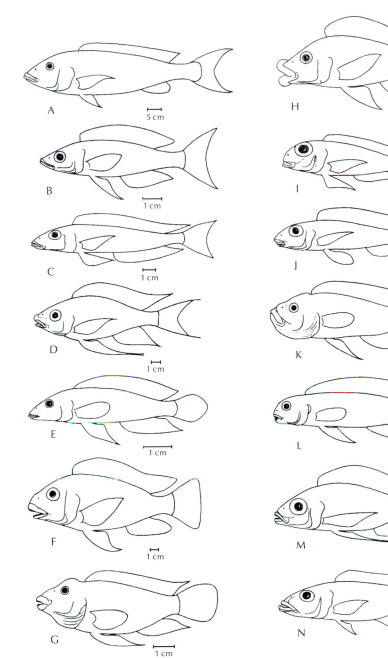

Figure 6.7. The diversity of body shape among cichlid fishes in Lake Tanganyika after 3.5–5 million years of trophic diversification. **A,** *Boulengerochromis microlepis;* **B,** *Limnochromis leptosoma;* **C,** *Xenotilapia melanogenys;* **D,** *Ophthalmochromis nasutus;* **E,** *Julidochromis transcriptus;* **F,** *Petrochromis polyodon;* **G,** *Spathodus marlieri;* **H,** *Lobochilotes labiatus;* **I,** *Xenotilapia sima;* **J,** *Asprotilapia leptura;* **K,** *Telmatochromis caninus;* **L,** *Telmatochromis vittatus;* **M,** *Cunningtonia longiventralis;* **N,** *Bathybates ferox.* Redrawn from Fryer and Iles (1972).

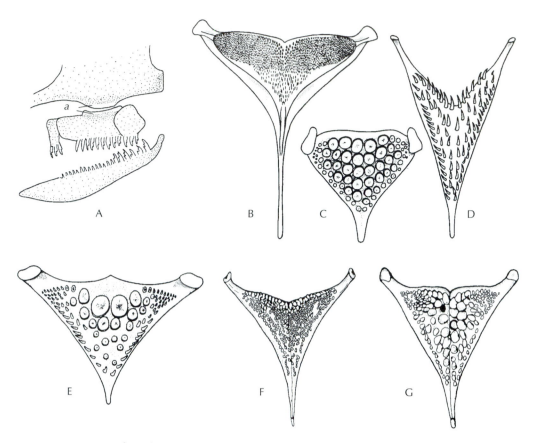

Figure 6.8. Pharyngeal jaws of cichlids. **A,** Upper and lower pharyngeal jaws of a piscivorous cichlid in lateral view; *a* = articulation between upper jaw and base of braincase (from Fig. 6.7 in *Cichlid Fishes: Behaviour, Ecology, and Evolution,* by K. F. Liem, Chapman & Hall, 1991). **B–E,** dorsal views of lower pharyngeal jaws in cichlids representing several feeding types: B, *Tilapia esculenta,* a phytoplankton eater, Lake Victoria; C, *Haplochromis placodon,* a mollusc crusher, Lake Malawi; D, *Bathybates leo,* a piscivore, Lake Tanganyika; E, *Lamprologus tretocephalus,* a mollusc crusher, Lake Tanganyika (from Fryer and Iles 1972). **F–G,** Lower pharyngeal jaws of *Haplochromis hiatus:* F, feed on a diet of soft food; G, feed on a diet of snails (from Hoogerhoud 1984).

bined ram and suction feeding, and a concentration on suction feeding (Fig. 6.9). The buccal cavity of a ram feeder is in the shape of a cylinder of nearly equal dimensions throughout its length. In suction feeders, which consume small prey, the cavity is much expanded posteriorly, but the mouth opening is very restricted. Fish using their oral jaws in forceful biting also have conically shaped buccal cavities because of their relatively short jaws (necessary to increase their mechanical advantage) and massive jaw musculature at the back of the skull. Predators with cylindrical buccal cavities concentrate on the pharyngeal jaws rather than the oral jaws in masticating their prey.

While field observations and study of gut contents show that cichlids commonly feed on a particular type of prey that appears to be most appropriate to skull structure, experiments show that they are also adept at feeding on many other food sources. This is possible because of the extremely varied ways in which the

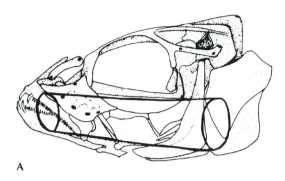

A   B

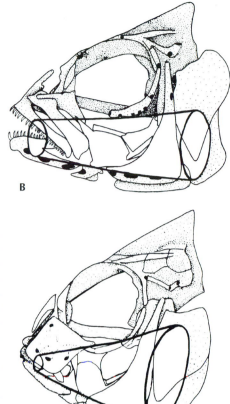

Figure 6.9. Models of differently designed buccal cavities (shown in heavy
lines) in three cichlids from Lake Tanganyika. **A,** a typical ram feeder with a cylin-
drically shaped buccal cavity, *Lamprologus elongatus.* **B,** a combined ram and
suction feeder, with an intermediate configuration of the buccal cavity, *Grammatoria lemairei.* **C,** a typical suction feeder
with the buccal cavity in the shape of a
truncated cone, *Tylochromis microlepis.*
From Fig. 6.1 in *Cichlid Fishes: Behaviour, Ecology, and Evolution,* by K. F.
Liem, Chapman & Hall, 1991.   C

muscles controlling the movements of the jaws and the shape of the buccal cavity
can be used. Both the force and the sequence of muscle contractions can be altered
extensively to accommodate prey as different as zooplankton, arthropods, worms,
and algae. This flexibility allows individual species to adjust readily to the availability of different food sources and competitive regimes.

Not only do cichlids have the capacity to vary their behavior and the neuromuscular activity of feeding and processing food, but some also show striking developmental plasticity in response to the availability of different food sources. This is
particularly evident in populations within species that have been raised on either
molluscs or soft food (Witte, Barel, and Hoogerhoud 1990). In populations that
feed habitually on soft food, such as insects, the pharyngeal jaws are delicate and
the teeth slender and numerous. In those feeding on snails, the jaws are massive,
especially where they articulate with the skull, and the teeth are heavy and blunt
(see Fig. 6.8F,G). The accommodation of bones and muscles to the differing
stresses they encounter is a common feature of vertebrates, but this ability is exaggerated in these small fish to the point that changes occurring during early growth
may influence their lifelong feeding pattern.

Clearly, the functional plasticity of the cichlids has enabled them to occupy a vast
array of different trophic levels, but such radiation is also associated with their capacity to form a great number of genetically isolated units, each of which can respond to different selective forces.

Cichlids are distinguished from other fish in the East African Great Lakes by their extremely rapid speciation and the maintenance of large species flocks. There were as many species in one 45-km² area of Lake Victoria (200) as in all of the twenty-nine fish families known throughout Europe (Barel et al. 1991).

Although the entire period available for evolutionary change in Lake Victoria may have been less than 13,000 years, the differentiation of individual species may have been even more rapid. This is suggested by the study of the fish in Lake Nabugabo (Greenwood 1965). This lake has been isolated from Lake Victoria for approximately 4,000 years, yet the five cichlid species that have arisen in Nabuga-bo during this time appear as different from putative ancestors in Victoria as are any of the Victoria species from one another. Cichlids in Lake Malawi may have attained the status of separate species in a much shorter time (Owen et al. 1990). Approximately twenty-five species are endemic to the waters around small rocky islands at the southern end of the lake that could not have been inhabited before about 200 years ago, when the water level rose to its present height. Although it is possible that these species migrated from other parts of the lake, the varying degrees of their differentiation correspond closely with the distances of their current habitats from those of other, related species, suggesting that they evolved around the same rocky islands they occupy today. The capacity for rapid speciation can be associated with several interrelated aspects of cichlid behavior: (1) intensive and long-term parental care; (2) limited tendency for dispersal; (3) complex courtship; and (4) sexual selection (Dominey 1984).

Cichlids are unique among fish in the importance of parental care (Keenleyside 1991b). In all known species, one or both parents protect their progeny, from spawning to independence. Primitive cichlids, including most of the riverine species and the most primitive species in Lake Tanganyika, lay their eggs on the substrate; the eggs and young are guarded by both parents until they are capable of feeding themselves and avoiding predators. The parents may use their mouths to free the young from the eggs, and to carry them from place to place. This is a logical bridge to the more advanced condition of mouth brooding, in which either the male or (much more commonly) the female carries the eggs and young in the mouth. Since there are both substrate spawners (60 species) and maternal mouth brooders (110 species) in Lake Tanganyika, the latter characteristic must have evolved within the lake and then spread to Lakes Malawi and Victoria, where all the cichlids are female mouth brooders.

Both substrate spawners and mouth brooders are strongly territorial. Many cichlid families guard the same territories generation after generation, and the young may show little tendency toward dispersal. The most extreme pattern can be seen among cichlids occupying rocky substrates in Lake Malawi, where Ribbink et al. (1983, p. 301) postulated that individuals might spend their entire lives within a few square meters.

In all cichlids, the possibility of reproduction between closely related species is strongly limited by their complex courtship, in which the females select particular males based on distinctive coloration and/or specific behavior. This is further restricted in mouth brooders, in which the eggs are picked up as soon as they are released. They may be fertilized in the water or within the mother's mouth; either way, there is little likelihood that males of another species could fertilize the eggs.

In some mouth brooders, the males remain with the females while the young are maturing, but in most they leave immediately after fertilization and may soon mate with other females. The polygamous behavior of most of the cichlids in the East African Great Lakes make them even more strongly subject to sexual selection than are the substrate spawners. Thousands of males may display before the females. If the characters the females use to select mates are both variable and strongly heritable, they may be greatly accentuated within a very few generations, provided they are not strongly opposed by environmental selection.

If, as is the case in the East African Great Lakes, the environment is geographically very heterogeneous, each local population of a species may occupy areas that are close to one another, but separated by inappropriate habitats. Under these conditions, each population may also differ in the particular features subject to sexual selection. Males belonging to the same species but a different population that disperse from one area to another will not be recognized as potential mates by the females; thus, reproductive isolation is achieved very rapidly. This is most clearly evident in Lake Malawi, in which the more than 200 species termed the Mbuna occupy the waters around isolated rocks and rocky islands.

The ease of speciation and the perpetuation of species in small, homogeneous habitats enable selection to favor any morphological or behavioral traits that improve survival and reproduction under a very limited variety of conditions. Hence, rapid speciation permits rapid evolutionary change in all the myriad local habitats. Any rapidly speciating group living within a relatively stable but geographically heterogenous environment might produce many species; but cichlids have the added feature of great structural and behavioral flexibility in the feeding apparatus that has permitted them to become adapted to an especially wide range of food sources and ways of feeding. It is this combination of feeding and reproductive traits that has made it possible for the cichlids to differentiate so widely and at such a high rate in the East African Great Lakes.

Species-level evolution among the cichlids of the East African Great Lakes

While the cichlids may be unique in their ease and rate of speciation, they nevertheless provide an informative example as to how evolution may proceed at the species level.

The most recently evolved fauna, that of Lake Victoria, presumably records the shortest period subsequent to speciation of any of the major lakes. The pattern seen there is markedly different from that expected from observations of the majority of longer-established living species.

Whereas most closely related species in other vertebrate groups are clearly distinct from one another in their degree of anatomical specialization, nearly all of the particular trophic levels of the Lake Victoria cichlids are represented by numerous species that appear as a graded series. For example, among the mollusc-crushing cichlids, the pharyngeal dentition shows progressively more thickened and blunted teeth from species to species. Frequently, more than one species of such series feed in the same area, potentially competing for the same type of prey. Greenwood

(1984) referred to the apparent pattern of evolution among these species as *cladistic gradualism,* suggesting that a series of cladogenetic events (speciations) each resulted in a small change in anatomy, mimicking the tiny intergenerational changes thought to result in phyletic gradualism.

However, in at least one well-studied case, the evolution of a species group did not occur as hypothesized by Greenwood. Dorit (1990) established relationships among the *Psammochromis–Macropleurodus* (mollusc-feeding) lineage on the basis of mitochondrial DNA restriction sites and then imposed on this phylogeny the morphological characters that had been observed. The resulting phylogenetic sequence did not show a graduated series of morphological changes; rather, it demonstrated a close common ancestry among forms showing more extreme patterns, and the subsequent appearance of forms filling in intermediate morphologies. If the pattern shown by Dorit applies to other species, it suggests an ever finer partitioning of the trophic levels, and explains how graded series of species are maintained in Lake Victoria.

However, without a fossil record, it is not possible to determine the pattern of evolution within the individual lineages. If the most derived lineages had undergone gradual change during their evolution, they may initially have been as similar to one another as are the intermediate species today. No matter the specific pattern of evolution, many very similar forms do cohabit within the modern fauna of Lake Victoria. This phenomenon suggested to Greenwood (1984) that selection through competition played a very small role in the evolution of cichlids, especially in the Lake Victoria flock, with which he was most familiar.

Competition, both between and within species, has long been thought to be a highly significant factor in natural selection; yet many cichlid species, very similar morphologically and seeming to have very similar environmental requirements, cohabit in all the East African Great Lakes. How is this possible?

Both Greenwood (1981, 1984) and Liem (1991) recognized that cichlids show a great deal of behavioral plasticity. Although it is possible to differentiate many trophic levels on the basis of observed behavior as well as of stomach contents, cichlids are observed, under both laboratory and field conditions, to feed on a great variety of prey items, from plants to small invertebrates to vertebrates, depending on their relative availability. Although some highly specialized species may be more efficient in making use of a particular food source, they are also capable of feeding on many others. This allows many species to cohabit through behavioral adjustment to avoid competition. Ribbink et al. (1983) found evidence from the Mbuna of Lake Malawi that species within a particular trophic group, which feed upon apparently identical food, collected it from different microhabitats.

The reason for the persistence of many cohabiting species within a particular trophic zone in Lake Victoria may also be attributed to another factor, one documented by Peter and Rosemary Grant in their studies of the Galápagos finches (see Chapter 3): Competition for particular resources does not necessarily result in continuous selection for either structural or behavioral traits. As long as resources were sufficient, no significant differential in the survival of birds with different-sized beaks or behavioral preference for one food or another was observed. However, in times of scarcity, such as result from droughts, selection is extremely severe and leads to measurable morphological differences within a single generation. In short-term studies of the finches, such severe conditions always ameliorated prior to species extinction and even before permanent changes occurred in the reservoir of genetic variability, so that the species have remained unaltered in the long run; on the other hand, some local populations have become extinct during historical times. Like the cichlids, individual species of Galápagos finches show a great deal of behavioral plasticity that enables them to take advantage of a wide range of food, including both plants and animals, so that only the most serious climatic changes have genetic impact. Selection clearly acts most of the time to maintain behavioral and structural variability. Strong directional selection to the degree of allele loss and fixation throughout a species is probably uncommon for structural traits over a scale of hundreds or thousands of years; once it occurs, however, it results in evolutionary change that can be reversed only by the appearance of new mutations.

The greater age of the Lake Tanganyika fauna increases the probability for the occurrence of rare but severe periods of selection that would favor more highly specialized species. This would explain the greater degree of distinction between species and trophic levels observed in that lake.

The cichlid radiation as a model to bridge the levels of microevolution and macroevolution

The evolution of cichlid fishes in the East African Great Lakes appears as a plausible model for the early stages of major radiations, such as those that occurred among placental mammals and the modern avian orders in the Late Cretaceous and Early Tertiary, the differentiation of a spectrum of amphibian orders in the Late Devonian and Early Carboniferous, and the radiation of metazoans at the end of the Precambrian and beginning of the Cambrian. Like the cichlid radiation, these macroevolutionary episodes involved the proliferation of a great many distinct lineages occupying a diversity of trophic levels over a geologically very short length of time.

Cichlids provide an example of how such radiations may be initiated at the species level. Although cichlid diversification has not yet achieved the degree of anatomical divergence of any of the major vertebrate radiations, we can determine the specific patterns and processes involved in changes up to the level of the genus in a great many different lineages. The cichlid example also provides a basis for determining what intrinsic factors of the organism and extrinsic factors of its environment are necessary for the occurrence of such macroevolutionary events.

The cichlid radiation has been emphasized because of the exceptionally rapid and extensive radiation at the species level. However, the prodigious capacity of the cichlids for speciation may be exceptional among vertebrates, and the particular combination of features of the head region that enabled them to adapt to a great many trophic levels with relatively minor anatomical change may be unique among fish. On the other hand, the cichlids are not unique within the biota of the East African Great Lakes in undergoing species- and generic-level evolutionary changes over a relatively short period of time.

Coulter (1991, chap. 9) reviewed the composition of the entire flora and fauna of Lake Tanganyika (Table 6.1). The single family Cichlidae has 172 species, 167 of which are endemic to the lake. There are nineteen families of noncichlid fish, with 115 species, but 52 of these are also endemic to the lake. Like the cichlids, but at a smaller scale, many lineages within the noncichlid families differentiated morphologically and behaviorally to accommodate to new environmental conditions that were common to the lakes but were absent from their previous, riverine habitats. None of the noncichlid families gave rise to large species flocks, although each of two genera includes six endemic species, and eleven of the twelve spiny eel species are endemic. Perhaps the most spectacular adaptive change is that of brood-parasitic behavior in one of the six endemic species of the catfish *Synodontis, S. multipuncatus,* whose eggs are incubated in the mouths of at least six species of inshore cichlids. The *Synodotis* eggs, which are the first to hatch, feed on the yolk sacs of the host's larvae. This reproductive behavior is unique among fish.

Going beyond vertebrates, there are nine families of ostracods known in the lake, with eighty-five species, seventy-four of which are endemic. There are also nine copepod families, with sixty-nine species, thirty-three of which are endemic. The most surprising feature of the fauna is the large number of extremely divergent molluscs, with thirty-four endemic species of gastropods, all but one of which belong to the prosobranch family Thiaridae. In fact, the highly specialized nature of the gastropods was recognized much earlier than that of the cichlids (Woodward 1859). The most striking are the thalassoid forms, which greatly resemble marine molluscs. The nature of the molluscan fauna was long thought to indicate a marine origin for the fauna of Lake Tanganyika, going back to the Jurassic. It was later suggested that marine features of these shells, such as their great thickness and extensive ornamentation, might result from unusual chemical constituents of the water; but it is now agreed that the marine features are the result of adaptation to the marinelike physical and biological characteristics of the lake. This is confirmed by the presence of similar thalassoid morphotypes in other very large lakes and inland seas, including Lakes Baikal and Malawi and the Caspian and Aral seas. Similar molluscs in the Zaire River, however, probably dispersed from Lake Tanganyika. Fossil shells with similar features are also known in sediments of Lake Kaiso, which, like Lake Tanganyika, formed in the late Miocene or early Pliocene, and lasted during approximately 6 to 2.5 million years ago. Boss (1978) thought that the radiation of the thalassoid prosobranchs occurred gradually over a long stretch of time. Van Damme (1984), in contrast, cited fossil evidence from both Lake Kaiso and a particular period in the existence of Lake Tanganyika to show that spectacular evolutionary change took place in geologically brief time spans early in the history of these lakes and were followed by long periods of stasis. The endemic gastropods are all live-bearing and, like the cichlids, apparently have limited dispersal of the young.

Although less striking than the gastropods, nine of the fifteen bivalves in Lake Tanganyika are also endemic. Stanley (1979) stressed the great longevity of bivalve molluscs, yet even this group has given rise to new species within the new environment provided by the East African Great Lakes. As emphasized by Coulter (1991), the spectrum of new environments, the long period of time available, and the isolation from other faunas have enabled all the species in Lake Tanganyika to undergo significant evolutionary change.

The cichlid fishes of the East African Great Lakes constitute the most informative and best-studied examples of species flocks, but numerous other species flocks

Table 6.1. *Number of families, genera, and species of a diversity of major taxa in Lake Tanganyika*

	Family	Genus	Species	Endemic species
Algae	49	160	759	
Aquatic plants	27	48	81	
Protozoa	25	33	71	
Cnidaria	(Orders 2)	2	2	
Porifera	1	6	9	7
Bryozoa	(Classes 2)	6	6	2
Platyhelminthes				
Cestoda	5	6	8	5
Trematoda	1	1	1	1
Turbellaria	2	2	2	1
Nematoda	7	12	20	7
Nematomorpha	3	3	9	
Acanthocephala	1	1	1	
Annelida (Oligochaeta)	5	7	8	5
(Hirudinea)	4	8	20	12
Pentastomida		1	1	
Rotifera	16	25	70	5
Mollusca (Gastropoda)	8	36	60	37
(Bivalvia)	5	10	15	9
Arthropoda				
Arachnida	12	21	46	17
Branchiopoda (Cladocera)	1	19	24	
Copepoda	9	23	69	33
Isopoda	1	1	3	3
Branchiura	1	3	13	7
Ostracoda	9	28	85	74
Bathynellacea	1	1	1	1
Decapoda (Caridea)	2	5	15	14
(Brachiura)	1	1	10	8
Insecta	23	107	155	
Pisces (Cichlidae)	1	50	172	167
(non-Cichlidae)	19	48	115	52
Amphibia	7	14	34	
Reptilia	6	16	29	2
Aves (aquatic birds)	37	92	171	
Mammalia	(Orders 2)	3	3	
Total fauna genera and species:		591	1,248	

Source: Coulter (1991), *Lake Tanganyika and Its Life*. Reproduced by permission of Oxford University Press.

have been identified in both modern and fossil lakes (Echelle and Kornfield 1984). The next most striking modern example consists of the sculpins in Lake Baikal, which at 1,620 m is the deepest lake in the world (Taliev 1955; Smith and Todd 1984). Thirty-six species and subspecies are recognized, within nine genera. The Russians have recognized two families, but other taxonomists include all the species in the family Cottidae. This species flock has apparently evolved within Lake Baikal for much of the late Cenozoic, approximately twice the age of the East African cichlid faunas, and produced highly derived abyssal forms.

Other species flocks are known among the cyprinids of Lake Lanao in the Philippines, the killifish of the Andes, the atherinids of the Mesa Central and the pupfishes of Laguna Chichancanab in Mexico, and the ciscoes in the North American Great Lakes. Few of these flocks are well documented or have more than a few species whose monophyly can be adequately demonstrated.

Although cichlids remain the best-known example of rapid and extensive evolution within modern lake systems, radiations of this type have occurred in other fish groups over a wide span of geological time. The most extensive and well-studied example is that provided by a series of rift valleys from the early Mesozoic of eastern North America that extends from North Carolina into Nova Scotia (McCune et al. 1984; McCune 1987, 1990, 1996). Although the entire sequence covers much of the Late Triassic and Early Jurassic, the individual lakes persisted for no more than 24,000 years. Again, there is a single dominant group of fish, in this case, the primitive neopterygian semionotids. There are about twenty species in the most thoroughly studied lakes, but they apparently differ only in body proportions and the configuration of the ridge scales. Most of the lineages appear near the base of the lake deposits and continue with only minor change until the lake begins to dry up. The very base of the lineages is not known in any of the lakes that have been studied, but it appears that change was most rapid early in the history of each lineage, then slowed to stasis long before their extinction. The limited period of evolution recorded in these lakes is most closely comparable to that of Lake Victoria.

Isolated lakes provide an especially informative environment for the study of evolutionary change because their fauna is typically confined to a particular area, all of which can be studied, rather than being influenced by immigration from outside the lake. In addition to providing the advantage of an essentially closed system, lakes have the potential for accurate dating over short time scales.

Unfortunately, lakes have serious limitations for analysis of larger-scale, longer-term evolution. Lakes, like oceanic islands, are ephemeral on a geological time scale, and their degree of isolation severely limits the possibility of their faunas contributing to wider-scale evolution. The Late Triassic and Early Jurassic lakes provide an impressive example of hundreds or even thousands of separate lacustrine radiations, all of which have terminated within less than 24,000 years, with little if any influence on subsequent events in the history of fish evolution. Many lakes have formed and then disappeared in East Africa since the beginning of tectonic activity 12 million years ago. Lake Victoria, with the second largest surface area of any lake in the world, was completely dried up as recently as 12,400 years ago.

Even if these lakes persist for tens of millions of years, they have little likelihood of contributing to the larger-scale evolution of fish because all the highly derived species are specifically adapted to lacustrine conditions, which are not present outside the lakes. Riverine fish originally colonized the lakes, and only fish that are still capable of living in a riverine environment can be expected to leave these lakes or to survive their desiccation. The sculpins in Lake Baikal have survived there for up to 20 million years, but the chances for the wider distribution of highly adapted abyssal forms into other lakes are vanishingly small.

In addition, isolated faunas, such as those that have evolved in Lakes Victoria, Malawi, and Tanganyika, appear especially subject to extinction as a result of competition and predation by immigrants from larger faunas outside the lakes. The majority of cichlid species in Lake Victoria were wiped out within less than twenty years by the introduced Nile perch. The cichlid faunas of Lakes Albert and Turkana, which have had intermittent contact with the Nile River drainage, are much smaller than those of the much more isolated lakes.

Despite the improbability of species flocks in large lakes contributing to larger-scale, longer-term evolutionary phenomena, they nevertheless provide a model for considering other aspects of macroevolution. What we can see in the East African lakes, over periods from tens of thousands to 5 million years, is an astonishing amount of trophic diversification and conspicuous morphological change. The rate of trophic diversification that occurred in Lake Victoria may be unique in vertebrate history. That such diversification transpired not just once but at least three times in the major lakes, and also occurred in such divergent taxa as the copepods and molluscs in Lake Tanganyika, suggests that it does provide a general model for rapid evolutionary change.

The common element in the evolution of the faunas of these lake systems, shared by many larger evolutionary radiations, is the sudden availability of an extensive new environment that is relatively depauperate in potential competitors and predators. This is certainly the case for the radiation of placental mammals in the early Cenozoic, for which a broad spectrum of terrestrial adaptive zones became available because of the sudden extinction of dinosaurs. The island continent of South America became available for the radiation of different groups of placental mammals at various times when intervening islands and finally a continuous land connection made it accessible to species from other continents. The arctic and tundra environments, appearing in the late Cenozoic as a result of changes in the world's climate, became the site of major changes in particular mammalian lineages.

Major evolutionary changes would have not occurred in any of these cases without prior alterations in the environment. Neither the cichlids nor any of the other aquatic groups living in East Africa would have undergone significant evolutionary modification during the late Cenozoic were it not for the formation of the East African Great Lakes. Their riverine relatives show no such dramatic changes during this time interval. The arctic and tundra fauna would not have evolved without the chilling of the Northern Hemisphere, and the history of placental mammals would have been far different were it not for the extinction of dinosaurs.

In all these examples, a host of new adaptive zones became available. In the cases of cichlids and placental mammals, these zones were very rapidly occupied by a diverse assemblage of lineages that shared a close common ancestry. The cichlids that radiated in the East African Great Lakes initially differed little from their immediate riverine ancestors. There is no evidence of significant new morphological changes that facilitated their differentiation into many trophic levels; rather, they capitalized on structural and behavioral flexibility that was evident in their ancestors. Similarly, the most primitive members of the derived mammalian groups initially show relatively few changes aside from dental specialization.

On the other hand, both the cichlids and the early placental mammals radiated trophically to a much greater extent than did other groups that occupied the same, newly available environments. Cichlids were certainly capable of a much higher rate of speciation than were other fish in the East African Great Lakes, and were able to differentiate into many different trophic levels with a minimum of morphological change. Similar factors may have favored the early Cenozoic placentals. Like small rodents, they may have been highly speciose, but the possible degree of behavioral and dietary plasticity can only be hinted at from the pattern in modern placentals.

Summary

In contrast with the pattern of evolution in Northern Hemisphere mammals of the late Cenozoic, within which phyletic evolution was pervasive in a broad spectrum of taxa, a much different picture emerges from the fossil record of nonmammalian vertebrates. The progressive changes in the climate and vegetation that were a driving force in the evolution of mammals had a much less systematic effect on fish, amphibians, and reptiles.

The cichlid fishes began one of the most rapid and large-scale trophic differentiations in the history of vertebrates in response to the formation of a series of gigantic lakes in eastern Africa. In contrast, there is little fossil evidence for significant skeletal change among the well-established lineages of amphibians, reptiles, and birds; most of the living genera had been established by the Miocene or early Pliocene.

The nature of the skeleton strongly influences our ability to study the evolutionary history of groups at the level of the genus and species. No characters of amphibians, reptiles, or birds are as useful as the teeth of mammals in identifying species or recording patterns of evolution. The skeletons of passerine birds, which radiated extensively at the level of the family and genus during the late Cenozoic, are so homogeneous that few fossil remains can be identified below the ordinal level. Extensive research on living species of amphibians, reptiles, and birds demonstrates significant variability that can be associated with adaptation to particular environments or ways of life, but most of the characters studied would not be preserved in the fossil record.

Rates of evolution are extremely variable from group to group and from one time period to another both at levels of species and of higher taxa. Both intrinsic aspects of organisms and extrinsic factors of the environment are important in enabling taxa to undergo major changes in structure and ways of life. This is clearly evident from the extensive radiation of the cichlid fishes in the East African Great Lakes, which would not have occurred if these lakes had not been formed, but was accentuated by anatomical and reproductive traits that are unique to the cichlids.

7 The influence of systems of classification on concepts of evolutionary patterns

[B]iological taxonomy must eventually outgrow the Linnaean system, for that system derives from an inappropriate theoretical context. Modern comparative biology requires a taxonomic system based on evolutionary principles.
                                        – de Queiroz and Gauthier (1992, p. 472).

Patterns and processes at the species level

In the previous chapters, we have concentrated on examples of evolution in living species and from the fossil record of the late Cenozoic. Geologically this is a short period of time, but it provides the best possible evidence of the patterns and processes of evolution at the level of populations and species.

All of those observations accord with the theory of natural selection as formulated by Darwin and further elaborated in the modern evolutionary synthesis. There is no evidence from vertebrates that the theory of punctuated equilibrium is necessary or appropriate to explain evolution at the level of genera or species in the modern fauna, nor (to the level that it can be evaluated) in earlier periods of vertebrate history.

What of Eldredge and Gould's other major argument, that patterns of evolution over much longer periods of time imply different factors acting on evolutionary change than those that are observed in modern populations? First, we should re-examine Darwin's hypothesis of the pattern of evolution at the level of populations and species over short periods of time.

Darwin's only illustration in the first edition of *On the Origin of Species* (see Fig. 1.1) was an attempt to diagram the patterns of evolution. That figure is matched in many respects by Greenwood's diagram of the evolution of cichlids in Lake Victoria (Fig. 7.1). The cichlid figure does not show the extinction of any lineages, since there is no fossil record from the lake, but there are a number of divergent and branching lineages and others that exhibit stasis throughout this period. A comparable diagram has not been generated for Lake Tanganyika, but it would show essentially the same pattern except for longer durations of all the lineages.

However, other studies of both living populations and the history of extant species and genera over 1–5 million years show that the patterns of short-term evolution are more diverse than those conceived by Darwin. He recognized that evolutionary rates might differ significantly from one trait to another within a species, and within a single trait over time, but greatly underestimated the scope of those differences. Darwin did illustrate stasis with many vertical lines in his diagram, but he felt that lineages exhibiting very little variability were not likely to persist for long periods of time. He certainly did not postulate groups such as the amphibians and reptiles, within which most species and genera show almost no change in their skeletal anatomy throughout the millions of years of the late Cenozoic; nor did he consider the possibility of extremely rapid trophic diversification as illustrated by the radiation of the cichlids and other species flocks.

As was emphasized in Chapter 1, the patterns of evolution over longer periods depart to an even greater degree from those postulated by Darwin. Most major groups appear to originate and diversify over geologically very short durations, and to persist for much longer periods without major morphological or trophic change. Conspicuous gaps in morphology and ways of life separate major groups, in contrast with the spectrum of intermediate forms that fill Darwin's diagram of both short- and long-term evolution. Eldredge, Gould, and Stanley have argued that the major differences in the patterns of evolution over short and long time scales indicate that different processes were involved (see Gould 1995a and references cited therein). Even if evolution at the level of populations and species is essentially similar throughout the history of vertebrates, it is logical to think that other factors may have been involved at higher taxonomic levels over the much longer periods of geological time.

For example, phenomena such as continental drift and all the environmental changes that it can engender only show their effects over scales of tens to hundreds of millions of years. Mass extinctions have greatly altered the history of life on Earth but can only be observed within a time frame of tens of millions of years. Major changes in developmental patterns probably occur only at extremely widely separated intervals within the history of phyla. Although some modern species and genera have undergone major trophic radiations and significant morphological and behavioral changes, there is no direct way that changes at this level can be extrapolated to the appearance of radically different body plans and ways of life. It is also impossible to conceive of a close analogy among living species of the phenomenon of stasis that has effected the total appearance of major groups for tens to hundreds of millions of years. These are questions that can only be answered through study of the entire fossil history of major lineages.

The remainder of this book is devoted to the question of the degree to which the patterns of large-scale, long-term evolution have been influenced by phenomena other than those that can be recognized or studied at the level of species, and whether this requires a separate theory of macroevolution, as argued by Gould (1995b).

Linnean classification

Before investigating possible differences in physical and biological processes influencing large-scale evolution, we should first consider the degree to which the very way animals are named and classified may affect the way we view their evolution. This is particularly important in light of radical changes in methods of establishing relationships and recognizing taxonomic groups that have been elaborated since Simpson's (1944, 1953) major works on large-scale evolution (Smith and Patterson 1989).

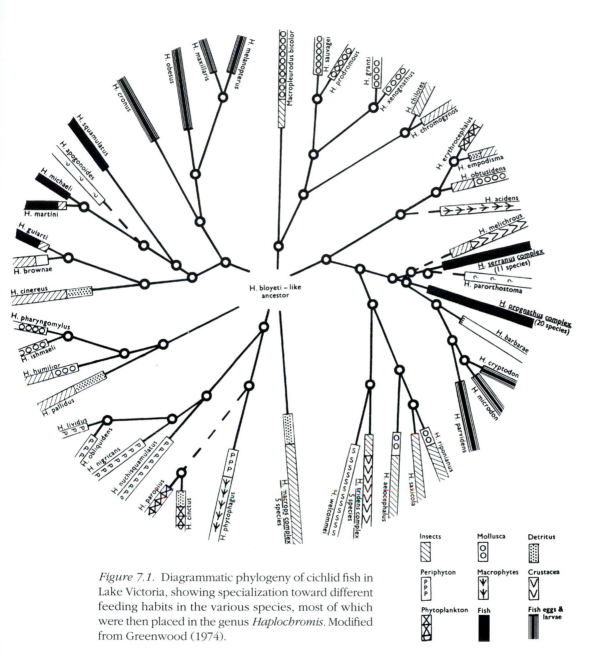

Figure 7.1. Diagrammatic phylogeny of cichlid fish in
Lake Victoria, showing specialization toward different
feeding habits in the various species, most of which
were then placed in the genus *Haplochromis.* Modified
from Greenwood (1974).

The means of determining relationships and classifying groups have not been
discussed until now because nearly all of the animals that have been considered
are either members of living species or genera, or so similar to them that there is
little question regarding their relationships or classification. As we attempt to compare the evolutionary patterns observed in these modern groups with those known
over much longer periods of geological time, establishing relationships and expressing their nature in an appropriate system of classification become much more
difficult.

Until recently, relationships among taxonomic groups have been based almost
entirely on the system of classification established by Linnaeus in the eighteenth

century. This system assumed that all taxa had an essentially unchanging morphology, determined by God. Linnaeus developed a hierarchical system of classification of all organisms based on their relative degree of morphological similarity. The basic unit of classification was the species, which encompassed plants or animals of a single type. The species was conceived as an unchanging entity, created in its present form. Similar species were grouped as members of a genus, similar genera were combined in families, families into orders, orders into classes, classes into phyla, and phyla into kingdoms.

Living species, as recognized by Linnaeus within a particular geographical area, have remained as clearly identifiable taxonomic units, with hybrids between species and significant ranges of morphological difference within species comparatively rare. This is because species have a biological reality, that of being reproductively integrated units, a feature that does not occur at any other taxonomic level. Living species are the only truly nonarbitrary taxonomic units. Unlike the species, the genus has proven to be a more labile taxonomic unit, with species frequently transferred from one genus to another, and new genera proposed to accommodate different species groupings. Many of the genera originally named by Linnaeus have since been elevated to the level of family. Although the family would appear to be as arbitrary a unit as the genus, it and higher taxa such as orders and classes have proven to be much more stable taxonomic units. Among living vertebrates, there has been little change in the concept of the families, orders, and classes until very recently.

The system of classification codified by Linnaeus made it possible for scientists throughout the world to use a single, readily communicated means of referring to every single species within a readily conceived hierarchy of relative similarity. Stability and ease of communication remain very important attributes of Linnean systematics.

Following the publication of Darwin's *Origin of Species,* the concept of classification changed fundamentally from being based on a God-given pattern to being based on evolutionary relationships. However, the hierarchical pattern of overall similarity among living species appeared to reflect evolutionary relationships among modern groups so accurately that there was no immediate need to modify either the content or the methodology of Linnaeus's system of classification. Higher taxa continued to be recognized on the basis of overall similarity, although they were accepted only if they were "natural" in the sense of having a common ancestry. For example, classification of monkeys from South America and Africa within the same taxonomic unit was dropped when it was recognized that the geographical isolation of these continents in the early Cenozoic made it nearly impossible for them to have had an immediate common ancestry.

As knowledge of the fossil record increased in the nineteenth century, extinct species and genera were either placed as distinct taxa within modern groups, as in the case of primitive elephants, sloths, and horses, or recognized as totally distinct forms of plants and animals, such as the seed ferns, ichthyosaurs, and dinosaurs. Until well into the twentieth century, the discovery of fossils did not greatly influence the classification of living organisms – partially because few fossils clearly indicated patterns of relationship that differed significantly from those recognized by Linnaeus, and partially because of the conventions of classification.

The concept of Linnean systematics is primarily typological. Species, whether living or fossil, are based on type specimens, with which other individuals are compared. Among fossils, as long as there is a continuity of morphological variability within a narrow range (about 2 standard deviation units [σ] for quantifiable characters), individuals have been placed within a particular species. For most vertebrate groups throughout the Paleozoic, Mesozoic, and much of the Cenozoic, species frequently are known from only a single stratigraphic horizon, leaving large morphological gaps that distinguish them clearly from other, related species. This was a common factor in the late Cenozoic as well, but with subsequent collecting, fossils from intermediate horizons have frequently been discovered that form a continuum between previously described species. This led initially to the application of a series of species names designating temporal segments of a continually evolving lineage, although there can be no nonarbitrary means of differentiating such successional "species." Robert Martin (1993; see Chapter 5) argued that a single species name should be applied to all such lineages, no matter how great the range of morphological change, if they cannot be subdivided on the basis of speciation events. This practice has the potential for recognizing much greater amounts of change at the species level than has been apparent from the more typological approach, but also demonstrates the incapacity of Linnean systematics to reflect evolutionary change within species.

Clearly, the typological manner of applying species names obscures our view of evolutionary change at the species level, and provides spurious support for the concept of punctuated equilibrium. This is most clearly seen in Stanley's (1979) effort to establish the pattern of evolution at the species level in mammals. He assumed that all animals assigned to a particular fossil species were as similar to one another as members of modern species, and hence that they reflect little evolutionary change over their duration. On this basis he reasoned that so little change was occurring *within* species that any significant change must occur *between* species, that is, during episodes of speciation, in order to account for the major morphological changes that occurred in most of the mammalian orders during the Cenozoic. He failed to consider that the designation of fossils as members of a single species does not preclude a significant spectrum of evolutionary change. The distinction between taxonomic usage and observed differences within and between species of Cenozoic mammals has been clearly demonstrated in studies by Maglio (1973), Krishtalka and Stucky (1985), and MacFadden (1992), and in many papers by Gingerich and his colleagues (Clyde and Gingerich 1994, and references therein).

The typological concept of defining groups continues at higher taxonomic levels. While an individual specimen is the type of a species, the first-named species within a genus is the type species. Subfamily, family, and superfamily names are based on the name of the first genus that is recognized as belonging to that assemblage. Names of orders and classes are not required to be based on names of included taxonomic groups, but they have, nevertheless, been conceived of in a typological manner.

Definitions of modern families and orders were initially based on living representatives, and fossils have been included in or excluded from those groups on the basis of those definitions. For example artiodactyls (cows, pigs, sheep, etc.) and perissodactyls (horses, tapirs, and rhinoceroses) have long been defined on the ba-

sis of the morphology of the ankle bones, and whales by their obvious aquatic specializations. Primitive fossil mammals lacking these specializations were automatically excluded from these orders and placed in separate groups, defined on the
basis of the absence of characters that distinguish their putative descendants. This
practice has been one of the major stumbling blocks to developing an integrated
view of evolutionary relationships. It may be recognized by paleontologists that
particular groups are related to one another, but until recently there was little effort to indicate this in the manner of their classification.

Among the large herbivorous mammals, this is epitomized by an assemblage of
Paleocene and Eocene genera termed the "Condylarthra," named by Cope (1881)
to include possible ancestors of perissodactyls and later expanded to include the
putative ancestors of all ungulates, including the South American hoofed mammals, artiodactyls, perissodactyls, elephants, sirenians, and whales, most of which
appeared in the early Eocene. The modern ungulate groups were all defined by
advanced characters not possessed by condylarths, which implied large morphological and adaptive gaps between the two groups. In fact, it was long recognized
that particular lineages among the condylarths shared features with specific modern orders, although this was not reflected in the classification of either group.

Many other major groups have been defined on the basis of primitive characters,
obscuring evolutionary relationships with their more advanced relatives. The most
primitive of all vertebrates are included in the Agnatha, defined by the absence of
jaws, which characterize all more advanced vertebrates. Amphibians are categorized as primitive land vertebrates that lack reproductive features common to reptiles, birds, and mammals. Reptiles have been defined on the basis of the absence
of advanced features of birds and mammals. All of these primitive assemblages are
assumed to include the ancestors of two or more descendant groups, but adher-
ence to the typological Linnean system of classification precludes designation of
specific relationships.

Phylogenetic systematics

It is only during the past thirty years that a new method of establishing relationships and naming taxonomic groups has begun to clarify these problems. Surprisingly, the need for a new system was initially recognized not by paleontologists but
by an entomologist, Willi Hennig (1950, 1966, 1981). Hennig developed a method
of determining relationships that depended explicitly on evolutionary principles.
The major tenets of his methodology were as follows:

1. Use derived characters rather than overall similarity in establishing relationships.
2. Reject taxonomic groups that include some but not all descendants of a
   common ancestral species.
3. Insist on a system of classification that directly reflects evolutionary relationships.

Hennig's most important contribution was to stress the use of derived, or advanced, characters in establishing relationships, as opposed to overall similarity. He

reasoned that the unique presence of derived characters, or **apomorphies**, demonstrates that the groups possessing them had evolved from an immediate common ancestor in which these characters had first appeared. In contrast, the common presence of other similarities of a primitive nature, or **symplesiomorphies**, cannot be used to establish specific relationships. For example, many of the characteristics of lizards and snakes are common to other living groups customarily classified as reptiles, and in fact to all vertebrates. On the other hand, they possess one character complex, a paired eversible penis, which is not possessed by *Sphenodon,* turtles, or crocodiles – or, for that matter, any other vertebrates. The common presence of this derived feature can be most logically explained by its origin in an immediate common ancestor that gave rise to lizards and snakes but to no other reptilian group.

Hennig used the expression **sister-groups** to refer to taxa whose derivation from a common ancestor can be demonstrated by the presence of one or more shared derived characters. Sister-groups may be recognized at all taxonomic levels: Two species or two orders may be sister-groups, or a single species may be the sister-group of all other members of an order or class. To emphasize the paramount importance of shared derived characters in establishing relationships, he coined a new term for them: **synapomorphies**. For instance, a paired eversible penis is a synapomorphy supporting the sister-group relationship of lizards and snakes, and demonstrating their phylogenetic divergence from all other reptiles.

Autapomorphies, characters unique to particular taxa, are the basis of distinguishing one taxon from another. Note that the terms *synapomorphy* and *autapomorphy* are not specific to different characters but may be applied to the same character at different levels of inclusiveness: An eversible penis is a synapomorphy that unites lizards and snakes in a common monophyletic group, the Squamata, but is an autapomorphy of the Squamata within a larger monophyletic assemblage.

Because of the clearly objective and evolutionary basis of this methodology, the use of shared derived characters in establishing relationships is now practiced routinely by nearly all taxonomists. Details of the application of this methodology are covered in many recent books (Wiley 1981; Ax 1987; Forey et al. 1992; A. B. Smith 1994).

Hennig referred to the explicit use of evolutionary relationships in taxonomic procedures as **phylogenetic systematics**, but this methodology is frequently termed **cladistics** because of the emphasis placed on the recognition of monophyletic groups or clades. In contrast with the use of overall similarity in establishing relationships, the use of a few, uniquely derived characters enables biologists to define the origin of a group or clade very specifically. This is especially important to biologists in determining patterns of evolution.

Monophyly and paraphyly

Although earlier taxonomists urged that all groups should have a monophyletic origin, that origin might be extremely broad, as exemplified by Simpson's definition (1961, p. 124): "Monophyly is the derivation of a taxon through one or more lineages from one immediately ancestral taxon of the same or lower rank." That is, the mammals would be accepted as monophyletic if they evolved from some-where among the reptiles, as then defined, or from any order of the Class Reptilia, although this might include origin from different infraorders. In fact, it was suggest-ed by Simpson (1959) and others that mammals might have arisen from two major subdivisions of the synapsid reptiles. Olson (1947) similarly suggested that reptiles might have arisen from two major groups of amphibians. One of the reasons for such a relaxed view of monophyly as late as the 1960s was the incomplete knowl-edge of the fossil record. There were many groups for which their origin was plau-sibly sought among another large assemblage, although no individual species or genus could be recognized as the ancestor.

Hennig and his followers have insisted that the origin of a group should be nar-rowed to a single species if possible. However, such a species would be viewed not as an ancestor of the group but as its most primitive (i.e., **plesiomorphic**) member. Cladistic methodology differs from the traditional practice of paleontolo-gists in not seeking ancestors. If all relationships are established by the recognition of shared derived features, ancestors cannot be recognized as such because they lack derived traits that are otherwise thought to characterize the group in question. They could, however, be recognized as belonging to the sister-group of the clade in question if they share synapomorphies with it. For example, the osteolepiform fish have long been suggested as ancestral to the tetrapods. No specific genus has been considered an ideal ancestor, but both *Eusthenopteron* and *Panderichthyes* share an important synapomorphy with tetrapods: an internal naris. Either genus might be a sister-group of more terrestrial choanates, such as the Devonian am-phibians. If osteolepiforms possessed autapomorphies not expressed in tetra-pods, they would be considered not ancestors of the tetrapods but a sister-group. If, on the other hand, all the synapomorphies that unite the osteolepiforms as a clade were also possessed by early tetrapods, tetrapods would be included in the same clade as the osteolepiforms.

This brings us to the second tenet of phylogenetic systematics: the rejection of taxonomic groups that include some but not all descendants of a common ances-tral species. Hennig (1966, p. 73) redefined the term **monophyletic group** as "a group of species descended from a single species, and which includes all species descended from this stem species." He coined a new term, **paraphyletic**, to ap-ply to groups, such as the Condylarthra as it was commonly constituted, that had a monophyletic origin but do not include all of its descendants.

Rejection of paraphyletic assemblages makes a striking difference in the way we view the evolution of the condylarths and the advanced ungulate orders (Fig. 7.2). If we concentrate on recognizing shared derived characters, as opposed to using overall similarity in assigning particular fossils to larger taxonomic groups, many of the genera and families that had been assigned to the Condylarthra can be placed in orders with living representatives. Whereas nearly all of the ungulate orders, as previously defined, appeared as clearly distinct groups at the base of the Eocene, sister-groups of several of these orders are known as early as the lower Paleocene, some 9 million years earlier. These include the Hyopsodontidae and Periptychidae, either of which may be the sister-group of some or all of the South American ungu-lates, and the triisodontine mesonychids, which are the sister-group of the whales. Two upper Paleocene genera, *Minchenella* and *Radinskya,* are sister-groups of an

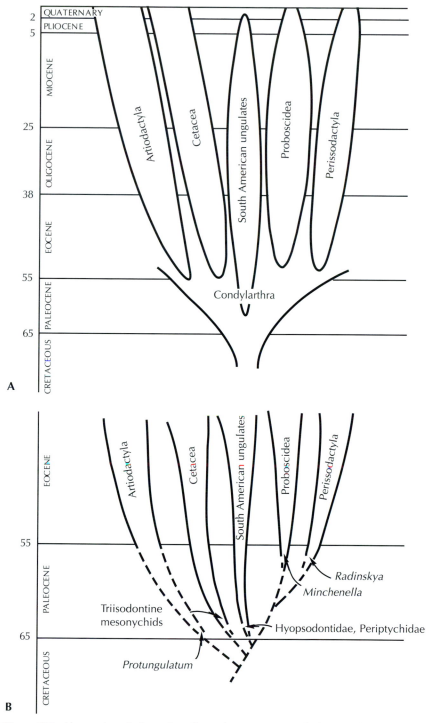

Figure 7.2. Alternative phylogenies of ungulate groups. **A,** Phylogeny showing the customary range of the major ungulate groups based on the achievement of new structural or adaptational features. **B,** Phylogeny based on recognized affinities with particular taxa typically classified as condylarths. Data from Prothero et al. (1988), Prothero (1994), and Janis et al. (in press-a).

assemblage including the Tethytheria and the perissodactyls. The Tethytheria includes the sirenians and proboscidians of the modern fauna, as well as two extinct orders, and the perissodactyls may include the hyracoids (Prothero, Manning, and Fischer 1988; Prothero 1994; Janis et al. in press-a,b). The one major group of ungulates for which a specific Paleocene sister-group has not been recognized is the Artiodactyla. Features of their dentition are more primitive than are those of any other ungulates, including early Paleocene genera such as *Protungulatum*. This indicates that artiodactyls must have diverged from the remaining ungulates at the very base of this radiation. There is no way of determining exactly when this was, although it must have been at least slightly before the appearance of the earliest known fossil record of any of the other orders.

Recognition of monophyletic clades as defined by Hennig gives an entirely different view of the origin of the ungulate orders. Instead of appearing clearly defined in the early Eocene, they diverged from one another over much of the Paleocene from a series of initially very similar lineages showing only a few recognizable features of their descendants. Whereas the origin of the modern orders had appeared to be the result of a major anatomical change – and/or a dramatic shift in habitat, as in the case of whales – in fact, the most primitive members of each order were all terrestrial animals that differed relatively little from their immediate common ancestors. Emphasis on monophyletic clades results in the distinction between ungulate orders at the time of their phylogenetic divergence from one another rather than at the time of major morphological and/or habitat changes within clades. This provides a close parallel with Robert Martin's suggestion that series of successional species be included in a single species if they cannot be separated by speciation events. At both taxonomic levels, much more change is recognized *within* groups, and much less is attributed to divergence *between* groups.

With the inclusion of their "condylarth" sister-groups, the fossil record of many of the ungulate orders can be traced back to the base of the Paleocene. Novacek's illustration (Fig. 7.3) further extends the range of these orders back into the Upper Cretaceous through what Norell (1993) refers to as **ghost lineages**. These are extrapolations of the temporal range of one of a pair of sister-groups that must have originated at the same time, although they may not be equally represented in the fossil record.

Although the concept of what constituted the origin of each of these orders must change, and the time of origin is extended several million years further back into the Late Cretaceous, the broad pattern of evolution within most of the orders remains similar to that previously suggested. Most diverged from one another during a relatively short time span during the Late Cretaceous and Early Tertiary. The distinct anatomy of the modern orders had become established by the early Cenozoic, and most retained their basic structural pattern from that time on.

The inclusion of the condylarth groups makes no difference in the classification of the modern ungulates. The ordinal name for archaic ungulates, Condylarthra, may either be abandoned if all species that had been assigned to this group can be related to one or the other living orders, or retained if some of the original species are found to belong to a separate monophyletic assemblage. Much greater changes are expected to take place in the classification and naming of other vertebrate groups as the ever-increasing fossil evidence is treated according to the methods of phylogenetic systematics.

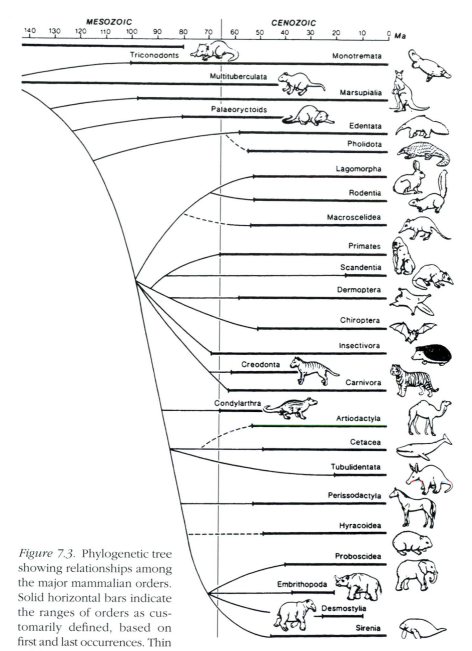

Figure 7.3. Phylogenetic tree showing relationships among the major mammalian orders. Solid horizontal bars indicate the ranges of orders as customarily defined, based on first and last occurrences. Thin solid lines indicate the branching sequence based on the relationship of clades and their known fossil record. Dashed lines indicate more ambiguous relationships. From Novacek (1994).

Monophyletic groups of vertebrates

The customary classification of vertebrates based on the methodology of Linnean systematics was shown in Figure 1.5. It is simplified in Figure 7.4 to emphasize the major groups. The cladogram in Figure 7.5 shows a nested hierarchy of monophyletic clades according to the methods of phylogenetic systematics. Figure 7.6 shows these clades diagramed as an evolutionary tree to provide more direct comparison with the Linnean scheme. Figures 7.4 and 7.6 differ both because of different approaches to classification and because of a great deal of new information from the fossil record.

When relatively few fossils were known, there appeared to be very large gaps between major taxonomic groups and most structural and adaptive changes, such as those occurring during the origin of mammals from reptiles and the origin of tetrapods from fish, were discussed in terms of transitions between higher taxonomic ranks. The emphasis on higher taxonomic ranks in the origin of major adaptive assemblages persisted in the writings of Simpson into the 1960s. As the fossil record improved, much more continuous sequences of morphological change could be established across major transitions, such as those between primitive amniotes and mammals, between dinosaurs and birds, and in the origin of many of the orders of Cenozoic mammals. At the same time, increasingly wide acceptance of phylogenetic systematics emphasized the need to modify the way major taxonomic groups were recognized and defined. This has led to proposals for significant changes in the way in which the major groups of vertebrates should be classified. Increased knowledge of the fossil record and changes in the manner of classification both influence how the patterns and processes of vertebrate evolution are conceived.

There are still many controversies regarding how major groups of vertebrates should be classified and named. For some taxa, the fossil record is not sufficiently

Figure 7.4. Simplified phylogeny of vertebrates emphasizing groups with living representatives, and showing the eight classes recognized in Linnean classification.

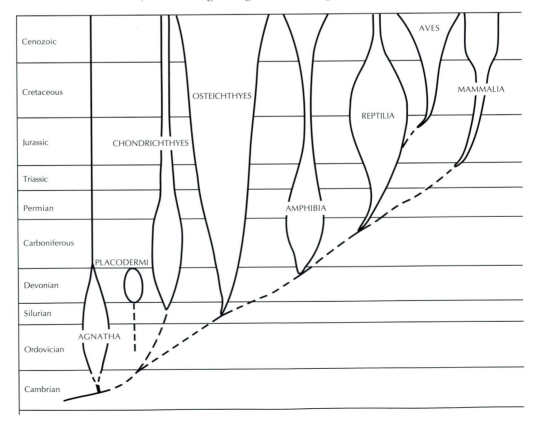

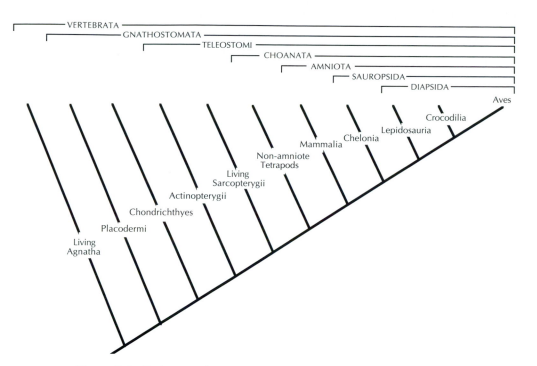

Figure 7.5. Cladogram showing a nested series of monophyletic groups among the vertebrates. Emphasis is placed on sister-group relationships of groups with living representatives.

Figure 7.6. Simplified phylogeny of vertebrates emphasizing groups with living representatives. Except for the Amphibia and Dinosauria, all named groups are monophyletic in the sense of Hennig.

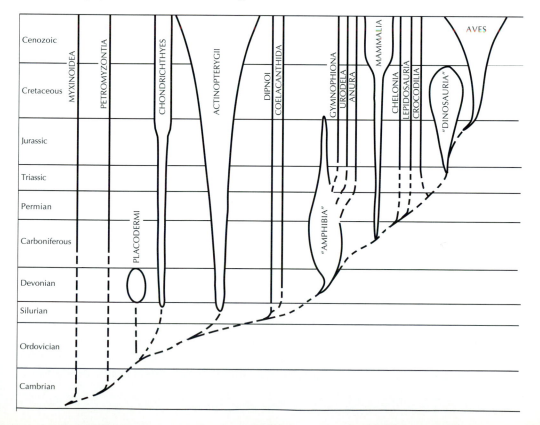

well known to determine specific relationships, but there is strong evidence that previously recognized groups are paraphyletic and should be broken up into monophyletic clades and named accordingly. In other cases, knowledge of the fossil record clearly demonstrates the need for specific changes in the way previously recognized groups were classified, but these require such major modifications in the names and concepts of long-accepted groups that an acceptable manner of classification has not yet been devised. These problems affect the highest taxonomic level, the vertebrate classes.

The eight customarily accepted vertebrate classes were originally established on the basis of clearly distinct anatomical patterns, with few or no intermediates known from the fossil record. However, since the major groups are represented as having evolved from one another, several of the previously accepted classes must be paraphyletic. Beginning with the most primitive known vertebrates, the Class Agnatha was originally distinguished by the absence of jaws. This character was initially recognized in the living lamprey and hagfish and later discovered to apply to a diverse assemblage of Paleozoic fish collectively referred to as ostracoderms because of their extensive dermal ossification. The term Agnatha is commonly applied to both the living and fossil forms. However, the absence of jaws must have been a characteristic of *all* primitive vertebrates – not just the ancestors of the living jawless fish but also the ancestors of all fish with jaws. There must have been an assemblage of primitive jawless fish that included the ancestors of both the living agnathans and the gnathostomes. Such an assemblage is paraphyletic by definition, but it must be capable of subdivision into lineages that were more closely related to one or the other of the subsequent groups. Nevertheless, it has not yet been determined whether the known Paleozoic agnathans are (1) part of a monophyletic group including all of the modern agnathans, (2) part of a monophyletic group including the lampreys but not the hagfish, or (3) a paraphyletic assemblage including some lineages that are more closely related to the jawed vertebrates.

If Schaeffer and Thomson (1980) are correct in their recognition of features of the jaws and gill arches as uniting all known Paleozoic and living agnathans into a single monophyletic clade, and other features as uniting all jawed vertebrates into a separate clade, no vertebrates are known that could be the common ancestor of both major groups. In this case, all known jawless vertebrates do belong to a single monophyletic assemblage; the class designation Agnatha can be retained while accepting that the absence of jaws, being a primitive character, cannot be used to define the group. On the other hand, if Forey and Janvier (1993) are correct in their hypothesis that some Paleozoic groups long accepted as agnathans are the sistergroups of jawed vertebrates, the living lampreys become the living sister-group of jawed vertebrates; the hagfish become the only living representative of a separate lineage that is the sister-group of all other vertebrates. These alternatives remain unresolved, and so the recognition and naming of the groups remain uncertain.

The exact relationship of placoderms (archaic jawed fish of the Devonian) to other jawed fish remains controversial, but their recognition as a monophyletic assemblage has not been challenged since there is no evidence that they gave rise to any subsequent group. Chondrichthyes (sharks, skates, rays, and chimaeroids) also retain their monophyletic status. The Class Osteichthyes (bony fish), as customarily defined, is paraphyletic because it specifically excludes a major descendant group: the terrestrial vertebrates. The more inclusive term Teleostomi may be used to include both the osteichthyian fish and the tetrapods. Osteichthyian fish were divided into two major sister-groups, the actinopterygians (ray-finned fish) and sarcopterygians (lobe-finned fish). The actinopterygians appear to be a monophyletic clade, but the sarcopterygians are but part of the assemblage including terrestrial vertebrates. The sarcopterygians are divisible into two sister-groups, one including the extinct porolepiforms as well as the living coelacanths and lungfish, and the other including the Paleozoic osteolepiformes and the tetrapods. The latter assemblage has been termed the Choanata.

The Amphibia, like the Agnatha, is conceptionally a primitive assemblage, including the ancestors of two or more derived groups. Amphibians were long considered as bridging the gap between fish on one hand and reptiles, birds, and mammals on the other. When primitive Paleozoic tetrapods were first discovered, they were classified among the amphibians, although it has long been assumed that they must also include the ancestors of reptiles. Paleozoic tetrapods must constitute a paraphyletic assemblage, but it is still not possible to determine their interrelationships or the specific affinities of particular lineages to the modern amphibians or the ancestry of reptiles (Carroll 1992, 1995). It is confidently established that the lineage leading to amniotes, including living reptiles, birds, and mammals, diverged early in the Carboniferous or perhaps even in the Devonian (Lebedev and Coates 1995), but the specific interrelationships of the modern amphibian groups are not adequately documented despite the apparent consensus of a common lissamphibian ancestry (Milner 1993). The three modern orders may have diverged from one another late in the Paleozoic from a single common ancestor, or their ancestral lineages may have separated early in the Carboniferous or possibly the Late Devonian. Most clades of Paleozoic tetrapods cannot be confidently related either to the lineage leading to amniotes, or to the one or more lineages leading to the modern amphibian orders (Carroll 1995). Until such relationships are established, it is premature to assign members of the Paleozoic amphibian assemblage to any higher taxonomic unit, and unjustified to place the three modern amphibian orders within a single higher category, distinct from that of the Paleozoic assemblage.

The situation in regard to the next long-accepted category of vertebrates, the Class Reptilia, differs in that phylogenetic relationships are now confidently established but require such major changes in customary taxonomic usage that an acceptable solution has not yet been achieved. The concept of reptiles originated with the modern lizards, snakes, crocodiles, turtles, and the tuatara, *Sphenodon*. Fossils of the modern groups were later included, as were a wide range of Mesozoic orders: dinosaurs, pterosaurs, and a number of aquatic lineages. The Paleozoic ancestors of these groups were subsequently added, as were members of a divergent lineage, the pelycosaurs and therapsids (Subclass Synapsida). It was later recognized that the pelycosaurs and therapsids were associated with the origin of mammals (Hopson 1994), and more recently theropod dinosaurs have been recognized as the sister-group of birds (Ostrom 1994).

All these groups had been accepted as reptiles on the basis of similar anatomical features and fossil evidence of their having a common ancestry, but as constituted, reptiles are clearly a paraphyletic assemblage. They do not include two descendant groups, the birds and mammals. A simplified phylogeny of this assemblage (Fig. 7.7) shows that crocodiles share a more recent common ancestry with birds than they do with any of the other living groups customarily included within the Reptilia. No nonarbitrary division is possible between the synapsid "reptiles" and the immediate ancestors of the major groups of living mammals, or between theropod dinosaurs and birds.

On the other hand, the entire assemblage of reptiles, birds, and mammals is clearly monophyletic. It can be differentiated from all other terrestrial vertebrates by the presence of a number of skeletal features common to the earliest fossils of all the major groups (Carroll 1991; Smithson et al. 1994), as well as the presence of extraembryonic membranes in all the living taxa that justifies the name Amniota for the entire assemblage. No descendant groups are omitted from the Amniota. The Amniota may be divided into two major sister-groups, one lineage leading to and including mammals, and the other including the classes customarily termed reptiles and birds, as well as all their extinct relatives.

We should no longer think of the groups customarily referred to as mammals and birds as having evolved from reptiles, with the assumption that a major change in anatomy and physiology has occurred in those transitions. Rather, it should be recognized that the assemblage including mammals diverged from other amniotes well back in the Carboniferous, and that the earliest members of each clade were no more different from one another than were different species. Similarly, the lineage leading to modern birds diverged from genera customarily classified as dinosaurs sometime in the early Mesozoic. At its inception, the lineage leading to living birds may not have had any of the morphological characters that differentiate their living descendants.

All of the factors that distinguish living mammals and birds arose *within* the monophyletic clades to which they belong, rather than during transitions between them and a clearly distinct ancestral taxon. The use of phylogenetic systematics to recognize monophyletic groups thus casts an entirely new light on the origin of these higher taxonomic groups.

Naming and defining clades and included groups

Most vertebrate paleontologists would agree with the general pattern of relationships illustrated in Figure 7.6 and with the evolutionary concepts arising from the emphasis on these monophyletic groups. On the other hand, the consequences of accepting these relationships may very much alter the formal classification and naming of major vertebrate groups. In order to demonstrate the evolutionary consequences of phylogenetic systematics, Hennig and his followers have argued that the systems of classifying and naming groups should also explicitly reflect their relationships. Acceptance of Hennig's arguments for the rejection of paraphyletic groups would necessitate the rejection of four of the eight vertebrate classes – the Agnatha, Osteichthyes, Amphibia, and Reptilia – as they are presently conceived.

Beyond altering the names and the current concepts of particular taxonomic groups, there is a fundamental difference as to how taxonomic names should be applied according to phylogenetic systematics. According to Linnean systematics, the long-recognized classes of vertebrates were considered as a series of essentially

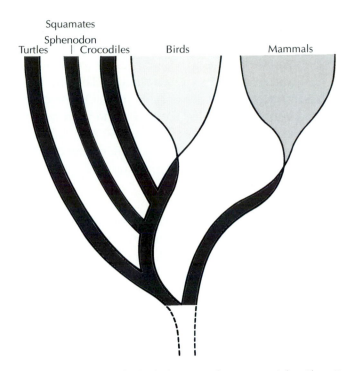

Figure 7.7. Simplified phylogeny of amniotes. The Class Reptilia, as customarily defined, includes all the darkly stippled lineages. Reptiles have a common ancestry but are considered paraphyletic since they do not include all their descendants. Birds and mammals can each be defined on the basis of unique, shared derived characters. Reptiles have been defined as amniotes that lack the specialized characters of birds and mammals. From Carroll (1987).

equivalent categories, and hence have the same taxonomic rank. If all recognized groups must be monophyletic, in the Hennigian sense of including all their descendants, successive subgroups within a larger monophyletic assemblage such as the Vertebrata are nested within one another like a series of Russian *matrioshka* dolls. The mammals and other amniotes are nested within the Amniota, the Amniota are nested within the Choanata, the Choanata within the Teleostomi, and the Teleostomi within the Gnathostomata. Each successive taxon in this series should thus be given a higher taxonomic rank. The Chondrichthyes would have a rank not equal to that of Mammalia but much higher.

Hennig reasoned that each successive dichotomous branching should be recognized by a successive taxonomic rank. Initially, this led to the proliferation of a great number of additional taxonomic ranks; subsequently, it has been suggested that the Linnean system of hierarchical ranking of taxa should not be used at all since it is both impractical and arbitrary (de Queiroz and Gauthier 1992, 1994). In practice, many taxonomists, even those fully committed to other aspects of phylogenetic systematics, have retained the Linnean system of taxonomic ranks because it still serves effectively in transmitting information regarding individual taxa. The recent compendium of the longevity of groups within all major taxa, *The Fossil Record 2* (Benton 1993), relied throughout on the duration of a particular rank, the family. Novacek's (1994) recent review of the radiation of placental mammals retained the customary series of orders with no effort to nest them in a series of subordinate ranks. In both of these cases, there is still only limited information regarding the specific interrelationships of these taxa within larger monophyletic groups, which precludes the recognition of series of nested taxa. This, in fact, applies to many taxonomic levels in most groups at the present time. On the other hand, authors writing on particular vertebrate groups for which interrelationships are better known were able to nest them in a hierarchical series, and did so without use of taxonomic ranks (Prothero and Schoch 1994). At present, the application of the principles of phylogenetic systematics to the names and ranks of taxa is in a state of flux (Minelli 1993).

While recognizing the logic of Hennig's arguments for making classification directly reflect relationships, our knowledge of many groups is too limited to apply this procedure in a consistent manner. In this time of transition, communication is frequently better served by retaining Linnean ranks and names unless they are very clearly misleading.

In addition to the need to apply new names and concepts to previously recognized groups, a problem unique to phylogenetic systematics is the difficulty of differentiating and designating subgroups within monophyletic assemblages. It is easy to recognize two major sister-groups among the Amniota, but how is one to divide up the large assemblages that include the birds and mammals? One may recognize living birds and all their fossil relatives as members of a monophyletic group on the basis of shared derived characters, but there is no formal way to recognize or name the more primitive taxa within the larger monophyletic assemblage that are distinguished from birds only by primitive characters such as the absence of feathers. If birds are descended from dinosaurs, paleontologists have suggested either that birds be classified with dinosaurs, or that dinosaurs be considered as nonflying birds. Similarly, mammals have long been recognized by derived features of the jaw, dentition, and middle ear, but how is one to handle the more primitive members of the same monophyletic assemblage? One could apply the term Mammalia to the entire assemblage, but this would appear very misleading to persons who study living mammals, which certainly differ fundamentally from their Paleozoic ancestors in their anatomy and physiology.

One way to handle this problem is to recognize two subdivisions within large monophyletic groups: the **crown group**, including all taxa with living representatives, and the **stem group**, which includes more primitive members of the same monophyletic assemblage (Fig. 7.8). De Queiroz (1994) argues that the most appropriate way to differentiate subgroups within a monophyletic assemblage is by a phylogenetic definition based on their relationship with other monophyletic groups (Fig. 7.9). He thus defines the Mammalia as "the clade stemming from the most recent common ancestor of monotremes and therians." This is referred to as a **node-based definition**, since the content of the group is defined by its membership in the two lineages that separate at a particular node in the cladogram. Paleozoic and early Mesozoic synapsids and many Mesozoic species formerly classified as mammals would be considered members of the stem group. The term Synapsida, long used by paleontologists for the lineage leading to mammals, could be applied to the larger monophyletic group including both the Mammalia and their stem group.

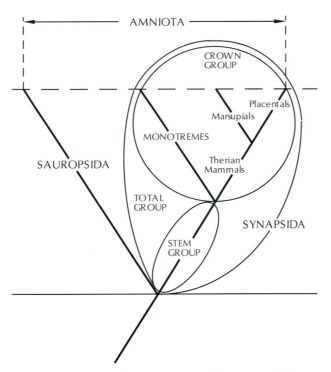

Figure 7.8. Stem and crown groups. A large monophyletic group such as the Synapsida may be divided into a crown group, including all the living representatives and their immediate common ancestors, and a stem group including all other members of the total group.

The term Sauropsida has long been used by paleontologists in reference to the large assemblage including modern reptiles and the ancestors of birds. This could be applied to the entire sister-group of the Synapsida. The first major lineage to diverge from the Sauropsida was that leading to the turtles (Chelonia or Testudinata). This lineage is primitive in retaining the solidly roofed, or anapsid, condition of the skull of all early tetrapods. Other sauropsids are derived in the presence of two pairs of openings in the temporal region, the diapsid condition. The derivatives of the diapsid assemblage include two clades with modern representatives: the lepidosaurs (including *Sphenodon*, lizards, and snakes) and the archosaurs (including crocodiles, dinosaurs, and birds).

By analogy with the node-based definition of Mammalia, the Class Aves may be restricted to living birds and other members of the clade stemming from the most recent common ancestor of the modern orders. This, however, would exclude *Archaeopteryx* and most other Mesozoic birds from the Aves. Alternatively, one might define the crown group Aves as including the clades stemming from the most recent common ancestor of modern bird orders and *Archaeopteryx*, thus retaining all currently known feathered birds within the Aves. Other alternatives would be to define Aves as including all species that share a more recent common ancestry with modern birds than they do with nontheropod dinosaurs, or with crocodiles. These latter two ways of distinguishing Aves are termed **stem-based definitions**; a further way to distinguish crown taxa is via **apomorphy-based definitions** (Fig. 7.9A,C).

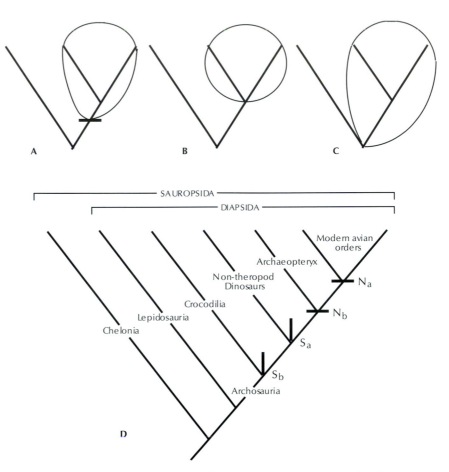

Figure 7.9. Three possible ways of defining taxonomic groups within larger, mono-phyletic assemblages. **A,** Apomorphy-based definition, defining the taxon as the clade derived from the first ancestor to possess a particular apomorphy. For example, the an-cestral birds may be recognized on the basis of the first occurrence of feathers. **B,** Node-based definition, defining the name of a taxon as the clade stemming from the most recent common ancestor of two other taxa. For example, the Mammalia may be defined as the clade stemming from the most recent common ancestor of monotremes and therians. **C,** Stem-based definition, defining the name of a taxon as all those species sharing a more recent common ancestor with one recognized taxon than with another. For example, defining birds as including all species that share a more recent common ancestry with modern birds than they do with nontheropod dinosaurs. Redrawn from de Queiroz and Gauthier (1990). **D,** Phylogeny of the Sauropsida, illustrating various ways in which the crown group Aves might be defined. N_a, N_b, two possible node-based definitions; S_a, S_b, two possible stem-based definitions.

These alternative ways of defining Aves demonstrate that although phyloge-netic systematics is nonarbitrary in the use of dichotomies to distinguish clades, the choice of which dichotomy to use is arbitrary. Actual phylogenies are almost always much more complicated than those shown in these figures, with many mi-nor lineages diverging from a main stem. Any of these might be used as a point of division within a larger clade.

Whatever sauropsids are not included in Aves become the stem group of the larger clade. Unfortunately, there is no single name that can be applied to the three modern lineages that have been classified as reptiles, since they do not have a common ancestry separate from the lineage leading to modern birds. Rather, the Chelonia, the Lepidosauria (including *Sphenodon,* lizards, and snakes), and the Crocodylia are all crown groups, each of which may be defined at the point of divergence from their more derived sister-group. For example, the Chelonia would be defined as all species sharing a more common ancestor with living turtles than they do with the ancestors of lepidosaurs, crocodiles, and birds.

The impact of phylogenetic systematics on the study of evolution

The application of phylogenetic systematics has resulted in extending the duration of many lineages through the more accurate determination of the content of monophyletic groups. In many cases, the inclusion of earlier members within a lineage demonstrates that the achievement of the definitive characters of the group occurred over a longer period of time than had been assumed, changing our perception of the actual pattern of evolution.

Of equal importance, the insistence that classification directly reflects evolutionary relationships alters the way that we conceive of the origin of major taxa. We have long been concerned with transitions between major assemblages distinguished by significant differences in their anatomy, physiology, and/or way of life: for example, the fish–amphibian transition, the amphibian–reptilian transition, and the reptilian–mammalian transition. These have been held to correspond with the origin of new taxonomic groups, which has frequently been interpreted as implying the existence of special evolutionary processes. Extensive study of evolutionary transitions, whether at the level of species or classes, demonstrates that it is impossible to establish nonarbitrary points of division within evolving lineages. For this reason, it is very difficult to argue that customary points of division between fish and amphibians, amphibians and reptiles, and reptiles and mammals necessarily result from a distinct category of biological processes.

Emphasis on the recognition of strictly monophyletic groups demonstrates that major clades, like species, are separated on the basis of nothing more dramatic than speciation events. There is no reason to think that the phenomenon of speciation differs in accordance with the subsequent history of the descendants of the daughter species. The origin of an individual species that soon becomes extinct is no different than the origin of a species that survives to give rise to thousands or tens of thousands of species that may dominate Earth for millions of years. Hence, the divergence of all clades is essentially equivalent, no matter what taxonomic rank is given to them. The focus of the study of macroevolutionary phenomena then shifts to survival and change within clades. The basic aspect of cladogenesis, speciation, goes on continuously within all more inclusive clades. What differentiates larger-scale evolutionary phenomena is the relative survival and degree of change within clades. For example, we have seen that riverine cichlids show the capacity for extensive speciation but did not give rise to a great variety of trophic levels. What distinguishes the cichlids in the East Africa Great Lakes is the persistence of species- and generic-level clades and their structural and trophic specialization for periods of up to 5 million years. This forms a plausible model for much longer-term and larger-scale radiations, such as that of the Late Cretaceous and Cenozoic mammals.

Summary

The processes of evolution at the level of populations and species postulated by Darwin provide a sufficient explanation for the patterns of evolution observed in modern populations and from the fossil record throughout the past 5 million years. The patterns of evolution during this time do differ from those hypothesized by Darwin in the greater range of evolutionary rates, the predominance of stasis in some groups, and the great amount of trophic radiation in others. Phenomena not considered by Darwin, such as continental drift and mass extinction, may be responsible for patterns of evolution much different than those he postulated as occurring over long periods of time.

The patterns of evolution over long periods must be reevaluated in light of the methodology of phylogenetic systematics, which provides an objective means of establishing relationships based explicitly on evolutionary principles. Strict definition of monophyly specifies the point of origin of clades, which may be significantly earlier than the time of morphological change previously accepted as indicative of the origin of major groups. The practice of phylogenetic systematics offers an objective way to compare patterns of large-scale evolution from group to group and within groups over time, through a consistent way of determining the time of clade origination.

The pattern of relationships established by phylogenetic systematics substantiates broad aspects of the pattern of long-term evolution established through study of the fossil record. Within many groups, major episodes of cladogenesis are limited to relatively short lengths of time, after which the individual clades persist for long stretches with relatively little morphological or trophic change. This pattern certainly requires explanations beyond those possible by extrapolation from the pattern of evolution common to most living species and genera.

8 Evolutionary constraints

Introduction

Although the pattern of nearly continuous change seen in late Cenozoic mammals and cichlids resembles that hypothesized by Darwin for all evolutionary time scales, the pattern of vertebrate evolution over tens and hundreds of millions of years differs significantly in the long-term retention of particular body forms and ways of life. Instead of new families, orders, and classes evolving from one another over long periods of time, most had attained their most distinctive characteristics when they first appeared in the fossil record and have retained this basic pattern for the remainder of their duration.

The temporal range of families provides a broad basis for examining the persistence of morphological patterns. These data have been compiled in Benton's *The Fossil Record 2* (1993) by persons knowledgeable in all the major groups of organisms, according to both morphological and phylogenetic criteria. Examples of family longevity within various vertebrate groups are given in Table 8.1. In order

Table 8.1. *Average longevity of living families of vertebrates*

Group	Average family longevity (10^6 yr)	Group	Average family longevity (10^6 yr)
Aquatic vertebrates		*Amphibians and reptiles*	
Squalomorphii	108.6	Amphibians	68
Batomorphii	73.6	Chelonia	75.6
Chimaeriformi	107	Sphenodontidae	235
Basal Actinopterygii	103	Lacertilia	72.3
Teleosts	45.3	Serpentes	40.8
Sarcopterygians	177.3	Crocodilia	69.7
Mammals			
Xenarthra	35	Chiroptera	25.5
Lagomorpha	38.5	Artiodactyla	28.8
Rodentia	26.3	Cetacea	16.8
Carnivora	27.8	Perissodactyla	44.7
Insectivora	30.5		

Source: Data from Benton (1993).

to maximize the completeness of the fossil evidence and minimize the possibility of paraphyletic groups, only families with modern representatives have been tabulated, although we cannot determine how much longer they may survive. Nevertheless, nearly all have maintained a recognizably similar morphology for tens to hundreds of millions of years.

The oldest vertebrate family with living representatives is the lungfish lineage Neoceratodontidae, which is recognizable from deposits 245 million years old. The Latimeriidae go back 204 million years. The average duration of seventeen modern shark families extends more than 128 million years. Batomorphii (skates and rays) average 73.6 million years, and living chimaeriforms 107 million. The few surviving families of basal ray-finned fish average more than 100 million years' duration; the average longevity of 250 modern teleost families with fossil records is approximately 45 million years. Among tetrapods, the oldest recognized family is the Sphenodontidae, which has endured for 235 million years. The modern amphibian, turtle, lizard, and crocodilian families for which there is a fossil record average about 70 million years' duration.

The average longevity for families in selected mammalian orders ranges from 16.8 to 44.7 million years. Birds have a much less informative fossil record, but Feduccia (1995) argues that most of the modern families, aside from the passeriforms, probably date from the late Eocene or early Oligocene, indicating an average longevity of about 35 million years. The fossil record demonstrates that many modern avian genera go back to the mid-Miocene, about 15 million years.

As at the level of species, there is great variation in the amount of change both within and between families. On the other hand, the relatively small number of families that are recognized on the basis of either morphological or phylogenetic criteria shows that the appearance of new morphological patterns distinguished at the family level is a comparatively rare phenomenon. Of equal importance is the great relative stability of these patterns once they are established. There was certainly much less change during most of the duration of the majority of vertebrate families than was implied by Darwin. This requires an explanation that is not provided by the evolutionary synthesis.

Within families, individual taxa, termed **living fossils** since the time of Darwin, further emphasize the retention of particular anatomical forms and ways of life for extremely long periods. *Lingula* and *Limulus* (Fisher 1984) illustrate particularly conspicuous examples among the invertebrates that have persisted with almost no detectable change since the Paleozoic. The vertebrates *Latimeria* and *Sphenodon* are excellent examples of modern genera that resemble quite closely animals known from the fossil record of more than 200 mya. Members of the Latimeriidae have changed their habitat from shallow to deep marine waters, but their superficial morphology remains remarkably similar. *Sphenodon* differs from its Late Triassic relatives in only a few skeletal traits (Whiteside 1986; Wu 1994). Other examples are provided by bats (see Fig. 11.7), whose general body form has remained nearly unchanged for almost 50 million years. Even the common tree squirrel, *Sciurus,* has been with us for roughly 35 million years (Emry and Thorington 1984). This is one of the longest geological ranges of any modern mammalian genus.

Eldredge and Stanley (1984) used the phenomenon of living fossils to demonstrate the importance of stasis in the history of life, but admitted that they may be but extreme examples within a normal distribution of evolutionary rates at the level of the family. They do provide a basis for the evaluation of factors resulting in the maintenance of extreme stasis. According to Stanley (1984), many living fossil groups:

1. are depauperate in species,
2. retain a large number of plesiomorphies,
3. are ecologically eurytopic in many physiological and behavioral attributes, and
4. show great individual species longevity with
5. broad areal and habitat distribution.

On the other hand, he admitted that there are exceptions to all of these attributes. For example, *Sphenodon* has an extremely limited distribution, and *Sciurus* belongs to a moderately speciose assemblage.

Of course, none of these so-called living fossils has been absolutely static for all this time, and evolution at the level of species and genera can be documented in most families that have an adequate fossil record; but in nearly all major groups, one must accept a period of rapid change during origin and relatively much slower change subsequently. Change at a very rapid rate has been reported in many late Cenozoic species and can be readily explained by natural selection. What needs a different type of explanation is the much slower change within many, or perhaps most, well-established groups.

The writings of Darwin give the impression that evolutionary change has almost no limits: Given sufficient time, nearly anything is possible. If we look at what evolution has produced in the modern fauna, this seems logical; but if we look at the pattern of evolution over tens and hundreds of millions of years, it is evident that directional change is always limited. Certain types of change appear to be possible in some groups but not others, and rapid morphological change and extensive radiations are almost always limited to short periods of time. Observations of what appear to be limits to the extent of evolution have led to the concept of **evolutionary constraints**: limits to the direction, nature, rate, and amount of change that is possible.

Constraints may be categorized as being (1) historical, (2) chemical, (3) material, (4) genetic, (5) developmental, or (6) physical. (These last three are discussed in Chapters 9–11, respectively.) Chemical, genetic, and developmental constraints show considerable overlap, but we can examine individual examples that help to explain the degree to which each may influence the patterns of evolution seen in the fossil record. Phenomena such as competition, predation, and availability of metabolic necessities and space, all of which were discussed by Darwin, are not considered evolutionary constraints: They do not constitute specific limitations to evolutionary change, and they can be overcome by selection if appropriate genetic variation is present in the species.

Historical constraints

Historical constraints are the most all-encompassing. All of the inherent attributes of an organism – its genetics, developmental patterns, physiology, and anatomy – limit the capacity for evolutionary change in its immediate descendants. No lineage can depart significantly from the properties established by its phylogenetic heritage. Not even the most highly adapted aquatic mammals or reptiles can forgo the need for breathing atmospheric oxygen, evolved among the earliest tetrapods of the Paleozoic. All mammals, except for the few that are capable of hibernating, are constrained by the requirements of their endothermic heritage to consume approximately ten times the amount of nutrients necessary for amphibians and reptiles of comparable size.

The path of adaptation within a lineage is frequently constrained or directed by the nature of its immediate ancestors. Many groups of land vertebrates have become secondarily aquatic. Those that evolved from animals that retained the primitive, fishlike pattern of lateral undulation for terrestrial locomotion depend primarily on the trunk and tail for aquatic propulsion. Examples include the marine iguana, the Upper Cretaceous mosasaurs, and crocodiles. Animals such as turtles and birds, in which, for very different reasons, the trunk region is short and rigid, gave rise to descendants that use their limbs for aquatic locomotion. In both sea turtles and penguins, the forelimbs are used to "fly" underwater. Cetaceans use the trunk and tail for propulsion, but the force is supplied by vertical undulation, as was the case in the earliest mammals, rather than the horizontal undulation common to fish and aquatic reptiles such as ichthyosaurs (Carroll 1985).

A more specific instance of ancestral constraint is exemplified by the feeding mechanics of carnivores. Werdelin (1989) contrasted two groups of bone-cracking scavengers: the well-known hyenas, of African origin, and a lineage of specialized canids, the borophagines, which lived in North America from the Miocene into the Pleistocene. In both groups, upper and lower cheek teeth were specialized to crack long bones so that they could feed on the marrow, but the position of the modified teeth differs in the two groups as a result of different ancestry. Hyaenids evolved from primitive relatives of the cat family, Felidae, in which the posterior cheek teeth were greatly reduced and the facial region of the skull was shortened (Fig. 8.1). As in all other carnivores, the last upper molar (P^4) and first lower molar (M_1), together termed *carnassials,* are modified as narrow blades to cut meat. The next, more anterior teeth, the third upper premolar P^3 and third lower premolar P_3 are pointed and mediolaterally thickened to crack bones.

Among borophagines, like most other canids, the molar teeth are elongate and the entire tooth row is much longer. Instead of specializing the third premolars as bone-cracking teeth, they specialized the fourth premolars. The reason for this can be established by considering the general geometry of the skull (Fig. 8.2) In hyaenids the region of maximum bite encompasses the third premolars; but in canids, including the borophagines with their longer tooth row, it extends no further forward than the fourth premolar. The borophagines must hence specialize more posterior teeth. The jaw mechanics of the borophagid skull suggest that the force of their bite on the more posterior teeth may have been greater than that in hyaenids; but rapid wear on the carnassial teeth, which were modified for bone crushing, made them less effective in cutting fresh meat. This suggests that borophagines were more committed to scavenging than modern hyaenas. This is partially confirmed by their shorter limbs, relative to hyaenids, suggesting that they were not adept as pursuit predators. Hyaenids appear to be more ecologically versatile, capable of subsisting both as pack-hunting pursuit predators and as scavengers.

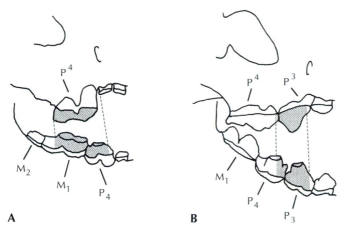

Figure 8.1. The bone-crushing region of **A,** the borophagine carnivore *Osteoborus cyonoides,* and **B,** the modern hyena *Crocuta.* Redrawn from Werdelin (1989).

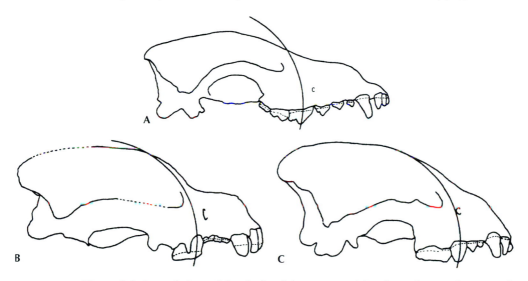

Figure 8.2. Lateral view of the skulls of **A,** an unspecialized canid *Canis lupus;* **B,** the borophagine *Osteoborus cyonoides;* and **C,** the modern hyena *Crocuta.* In B and C an arc has been drawn with its center at the center of the rotation of the lower jaw and a radius equal to the distance to the bone-crushing tooth. In each case the arch passes just in front of the orbit. The arc passes through a thickened area that extends toward the skull roof to dissipate the compressive force of bone crushing. In A, the arc passes through the unmodified upper carnassial, but the skull is not thickened above it. Note the different relative position of the upper carnassial tooth P^4 in the hyena and the canid species. Redrawn from Werdelin (1989).

Werdelin concentrated his study on the relatively primitive borophagine *Osteoborus cyonoides,* but pointed out that the more specialized species *Borophagus diversidens* further elaborated the same characters, showing how constraints are perpetuated to be expressed in evolutionary trends within lineages.

It is hardly surprising that, once well-established, a particular adaptive morphotype will be further accentuated. Hence, most orders of placental mammals retain their original adaptive zone, whether that of insectivore, carnivore, or large or small herbivore. This is particularly true of highly derived groups such as the Chiroptera and the Proboscidia. Some notable exceptions can be cited, especially among secondarily aquatic lineages; these include the seals, walruses, and sea lions, which descended from terrestrial carnivores, and the sirenians and extinct desmostylians, which diverged from near the proboscidians. The cetacean lineage underwent two major adaptive changes, beginning as small terrestrial omnivores related to ancestral ungulates and then going through a stage as large terrestrial carnivores before becoming adapted to an aquatic existence. Nevertheless, once they became committed to life in the water in the early Eocene, they perpetuated this way of life for the next 55 million years.

Although morphological and behavioral changes may be favored under changing environmental conditions or as a result of an adaptive shift within a group, other characters may exhibit stasis for long periods because they are optimal for the conditions under which they live. This is almost certainly the case for basic biochemical and physiological processes that have remained relatively static throughout vertebrate evolution.

Chemical constraints

The most ubiquitous constraints are those of the molecules that make up the body and constitute its metabolic and reproductive machinery. All vertebrates, like other organisms, comprise the same nucleic acids, amino acids, fatty acids, lipids, and sugars. The genetic code is identical throughout, as are the basic proteins and polysaccharides. Change is evident only at the level of specific amino acids in proteins. The degree of constancy at this level is illustrated by the fact that the total genome of humans, which codes for all the proteins, differs by no more than 1 percent from that of chimpanzees. Minor changes may occur, but there is little evidence that they are controlled by environmental selection. We may think of the range of variability of these basic molecules as being an evolutionary constraint, or conceive of variability as strongly limited by stabilizing selection, which results in the death of any organisms that vary from the optimum common to all vertebrates.

All of the basic metabolic pathways, such as energy transformation, synthesis of proteins, carbohydrates, and fats, and elimination of nitrogenous wastes, are fundamentally similar throughout vertebrates (Hoar 1983). Wilson, Carlson, and White (1977) cited extensive evidence that most differences in these processes that are evident between members of particular groups are attributable to differences in regulatory genes controlling the production and availability of rate-limiting proteins. In the processes that have been studied, the same suite of enzymes is common to all vertebrates, but their concentration may differ 10–1,000-fold from one group to another. For example, ruminating artiodactyls that utilize microorganisms to break down cellulose have altered the relative frequency of a number of enzymes. This enables them to make use of the large amount of propionic and acetic acids produced by the microbes, as well as the high concentration of ribosomes in the microbes themselves. The concentration of pancreatic ribonuclease in ruminants is a thousand times that of humans and most other primates.

Another biochemical pathway that shows considerable variability among vertebrates is the excretion of nitrogenous wastes. All vertebrates produce ammonia as a result of the breakdown of proteins. Many aquatic vertebrates excrete ammonia directly, whereas more terrestrial genera excrete nitrogen in the form of urea or uric acid. The different pathways have a strong phylogenetic component, but probably all vertebrates have the enzymes necessary for formation of urea and uric acid; those that excrete ammonia simply have much lower concentrations. Mammals, most of which excrete their excess amino groups as urea, have high levels of the urea-pathway enzymes, especially arginase, and low levels of the enzymes for synthesis of purines and uric acid. In contrast, birds, which form uric acid, have high levels of enzymes for purine biosynthesis and low levels of arginase (Wilson et al. 1977). It is not necessary for selection to direct the evolution of entire new enzymes, but only to alter the rates of production of those already present. Although one or the other pathway is followed by most members of large taxonomic groups, individual genera may utilize another, showing that this alternative is not strongly constrained. For example, the arid-land toads *Chiromantis* and *Phyllomedusa* both excrete uric acid rather than ammonia, as is the case for most other amphibians; this greatly reduces the amount of water they must use to eliminate their nitrogenous wastes (Eckert and Randall 1983). Many vertebrates change the nature of their excretory products during development.

Metabolic rates differ greatly between major vertebrate groups, with modern birds and mammals having approximately ten times the basal metabolism of fish, amphibians, and the animals customarily referred to as reptiles. This increased metabolism also is not the result of the evolution of new enzymes and metabolic pathways, but is related to the great increase in the number of mitochondria in many body tissues. It is also associated with the preponderance of red rather than white muscle fibers, with their much higher number of mitochondria.

The limitation of most fish, amphibians, and modern reptiles to a lower metabolic rate results not from chemical or metabolic constraints, but rather from the combination of a large suite of structural and behavioral factors involving the respiratory, digestive, and circulatory systems, as well as posture and locomotion. In fact, fast-swimming sharks, the tuna and its relatives, and sea turtles maintain a high temperature in their muscles even in cold water (Schmidt-Nielsen 1975). In these cases, the secret is the retention, through countercurrent exchangers in blood vessels associated with the muscles, of heat generated by muscular activity. Among snakes, brooding pythons also maintain a high body temperature through muscular activity. These examples show that physiological constraints can be overcome under strong selection regimes.

Material constraints

Bone and cartilage

The basic organic molecules and metabolic pathways are common to all eukaryotic organisms. In contrast, major taxonomic groups may have unique hard tissues: cellulose in the case of plants, chitin for terrestrial arthropods, and calcium carbonate for many aquatic metazoans. Most living vertebrates possess two types of hard tissue: bone and cartilage. These tissues are developmentally related in being formed from the same precursors among mesenchymal cells, which may mature as either chondroblasts or osteoblasts. Bone that forms superficially – as does much of the braincase in mammals and the external bones of the skull and scales in bony fish – develops directly from connective tissue into **intramembranous** or **dermal bone**. Bone that forms deep in the body, such as the vertebrae and long bones – termed **endochondral** or **cartilage replacement bone** – is preformed in cartilage, which is secondarily replaced by bone.

Bone and cartilage differ greatly in their histology, chemical composition, strength, and weight. Bone has a specific gravity of approximately 2.2, more than twice that of water, whereas cartilage has a specific gravity of about 1.1. Bone is rigid and will bend or break only under very strong pressure; cartilage is more capable of yielding to compression without fracture. Several types of cartilage are recognized, each with a different degree of resistance to deformation.

Bone is a composite of organic and inorganic components that give it a strength and resistance to fracture far above that of either of these substances in isolation. A mucopolysaccharide ground substance and a network of collagen fibers make up the organic constituents, within which is deposited a calcium phosphate salt similar to the naturally occurring mineral apatite. Unlike cartilage, which has uniform properties in all directions, bone, like wood, differs greatly in how it responds to force in different directions because of the specific orientation of its internal crystalline structure. This is particularly evident in regard to shear, in which bone will resist loads up to 500 kg/cm² parallel to the grain but 1,176 kg/cm² across the grain. Long bones stressed parallel to the grain will resist compression forces of 1,330–2,100 kg/cm² and tensile forces of 620–1,050 kg/cm² (Hildebrand 1995).

The geometry of the internal structure is such as to maximize the resistance to distortion or fracture when the bones are in their normal orientation. This is particularly clear in the complex orientation of trabecular bone in joints that are subject to stress from several different directions. The strength of bones and the muscles that move them must be an ultimate constraint to the size of terrestrial vertebrates. It may have been closely approached in the largest dinosaurs, but is not in any living mammals (Farlow, Smith, and Robinson 1995; Alexander 1996).

There is little variation in the chemistry or physical properties of bone among vertebrates. Bone can be made stronger by more extensive mineralization, but this makes it more brittle (Wainwright et al. 1982). On the other hand, the general configuration of bones can be extensively modified so that they can most effectively resist the forces acting upon them both within the body and externally. The long bones of terrestrial vertebrates are basically cylindrical, since the forces acting upon them by the attached muscles are nearly symmetrical. The support capacity of the nearly vertical limbs of large mammals and dinosaurs is proportional to the square of their cross section. Because the weight of the body they are supporting is proportional to the cube of linear dimensions, the diameter increases at a disproportionally rapid rate as body size is increased. If a long cylindrical bone, such as the femur in crocodilians, remains nearly horizontal while the body is lifted on the lower limb, the resistance to bending is proportional to the diameter divided by the square of the length ($R \propto d/l^2$).

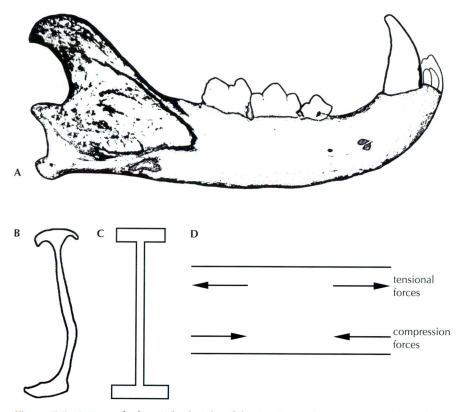

Figure 8.3. **A,** Jaw of a lion. The height of the jaw is much greater than the width to maximimize its resistance to tension and compression in the vertical plane. **B,** Cross section of the back of the jaw showing the much thicker bone on the dorsal and ventral margins to resist the maximum forces of tension and compression. **C,** I beam, showing the thicker dorsal and ventral portions that resist compression and tension. **D,** Directions of the forces of tension and compression on the jaw.

Other bones are subject to stress primarily in a single plane. This is seen in the structure of the jaw of a terrestrial vertebrate such as a mammal (Fig. 8.3), which must resist forces of tension and compression – the bite, the jaw articulation, and the force of the adductor muscles – all acting more or less in the vertical plane. The resistance of a horizontal beam to bending is proportional to its width times the square of its height ($R \propto wh^2$). A doubling of its width doubles its resistance, but a doubling of its height increases the resistance by a factor of 4. Hence the jaw ramus is much higher than it is wide.

In related animals of similar size and habits, there is an optimal shape of the bones that results in the maintenance of a similar morphology for tens of millions of years, as we see in the stasis of the appendicular skeleton of tree squirrels for 35 million years. This is attributable to the stability of the chemical composition of the bone, as well as to the conservation of the physical laws governing how force acts on the rigid bony skeleton. Since the general size and shape of the bone are genetically variable, one may also attribute the constancy of form to stabilizing selection.

On another level, the degree of ossification and details of internal and external structure are also controlled by physiological factors acting during the growth of the animals. This is evident early in embryological development, when the stress of developing muscles results in differential ossification of the bones to which they are attached. This process continues in the adult, such as when the amount of ossification is reduced in persons who are bedridden or live under the reduced gravity of outer space.

The greatest consistent difference in supporting tissues among vertebrates is seen between the actinopterygians and the chondrichthyans. In actinopterygians, bone is the dominant tissue for both the endoskeleton and the exoskeleton of the skull for most orders throughout the history of the group. In sharks and their kin, the endochondral bones do not ossify but remain cartilaginous, and no dermal bones are formed. The predominance of bone or cartilage presumably stems from the different methods of buoyancy control in the two groups. Bony fish apparently evolved a swim bladder early in their history, which had the potential for providing neutral or positive buoyancy even with a heavy, bony skeleton. In contrast, a swim bladder never evolved among cartilaginous fish, which would have placed a selective advantage on a lighter skeletal material. Cartilage, as we have seen, has only about half the specific gravity of bone, but it is also a much softer and more breakable material. Cartilage cannot be used for the complex mechanical roles that bone plays in actinopterygians. No sharks, skates, or rays have the complicated interconnections between elements of the feeding and respiratory apparatus that characterize bony fish. This may be one of the factors that has limited the morphological and adaptive diversity of the Chondrichthyes, which today are limited to only 793 species, compared to 20,839 species of actinopterygian fish.

Muscles

All voluntary movements of the vertebrate body are produced by the skeletal, or striated, muscles. The molecular structure and function of skeletal muscles are essentially uniform throughout vertebrates and may be considered as constraints that cannot be significantly changed by selection. The functional elements of striated muscles consist of interdigitating filaments of the proteins myosin (forming the thick filaments) and actin, troponin, and trophomyosin (together forming the thin filaments), which are drawn across one another to produce the force of muscle contraction (Fig. 8.4). The length of the filaments limits the distance over which the muscle may contract. In simple muscles, in which all the muscle fibers are parallel to one another, a muscle may contract as much as 30 percent of its resting length, or nearly 60 percent of its stretched length (Hildebrand 1995). The force of muscle contraction ranges from 1 to 8 kg/cm^2, with an average of approximately 3 kg/cm^2, as determined by the cross section of the muscle fibers measured at right angles to their length. Selection cannot significantly alter these limits. Muscles of a particular dimension, with parallel fibers, will always have about the same degree of possible expansion or contraction. This will tend to act as a constraint on changes in the dimensions of skeletal elements acted on by the muscles.

On the other hand, selection can act to change other properties of muscles that can alter their function. Although expansion and contraction of muscle fibers are limited to a particular percentage of the length of the muscle, the distance of contraction or expansion can be altered significantly through increase in the length of

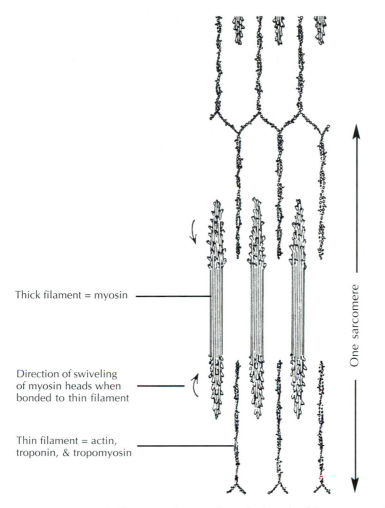

Thick filament = myosin

Direction of swiveling
of myosin heads when
bonded to thin filament

Thin filament = actin,
troponin, & tropomyosin

One sarcomere

Figure 8.4. Muscle filaments, showing how the length of their region of overlap constrains the distance over which muscles can expand or contract. From *Analysis of Vertebrate Structure* by Hildebrand, 4th ed. Copyright © 1995. Reprinted by permission of John Wiley & Sons, Inc.

the whole muscle. For example, if the muscles that close the jaw are confined to a small chamber, as they were in primitive bony fish, the gape of the mouth is narrowly limited. If the bones surrounding the adductor chamber are reduced, the length of the muscles can be greater and the mouth can be opened wider. Most genera from the Early Devonian through the Permian retained a small adductor chamber and had a limited gape. Among the many changes in the skull leading from primitive to modern ray-finned fish is the opening of the adductor chamber. Changes in the configuration of the adductor chamber that permitted increase in the length of the jaw muscles and the gape of the jaw also occurred in very early stages in the evolution of lungfish, several lineages of salamanders, and in the origin of mammals and many reptilian groups.

Although the force that muscles are capable of generating is limited by the cross section of their fibers, a greater force of muscular contraction can be achieved with the same muscle mass by altering the geometry of the fibers. Although the simplest

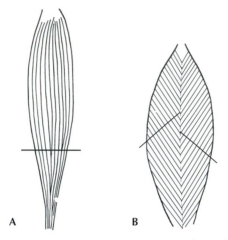

Figure 8.5. **A,** Parallel orientation of muscle fibers in a strap-shaped muscle. **B,** Diagonal orientation of fibers in a pinnate muscle. Heavy lines indicate the cross section, whose area determines the force of their contraction.

arrangement of the fibers is parallel to the long axis of the muscle, they are more commonly aligned obliquely, with attachment to an internal tendon – a pattern termed *pinnate* (Fig. 8.5). Because the sum of the cross sections of the fibers is then greater than the cross section of the muscle as a whole, a greater force can be generated, although the distance that the muscle as a whole can expand or contract is reduced. The force of muscles can also be increased through leverage involving the bones to which they are attached. Like bones, the size of muscles must increase disproportionately in larger animals, since their strength is proportional to their cross-sectional area, whereas the weight they are supporting or moving increases as the cube of linear dimensions.

The relative speed of muscle contraction and the susceptibility of muscles to fatigue are governed by the proportions of three types of fibers: fast-twitch fatigable fibers, slow-twitch fibers, and fast-twitch, fatigue-resistant fibers. Fast-twitch fatigable fibers contract rapidly with great force but fatigue quickly because they depend on anaerobic metabolism, which results in the rapid buildup of lactic acid. Fatigable fibers are white in color because they have few mitochondria or blood vessels; the other two types of fibers are darker because of numerous mitochondria, contributing to oxidative metabolism, which permits long use without fatigue. Slow-twitch fibers contract and fatigue the slowest but have the least force; they maintain slow, repetitive movements, as on the sides of slow-cruising fish. Most vertebrates have all three types of fibers but in differing proportions. Most fish, amphibians, and modern reptiles have a preponderance of white fibers: They can respond quickly but cannot remain active for more than a few minutes. Mammals and birds have a much greater proportion of fast, fatigue-resistant fibers, which enable them to be active for long periods of time, as in the case of migratory birds, which may fly thousands of miles over open water. The relative amount of red and white fibers may be nearly constant within major taxonomic groups, but it cannot be considered a fixed constraint. Larger amounts of red fibers are found in both sharks and some bony fish, which are exceptionally powerful long-distance swimmers.

The overall pattern of the muscles in vertebrates can be readily derived from that seen in the cephalochordate amphioxus, in which they are arranged as a series of nearly identical, paired myotomes running along either side of the trunk from the head region to the tip of the tail. The muscles in the head region became specialized during the origin of vertebrates to control the movement of the eyes, mouth, and gills. With the origin of jaws and paired fins, muscles of the head and body wall became specialized to enable their controlled movement. The muscles controlling the fins in choanate fishes became further elaborated as the tetrapod limbs evolved. Although the girdles and limbs are much altered among terrestrial vertebrates, the number, general position, and innervation of the muscles controlling the limbs have remained conservative throughout the evolution of terrestrial locomotion and flight. Nearly all the muscles can be homologized from amphibians through reptiles, birds, and mammals. Some have been lost, others split or fused, but their number has remained generally similar. On the other hand, the way in which they function, such as adduction, abduction, supination, pronation, or rotation of the limb, may change in relationship to the geometry of the bony elements of the limbs. It is clear from developmental studies that the muscles, along with the nerves and blood vessels, follow the bones in their development and show very extensive plasticity. Although muscle filaments have a fixed molecular pattern throughout the evolution of the vertebrates, constraints on entire muscles are much less stringent and are not a narrowly limiting factor in vertebrate evolution.

Summary

The pattern of large-scale evolution shows that all major taxa retain a basically similar morphological pattern for much longer periods of time than had been assumed by Darwin. Among vertebrates, many families, defined on both morphological and phylogenetic criteria, retain a similar structure for 100 million years or longer. This constancy of form may be attributed to evolutionary constraints: aspects of their biochemistry, genetics, development, and/or physical aspects of the environment.

Inherent aspects of structure, development, and physiology limit the extent to which lineages can diverge from the pattern of their immediate ancestors. Similar ways of life tend to be perpetuated and accentuated within most groups. Despite the constraints of the basic chemical components and metabolic pathways, physiological processes show considerable latitude because of the capacity for regulatory genes to control the relative quantity of different enzymes that can be produced. The strength of bone can be little altered by selection, but the shape of bones can be varied substantially to accord with the forces that they must resist. The molecular structure of vertebrate striated muscles limits the capacity of their basic properties to respond to selection, but other aspects of their structure and arrangement can respond significantly in relationship to changes in the bony elements to which they are attached.

9 Evolutionary genetics

Introduction

The structure and function of all organisms ultimately depend on information coded by their genes. In turn, all aspects of evolutionary change result from modifications in the nature and sequence of the DNA that makes up the genes. Knowledge of the rates and patterns of change at the level of the genes is thus vital to understanding the basic force underlying evolution. However, understanding the consequences of genetic change in controlling the rate and direction of evolution in species and higher taxa also requires quantitative evaluation of other forces, including selection and stochastic (probabilistic or random) phenomena that influence the presence and relative frequency of different genes in populations.

Readers who are not familiar with the basic principles of Mendelian genetics may wish to consult a general genetics text such as *An Introduction to Genetic Analysis* by Griffiths et al. (1993), or the appropriate chapters in evolutionary biology texts such as *Evolution* (Ridley 1996), *Evolutionary Biology* (Futuyma 1986), or *Human Heredity* (Cummings 1994). Other publications covering the broad field of population genetics include Falconer (1989) and Hartl and Clark (1989). Knowledge of the history of any science contributes to an understanding of the formulation of current ideas, but this seems especially important in the case of the interrelationships of genetics and evolutionary theory. In particular, such a historical perspective shows how periodic lack of communication has led to the elaboration of strongly divergent hypotheses, both within and between fields. This is ably demonstrated in the publications *The Growth of Biological Thought* by Mayr (1982), *Sewall Wright and Evolutionary Biology* by Provine (1986), and the review by Gilbert, Opitz, and Raff (1996).

Basic models of genetics

Mendelian inheritance

One of the most serious inadequacies of selection theory as elaborated by Darwin was the absence of information regarding the manner of inheritance. Darwin either did not know of the work of Mendel or did not appreciate its significance. It was not until the early years of the twentieth century that Mendel's work was rediscovered and the science of genetics emerged. Ironically, knowledge of the mechanism of inheritance did not initially serve to support selection theory; rather, it formed the basis of an alternative evolutionary theory, mutation theory, which argued that the spread of new mutations, and not the control of variability by selection, was the main force determining the direction of evolution.

The integration of genetics and selection theory was achieved through the development of population genetics in the first thirty years of the twentieth century (Haldane 1924, 1932; Fisher 1930; Wright 1931). This integration was made possible through mathematical analysis of the alternative forces of mutation and selection and appreciation of the significance of population size. Subsequent research has provided additional quantitative information regarding mutation rates, the prevalence of genetic variability in natural populations, and the efficacy of natural selection in both natural and laboratory populations.

The fundamental rules of inheritance were discovered by Mendel (1866) on the basis of the study of clearly distinguishable traits in sweet peas that occurred as simple alternatives. The plants were tall or short, without any of intermediate stature; the seeds were yellow or green, without intermediate colors; and the seed coats were either wrinkled or smooth. We now recognize that each pair of traits studied by Mendel was governed by alternative alleles at the same locus, or position, on a single pair of chromosomes, and that genes governing different characteristics were carried on different chromosomes. The genes controlling homologous traits segregate at the time of meiosis, one allele going to each gamete. Traits carried on different chromosomes undergo independent assortment. The color and texture of the seed coat and the height of the plant are all expressed independently.

Thousands of genes controlling specific traits have been recognized in all eukaryotic organisms that have been studied. The genes responsible for clearly distinguishable traits can be studied individually to establish their specific molecular structure and function, mutation rate, and (within limits) the selection coefficients that contribute to determining their frequency in populations. Extensive data bases for traits that are influenced by specific loci have been assembled for humans (McKusick 1994), mice (Lyon and Searle 1989), and fruit flies (Lindsley and Zimm 1992) and are being compiled for many other organisms.

The early development of population genetics concentrated on traits of this nature. There may be numerous alternative alleles at a single locus, but the frequently used models are based on a single pair. The alternative alleles are commonly designated A and a – the capital letter indicating dominance and the lowercase one the recessive alternative. The frequencies of the two alleles (typically indicated by p and q) may occur in any proportion, adding up to 100 percent or unity.

$$p + q = 1 \quad p = 1 - q$$

In large, randomly mating populations with no mutation, no migration, and no selection, the frequencies of the alleles will remain in equilibrium from generation to generation. So too will the frequency of the genotypes – the homozygotes expressed as p^2 and q^2, and the heterozygotes as $2pq$. The relationship of the frequencies is expressed mathematically as the Hardy–Weinberg equation:

$$p^2 + 2pq + q^2 = 1$$

If the actual frequencies of the genotypes do not accord with this equation, or if they change from generation to generation, selection, nonrandom mating, migration, and/or mutations are assumed to be acting to change the frequencies of the alleles, or the population size may be so small that random factors become important in determining allele frequency.

In fact, all populations must be influenced to some degree by both mutation and selection. Looking first at mutations, if A_1 is mutating to A_2 at a rate of u per generation without reverse mutations, the frequency of A_2 will eventually reach 100 percent. If A_2 is originally rare, its frequency will initially increase rapidly; but as the frequency of A_1 is reduced, the rate of increase in A_2 will be progressively lessened.

If mutations are acting in both directions, an equilibrium will be reached that is determined by the ratio of the different mutation rates. The frequency of A_1 will be equal to the rate of mutation from A_2 to A_1, divided by the sum of the mutation rates in both directions. If q and p are, respectively, the frequency of A_1 and A_2, u is the mutation rate of $A_1 \rightarrow A_2$, and v is the mutation rate from $A_2 \rightarrow A_1$, then the equilibrium frequency $q_e = v/(u + v)$.

Genes are acted upon individually by mutations, but selection acts on the entire organism. The force of selection on alternative alleles is approximated by assuming that the remainder of the genome is essentially constant. In diploid organisms, all alleles (except those on the sex chromosomes in males) occur as pairs, so that the effect of both alleles must be considered together, unless one is fully dominant. Selection on a fully recessive trait can occur only when it is in the homozygous state. Selection favoring a rare dominant allele will result in rapid increase in its frequency until the frequency of the recessive is much reduced. Thereafter, selection will have very little effect because the recessive will rarely appear in the homozygous state; its frequency will approach zero in an asymptotic manner.

The force of selection is expressed in terms of the **selection coefficient**, which compares the fitness of one phenotype relative to the other. For example, if 100 fertile progeny are produced by the favored phenotype and only 90 are produced by the phenotype selected against, the selection coefficient s is equal to 10 percent or 0.1. Alternatively, one may refer to the fitness of the alternative phenotypes – for example, AA and Aa (if A is dominant) relative to aa – as 1.0 and 0.9.

The change in frequency of a recessive allele subject to selection alone is calculated by the expression

$$\Delta p = \frac{spq^2}{1 - sq^2} = \frac{sp(1-p)^2}{1 - s(1-p)^2}$$

where Δp is the change in frequency of a from one generation to the next.

To establish the amount of change over a number of generations, this equation must be solved for each in succession. One can also use this equation to determine the selection coefficient when the number of generations and amount of change are known, as in the case of the peppered moth, *Biston betularia*. In the fifty generations between 1848 and 1898, the frequency of the melanic gene increased from 10^{-5} to 0.8. This could be accounted for by a selection coefficient of 0.33 (Ridley 1996). In any generation, only two nonmutant moths would reproduce successfully for every three melanics. Application of this formula at lower selection coefficients shows that both alleles may persist from tens of thousands to millions of generations (Fig. 9.1). These examples show how very low selection coefficients might give rates of change as slow as those that seem to be common in the fossil

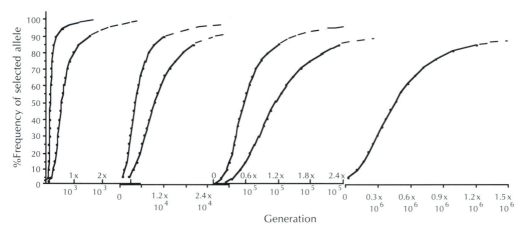

Figure 9.1. The effects of selection favoring a dominant allele over time. Starting at the left, the selection coefficients against the alternative, recessive allele are 0.05, 0.01, 0.001, 0.0005, 0.0001, 0.00005, and 0.00001. The extreme differences in time involved with changing selection coefficients require a sequence of different time scales. Curves calculated by David Carroll.

record. It seems very unlikely, however, that selection would ever proceed at such a steady rate for such a long period of time.

If selection and mutation are both acting to increase the frequency of one allele, their effects are essentially additive. If they act in opposite directions, the frequencies of the alleles will reach equilibrium. Two equations are necessary to describe the equilibrium frequency: When the allele in question is dominant, the equation is $P = u/s$; if the allele is recessive, the equation is $p = \sqrt{u/s}$, since selection will act on a recessive allele only when it occurs in a homozygote.

The relative significance of mutation and selection becomes evident when actual mutation rates and measured selection coefficients are entered into these equations.

Mutation rates

The term *mutation* was used by early geneticists to describe rare but relatively conspicuous differences in the structure, physiology, and/or behavior of organisms that were inherited according to a Mendelian pattern. We now recognize that such changes may result from a broad range of phenomena, from the loss or substitution of individual bases in the DNA, to loss or duplication of entire genes and diverse modifications in the structure and number of chromosomes.

Most mutation rates that have been calculated are associated with changes in clearly recognized traits whose expression is controlled by limited regions of the chromosome, recognized as individual genes. Although the correlation is not exact, many genes code for individual proteins – typically, enzymes associated with particular metabolic or developmental processes. Among the best-documented mutation rates are those studied in humans, in which it is possible to determine the exact number of new mutations that appear per generation within a population of known size. The most direct evidence comes from traits that either are dominant or that occur on the sex chromosomes, so that they are expressed in all carriers, at least of one sex, in each generation. The examples given in Table 9.1 show a wide spectrum of rates, from $6–12 \times 10^{-4}$ in the case of polycystic kidney disease to less than one in a million for the von Hippel–Landau syndrome. Additional examples from other eukaryotes suggest that most mutation rates at the level of the gene occur at a rate ranging from 10^{-5} to 10^{-6} per locus, per generation. For example, 25 new coat-color mutations at five loci were found in a study of over 2 million mice tested, with an average rate of visible mutants of 1.1×10^{-5} per locus, per generation (Schleger and Dickie 1971). Since most published mutation rates are based on such clearly distinguishable traits, they have played an important role in developing models of population genetics. Although the human examples are primarily highly deleterious, the traits in mice and fruit flies are the sort that might provide a guide to the relative roles of selection and mutation in natural populations.

Prior to determination of the specific structure of DNA, it was not possible to establish the actual molecular changes that resulted in differences in the expression of traits at the level of the gene. We now recognize that these may be as small as the substitution of a single base within a codon, as in the case for sickle-cell anemia, wherein a U is substituted for an A, changing a codon from GAG to GUG, which results in the substitution of valine in the place of glutamic acid in beta globin. This results in altering the electrical charge, thus reducing the solubility of the protein so that the red blood cells assume a characteristic sickle shape. In other cases, what is recognized as a single mutation may result from the loss of an entire gene, which may consist of hundreds of nucleotides.

Mutation rates at the level of the gene are thus recording many different phenomena. The most ubiquitous type of mutation is that of base substitution, which affects single bases in all parts of the genome. The nature and amount of base substitution can be determined directly by sequencing entire genes and ultimately entire genomes. In prokaryotes, the rate of most base substitutions has been found to range from 10^{-9} to 10^{-10} per base, per cell division, the specific rate depending on the neighboring bases (Maynard Smith 1989).

Base substitution is fundamental to the gradual evolution of the genome, but the effect on the phenotype and thus on selection differs greatly depending on the position of the substitution. In complex eukaryotes such as vertebrates, much of the genome does not code for proteins, nor does it appear to carry any information that directly affects the form or function of the organism. We still know little of the function of the noncoding regions of the genome, but individual base substitution in these areas would probably not appear as an observable phenotype, nor affect the fitness of the organism. Even within the coding portion of the genes, much replacement in the third position of the codon is "silent" in the sense that it results in no change in the amino acid that is coded for. Change in bases that do result in coding for different amino acids may alter the electrical charge of the protein (as in the case of sickle-cell anemia) or otherwise result in changes in the physiology of the organism. On the other hand, nearly all populations show some variation in individual amino acids within proteins that do not appear to result in any differences in viability or fecundity.

One can see from the wide range of effects of single base substitutions that the rate of change at this level does not in itself provide direct information regarding

Table 9.1. *Mutation rates of specific genes in various organisms*

Organism and trait	Mutations per cell or gamete
Bacteriophage *T2* (virus)	
Host range	3×10^{-9}
Lysis inhibition	1×10^{-8}
Escherichia coli (bacterium)	
Streptomycin resistance	4×10^{-10}
Streptomycin dependence	1×10^{-9}
Sensitivity to phage *T1*	2×10^{-8}
Lactose fermentation	2×10^{-7}
Salmonella typhimurium (bacterium)	
Tryptophan independence	5×10^{-8}
Chlamydomonas reinhardi (alga)	
Streptomycin resistance	1×10^{-6}
Neurospora crassa (fungus)	
Adenine independence	4×10^{-8}
Inositol independence	8×10^{-8}
Zea mays (corn)	
Shrunken seeds	1×10^{-6}
Purple seeds	1×10^{-5}
Drosophila melanogaster (fruit fly)	
Electrophoretic variants	4×10^{-6}
White eye	4×10^{-5}
Yellow body	1×10^{-4}
Mus musculus (mouse)	
Brown coat	8×10^{-6}
Prebald coat	3×10^{-5}
Homo sapiens (human)	
Achondroplasia	1×10^{-5}
Aniridia	2.6×10^{-6}
Retinoblastoma	6×10^{-6}
Osteogenesis imperfecta	1×10^{-5}
Neurofibromatosis	$0.5–1 \times 10^{-4}$
Polycystic kidney disease	$6–12 \times 10^{-4}$
Marfan syndrome	$4–6 \times 10^{-6}$
Von Hippel–Landau syndrome	1.8×10^{-7}
Duchenne muscular dystrophy	$0.5–1 \times 10^{-4}$

Source: Human mutation rates from Cummings (1994, Table 11.1); other organisms from Dobzhansky, Ayala, Stebbins, and Valentine (1977, Table 3-2).

rates or patterns of evolution that are studied in either living or fossil populations. However, it can be observed that the rates of evolution at the level of the gene and at the level of the individual base are roughly comparable. Proteins typically comprise several hundred amino acids. Each amino acid is coded for by three bases, so a complete protein requires up to several thousand bases. Hence a mutation rate of 10^{-9}–10^{-10} at the level of individual bases is similar, if not exactly equal, to a mutation rate of 10^{-5}–10^{-6} at the level of the gene. This suggests a rough equivalence between the potential for genetic change at a generation-to-generation level throughout the range of organisms from prokaryotes to vertebrates.

Sequencing the entire genome, or even selected genes, is still far too expensive and time consuming for this to be a practical approach to the study of the nature and rate of mutation at the population level. However, other techniques do provide a practical means of sampling differences in particular enzymes among large numbers of individuals, yielding insight into the significance of evolutionary change at this level. One of the most widely used methods is **gel electrophoresis**, in which tissue samples in a gel are placed in an electrical field wherein proteins move at different rates depending on their electrical charge. Protein can be identified by specific stains that enable one to recognize different allelic forms depending on the relative speed of their movement. Within populations, it is possible to recognize the existence of different alleles at a locus by the presence of adjacent bands representing homologous proteins with different electrical charges. Estimates of the rate of appearance of electrophoretically detectable mutations in *Drosophila* are close to 4×10^{-6} per locus, per generation (Maynard Smith 1989). Since only about 25 percent of amino acid substitutions will be detected electrophoretically, this implies a higher mutation rate than that of most phenotypically conspicuous traits in vertebrates; but this may more accurately reflect the rate of appearance of subtler genetic differences that are commonly acted upon by selection.

Although mutations are the ultimate source of all genetic variability, the rate of appearance of *new* mutations is so slow, compared with selection coefficients high enough to be measurable, that mutation rates are not included in many commonly used equations illustrating change in allele frequency within populations. With a mutation rate of 10^{-5} to 10^{-6} per locus, per generation, and up to 10^5 loci in the vertebrate genome, we can expect one or two new mutations in every individual. What is much more important for the course of evolutionary change within populations and species is the *total number* of alternative alleles present in the population at one time.

Even though not all mutations will be detected, the most practical means of estimating the total amount of variability is through electrophoresis. Table 9.2 shows the results of studies tabulated by Ward, Skibinski, and Woodwark (1992) and Merola (1994) of many eukaryote species. Among vertebrates, 14–27 percent of the loci studied were polymorphic within populations. The average proportion of loci for these traits that were heterozygous within individuals was in the range 5–11 percent. This method detects only mutations that result in differences in the electrical charge of the enzymes studied, so approximately 75 percent of the variability goes undetected. Perhaps as much as a third of all enzyme loci within each individual may be heterozygous, but only a small proportion of these traits would be visually apparent. What proportion would be acted upon by selection is impossible to quantify, and would differ depending on the force and nature of selection. At this level, the amount of variability present in natural populations is certainly large enough to account for the extremely rapid rates of evolution that have been reported in living species (see Chapter 3). This rate is far above that which appears to be reflected in morphological change in any groups known from the fossil record.

Table 9.2 *Comparison of levels of genetic variation among different animal groups*

	$\bar{H}$		$\bar{L}$		n	$P(\%)$
Vertebrates						
Total	0.071	± 0.002	24.76	± 0.30	648	
Mammals	0.067	± 0.004	24.48	± 0.52	172	14
Birds	0.068	± 0.005	28.70	± 0.90	80	15
Reptiles	0.078	± 0.007	22.72	± 0.54	85	22
Amphibians	0.109	± 0.006	21.93	± 0.44	116	27
Fish	0.051	± 0.003	25.97	± 0.65	195	16
Invertebrates						
Total	0.122	± 0.004	21.71	± 0.39	370	
Insects	0.137	± 0.005	21.18	± 0.70	170	
Crustaceans	0.052	± 0.005	23.01	± 0.82	80	
Molluscs	0.145	± 0.010	21.78	± 0.36	105	
Others	0.160	± 0.016	20.33	± 1.77	15	

Note: $\bar{H}$, average heterozygosity per species; $\bar{L}$, average number of loci screened per species; n, number of species; P, percentage of loci examined that are polymorphic per species within vertebrate groups.
Sources: $\bar{H}$, $\bar{L}$, and n from Ward et al. (1992, Table III); P from Merola (1994).

Selection

Endler (1986) has provided a very useful compendium and analysis of data derived from a large number of selection studies of natural populations representing a wide variety of multicellular plants and animals. This information includes both polymorphic traits determined by single loci and quantitative traits that may be controlled by genes at many loci. All show a broad spectrum of selection forces.

Studies of traits encoded by a single locus that can be statistically analyzed show selection coefficients ranging from zero (neutrality) to one (lethality). In populations that are not in stressful conditions, they are concentrated toward the lower end, with the highest proportions lying between 0.0 and 0.1 (Fig. 9.2). Presumably most well-established populations have relatively few highly deleterious traits and many for which selection coefficients are small and variable. Endler commented that there may be many more traits that are near neutrality, but these are virtually impossible to recognize over short periods of time in wild populations. Very few of the lower selection coefficients are statistically significant.

In stressful environments, the selection coefficients are more equally distributed, with a higher proportion of traits that are strongly affected by selection (Fig. 9.2B). Endler emphasized that selection coefficients greater than 0.1 are not rare, in contradiction to the common assumption of theoretical population genetics and ecological genetics that most coefficients will be less than 0.1. All selection coefficients that can be identified and quantified are far larger than necessary to counter random phenomena (see next section).

Selection coefficients a great deal lower have been postulated for long-term change in morphological characters. Lande (1976) calculated rates as low as 10^{-5}–10^{-7} selective deaths per generation among Tertiary mammals. The significance of such very low selection coefficients are discussed later in this chapter.

The significance of small population size

Another factor that must be considered in evaluating the causes of evolutionary change is chance. This becomes significant when the differential between mutation rate and selection coefficient is very small, when population size is small, and when long periods of time are being considered. The significance of small population size can be seen if we take the most extreme case in which there is only a single mating pair, one of which is heterozygous for a trait, while the other is homozygous. In a single birth, there is a 50 percent chance of the zygote receiving both of the two alternative alleles, and a 50 percent chance of one allele being lost within the family if no more progeny are produced:

parental generation   *Aa*   ×   *AA*
gametes   *A* and *a*   *A* and *A*
zygotes   50% *Aa* and 50% *AA*

In more general terms, the chance of an allele being lost is roughly proportional to $1/2N_e$ (Crow and Kimura 1970), where N_e is the effective population size, the number of individuals that would be expected to participate in reproduction at a given time. This may be significantly smaller than the total population, omitting, for example, those that are either too old or too young to reproduce and those precluded from doing so by a major imbalance in the sex ratio. The number may also be reduced if polygyny or polyandry is common in the species – for example, if a

Figure 9.2. The distribution of selection coefficients *s* for alternative alleles at a single locus. The total height of each bar indicates the percentage of *s* values in each interval, and the shaded portion indicates the percentage of *s* values that are significantly different from zero at the 0.05 level. **A,** Data from undisturbed populations; thirty-six species, 239 of 566 *s* values significant. **B,** Data from perturbations, field cages, or stressful environments; twelve species, 70 of 144 *s* values significant. *s* = 0, neutrality; *s* = 1, lethality. From John A. Endler (1986), *Natural Selection in the Wild*. Reproduced by permission of Princeton University Press.

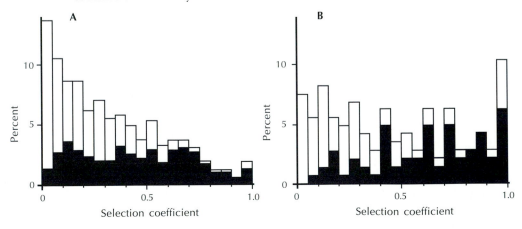

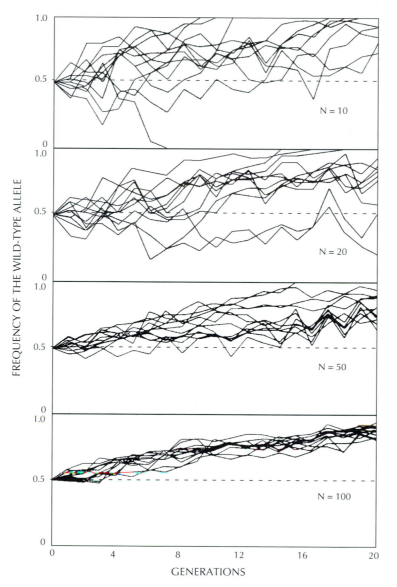

Figure 9.3. Random drift. Change in frequency of alternative alleles for body color in the flour beetle *Tribolium*. *N* = population size. The largest populations show gradual response to selection over twenty generations with the frequency of the wild-type allele increasing from 0.5 to approximately 0.9 at a nearly equal rate in twelve populations. In the smallest population, the frequency varies much more significantly from generation to generation. In one lineage, the favored allele was eliminated after seven generations. In half the populations, the frequency of this allele reached 100 percent before the end of the experiment. Redrawn from Rich, Bell, and Wilson (1979).

small number of males mate with a large number of the females, as occurs in many mammals living in herds. Change in allele frequency and loss of alleles in populations by chance alone is referred to as **genetic drift**.

Figure 9.3 shows an example of both selection and genetic drift acting on different sized populations of the flour beetle *Tribolium*. Selection favors a gene for dark color that is expressed in both homozygotes and heterozygotes. Twelve lines each of populations of ten and a hundred were established and followed for twenty generations. Even when favored by selection, one of the small populations lost the dark allele after seven generations. In half of the lineages, the allele for light color was lost by the end of the experiment. In the large populations, there was a gradual increase in the frequency of the dark allele, but in no lines did it become fixed.

Although the random loss of an allele is unlikely even in fairly small populations over short periods of time, loss becomes inevitable over long periods unless it is countered by selection, as calculated from the expression (from Falconer 1989)

$$C = q_0 - 3p_0q_0(1 - 1/2n)^t$$

where C is the chance of one allele being either lost or fixed over long periods of time; p_0, q_0 are the original frequencies of alleles; n is the population size; and $t =$ time in generations.

Note that neither mutation nor selection is included in this equation, although they may be reflected in the relative frequencies of the alleles. The more disparate the two frequencies, the more likely is the loss of the rare allele. It is unlikely that species that survive for thousands or tens of thousands of generations ever have long periods in which the population size is so small that chance is a significant factor; however, short intervals in which the population size is very small may be enough to change the frequency of rare alleles to a significant extent.

Another condition in which random factors would definitely be important in determining the presence and frequency of particular alleles would be during the founding of new populations by small numbers of individuals (ultimately, only a single gravid female). Termed the **founder effect**, this would be particularly important in the colonization of oceanic islands or other isolated localities. Presumably the entire biotas of the Hawaiian and Galápagos islands were established in this manner. Those individuals that chance to reach these localities – whether by flying or being carried by the wind or floating debris – bring with them the entire genetic component from which the ensuing fauna or flora will evolve. The amount of polymorphism expected in a small number of individuals is far lower than that of the populations at large (see Table 9.2), and the frequencies of alleles are certain to differ from that in the parental populations. Alleles that had been very rare (but not disadvantageous) in the parental population might dominate the new locality and could enable selection to take a much different course.

The shifting balance theory of evolution

Wright (1931, 1932, 1977) argued that genetic drift might also be a significant factor in other, more common situations, such as shifts from one habitat or specific way of life to another. To explain such changes, he elaborated the **shifting balance theory** of evolution. Under most circumstances, individual species are confined to particular habitats and specific ways of life. This is most evident in restricted geographical areas in which all species are clearly differentiated from one another by structure and behavior. The primary mode of selection is to stabilize the characters of the population around well-defined means. Most directional change will further accentuate the degree of adaptation to the conditions under which the population has lived. Wright illustrated this situation by an **adaptive topography**, in which each adaptive zone is represented by a peak, and areas for which adaptation was less favorable are represented by valleys. Selection would drive each species to an isolated peak but could not explain how species passed from peak to peak. How, then, can species evolve from one adaptive regime to another? Wright, followed by Simpson (1944, 1953) and Mayr (1963), argued that this might be accomplished over very short periods of time in small populations.

We have seen that, in very small populations, chance phenomena may have a stronger influence on allele frequencies than selection. Alleles that would otherwise rarely be expressed might enable individuals to invade adjacent habitats in where they were advantageous. If selection were to favor this alternative phenotype, the new population could spread into a distinct habitat. Mayr (followed by Eldredge and Gould 1972) emphasized this phenomenon as important in the origin of new species, and Simpson used it to explain how macroevolutionary change might originate. This hypothesis has since been hotly contested (e.g., Fisher and Ford 1947; Fisher 1958; Ford 1964; Barton and Charlesworth 1984).

Lande (1976) provided a mathematical model to establish the specific relationships between differing population size and selection pressures that would enable change to occur between adaptive zones of variable distance from one another. With no adverse selection against the trait in question, a population with an average size of ten individuals would require roughly a hundred generations for the mean of a character to change by three units of standard deviation (3σ) – approximately the difference between two species within a genus. With a hundred or a thousand individuals, the same amount of change would require about a thousand or ten thousand generations, respectively. Against a plausible selection pressure, 3σ of change would still require a hundred generations for a population of ten individuals and a thousand generations for a population of a hundred, but more than a billion generations for a population of a thousand. Lande (1986) again discussed the importance of small population size in the transition between adaptive peaks, pointing out that this model mirrors the idea of punctuated equilibrium in that evolution may be extremely fast in small populations but nearly static in large ones.

There are two problems with the approach taken by Wright and Lande: They concentrate on traits influenced by a single selective regime, and they ignore nongenetic phenomena that must have a major impact on adaptive change in nature. Lofsvold (1988) and others have pointed out that an adaptive topography is actually much more complex than envisaged by Lande in that potentially every character that varies within a species might have its own adaptive optimum. All individuals are subject to a great many different selective forces, and only a few of the multitude of characters they express can be optimal for a specific environment at any one time. This is very clearly shown in organisms living in environments that change rapidly over time, but applies to all species in which selective pressures on individuals change between birth and death, and from one part of the species range to another.

Although small population size may increase the importance of random fluctuation of allele frequency, large populations provide a much greater source of genetic variability. The larger a population is, the greater the potential there is for the accumulation of new mutations and the retention of previously generated variability. This provides a much greater potential for adaptation to different environments than is the case in small populations with limited total variability and much greater vulnerability to chance extinction. Enlarging populations are also more likely to extend their range into new habitats as competitional selection for available resources increases in the original range. In the summary of their review of founder effects and speciation, Barton and Charlesworth (1984, p. 157) state: "There are no empirical or theoretical grounds for supposing that rapid evolutionary divergence usually takes place in extremely small populations."

In addition, most vertebrate species show considerable behavioral plasticity that enables them to enter new habitats even if they are genetically similar to other members of the species. On the other hand, no matter the population size, competition or predation from species already occupying adjacent habitats may make it impossible for new species to adapt to them.

These models of population biology demonstrate the mathematical relationships of population size, mutation rate, and selection coefficients to the frequencies of alternative alleles. Knowledge of mutation rates and selection coefficients in natural populations shows that selection plays a dominant role in determining the frequencies of alleles except where population size or selection coefficients are extremely small. Most modern populations maintain sufficient genetic variability that they can evolve much more rapidly than has been observed for species known from the fossil record. However, despite the importance that most basic textbooks on genetics and evolution place on the study of clearly defined traits controlled by a single locus, such traits probably play a relatively minor role in large-scale changes in morphology and physiology over geological time.

Polygenic or quantitative inheritance

Conspicuous traits controlled by alternative alleles at a single locus have been discussed first in this chapter because they provide a readily visualized basis for understanding the mathematical relationships of various forces acting on allele frequency. Study of such traits has also shown how altered phenotypes result from specific changes in the genome.

Although a large number of traits have been discovered whose expression appears to be controlled by alleles at a single locus, most genes appear to have more complex interactions. Many genes influence several different aspects of the developing embryo and the structure and function of the adult organism, a phenomenon termed **pleiotrophy**. **Epistasis** refers to the circumstance in which two or more genes acting together produce phenotypes that differ from those expected when either locus is considered individually.

Other genes act together in an additive fashion to modify the degree of expression of a single characteristic. Characters so controlled are termed **quantitative traits**, as opposed to the **qualitative traits** that we have been considering. The terms **polygenic** and **multifactorial** are also used for traits controlled by a number of genes with additive effects. This last term refers to that fact that the study of quantitative traits is often complicated by environmental factors as well as genetic differences; for example, the size of a plant or animal may be as significantly affected by the fertility of the soil or the availability of food as by its genetic heritage.

Although study of traits controlled by a single locus will continue to contribute to understanding the nature of inheritance and the distribution of particular genes in populations, such traits may be of limited significance in determining the pattern and rate of evolutionary change over long periods of time. Darwin, in his earliest writing on natural selection (1859), distinguished between two kinds of traits: "sports," which were rare but conspicuous differences that appeared from generation to generation; and "fluctuating variability," smaller quantitative differences that were expressed in all populations – for example, differences in coat color, body weight, and number of progeny. He argued that fluctuating variability was much more important in long-term evolution. Unfortunately, it was also much more difficult to study genetically. Casual observations gave the impression that quantitative differences followed the then-prevalent view of blending inheritance. In general, progeny tend to be intermediate between the parents in most quantitative traits. Soon after the rediscovery of Mendel's work, genetic studies showed that quantitative traits are governed by the same rules of inheritance that apply to qualitative traits; the difference lies in the fact that many genes at different loci, frequently on different chromosomes, may contribute to the same trait (Fig. 9.4). Human stature, for example, may be controlled by scores of different genes, with more or less additive effects, and skin color may be attributed to four. An extremely useful review of the evolutionary aspects of quantitative genetics, with a minimum of formal mathematics, is provided by Barton and Turelli (1989).

In contrast with qualitative traits, it is rarely if ever possible to know the true genotype of quantitative traits. We still have very limited knowledge as to the actual identity of the individual genes involved, and even less of their mutation rate. All that can be established with reasonable assurance is the minimum number of genes involved. Selection coefficients cannot be established for each individual gene or allele, but the force of selection can be measured in terms of the amount of change in characters over time. All measures are based on phenotype.

Analysis of quantitative traits

Quantitative traits are characterized by continuous variability of expression within a population that typically shows a normal, or bell-shaped, pattern of distribution. Even four or five genes, represented by as few as two alternative alleles each, will give a nearly smooth curve, and many traits may be controlled by tens or even hundreds of genes. Even with only three or four genes, it is difficult to distinguish individual size classes because the expression of quantitative traits is also influenced by environmental factors. These may also influence the degree or nature of expression of alternative alleles for single genes, but this is a much more serious problem when dealing with quantitative inheritance.

Quantitative traits are described in terms of the mean and standard deviation of their expression in populations. The **mean** is the expression of a quantitative trait, averaged over all members of a population. The value of a character in an individual is expressed as a deviation from the population mean. The evolution of quantitative traits is measured by differences in the mean and variance of their expression in populations. **Variance** (commonly abbreviated as σ^2) is a measure of the

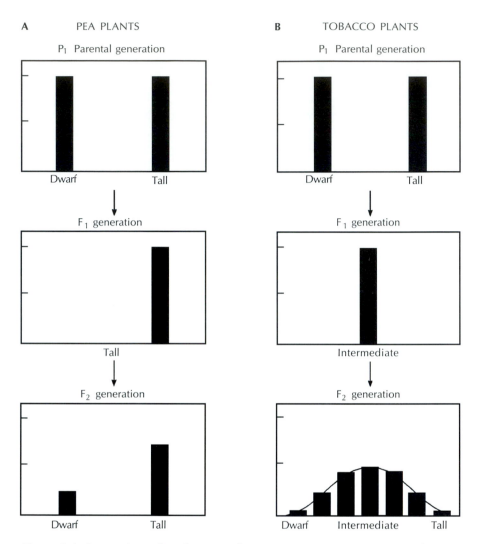

Figure 9.4. Comparison of qualitative and quantitative traits. **A,** Histograms showing the percentage of plants of different height in crosses between tall and short pea plants carried to the F_2 generation. In the pea plant, height is controlled by two alleles at a single locus, one of which is fully dominant. **B,** Histograms showing the percentage of plants of different height in crosses between tall and short tobacco plants carried to the F_2 generation. In tobacco, there are several genes, each with alternative alleles, all of which give additive effects. Reprinted by permission from page 403 of *Human Heredity: Principles and Issues* by Cummings, 3rd ed. Copyright © 1994 by West Publishing Company. All rights reserved.

deviation of the individual members of a population from the population mean. It is calculated as the sum of the squares of the deviation of each member from the mean, divided by the number of individuals.

$$V = \frac{\sum (x_i - \bar{x}_i)^2}{N}$$

The square root of the variance is termed the **standard deviation** σ. This is a fixed fraction of the area under the curve of a normal distribution, so that 68 per-

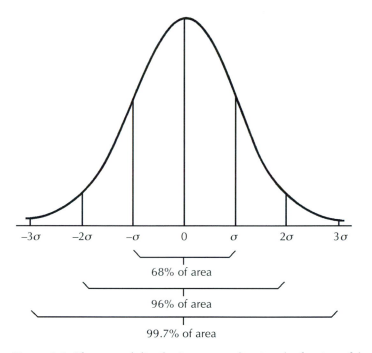

Figure 9.5. The normal distribution curve, showing the fraction of the area embraced by 1, 2, and 3 σ (standard deviations) on either side of the mean. From Futuyma (1986).

cent of the observations fall within 1 σ on either side of the mean, 96 percent fall within 2 σ, and 99.7 percent within 3 σ (Fig. 9.5).

Variance may result from several factors. The total phenotypic variance may be written V_P. The portion that is of particular interest to an evolutionary biologist is the genetic variance V_G, but variance nearly always has an environmental component V_E as well. The genetic variance itself will have two components if there is any degree of dominance expressed by alternative alleles. If one of two alleles, A, is fully dominant relative to the other, a, an individual that is phenotypically A may have a genotype of either Aa or AA. A parent with a genotype of AA will pass on the A allele to all of its progeny; a parent with a genotype of Aa will pass the dominant allele to only half of its progeny. The amount by which Aa heterozygotes deviate from the population mean is due to dominance and is not inherited by their offspring. The two components of the genetic contribution to the variance of the phenotype are termed the **dominance effect** V_D, and the **additive effect** V_A. Only the additive effect is significant for evolutionary change, and it is highest when alternative alleles have an additive effect rather than showing dominance.

In order to establish the effect that selection has on variance within a population, it is necessary to determine the relative amount of the total phenotypic variance V_P that is caused by the additive effect of the genetic component. This is termed the **heritability** of a character, notated as h^2 and expressed by the equation

$$h^2 = V_A / V_P$$

Heritability can be measured by correlation of traits between relatives, or by *response to selection*. Fig. 9.6 shows the correlation between beak size in the parents and offspring of the Galápagos finch *Geospiza fortis* during two different periods. Heritability is measured by the slope of the line. In this case the slopes of the two lines average 0.79; hence 79 percent of the variance in the depth of the beak is the result of additive genetic factors. With such a high heritability, selection should be very effective in promoting evolutionary change.

The effect of selection on quantitative traits

When selection is acting on a population, as in the case of the Galápagos finches, only a small portion of the population may survive to give rise to the next generation. The **selection differential** *S* is defined as the difference between the mean of the parents giving rise to the next generation and the mean for the entire parental population. The **response to selection** *R* is the difference between the mean phenotypic value of the offspring population and the mean of the entire parental population before selection. It can be calculated as $R = h^2S$.

In the study of *Geospiza fortis,* several characters were used to estimate selection differentials (Fig. 9.7). Following the drought in 1976–7, beak dimensions av-

Figure 9.6. The relationship between the beak depth of offspring and their parents in the medium ground finch *Geospiza fortis* on Daphne Major. The slope of the relationship is the heritability. In two years the slopes were nearly the same; 0.82 in 1976 (○) and 0.74 in 1978 (●). Offspring reached larger sizes in 1978 than in 1976. Crosses (+) indicate bivariate means. Redrawn from Peter R. Grant (1986), *Ecology and Evolution of Darwin's Finches*. Reproduced by permission of Princeton University Press.

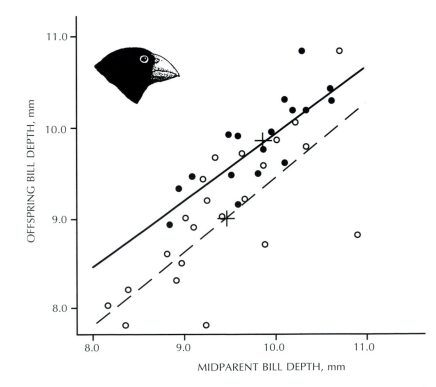

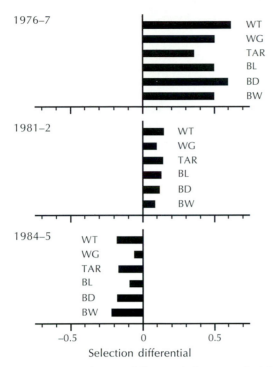

Figure 9.7. Selection differentials for traits of adult *Geospiza fortis* during two years of drought and low food supply, 1976–7 and 1981–2, and during the years that followed the El Niño event of 1982–3. Traits measured are abbreviated as follows: WT, weight; WG, wing length; TAR, tarsus size; BL, bill length; BD, bill depth; BW, bill width. Redrawn from *Nature* (Gibbs & Grant, vol. 327). Copyright © 1987, Macmillan Magazines Limited.

eraged 4 percent larger than the previous mean; this is the response to selection. The response to selection may also be expressed in terms of units of standard deviation – in this example, 0.47. Knowing the heritability – for beak size, 0.79 – and the response to selection, the selection differential can be calculated as

$$S = R/h^2 = 0.47/0.79 = 0.59 \ \sigma$$

After the extensive rains in 1983, selection acted in the opposite direction with a selection differential of $-0.2 \ \sigma$. If the response to selection and the selection differential are known, the same equation can be used to establish the heritability of the character.

A more general view of the effect of selection on quantitative traits can be gained from the data of Endler (1986). For both undisturbed and disturbed environments, these data show a predominance of small changes, over the range 0–2 σ (Fig. 9.8). Five examples taken from studies of fossil populations showed larger changes, but the distribution of values that were statistically significant did not differ from those in modern populations. This may be attributed to the smaller sample size of the fossil species.

Endler pointed out that there is extensive overlap between the range of values taken from animal breeding and artificial selection experiments and those found in natural populations. The geometric mean of the departures from the mean in natural populations is 0.59 σ, whereas that in selection experiments is 0.71 σ. This indicates that selection on quantitative traits in natural populations may be nearly as strong as that under experimental conditions. Clearly, natural selection has the potential for producing significant change in wild populations, at least in the short term.

The well-studied cases of change in the means of quantitative characters were generally over very short periods of time. Those of the Galápagos finches were limited by the duration of extreme climatic conditions. Larger-scale changes may ultimately be limited by the mechanical nature of the beak and the long-term availability of different food supplies. An additional factor that might limit the scope of change in populations is the availability of genes that can respond to selection. Under continuous strong directional selection, one would expect variability to be reduced at a higher rate than it could be replaced by new mutations, and the possibility for change would eventually become exhausted. In addition, other characters that are affected by strong selection may lead to reduced viability or fertility (Falconer 1989).

The power of artificial selection to act on quantitative traits is documented in a host of papers from two symposia on quantitative genetics (Pollak, Kempthorne, and Bailey 1977; Weir et al. 1988). In most of these examples, selection is narrowly directed and much more stringent than under any natural conditions, but they demonstrate the potential for extremely rapid change. One of the most striking features of quantitative traits is that populations under both natural and artificial selection can rapidly evolve far beyond the limits of variation in the original base population, not uncommonly more than 4 σ.

One of the most impressive and oft-quoted examples is that of selection in corn, *Zea mays* (Dudley 1977). For seventy-six generations, starting in 1896, separate lines for high and low oil and high and low protein were selected (Fig. 9.9). By the end of this period, the oil content had been increased by 279 percent or reduced

Figure 9.8. The distribution of directional selection differentials in units of standard deviation. The total height of each bar indicates the percentages of values in each interval, and the shaded portion indicates the percentage that are significantly different from zero at the 0.05 level. **A,** Undisturbed; twenty-five species, 102 of 262 values significantly different from zero. **B,** Disturbed; five species, 27 of 62 significantly different from zero. **C,** The distribution of selection differentials in studies of fossil and subfossil data; five species, 15 of 46 significantly different from zero. Redrawn from John A. Endler (1986), *Natural Selection in the Wild.* Reproduced by permission of Princeton University Press.

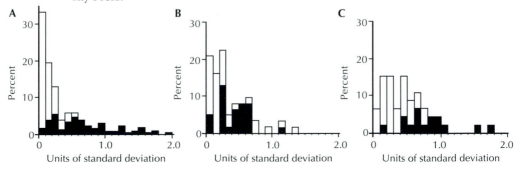

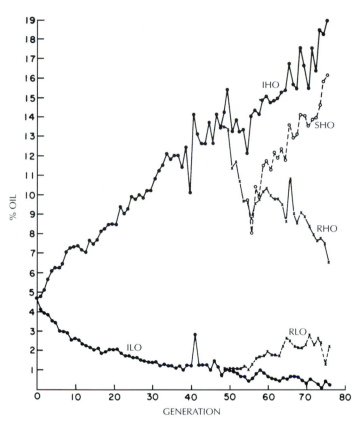

Figure 9.9. The results of seventy-six generations of selection for oil content in maize. IHO = line selected for high oil; ILO = line selected for low oil; RHO = reverse high oil; RLO = reverse low oil; SHO = switchback high oil. From Dudley (1977).

by 92 percent, and protein content increased by 133 percent or reduced by 78 percent. In terms of units of standard deviation, these changes can be expressed as 20 σ for oil gain, 6 σ for oil loss, 20 σ for protein gain, and 12 σ for protein loss.

One of the most important observations is that the capacity for change was apparently not diminished, even at the very end of this experiment, except in the case of the reduction in oil content, which was approaching zero. Reverse selection for each of these traits in subpopulations of the main lines showed that considerable allelic variability was retained, at least after fifty-two generations. The frequency of alternative alleles at this point is thought to have ranged from approximately 0.25 to 0.50. On the assumption that limits had been reached in the selected traits and that most loci were homozygous for the + alleles at the end of the experiment, it was estimated that 122 genes controlled the percentage of protein, and 54 the oil.

It is difficult to determine the degree to which linkage affected the response to selection because change proceeded at an almost regular rate throughout the experiment. Based on the number of chromosomes and the probability of crossing over, recombination would have little effect for the first seven to ten generations; by the end of seventy-six generations, however, all the genes would show essentially free recombination. That is, even over such a short time span, geologically speaking, linkage is not a constraint to stringent selection.

Given the importance of quantitative traits to large-scale evolution, it is vital that we learn more of their attributes. Unfortunately, many key aspects of quantitative traits remain poorly known, including the number of loci that contribute to variation, the rate of polygenic mutation, the extent of pleiotrophy, and the mechanisms that maintain additive variance (Barton and Turelli 1989).

Determining the number of genes contributing to quantitative variation between populations

Wright ([uncredited] in Castle 1921; Wright 1968) established a means to approximate the number of genes that determine the expression of a quantitative character in closely related lineages in which selection had favored either increasing or decreasing the character in question. Lande (1981) showed that this method could be extended to natural populations and even different but closely related species. Estimates are obtained by comparing the phenotypic means and variances of the character in parental, F_1, F_2, and backcross populations according to the equation

$$n_E = (\mu p_2 - \mu p_1)^2 / 8 \sigma_s^2$$

in which n_E is the effective number of genes, μp_1 and μp_2 are the differences in means of the parental populations, and σ_s^2 is the extra genetic variance segregating in the F_2 population beyond that in the F_1 hybrids. It is assumed that all genetic variance is additive.

Over short periods, such as those covered by breeding experiments, the effective number n_E of elements (or freely segregating genetic factors) that can be detected through comparison of variance is limited by the number of chromosome segments that are regularly recombining. This equals the haploid number of chromosomes plus the mean number of crossovers per gamete.

Lande (1981) provided data on estimates from several studies: mean weight of tomato fruit, percentage of oil in maize kernel, female head shape in two species of Hawaiian *Drosophila,* anthesis in goldenrod, human skin color, and the diameter of the eyes of fish living in a normal and a cave environment. The estimated number of effective genes ranges from approximately four to twenty-two (Table 9.3). All involved large changes in the mean of the parental populations, ranging from six to thirty phenotypic standard deviations. The cave populations may be as little as 10,000 years old. Lande used these results to support the neo-Darwinian theory that large evolutionary changes usually occur by the accumulation of multiple genetic factors, each with relatively small effects.

Much larger numbers of loci controlling quantitative traits were hypothesized by Bailey (1985, 1986), who argued that more than a hundred genes would be required to account for morphometric differences in the configuration of the mandible in different strains of the laboratory mouse. Atchley (1993) pointed out that this number may include many genes that also act on other parts of the skeleton. Only one of these putative genes has been recognized as influencing the shape of the lower jaw (*Mdmg-1*), compared with about a hundred genes that have been identified as influencing all other aspects of the skeleton in the mouse. Whether the average number of loci affecting individual quantitative traits is in the range of five to twenty or is over a hundred has not been resolved (Barton and Turelli 1989). The

Table 9.3. *Number of loci for quantitative traits in various organisms*

Organism	Character	No. of loci	Units of std. dev.
Maize	Percent oil in kernels	16–22	9
Drosophila	Female head shape	6–8	8
Goldenrod	Date of anthesis	5–	15
Human skin	Skin color	~4–6	6
Tomato	Fruit weight	10	14
Fish	Eye diameter	~6	30

Source: Data from Lande (1981, Table 3).

higher the number of loci, the more responsive a species is to selection, but the more difficult both mathematical or genetic analysis would be.

Rate of accumulation of quantitative traits

The mutation rates for genes that affect quantitative traits are nearly impossible to measure directly, but it is possible to estimate the rate of accumulation of genetic variability that must include loci involved with quantitative traits. This can be achieved most effectively in *Drosophila,* in which chromosome manipulation and inbreeding can result in the production of nearly homozygous populations. These can then be examined visually and analyzed by electrophoresis to determine how rapidly total variability returns to a normal level. This occurs amazingly rapidly, over as short a period as 500–1,000 generations (Maynard Smith 1989).

Lande (1988) cited evidence that spontaneous polygenic mutations are thought to make an important contribution to the response to selection within what are described as "long-term" programs of artificial selection; "long-term" meaning less than 100 years! This implies either very large numbers of quantitative genes or extremely high rates of spontaneous mutation, or both. The latter are suggested as being associated with transposon activity in particular studies of *Drosophila melanogaster* (Lynch 1988).

The total amount of variability that can be attributed to mutations (V_m) is measured in comparison with the variability that results from environmental differences (V_e). Lynch (1988) stated that the ratio of these two factors, $V_m : V_e$, may commonly range from approximately 10^{-4} to 5×10^{-2}, with exceptional cases as high as 10^{-1}, based on a diversity of organisms including *Drosophila, Tribolium,* mice, and several crop plants. That is, the amount of *new* genetic variability involving quantitative traits that accumulates each generation may be as high as 1 percent, or even 10 percent, of the variability resulting from environmental factors. This provides an extensive amount of genetic variability on which even stringent selection can act to produce the very high rates of continuing change observed under experimental conditions. These rates are several orders of magnitude higher than the rates of evolution reported in groups known from the fossil record.

Nature of genes for quantitative traits

Despite the apparently great importance of genes controlling quantitative characters to both short- and long-term evolutionary change, the actual nature and identity of such genes remain uncertain. Changes in the size, proportions, and configuration of skeletal elements provide most of the evidence for the pattern of
vertebrate evolution, yet only a very small portion of the more than a hundred
genetic traits that have been recognized as affecting the mouse skeleton are of the
sort that are recognized in fossil lineages (Lyon and Searle 1989). A few produce
conditions that broadly resemble traits occurring in related taxa, such as differences
in the number and configuration of the vertebrae, change in shape and proportions
of limbs, and greater or lesser numbers of digits; but the vast majority (nearly all of
the 114 listed genes) are elements of broad systemic abnormalities, such as dys-
function in cell aggregation, migration of the neural crest cells, chondrification, and
ossification, nearly all of which are either fatal or lead to greatly reduced viability
and/or fertility. Clearly, none of these mutations could be responsible for the types
of change that are encountered in the evolution of vertebrates. They simply inter-
fere with the normal function of well-established developmental processes.

On the other hand, Grüneberg (1963), concentrating on morphological abnormalities rather than specific mutant alleles, identified a number of skeletal differences of the sort that are typical of evolutionary changes. One, in guinea pigs, was
originally studied by Wright (1934): Whether there are three or four toes is governed by three or four genes with additive effects that determine whether an animal crosses or fails to cross the threshold to polydactylism. Similar genetic control
affects the size and presence or absence of the third molar tooth in mice and the
number of presacral vertebrae.

Falconer (1989) argued that genes that cause variation of metric characters are
not a special category of genes. Rather, their effects may be secondary to their primary function in development or metabolism. Study of allelic series of genes affecting blood groups in cows, for example, showed significant differences in milk
fat percentage. Another study showed association of change in vertebral shape
with selection for body weight in mice (Johnson, O'Higgins, and McAndrew 1988).
Regulatory genes, which control the amount of gene-product, may be a major
source of the variation of metric characters, but there is little concrete evidence.
Some of the genes that have many multiple copies may be associated with quantitative traits, but this too has not been documented.

Although it remains very difficult to establish the nature and number of genes
that are responsible for quantitative traits, their existence and polymorphism are
amply demonstrated by the extremely rapid response to selection that is possible
under both natural and artificial selection over short periods of time. They must
play an equally important role in large-scale, long-term evolutionary change.

The enigma of low selection coefficients for long-term
evolutionary change

Quantitative traits appear to have all the attributes required by Darwin's theory of
evolution to explain the manner in which organisms respond to long-term selec-

tion through the modification of all aspects of their anatomy, physiology, and behavior. Ironically, although the rates and patterns of evolutionary change in modern populations accord well with the genetic potential for rapid change, most of the evidence available from the fossil record indicates much lower rates of evolution. As demonstrated by Gingerich (1983), measured rates of evolution appear several orders of magnitude slower. The data he analyzed included 228 vertebrate examples covering time spans from 8,000 to 98 million years. They showed a geometric mean of 0.08 darwins, compared with post-Pleistocene mammals (3.7 d) and much higher rates in living populations.

Lande (1976) pointed out that the rate of evolution measured over geological time was so slow that it was difficult to differentiate the pattern of change from that which might result from genetic drift. He used the expression

$$N^* = (1.96)^2 \, h^2 t / (z/\sigma)^2$$

where N^* is the effective population size at which there is a 5 percent chance of randomly drifting a distance at least z in either direction in t generations; and h^2 is the heritability of the trait being considered. Here, σ is the number of units of standard deviation used to establish the maximum effective population size that would allow genetic drift to produce the same phenotypic change as that expected from the selection coefficient postulated to explain the rate of morphological change.

Lande used published accounts of the evolutionary rates among Tertiary mammals, including horses, the primitive ungulate *Hyopsodus,* and oreodonts, to show that they could be accounted for by random change in fairly large populations. Among the horses, he used the amount of change in two dental features, paracone height and ectoloph length, in four transitions between genera (Table 9.4). The force of selection was measured as the proportion of individuals in a population that were selected against per generation. The most stringent selection removed two per million; the weakest, two per ten million. In order for these to exceed the rate of change expected as a result of random drift, the population sizes would have had to be greater than 10,000 in the first example and over 5 million in the second. These results suggest that most rates of morphological change that have been determined from the fossil record are so slow that they could be attributed to random drift.

Turelli, Gillespie, and Lande (1988) provided an alternative equation that took into consideration the complication of new mutations that would be expected to accumulate over geologically long periods of time. The data on Tertiary horses previously compiled by Lande were recalculated, but again they indicated that the observed rate of evolution was not too rapid to be explained by drift in populations of reasonable size.

This problem was later tackled by Lynch (1990). He too concluded that the mean rate of morphological evolution in all Cenozoic mammals, with the possible exception of *Homo sapiens,* was well below the rate expected by random drift acting on neutral mutations. The only role he accorded to selection was stabilization, when the rate of change was below that expected from random drift.

The conclusions reached by Lande and his colleagues imply that the forces of selection necessary to account for the rates of evolution observed from the fossil record were five- to sevenfold weaker than those observed for morphological changes in modern populations. This seems extremely unlikely. In fact, it is simply

Table 9.4. *Evolutionary changes in dental features among Cenozoic horses*

(A) Mean and standard deviation transformed to natural log scale

Species	ln paracone height		ln ectoloph length	
	$\bar{z}$	σ	$\bar{z}$	σ
A. *Hyracotherium borealis*	1.54	0.062	2.11	0.056
B. *Mesohippus bairdi*	2.12	0.048	2.48	0.046
C. *Merychippus paniensis*	3.53	0.059	2.99	0.053
D. *Neohipparion occidentale*	3.96	0.046	3.03	0.053

(B) Amount of morphological changes in units of phenotypic standard deviation (avg. %), from (A)

Transition	z/σ		t generations
	ln paracone height	ln ectoloph length	
A–B	10.6	7.1	10×10^6
B–C	25.6	9.8	5×10^6
C–D	7.8	0.8	1.75×10^6
A–D	44.0	17.7	16.75×10^6

(C) Minimum amount of selection (proportion culled per generation) needed to explain the transition in (B) using $h^2 = 0.5$, and assuming no genetic drift

Transition	Minimum selective mortality	
	ln paracone height	ln ectoloph length
A–B	4×10^{-7}	3×10^{-7}
B–C	2×10^{-6}	8×10^{-7}
C–D	2×10^{-6}	2×10^{-7}
A–D	1×10^{-6}	4×10^{-7}

(D) Effective population sizes for rejection of the neutral hypothesis at the 95% level of confidence for the transitions in (B)

Transition	N^*	
	ln paracone height	ln ectoloph length
A–B	2×10^5	4×10^5
B–C	1×104	1×10^5
C–D	6×10^5	5×106
A–D	2×10^5	1×10^5

Source: Modified from Lande (1976, Table 1a–d).

Table 9.5. *Estimated force of selection on quantitative traits distinguishing subspecies of the rodent* Peromyscus, *over four time intervals*

| | Taxon transition | | | | | |
| | PMB → PMN | | PMB → PLN | | PMN → PLN | |
t	b	$Q(b)$	b	$Q(b)$	b	$Q(b)$
10^3	2.9	1.9×10^{-3}	2.7	3.8×10^{-3}	2.7	3.7×10^{-3}
10^4	3.6	6.0×10^{-4}	3.4	3.1×10^{-4}	3.4	2.9×10^{-4}
10^5	4.2	1.4×10^{-5}	4	2.6×10^{-5}	4	2.5×10^{-5}
10^6	4.7	1.2×10^{-6}	4.6	2.3×10^{-6}	4.6	2.2×10^{-6}

Abbreviations: PMB = *Peromyscus maniculatus bairdii;* PMN = *P. maniculatus nebras-censis;* PLN = *P. leucopus noveborascensis.*
Source: Modified from Lofsvold (1988, Table 6).

another aspect of the problem that Gingerich (1983, 1993b) faced when analyzing evolutionary rates (see Chapter 4). He demonstrated that rates of evolution depend on the time scale over which they are measured: They are very rapid over short intervals but appear very slow over longer ones. Gingerich argued that the apparent difference between rates of evolution for modern populations and those for fossil lineages was the result of averaging changes over time. Clearly, the selection coefficients that are associated with evolutionary changes are subject to the same averaging effect. Selection coefficients and selection differentials calculated by averaging change over hundreds of thousands or millions of years will give the same spurious impression of very low values as those for rates in darwins, or other morphological measures.

Lofsvold (1986, 1988) provided an informative study of how widely differing rates of evolution could be used to explain species-level changes in a group of living rodents. He studied differences in skeletal characters in three subspecies of the cricetid rodent *Peromyscus,* undertaking multivariate analysis of variance of fifteen divergent characters of the skull and lower jaw and estimating their additive genetic variance and covariance by regression of offspring means on sire values. He had no evidence for the actual times of phylogenetic divergence, but established selective mortalities per generation as a measure of the force of selection, over a range of time intervals of 1,000–1,000,000 generations (Table 9.5). For the observed means of the differences in phenotype of the three possible transitions between the subspecies, the selective mortalities per generation varied from 1 in 1,900 to 1 in 2.3 million. Even the most rapid of these measures is much slower than some observed in modern populations. Using the same measure of selection, *truncation mortality* (i.e., all animals above or below a certain limit fail to reproduce), the equivalent number for the force of selection acting on the Galápagos finch *Geospiza fortis* from a drought in 1977 is 1 in 260; for other studies of modern populations, it is 1 in 100 and 1 in 4,000. He concluded that it was much more plausible to attribute changes of the magnitude observed in rodent species to short episodes of intensive selection than to continued selection at very low rates over long periods of time. Lofsvold showed that differences great enough to distinguish subspecies and species, which may have required half a million years, could have been carried out in a thousand years at a rate of selection comparable to those observed in modern populations.

Very rapid rates of evolution have been observed and measured in a small number of modern species, including the immigration and radiation of house sparrows in North America, the response to climatic changes among the Galápagos finches, and the response to different predators among stickleback fishes. Certainly, comparable situations, although rarely observed in the fossil record, must have occurred throughout evolutionary history. Terrestrial vertebrates are almost never known from sufficiently complete remains and from close enough sampling intervals to determine how general this phenomenon may be. On the other hand, unicellular organisms preserved in deep-sea sediments accumulated at relatively constant rates do provide plausible examples of rapid changes within long-lived lineages.

Rapid variation over time is clearly shown by foraminifera and radiolarians (Kellogg 1983; Malmgren, Berggren, and Lohmann 1983). In the foraminifera *Globorotalia* (Fig. 9.10), change over a period of 500,000 years adjacent to the Miocene–Pliocene boundary is sufficiently great for the recognition of two species; but throughout both the ancestral and parental lineages, as well as during the intermediate period, there are also large-magnitude changes from sample to sample, taken at 5,000–15,000-year intervals.

On the basis of differences in their means, it would be difficult to distinguish whether individual pairs of successive samples had been taken from within the period of transition between the two species, or within the ancestral or descendant lineages. Selection coefficients may have been of nearly equal magnitude. What differentiates the transitional period is the relative persistence of selection in a single direction.

Malmgren et al. (1983) referred to the overall pattern of evolution in this example as *punctuated gradualism,* in reference to the long-term absence of evolutionary trends within each of the species, separated by a relatively short period of rapid phyletic change without lineage splitting. I would like to emphasize another aspect of this record: the continuing pattern of oscillatory change throughout the entire

Figure 9.10 (facing). Oscillatory evolution in late Cenozoic marine unicellular organisms. **A,** The foraminifera *Globorotalia*. Variation in shape (second eigenshape) of the *Globorotalia* lineage in site 214. Mean and 95 percent confidence intervals shown by horizontal lines. Size shown by outline of tests. M, Late Miocene; P, Pliocene; Q, Quaternary. Inset is a close-up of the changes across the Miocene–Pliocene boundary showing in detail the *G. plesiotumida–G. tumida* transition. Sampling resolution is here in the range $5–15 \times 10^3$ (from Malmgren et al. 1983). **B,** The radiolarian *Eucyrtidium*. Mean width in microns of the fourth segment of *Eucyrtidium* versus depth in core V20-105. In the magnetic stratigraphy, black indicates times of normal polarity and white times of reversed polarity. Dotted lines connect points at which samples consisted of five or fewer specimens, in order to indicate that caution should be exercised in the statistical interpretation of these extremely small samples (from Kellogg 1983). The size of each sample and 95 percent confidence intervals around their means were published as Kellogg (1976, fig. 4).

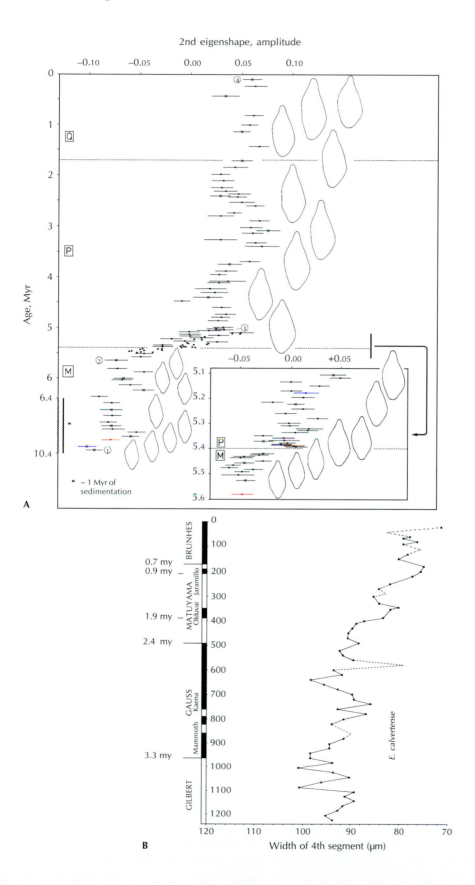

A

B

evolutionary sequence. Eldredge and Gould (1972 and later papers by each) argued that species should be considered as showing stasis even if they undergo oscillatory change throughout their duration. If they do not exhibit net directional change over time, they are not considered to have evolved. Nevertheless, the same phenomenon observed over seasons, years, or decades among living organisms would be accorded selective values and evolutionary rates. Measurable change during the duration of species, even without overall directional change in mean (see Figs. 5.2, 5.5), shows a pattern of evolution that is probably common to all groups. There is no reason not to assume that such oscillations also occur during long-term directional change, as suggested in Gingerich's (1976, 1987) data on Tertiary mammals. Rates averaged over very long time spans may appear very slow, but they almost certainly comprise much more complicated patterns over shorter time scales. Such patterns of evolution may appear as random walk but can be interpreted as continuous tracking of small environmental changes.

Emphasis has long been placed on three modes of selection: stabilizing, directional, and disruptive. It is important to recognize a fourth pattern – oscillatory – that probably characterizes subunits of time within the evolution of most species. The periodicity and amplitude of oscillation may vary both within and between groups, and might be either regular, as would be expected in response to regular environmental oscillation (e.g., Milankovich cycles), or highly irregular. The pattern of racemic or iterative evolution in freshwater populations of the threespine stickleback provides a further example, in which selective forces on a species may differ greatly under differing environmental conditions (Bell and Foster 1994b).

In contrast with many mathematical models proposed for analysis of evolutionary patterns, there is no justification for the assumption that selection coefficients or selection differentials are constant over time. In contrast, one can assume that the range of selection coefficients seen in modern populations applies to most periods in the history of individual species and throughout vertebrate evolution.

Continual changes in the intensity and direction of selection are in strong contrast with the relative constancy of mutation rates for particular alleles. This indicates a much different pattern of expression of evolutionary forces mediated by the organism itself (mutations) compared to that resulting from environmental interactions (selection).

Genetic constraints

To what degree does the evidence discussed in this chapter support the hypothesis that processes inherent to the level of the gene constrain the rate, direction, and pattern of evolution?

Genetic constraints were defined by Loeschcke (1987, p. 1) in his introduction to the book *Genetic Constraints on Adaptive Evolution* in the following manner: "Genetic constraints on adaptive evolution can then be understood as those genetic aspects that prevent or reduce the potential for natural selection to result in the most direct ascent of the mean phenotype to an optimum." Whereas *adaptive evolution* may be used very broadly to include nearly all aspects of evolutionary change and radiation of vertebrates, the phrase *genetic aspects* is here restricted to exclude developmental constraints (which are considered in Chapter 10).

Genetic constraints are limited here to those factors involving the molecular and chromosomal structure of the hereditary material and the changes that it undergoes during evolution. The basic structure of DNA, the manner of coding for amino acids by nucleotide triplets, and other aspects of protein synthesis specifically controlled by the genes are very strongly constrained and presumably have been so since the origin of life as we know it today. The nature of these molecules and the chemical processes that they undergo were presumably subject to natural selection during the early stages in the origin of life and during the origin of eukaryotes, but are essentially constant among all living eukaryotes. This is emphasized by the fact that all metabolic processes among eukaryotes appear to be associated with fewer than 900 basic proteins or protein domains (Green et al. 1993). Changes at this level are presumably extremely rare among vertebrates and may be considered an almost unalterable constraint. In fact, the constancy of these basic genetic elements and processes is so great that it cannot contribute at all to understanding differences in evolutionary patterns and processes that are observed among vertebrates.

There are many levels of mutational change that may be considered. The frequency of substitution of the different bases is constrained, both individually and in the context of surrounding bases (Golding 1987). This constraint probably acts in a similar fashion throughout eukaryotes and would not be expected to be reflected in the different patterns and rates of evolution among the various vertebrate groups.

There is much variation in mutation rates at the level of enzymes and other proteins. Among species, the mutation rates differ greatly from gene to gene, and to a lesser degree for specific genes, both of which may be subject to some control by selection. Limits to the range of possible mutation rates certainly act to slow or prevent changes that would lead to improved adaptation. There are almost certainly limits to the nature of mutational changes that may occur. It may be impossible for mutational changes to result in the synthesis of particular enzymes or other proteins within some groups; for example, no eukaryotes are capable of synthesizing cellulase, although it is produced by some prokaryotes. All vertebrates that digest cellulose do so as a result of a symbiotic relationship with bacteria that produce cellulase. Is there a genetic constraint that precludes eukaryotes from evolving a gene to synthesize cellulase, or is it simply easier or metabolically cheaper to form a symbiotic relationship with bacteria than it is to evolve the means to produce a novel substance?

Mutation rates appear very low compared with the forces of selection, but the total genetic variability in populations is sufficient to support sustained periods of change for hundreds or thousands of years at rates far above the average rates of evolution within most groups. The rate of mutation of genes for morphological traits is certainly too high to explain the existence of so-called living fossils or the long-term stasis that characterizes the basic body plans in most taxonomic groups.

Pleiotrophy, the phenomenon in which particular loci control many aspects of development and metabolism, may significantly constrain the rate of change in individual traits. Selection may act in different directions on different effects of the same allele, limiting the direction of evolution to that which is most strongly selected. The reduction in viability and fertility often observed under stringent artificial selection may be the result of selection for quantitative traits that are influenced by genes that also contribute to basic metabolic pathways.

Other constraints occur at the chromosomal level. The fact that genes are attached to one another in chromosomes is certainly a constraint, since the closer they are linked, the less likely they are to assort independently during reproduction. Long-term selection may act to conserve or alter different arrangements of genes within and between chromosomes, but these arrangements are resistant to change over short periods of time. For example, the full range of expression of multifactorial traits may appear only after scores of generations of stringent selection, yet some alternative alleles will commonly remain available to respond to changing conditions. This is clearly a case in which the potential for natural selection is precluded from leading to the most direct ascent of the mean phenotype to an optimum. On the other hand, this phenomenon is presumably so widespread among vertebrates that it does not form a basis for explaining different rates and patterns of evolution from group to group, or within groups over time.

In fact, this appears to be the case for nearly all genetic constraints. They may slow or even preclude particular paths of evolution, but most changes that do occur take place sufficiently rapidly that genetic constraints cannot be used to explain the much slower rates of evolution of the same characters in other groups, or during different periods of time.

A striking example of the lack of concordance between the rates of genetic versus anatomical change is evidenced by studies of the rates of evolution at the molecular level in living fossils. Avise, Nelson, and Sugita (1994) compared the rate of change in 16S rRNA gene sequences from mtDNA in two groups of arthropods: horseshoe crabs, noted for the extremely slow rate of morphological evolution, and king crabs and hermit crabs, which show marked change in body size and proportions (Fig. 9.11). The rates of molecular change were essentially the same in the two groups. All other groups of living fossils in which molecular changes have been studied in this fashion have rates of change comparable with related taxa that show much more rapid rates of morphological evolution. To the degree that comparisons can be made, the rate of change at the level of the gene and the basic enzymes controlling metabolic processes is essentially constant in all groups, regardless of the amount or rate of morphological change. The two processes are, in essence, uncoupled.

In conclusion, there are many phenomena that may be considered genetic constraints, but they provide relatively little help in understanding the directions and rates of evolution that actually occur over millions and hundreds of millions of years. Evidence from both qualitative and quantitative traits shows that mutation rates, the amount of variability in populations, and selection coefficients that control gene frequencies in modern populations are sufficient for evolutionary change to occur at rates that are far in excess of those seen in the fossil record. Rather than acting as a constraint to evolution, genetic phenomena appear as a means of facilitating change at a very rapid rate at the level of populations and species while retaining sufficient variability for adaptation to occur in a diversity of directions for millions of years. No aspects of genetics appear to explain either long-term stasis of anatomical features at any taxonomic level or the particular patterns of evolution that are characteristic of the origin, radiation, and persistence of major taxonomic assemblages.

As summarized by Barton and Turelli (1989, p. 362):

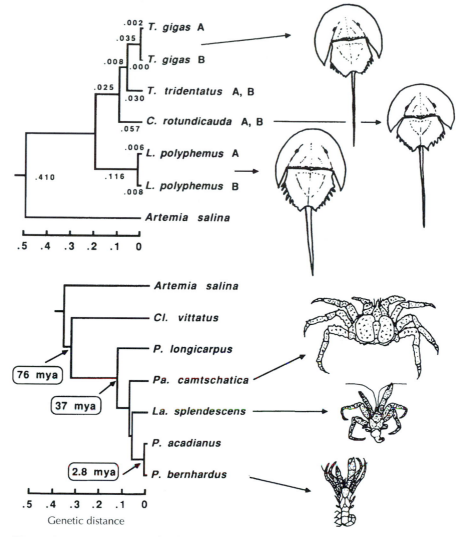

Figure 9.11. Comparative dendrograms for horseshoe crabs *(top)* and selected king crabs and hermit crabs *(bottom)* based on 16S rRNA gene sequences from mtDNA. Dendrograms based on unweighted pair-group method with arithmetic means. The genetic distance scale is the same for both phenograms, and is drawn to indicate mean clustering levels (average sum of branch lengths connecting relevant nodes and extant taxa). Also shown for the horseshoe crab lineages are individual branch lengths in a neighbor-joining tree that produced the same branching structure. The indicated dates for particular nodes (mya, millions of years ago) stem from vicariant geological evidence. *T., Tachypleus; C., Carcinoscorpius; L., Limulus; Cl., Ciblanarius; P., Pagurus; Pa., Paralithodes; La., Labidochirus.* The brine shrimp, *Artemia,* is an outgroup common to both studies. From Avise et al. (1994).

Even very weak selection can produce rates of phenotypic evolution that are extremely fast on the geological time scale. Thus, there is no basis for claims that developmental constraints (in the accepted sense of genetic correlations) cause stasis and no support for the view that population bottlenecks are necessary for rapid morphological evolution.

10 Development and evolution

The turn of the century saw the emergence of genetics, embryology, and evolutionary biology as three separate fields. Before that time, they were approached as a single discipline, and the term *development* applied to them all. By 1926, however, the three sciences had their own journals, their own textbooks, and their own jargon. Moreover, each discipline had largely ceased referencing papers in the other two fields. The geneticists studied the transmission of inherited traits, the embryologists studied the expression of these traits, and the evolutionary biologists studied the changes, selection, and propagation of these traits over many generations.
– Gilbert (1988, p. 631)

Genetics and development

Embryological development has long been considered key to understanding major features of evolution. Following von Baer (1828), Darwin emphasized the similarity of early developmental stages as evidence of the interrelationships of the major groups of vertebrates, and many other scientists have argued that the progressive changes that are seen in embryogenesis reflect the changes that occurred during evolution (e.g., Haeckel 1866; Gould 1977; Thomson 1988; Hall 1992). Despite the obvious role that changes in developmental patterns and processes must have played throughout the history of life, developmental biology did not contribute significantly to the formulation of the evolutionary synthesis (Mayr and Provine 1980; Gilbert et al. 1996). The major new support for the validity of Darwinian selection theory came not from embryology but from an understanding of the nature of inheritance, knowledge of the quantitative importance of mutation and selection, and appreciation of the significance of speciation.

Since modifications in developmental patterns and processes presumably reflect changes in genes comparable to those that have long been studied by geneticists and population biologists, it seems ironic that genetics has long overshadowed developmental biology in studies of the mechanics of evolutionary change. In fact, there are significant differences between the nature of the majority of genetic differences studied by evolutionary biologists and of those that contribute to understanding the embryological patterns and processes of interest to developmental biologists.

Developmental biology is the study of the factors that control the differentiation of individual cells and their organization into the many different tissues, organs, and organ systems of multicellular organisms. Development in vertebrates begins with a single fertilized egg, bearing the combined genetic components of the two

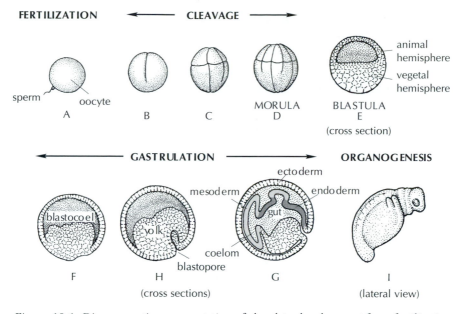

Figure 10.1. Diagrammatic representation of chordate development from fertilization to organogenesis. By the morula stage, the cells form a hollow sphere. During gastrulation, three types of embryonic tissues, termed the *germ layers,* develop. The external surface of the gastrula is made up of the *ectoderm,* which later forms the skin and much of the central nervous system. The gut develops by invagination at the blastopore; its lining is formed by the *endoderm.* A third tissue, the *mesoderm,* forms by outpocketing of the gut. It has two layers, separated by an open space, the *coelom.* The mesoderm forms the muscles, most of the bones, and other internal structures of the adult.

parents. Cell division that begins in the zygote eventually produces all the cells of the adult organism, all of which carry the same genetic information whether they are brain cells, muscle cells, or the cells that produce bone and cartilage. The key to cell differentiation is the activation and regulation of particular genes in the precursors of each different kind of cell. Cell fate is also influenced by where cells are located in the body, their relationship to other cell types and tissues, and the timing of their genetic activation.

Determination of the anterior–posterior and dorsal–ventral axes of the body is among the earliest developmental events, closely followed by the establishment of the body regions and germ layers (Fig. 10.1). This determines the framework or body plan within which organs and organ systems develop. From this pattern, structural features that distinguish each of the major vertebrate groups are rapidly elaborated. It is only late in development, when their anatomical patterns are nearly fully formed and the young are close to hatching or birth, that the traits commonly studied in classical genetics become evident. Many, such as the color of the flowers or the stature of the plant or animal, are manifest only when the organism reaches maturity.

In general, the earlier a gene is expressed, the greater influence it can have on subsequent events during development and the more likely that its disruption will result in malformation and death. According to studies reported by Gilbert (1988, p. 187), at least 50 percent of human conceptions abort spontaneously within the first month, and 90 percent of these show evidence of developmental abnormalities. It is hardly surprising that traits that control major features of the organism are not susceptible to study through procedures of population genetics. If nearly all mutational changes in the genes that control these traits are lethal, they will not be expressed as alternative alleles in living organisms and thus will not provide the opportunity for positive selection leading to evolutionary change.

There are a few mutations among well-studied organisms that are responsible for the gain or loss of a finger or toe in otherwise normal individuals, but most of the changes that produce skeletal abnormalities are associated with grave disturbances of developmental processes leading to malformations in many organ systems (Lyon and Searle 1989; McKusick 1994). Thus, it is rarely possible to speak in terms of mutation rates and selection coefficients of genes that might lead to changes in basic patterns or processes of development. Most mutations known to cause changes in developmental processes are too damaging to provide a model for the manner in which evolutionary advances may have occurred. There must be many other mutations that result in minor changes to individual parts of the body, but few have been recognized as related to specific embryological processes.

Heterochrony

In addition to the origin of totally new structures or major reorganization of existing patterns, much of evolution can be seen as proceeding by incremental modification of the size and proportions of already existing structures. This can be broadly attributed to modification in the timing of developmental processes. The development of some parts of the body may be either accelerated or delayed compared with others, a process known by the general term **heterochrony**. For example, the skulls of apes and humans closely resemble one another early in development; subsequently the facial region develops more slowly in humans, retaining the proportions of juvenile apes, while the period of growth and expansion of the braincase is greatly prolonged, allowing for a brain size at least three times the volume of that in apes of comparable body size.

Among the most clearly documented examples of heterochrony are seen in the development of salamanders (Duellman and Trueb 1986). Although most groups of salamanders pass through an aquatic larval stage that matures into a terrestrial adult, all the members of four families – Sirenidae, Amphiumidae, Proteidae, and Cryptobranchidae – are permanently aquatic. In other families, particular species may be obligatorily aquatic, or have the capacity to metamorphose into terrestrial adults, depending on environmental conditions. In species that undergo metamorphosis, the reproductive organs complete development in the terrestrial stage. In contrast, the permanently aquatic forms exhibit a particular category of heterochrony, termed **neoteny,** in which the reproductive system matures while other aspects of the body retain a level of development comparable to that of the larval stage of other genera.

The most famous neotenic salamander is the axolotl, *Ambystoma mexicanum,* which does not metamorphose in its natural environment but can be induced to do so by injections of the hormone thyroxin. In obligatorily neotenic salamanders, the tissues have little or no ability to respond to thyroxin, indicating that there has been a genetic alteration that precludes metamorphosis. The adult morphology and life-style of these animals is hence radically altered by what may have been a very simple genetic change.

Other aspects of heterochrony have been attributed to a particular category of genes, termed **regulatory genes**, that control the timing and degree of expression of other genes that code for the formation of structures.

Heterochrony was extensively discussed by Gould (1977), and many informative examples were provided in the recent volume edited by McNamara (1995). This concept provides a very useful way in which to view significant changes involving particular structures and even major modification of the entire body form and way of life. The remainder of this chapter, however, is devoted to other aspects of development that have the potential for providing detailed molecular explanation of both heterochrony and all other changes in the physical form of the body throughout evolution.

Until recently, it has been extremely difficult to investigate how changes in developmental processes might produce new characters and new body plans; however, the capacity to do so is increasing rapidly as a result of new tools of biotechnology. It is now practical to isolate genes defined by mutations and determine their DNA sequences; this, in turn, establishes the sequence of the amino acids in the proteins coded for by the genes, which often gives hints as to their probable function. The function of a gene can also be investigated by establishing the spatial and temporal regulation of its expression throughout development in both normal and mutant individuals. Specific genes can also be disrupted or deleted, and their normal function be inferred by what happens when they are absent.

Among the most important families of genes for studying the development of major structural elements are *homeotic genes,* discussed in the next section. Knowledge of these genes has already provided significant insights into the appearance of major new structures associated with the origin of vertebrates from nonvertebrate chordates and into the processes that control the way in which the vertebrate head, trunk, and limbs originated and evolved.

Homeobox genes

Homeotic mutants were first recognized and named by Bateson (1894) more than a century ago. In the fruit fly *Drosophila,* such mutations result in the appearance of structures in inappropriate segments of the head and thorax. One of the earliest mutants to be identified was *Antennapedia,* in which a pair of legs, similar to those of the second thoracic segment, developed in place of the normal antennae on the head. Goldschmidt (1938, 1940) used these and comparable observations to argue that new developmental patterns and adult structures might have appeared as the result of single, major chromosomal changes termed *systemic mutations.* Since most changes in the expression of major body structures are not advantageous, Goldschmidt coined the term **hopeful monsters**, indicating that out of thousands (or perhaps millions) of maladaptive mutations, a very few might confer a significant advantage and lead suddenly to major new structures or body plans. Simpson (1944, 1953) and more recently Wallace (1985) strongly attacked Goldschmidt's hypothesis, pointing out that it was not substantiated by knowledge of the nature of the genetic material and that neither the fossil record nor any modern species provided evidence for the instantaneous appearance of totally new structures from one generation to the next. Since patterns of development and adult structures both require extremely close integration of all their component elements and relationships with other aspects of the structure, physiology, and behavior of the organism, it is nearly impossible to conceive of the evolution of new functional complexes occurring through an instantaneous, global restructuring of the genetic material.

Ironically, it is exactly these homeotic mutants that are now providing the strongest evidence of how developmental processes control the body plan of all multicellular organisms. Little more than a decade ago, DNA sequencing showed that **homeotic genes** – a particular family of a larger group of genes termed **homeobox genes** – are characterized by a highly conserved region, the **homeobox**, consisting of 180 bases that code for a particular sequence of amino acids termed the **homeodomain** in regulatory proteins (McGinnis et al. 1984; Scott and Weiner 1984) (Fig. 10.2). The homeodomain serves as a sequence-specific, DNA-binding motif that allows particular proteins to bind to genes that control other aspects of development.

Not only were all homeotic genes in *Drosophila* shown to code for a similar homeodomain, but also comparable genes were recognized among vertebrates and later in all major metazoan phyla (Ruddle et al. 1994a,b; S. Carroll 1995) (Figs. 10.3, 10.4). The conservative DNA sequence of these genes provides an extremely

Figure 10.2. Homeodomain sequences of five proteins encoded by homeobox-containing genes. *Antp, ftz,* and *Ubx* are from *Drosophila; Hox-1* from the mouse; and *MM3* from *Xenopus.* The light boxes enclose homologous sequences, using *Antp* (i.e., *Antennapedia*) as the standard. The invariant sequence in the dark box is the DNA-binding site. From Gehring (1985).

	1																			20
Hox-1	Ser	Lys	Arg	Gly	Arg	Thr	Ala	Tyr	Thr	Arg	Pro	Gln	Leu	Val	Glu	Leu	Glu	Lys	Glu	Phe
MM3	Arg	Lys	Arg	Gly	Arg	Gln	Thr	Tyr	Thr	Arg	Tyr	Gln	Thr	Leu	Glu	Leu	Glu	Lys	Glu	Phe
Antp	Arg	Lys	Arg	Gly	Arg	Gln	Thr	Tyr	Thr	Arg	Tyr	Gln	Thr	Leu	Glu	Leu	Glu	Lys	Glu	Phe
ftz	Ser	Lys	Arg	Thr	Arg	Gln	Thr	Tyr	Thr	Arg	Tyr	Gln	Thr	Leu	Glu	Leu	Glu	Lys	Glu	Phe
Ubx	Arg	Arg	Arg	Gly	Arg	Gln	Thr	Tyr	Thr	Arg	Tyr	Gln	Thr	Leu	Glu	Leu	Glu	Lys	Glu	Phe

	21																			40
Hox-1	His	Phe	Asn	Arg	Tyr	Leu	Met	Arg	Pro	Arg	Arg	Val	Glu	Met	Ala	Asn	Leu	Leu	Asn	Leu
MM3	His	Phe	Asn	Arg	Tyr	Leu	Thr	Arg	Arg	Arg	Arg	Ile	Glu	Ile	Ala	His	Val	Leu	Cys	Leu
Antp	His	Phe	Asn	Arg	Tyr	Leu	Thr	Arg	Arg	Arg	Arg	Ile	Glu	Ile	Ala	His	Ala	Leu	Cys	Leu
ftz	His	Phe	Asn	Arg	Tyr	Ile	Thr	Arg	Arg	Arg	Arg	Ile	Asp	Ile	Ala	Asn	Ala	Leu	Ser	Leu
Ubx	His	Thr	Asn	His	Tyr	Leu	Thr	Arg	Arg	Arg	Arg	Ile	Glu	Met	Ala	Tyr	Ala	Leu	Cys	Leu

	41																			60
Hox-1	Thr	Glu	Arg	Gln	Ile	Lys	Ile	Trp	Phe	Gln	Asn	Arg	Arg	Met	Lys	Tyr	Lys	Lys	Asp	Gln
MM3	Thr	Glu	Arg	Gln	Ile	Lys	Ile	Trp	Phe	Gln	Asn	Arg	Arg	Met	Lys	Trp	Lys	Lys	Glu	Asn
Antp	Thr	Glu	Arg	Gln	Ile	Lys	Ile	Trp	Phe	Gln	Asn	Arg	Arg	Met	Lys	Trp	Lys	Lys	Glu	Asn
ftz	Ser	Glu	Arg	Gln	Ile	Lys	Ile	Trp	Phe	Gln	Asn	Arg	Arg	Met	Lys	Ser	Lys	Lys	Asp	Arg
Ubx	Thr	Glu	Arg	Gln	Ile	Lys	Ile	Trp	Phe	Gln	Asn	Arg	Arg	Met	Lys	Leu	Lys	Lys	Glu	Ile

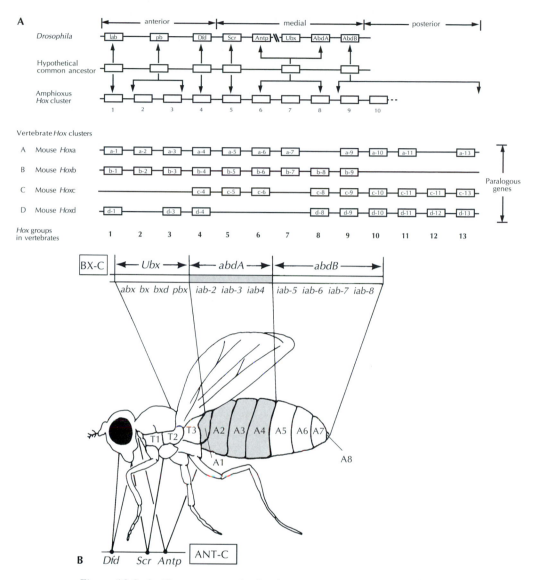

Figure 10.3. **A,** *Hox* gene organization in metazoans. Beginning at the top, the anterior-to-posterior sequence in *Drosophila;* hypothetical *Hox* genes of the common ancestor of *Drosophila* and other metazoans; amphioxus *Hox* cluster; and the four *Hox* clusters of a mouse. Based on data from S. Carroll (1995) and Ruddel et al. (1994a). The laterally directed arrows indicate the origin of new genes by tandem duplication. Chordates and *Drosophila* have independently duplicated the *Ubx* gene in primitive metazoans. **B,** Regions of expression of the bithorax complex and the antennapedia complex genes in *Drosophila,* showing the colinearity of genes within the *Hox* cluster and their expression along the anterior–posterior axis of the animal. From Gilbert (1988).

powerful tool for establishing the interrelationships of metazoan groups, as well as offering a basis for investigating the genetic and developmental changes that occurred in the transformation of one major group to another (Philippe, Chenuil, and Adoutte 1994). With the changing concept of the nature of homeotic genes, the term **Hox genes**, originally restricted to vertebrates, is now used for a particular family of homeobox genes that are universal among metazoans.

Hox and related genes do not code directly for structural elements; rather, they encode proteins regulating other genes that, in turn, influence processes – such as cell proliferation, differentiation, adhesion, movement, and cell death – that result in the formation of structures (Fig. 10.5). *Hox* genes act like switches that control whether or not structures are formed and where in the body they are expressed. For example, the wild-type *Antennapedia* gene is normally expressed in the second thoracic segment, where it is thought to repress the development of head structures. In the classical *Antennapedia* mutant, the gene is expressed in the head, causing a pair of thoracic legs, rather than antennae, to develop. This change in the function of the gene is caused by a shift in its position on the chromosome, so that its expression is controlled by a promoter that influences the head rather than the thorax (Gilbert 1994).

Mutations of *Hox* genes do not result in the sudden initiation of new structures, as argued by Goldschmidt, but instead alter the expression of structures that were already present and whose origin resulted from changes in other, still not fully understood, genes. However, since the position and expression of limbs and other major body structures are regulated by *Hox* genes, knowledge of their area of expression and mode of activation provides a readily identifiable starting point for determining the character of other developmental processes that more directly regulate the nature and localization of the cells and tissues that form the structures.

The current, explosive growth of knowledge of the function and phylogenetic distribution of *Hox* genes results from the ease of recognizing the highly conservative homeobox region. This makes it practical to screen genomes of many different organisms and so locate all genes in which this particular base sequence is present. Krumlauf (1994) points out that progress in identifying and establishing the specific functions of genes more directly responsible for the formation of structures, but whose activities are regulated by the homeobox genes, has been much slower.

In nearly all invertebrate groups that have been studied, most *Hox* genes are arranged in a continuous linear sequence on a single chromosome. This is called the **Hox cluster**. The close similarity of each of the sequential genes suggests that they evolved from a single ancestral gene that underwent *tandem duplication,* which may be attributed to unequal crossing-over during meiosis. In the fruit fly, the eight genes can be differentiated into three series, anterior, medial, and posterior (or head, thorax, and abdomen), which correspond to the position along the body where they are expressed.

Vertebrates differ in having not just one but as many as four homeobox or *Hox* clusters, designated by the letters *A, B, C,* and *D* (or *a, b, c,* and *d*), each occupying a different chromosome. Each cluster has a different number of *Hox* genes, but all can be compared with those present in *Drosophila.* Thirteen positions along the four chromosomes are occupied by different **gene groups**, although not all positions are occupied on particular chromosomes. Clusters *A, C,* and *D* have four to five genes in the posterior series, all of which resemble a single gene, *Abdominal-B,* in *Drosophila,* suggesting that they resulted from tandem duplication. The genes in groups 2 and 3 are comparable to a single gene in the anterior series of *Drosophila.* Since these duplicated genes occur in more than one cluster in vertebrates, it is almost certain that they initially arose within a single cluster, followed by two du-

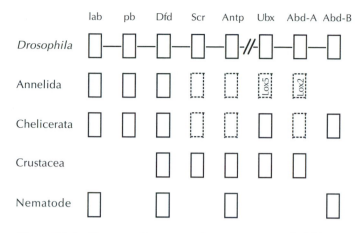

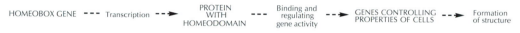

Figure 10.4. *Hox* genes in nonvertebrate metazoans. Solid boxes represent confirmed *Hox* genes; dotted boxes are genes that are too similar to another to be distinguished on the basis of available data; missing boxes represent the absence of information. Modified from *Nature* (Sean Carroll, vol. 376). Copyright © 1995, Macmillan Magazines Limited.

Figure 10.5. Sequence of activation involving *Hox* genes.

plication events involving the entire genome. If all thirteen positions were occupied prior to cluster duplication, either some of the duplication events did not copy the entire cluster, or up to fourteen genes have subsequently been lost. The *Hox* clusters of mice and humans have now been studied in sufficient detail that it is unlikely that additional genes will be discovered (Ruddle et al. 1994a).

The current use of letters to identify each of the *Hox* clusters and the numbers 1–13 to designate the gene groups dates from 1992 (Scott 1992); most papers published before 1993 used different ways to indicate the position of genes on a particular chromosome and to differentiate the four clusters. The term *Hox* is followed by a letter indicating which of the four clusters it occupies (e.g., *Hoxb*). This, in turn, is followed by a hyphen and a number that indicates the gene's linear position on the chromosome (e.g., *Hoxb-5*).

The greatest similarity in the nucleotide sequence of *Hox* genes is evident among those that occupy comparable positions in the linear sequence exhibited by each of the vertebrate clusters – for example, *Hoxa-4, Hoxb-4, Hoxc-4, Hoxd-4*. These genes are said to be **paralogous** with one another, in analogy with the term *homologous*. As is the case for the genes in the single cluster in *Drosophila* and other nonvertebrate metazoans, the linear arrangement evident along each chromosome is reflected in a comparably linear expression in the developing organism. As might be expected by their similar structure and position, paralogous genes within a particular group may have similar, but not identical, functions in particular tissues or structures. In some cases, mutational disruption of a single gene can, to a degree, be compensated by its paralogues. In other cases, only one of the three or four paralogous genes may be involved in the development of particular structures.

The phylotypic stage

One of the most striking features of the *Hox* complex, beyond its expression in all adequately known metazoan groups, is the colinearity of the genes within the complex and their expression along the body axis of the developing embryo. Although the number of *Hox* genes differs from phylum to phylum as well as within phyla, the sequence of genes that are recognized as homologous between one phylum and another is nearly constant. With the exception of some insects, including *Drosophila*, few non-*Hox* genes interrupt the sequence within each cluster.

Duboule (1994) provided a simply functional reason for this phenomenon that simultaneously explains the constraint expressed in the basic body plan within each of the metazoan groups. Although there may be significant differences in the early stages of their cell division – through the formation of the gastrula among vertebrate groups and the larval stages in many other metazoan phyla – all metazoans go through a developmental period termed the **phylotypic stage**, during which embryos within each phylum are strikingly similar to one another (Slack, Holland, and Graham 1993). This is particularly evident among vertebrates, in which the early embryos of fish, amphibians, reptiles, birds, and mammals appear almost identical (Fig. 10.6).

Duboule argued that the constancy of body form at the beginning of morphogenesis is necessary to correlate the timing of gene expression with the anterior to posterior sequence of development in the embryo. *Hox* genes are activated in a linear sequence corresponding to their position along the chromosome and the anterior–posterior axis of the body. More specifically, they are activated in a sequence that corresponds with a short period of time during which cells, arranged along the axis of the body in an anterior to posterior direction, are actively proliferating. That is, there is a *temporal* colinearity of cell proliferation that must correlate with the *spatial* colinearity of the genes on the chromosomes and the appropriate areas of gene expression in the developing embryo.

The initial activation of DNA within the cells that will lead to the formation of the major structural features of the body is limited to a very narrow temporal window, corresponding to the early somite stage in development. It is at this time that the embryos of all vertebrate groups are most similar to one another. This is the period during which anterior to posterior progression of somite formation and closure of the neural tube are clearly evident. In the mouse, this takes about two days. Subsequent development, initiated by the activation that occurred during this short time frame, proceeds differently from this stage onward, modified by more specific developmental processes.

The constancy of the function and linear expression of *Hox* genes provides a stable framework about which other elements can change without jeopardizing the survival of the organism. Their stability among vertebrates and among each of the other metazoan phyla is equivalent to the stability of the genetic code among all organisms. They provide the most general level of explanation for the endurance of features common to each phylum (Hall 1996). Although individual *Hox* genes may continue to vary in details of their DNA composition, their basic function in controlling the degree, place, and nature of expression of other genes responsible for particular developmental processes has apparently remained nearly constant.

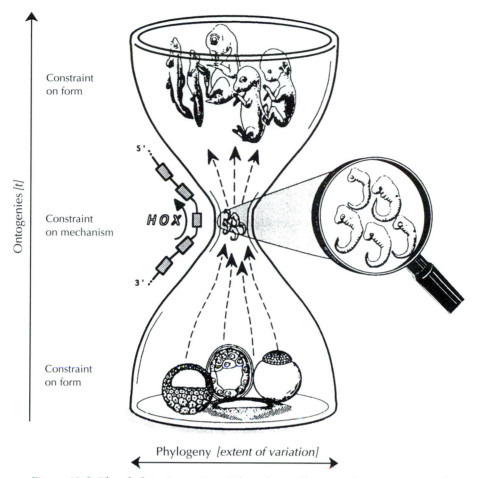

Figure 10.6. The phylotypic egg timer. This scheme illustrates the convergence of vertebrate developmental strategies toward the phylotypic progression, the acquisition of a stable *Bauplan*. Such a mechanism results in the progressive temporal activation of *Hox* genes along the axis of the chromosome and the anterior–posterior axis of the body. This coincides with the period of most active cell proliferation. Reprinted from *Development* (Duboule, Supplement 1994), Company of Biologists Ltd.

Hox genes in chordates

The number of *Hox* genes reflects the number and complexity of developmental pathways that can be controlled. This is clearly shown by the number of different cell types present in different groups of organisms. *Drosophila,* with eight genes in the homeobox cluster, has approximately fifty cell types, whereas jawed vertebrates, with thirty-eight *Hox* genes, have more than a hundred (Bonner 1988). All jawed vertebrates that have been studied to date have four clusters of *Hox* genes, although only mice and humans have been exhaustively mapped (Ruddle et al. 1994a). Not all craniates have been studied in detail, but *Hox A, B,* and *D* clusters contain the same set of genes in birds, mammals, and teleost fish, and in the same linear sequence (Holland and Garcia-Fernàndez 1996). This indicates a high degree of constancy in the primary system for the control of gene expression that governs the body plan of all terrestrial vertebrates as well as the extremely diverse assemblage of bony fish.

The difference in numbers of *Hox* groups and clusters between *Drosophila* and other invertebrates, on the one hand, and vertebrates, on the other, indicates a major increase in the degree of control over developmental processes between primitive metazoans and advanced chordates. Comparison between *Drosophila* and mammals suggests that a hypothetical common ancestor would have had approximately six *Hox* genes arranged in a single cluster. Other metazoan groups share a more recent common ancestry with vertebrates than they do with modern arthropods. Echinoderms appear to have nine genes in one cluster, but paralogues of groups 1–3, which control development of the head region in other metazoans, have not been recognized (Ruddle et al. 1994a). Among chordates, the hemichordate *Saccoglossus kowalsekii* has representatives from each of groups 1–9 among vertebrates, which are probably arranged in a single cluster (Pendleton et al. 1993). The urochordates, which have been thought to be more closely related to vertebrates than are hemichordates, actually have fewer *Hox* genes, although the number and the identity of these genes differs from one urochordate group to another. Ruddle et al. (1994a) argued that this probably reflects an irregular pattern of gene loss in this group, which suggests significant divergence of the urochordates from the lineage leading to vertebrates. Holland et al. (1994) determined that the cephalochordates, characterized by *Branchiostoma* (commonly referred to as amphioxus) has ten *Hox* genes arranged in a single cluster. The relative similarity of the genes in *Branchiostoma* to the first ten *Hox* groups in vertebrates shows that the linear sequence is the same in the two taxa. The three most posterior *Hox* groups present among vertebrates (*Hox 11–13*) have not been recognized in *Branchiostoma*.

There is thus a major difference in the number of gene clusters and the total number of *Hox* genes between the cephalochordates and most jawed vertebrates. This may be bridged to some extent by the condition in the living jawless vertebrates, the hagfish and lampreys. Nineteen *Hox* genes have been identified in the lamprey, and others would be expected from groups 11–13, for which a probe was not then available. The known *Hox* genes may be arranged in as few as two clusters, although a higher number cannot be ruled out (Pendleton et al. 1993).

Current evidence suggests two major episodes of *Hox* gene duplication in the early evolution of vertebrates: one between cephalochordates and jawless vertebrates, and another between jawless and jawed vertebrates (Holland et al. 1994). There was a much greater change in both developmental patterns and adult structures between cephalochordate and jawless vertebrates than occurred between jawless and jawed vertebrates. Presumably this was related to much more significant changes in the number and arrangement of the genes controlling developmental processes.

The origin of craniates

Although cephalochordates had achieved the primary characters of chordates – notochord, dorsal hollow nerve cord, and pharyngeal gill slits – as well as a post-anal tail and segmental muscles throughout the trunk and tail, they lack many other features that characterize vertebrates. The differences in the head region are particularly dramatic. Although *Branchiostoma* has a light-sensory structure and an organ of balance, these show few of the specialized characteristics of the vertebrate eyes or inner ear, and the anterior portion of the neural tube is only slightly expanded, without any of the derived features of the vertebrate brain. It is these characteristics of the head region that most clearly distinguish even the most primitive fossil and living vertebrates from nonvertebrate chordates. For this reason, many biologists prefer "craniate" rather than "vertebrate," since none of the primitive fish groups have vertebrae.

This view of the origin of vertebrates as being specifically associated with the origin of cranial structures was strongly emphasized by Northcutt and Gans (1983) and Gans (1989, 1993). They pointed out that nearly all of the specifically craniate features of the head region develop from a unique set of tissues: the sensory placodes and the neural crest cells. No comparable tissues are known in any of the nonvertebrate chordates. The *sensory placodes* are restricted to the head region (except for those that develop into the lateral line canals) and are primarily associated with the development of the olfactory, optic, and otic capsules, and chemical sensory structures within the oral cavity. The *neural crest cells* originally proliferate at the junction of the developing neural tube and the adjacent ectoderm, posterior to the skull; however, these cells are migratory and contribute significantly to structures throughout the head region as well as the trunk. Expression of the *premature death* mutation in both sensory placodes and neural crest tissue suggests that these tissues had a common phylogenetic and developmental history (Smith, Graveson, Hall 1994).

Gans and Northcutt (1983) argued that the brain and sensory structures of the head region in craniates evolved anterior to the end of the nerve cord in cephalochordates, forming what amounts to a new head. This was based on the fact that the notochord extends to the very end of the neural tube in *Branchiostoma,* in contrast to its more posterior termination in all vertebrates. In addition, the pattern of segmentation in the head region of craniates seems very different from that of the trunk, and much of the bone and muscle tissue in the head is derived from neural crest cells rather than from mesodermal somites, as is the case in the trunk region.

Subsequently, Holland et al. (1992) identified one of the *Hox* genes in the head region of *Branchiostoma, AmphiHox3,* as a homologue of the vertebrate gene *Hoxb-3.* In the chick, the anterior limit of expression of this gene is between rhombomeres four and five, close to the anterior margin of the hindbrain. In *Branchiostoma,* the homologous gene is expressed between somites four and five within the head region. This indicates that the anterior part of the neural tube occupies a broadly comparable position in cephalochordates and craniates, and hence the sensory and integrative structures in craniates evolved within the region of the cephalochordate "head" rather than in a more anterior position (Peterson 1994). This is further supported by electron microscopy of the cerebral vesicle of larval *Branchiostoma* that shows more vertebratelike features of the sensory structures and their neural circuitry than had been observed in earlier histological studies (Lacalli 1995).

The vertebrate head should not be considered an addition to the existing body of nonvertebrate chordates, as was argued by Gans and Northcutt (1993), but much of the tissue associated with the brain and paired sense organs develops from tissue that was not present in cephalochordates. The origin of neural crest and placode tissue thus become key to the origin of craniates (N. D. Holland et al. 1996).

Many of the primitive attributes of cephalochordates may be correlated with the absence of particular developmental genes. For example, neural crest cells are associated with the vertebrate *Hox* genes *Hoxa-1* and *Hoxa-3,* which are not present in *Branchiostoma,* although their paralogues, *Hoxb-1* and *Hoxb-3,* are (Holland 1992; Holland et al. 1994). The origin of neural crest may thus be linked with the duplication of gene clusters that produced these new *Hox* genes at the base of the vertebrate lineage (Peterson 1994). Several other genes also appear to have undergone duplication during the transition between cephalochordates and craniates. These include some homeobox genes that are not part of the *Hox* clusters (termed *non-Hox* or *diverged homeobox* genes by Ruddle et al. 1994a). Among these are members of the *Msx* gene family. Three are known in mammals: Two are expressed in neural crest–derived mesenchymal tissue involved with branchial arch development, palate formation, tooth morphogenesis, and development of the paired eyes; only a single gene is recognized in *Branchiostoma.* Others are members of the family of engrailed genes, within which the vertebrate gene *En-2,* not present in cephalochordates, has an important role in hindbrain formation (Holland 1992), and its expression extends as far forward as the midbrain (Langille and Hall 1993).

Although not limited to the head region, another major change between cephalochordates and craniates that can be associated with gene duplication is expressed in the development and adult structure of striated muscles. Among vertebrates, the development of striated muscles involves fusion of the precursors of muscle cells, the myoblasts, into much larger cells containing many nuclei. Cephalochordates, in contrast, have simple, mononucleate muscle cells resembling the embryonic myoblasts of vertebrates. The different patterns of development may be attributed to the duplication of genes within the family that controls formation of alkali myosin light chain molecules in vertebrate muscles. Holland et al. (1995) found only a single member of this gene family in *Branchiostoma.*

The duplication of *Hox* cluster genes and other genes controlling major developmental processes certainly played a key role in the origin of craniates. There was apparently another episode of gene duplication between primitive jawless fish and the modern groups of jawed vertebrates, although the magnitude and timing of the latter process cannot be established without additional knowledge of living agnathans and the sharks and their relatives. In contrast, there is no evidence for the appearance of new *Hox* genes among the vast assemblage of bony fish and terrestrial vertebrates during the past 400 million years, despite such radical morphological changes as the emergence of paired limbs and their continuing adaptation to a host of different environments. Sean Carroll (1995) argues that all the structural diversity within these groups may be attributed to alterations, not in the *Hox* genes themselves, but through modifications of regulatory genes that alter the timing, place, and manner of expression of the *Hox* genes. These fall into two large categories: changes in the regulation of *Hox* genes themselves, and changes in the interactions between *Hox* gene products and the genes they regulate. It is probably this latter process that led to most of the transformations that we can observe in the fossil record of vertebrates.

Developmental processes and the evolution of the skull and axial skeleton

The origin of new developmental processes, especially those involving the neural crest cells and placodes, played a major role in the establishment of the major features of the vertebrate skull. On the other hand, although much is known of how developmental processes affect the ontogeny of modern vertebrates (e.g., *The Skull*, Hanken and Hall 1993), it has been much more difficult to associate subsequent phylogenetic change with alterations in particular developmental processes. This may be partially due to the fact that genes of the *Hox* cluster, which are so useful in the study of developmental processes in other parts of the body, are expressed only in the most posterior part of the skull, associated with the hindbrain and the branchial arch region (Thorogood 1993). The remainder of the skull is completely remodeled through the activity of the neural crest cells and placodes, which appear to overlay any more primitive patterning that may have existed in earlier chordates. As recently as 1993, Hanken and Thorogood wrote: "Virtually nothing, however, is known about the role of homeobox genes in cranial evolution following the initial origin of the head" (p. 13).

Another reason that it has been difficult to associate changes in processes of development with the evolution of the head region among vertebrates is the sheer complexity of the system, with a three-dimensional expression of a host of different tissues, all of which have their own patterns of developmental control. In addition, most studies of cranial development have concentrated on particular species of birds and mammals, rather than investigating the broad spectrum of different cranial patterns illustrated by different major groups of fish, amphibians, and the modern reptile orders.

Two other areas of the vertebrate body, the axial skeleton and the limbs, provide much simpler, essentially one- and two-dimensional systems whose development has been studied in a wider range of taxa. They show in considerable detail how changes in the pattern of expression of particular genes, especially those of the *Hox* cluster, are associated with specific differences in developmental processes and adult morphology.

The thirteen *Hox* groups of all four clusters are expressed in a linear fashion in spatially restricted domains from the head to the end of the tail. They are expressed in a variety of tissues, including hindbrain segments, branchial arches, neural tube, neural crest, paraxial mesoderm, limbs, surface ectoderm, gut, and gonadal tissues (Krumlauf 1994). The regions of expression of different *Hox* groups and different paralogues within each group is clearly shown in different regions of the vertebral column (Burke et al. 1995).

Burke and her colleagues investigated the expression of *Hox* genes in several animals that exhibit clear differences in the relative number of vertebrae in the cervical, thoracic, lumbar, and sacral regions. They concentrated on the well-known laboratory mouse and the chick, but also studied the frog *Xenopus* and the goose. They found that the anterior limit of *Hox* expression was associated with the morphology of the vertebral regions rather than with the vertebral number (Fig. 10.7).

Hox groups 1–3 are expressed in the hindbrain and the occipital somites, whereas groups 4 and 5 have anterior boundaries of expression within the cervical region. Boundaries of expression of *Hox-4* genes are similar in mouse and chick, with *Hoxb-4* associated with the most anterior cervical vertebra and *Hoxa-4* and *Hoxc-4* toward the middle of the cervical series, regardless of the specific number of neck vertebrae. The anterior expression boundary of *Hoxc-5* (the only Hox-5 paralogue studied) marks the second-to-last cervical vertebra, which is number 13 in the chick and 6 in the mouse. *Hoxc-6* has an anterior boundary of expression at the level of the first thoracic vertebra (T-1) in all four animals; This is the eighth presacral vertebra in the mouse, the fifteenth in the chick, the eighteenth in the long-necked goose, and close behind the skull in the short-necked frog.

All of the paralogous genes of groups 7 and 8 have their anterior boundary of expression within the thoracic series. That of *Hoxa-9, Hoxb-9,* and *Hoxc-9* lies near the posterior end of the thoracic series, but the *Hoxd-9* boundary is near the lumbosacral transition, at somite level 29–30 in both chick and mouse. *Hox* genes *Hoxc-10* and *Hoxd-10* are expressed in the sacral region of the chick, but only the latter in the mouse. *Hoxd-11–Hoxd-13* are expressed in the tail of both species.

Burke et al. (1995) pointed out that the domains of *Hox* gene expression shift according to the functional anatomy of the vertebral regions in different vertebrate groups, rather than being tied to vertebral or somite number, suggesting that *Hox* genes may have played an important role in the evolution of axial variation. These changes might be explained either by different deployment of the upstream regulators of *Hox* expression, or through modification in the response to these regulators by the *Hox* genes themselves. The fact that individual members of paralogous groups have different sites of expression (e.g., the expression of *Hoxd-9* is more posterior than that of *Hoxa–c-9*) demonstrates that response to the same upstream regulators can differ, and suggests that the evolution of regulatory diversity within paralogous groups may have enabled major changes in morphology to occur without change in the basic function of the *Hox* genes.

The work by Burke and her colleagues shows the relationship between the position of *Hox* gene expression and regions of vertebral specialization, but does not show how their expression may influence the configuration of the vertebrae. This was investigated by Kostic and Capecchi (1994), based on gene disruption in the mouse. The most anterior limit of expression of *Hoxa-4* is at the level of the second cervical vertebra (C-2), the axis. This vertebra is characterized by a very large neural spine and an anterior extension of the centrum, the dens, that fits into the atlas. In all the mice that were homozygous for the disruption of *Hoxa-4,* the third cervical vertebra (C-3), which normally lacks a neural spine, develops one comparable to that normally expressed in the atlas; however, none of the other structures of C-3 is altered. Disruption of *Hoxb-4,* on the other hand, results in C-3 having a wide neural arch, like that of C-2, while C-2 develops a ventral tubercle that is normally expressed only in C-1, the atlas.

Kostic and Capecchi offered several alternatives to explain the different effects of disruption of *Hoxa* and *Hoxb*: action at different times during development (*Hoxb-4* before the neural arches have fused and *Hoxa-4* after fusion), regulation of different cell populations, or cell proliferation along different planes.

While disruption of the genes usually expressed in the second and third cervical vertebrae resulted in their characteristics being transformed anteriorly, disruption of *Hoxa-6,* which is usually expressed in the last cervical, resulted in its characteristics being transformed posteriorly. In normal mice, the configuration of the axial skeleton changes conspicuously between the last cervical (C-7) and the first thorac-

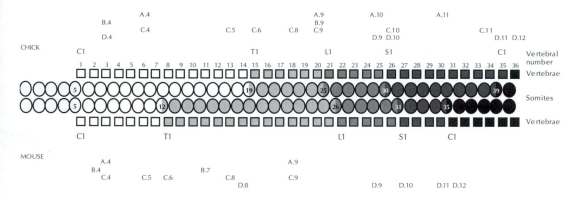

Figure 10.7. Representation of the vertebrae and somites of the chick *(top)* and the mouse *(bottom)* showing the anterior boundry of expression of *Hox* genes. C, T, L, S, and C refer respectively to the cervical, thoracic, lumbar, sacral, and caudal regions of the vertebral column. Modified from *Development* (Burke et al., vol. 121, 1995), Company of Biologists Ltd.

ic vertebra (T-1). Ribs are expressed in the thoracic but only rarely in the cervical region, and the transverse and dorsal processes are much larger. In thirty-one of the seventy-two homozygotes in which *Hoxa-6* was disrupted, a rib characteristic of the thoracic region was expressed on the last cervical. Other aspects of the last cervical vertebra were not modified, but the neural spine on the most anterior thoracic vertebra assumed the larger size normally seen in the second thoracic.

These studies provide extensive information on the specific areas of expression of the paralogues of various *Hox* groups. Each paralogue may control the expression of particular structural details within a single bone, but there is also considerable overlap and potential redundancy in more general aspects of bone formation. Loss of function in some *Hox* genes may result in only minor changes in the anatomy of particular bones without serious reduction in viability and fecundity, but loss of other genes, which also control other systems, may be fatal.

Although disruption of *Hox* genes may aid in determining the function of their normal counterparts, it is unlikely to contribute to understanding normal evolutionary processes. The presence of exactly the same number and arrangement of all the *Hox* genes in modern birds and mammals, despite these groups having separated from one another phylogenetically at least 300 million years ago, points to their extremely conservative nature. There is no evidence that any *Hox* genes have been lost since the much earlier divergence of actinopterygian and sarcopterygian fish, sometime in the Silurian. It is thus very unlikely that total loss of function of any *Hox* gene has played a role in small-scale anatomical changes among modern vertebrate groups. On the other hand, changes in the regulation of *Hox* genes, including altering or even eliminating their effect on other genes, might have results similar to those reported in these experiments.

The evolution of fins and limbs

Changes in the structure of the fins of fish and the limbs of tetrapods are among the most striking aspects of vertebrate evolution. They directly reflect changes in habitat and ways of life both within and between each major group. They are also among the most difficult evolutionary phenomena to explain. Darwinian selection theory can account for the modification of particular structures so that they are better suited for a given environment, but it is much more difficult to understand how the limbs of early tetrapods could have evolved from the fins of a fish or how the wings of birds could have evolved from the forelimbs of dinosaurs, since these changes involve radical shifts in function between completely different selective regimes. Paradoxically, an equally serious problem is to explain the stability of limb structure within individual groups or lineages, which may retain an extremely stereotyped pattern for hundreds of millions of years.

Changes of limb structures between adaptive zones are frequently both radical and relatively rapid, but they can be attributed to strong, persistent, and unidirectional forces of natural selection. *Within* each major adaptive zone, on the other hand, particular features of limb structure may be so extremely conservative that they seem immune from directional selection or even random fluctuation. It is the latter phenomenon that has led to the concept of *developmental stasis* (Maynard Smith et al. 1985). For instance, among the primates, modern humans have retained five digits on both the forelimbs and hind limbs – the same number as were present in the earliest primates from the early Cenozoic, the earliest mammals from the Late Triassic, and members of the stem group of amniotes from the mid-Carboniferous, nearly 340 mya. Nearly all lizards also have five digits, although their ancestry diverged from that of mammals more than 300 mya. Frogs and most salamanders also had five digits on the hind limb, but four or less on the forelimb, at least since the Early Jurassic, and one or both of these groups may trace its ancestry to lineages that had already lost one of the digits in the forelimb by the mid-Carboniferous. At an even more detailed level, the number of phalanges in each digit, expressed as the phalangeal formula, has been conserved within the mammalian lineage since at least the Early Triassic, when the primitive amniote number of 2,3,4,5,3 (counting from the thumb to the fifth finger) in the front limb and 2,3,4,5,4 in the hind limb was reduced to 2,3,3,3,3 in both hands and feet (Fig. 10.8).

Many amniote lineages have reduced the number of digits and phalanges, but none has evolved more than five true digits, and many have retained the original phalangeal formula, at least in the digits that remain. Except in taxa that have completely lost their limbs, the pattern of the major bones – the humerus, ulna, and radius in the forelimb, and the femur, tibia, and fibula in the hind limb – remains constant, even in groups that have adapted to such different forms of locomotion as swimming or flying. There must be some force acting to retain such stability, beyond selection for particular environments or ways of life.

Understanding how development of the limbs is controlled should contribute substantially to explaining why some aspects of their structure can remain so extremely conservative over hundreds of millions of years in particular taxa, whereas in other lineages comparable features change radically over far shorter periods.

In common with the patterning of the trunk region in both nonvertebrate metazoans and vertebrates, the *Hox* cluster genes are expressed in a linear sequence within the fins of bony fish and the limbs of tetrapods. Neither cephalochordates nor the most primitive jawless vertebrates have paired fins, but the potential for the regulation of their development may have begun with the first stage in the tan-

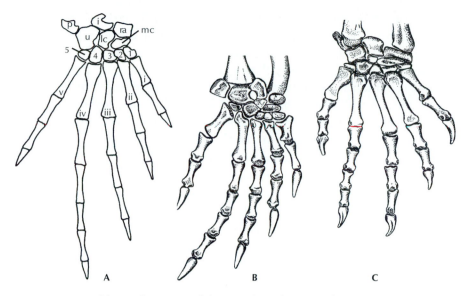

Figure 10.8. Stable configuration of the carpals and manus of amniotes with a lizard-like body form over the past 300 million years. **A,** The primitive amniote *Paleothyris,* from the Upper Carboniferous. **B,** The primitive diapsid *Thadeosaurus* from the Upper Permian. **C,** The living genus *Sphenodon*. From Carroll (1987). Abbreviations used in this and subsequent illustrations involving limb structures: a, astragalus (incorporates intermedium, tibiale, and proximal centrale of primitive tetrapods); c, centrale; cal, calcaneum (= fibulare of amphibians); F, femur; Fi, fibula; fib, fibulare; H, humerus; i, intermedium; lc, lateral centrale; mc, medial centrale; p, pisiform; R, radius; ra, radiale; T, tibia; ti, tibiale; U, ulna; u, ulnare; 1–5, distal carpals and tarsals; i–v, metacarpals and metatarsals.

dem duplication of the gene *Abdominal-B,* a homologue of the vertebrate limb-patterning genes, which is represented by a single copy in *Drosophila* but two copies in *Branchiostoma*. In jawed vertebrates, this area of the genome is represented by five successive gene groups, *Hox9–13*. These *Hox* groups have not yet been studied in the Agnatha (Pendleton et al. 1993), and no information is yet available regarding the number or pattern of expression of *Hox* cluster genes among the Chondrichthyes.

Sordino, van der Hoeven, and Duboule (1995) demonstrated that the same general pattern of expression of *Hoxa* and *Hoxd* genes is found in the pectoral fin of bony fish, represented by the zebra fish, as that described for tetrapods. *Hoxa-9–13* and *Hoxd-9–13* appear in sequence from the proximal to distal end of the appendage. *Hox* expression in the pelvic fin has not yet been described in bony fish but is presumably comparable. This raises a difficult problem regarding the origin of paired appendages: Pectoral and pelvic appendages arise in both bony fish and tetrapods as outgrowths from the trunk; it would be natural to think that they would develop in response to the expression of *Hox* genes associated with the area of the trunk from which they arise – that is, that the pectoral appendage would be regulated by *Hox* paralogues of groups 4–7 and the pelvic appendage by groups 9–10.

The fact that the forelimbs and hind limbs are both regulated by the same set of posterior groups, 9–13, led Tabin and Laufer (1993) to suggest that the pelvic fin might have been the first to arise in fish, and that the pectoral fin arose by ectopic expression of *Hox* genes originally regulating rear-limb development. Alternatively, both forelimbs and hind limbs might have evolved from a single, continuous lateral fin fold, which was once hypothesized as being the primitive condition for the ancestors of vertebrates with paired fins. Coates (1993, 1994) responded with extensive evidence from the fossil record demonstrating that the pectoral girdle evolved in several groups of primitive fish prior to the appearance of the pelvic fin, and that no vertebrates were known to have had a continuous lateral fin fold.

The oldest adequately known fossils of craniates, from the Middle Ordovician (Gagnier 1993), have a continuous covering of dermal bone over the head and trunk, back to the base of the tail, clearly precluding the presence of any paired fins. A large number of craniate lineages are known by the beginning of the Devonian, representing all the major groups of aquatic vertebrates. These show many different stages in the origin of paired fins. Unfortunately, the interrelationships of these groups are not well established, and it is still very difficult to distinguish whether different stages in the elaboration of paired fins can be attributed to a succession of changes within a particular lineage, or if they have evolved separately in different groups. The general picture is that the earliest craniates lacked any trace of paired fins, and that the pectoral and pelvic fins evolved separately but in that order: Pelvic fins are never found in the absence of pectoral fins; nor is there any fossil evidence that once-continuous lateral fin folds may have divided to form successive anterior and posterior fins (Carroll 1987).

Although most of the well-known genes that control limb development are the same as those expressed in the posterior trunk region, a few are not. *Hoxb-8* is expressed in both the anterior and posterior limbs early in development in the mouse, but it is involved with further development only in the pectoral limb (Charité et al 1994). *Hoxc-6* is expressed in the formation of the pectoral limb field in mice, the frog *Xenopus,* and zebra fish; it is also involved in regeneration of both the front and hind limbs in the salamander *Notophthalmus* (Savard and Tremblay 1995), and is expressed in the front and hind limbs of the frog *Xenopus,* although in low concentrations. In addition, *Hoxb-5* is involved in determining the position of the shoulder girdle in mice (Rancourt, Teruhisa, and Capecchi 1995).

The pattern of expression of these genes in modern bony fish and tetrapods suggests that development of paired appendages may originally have been regulated by *Hox* genes common to the appropriate portion of the trunk, but that once both anterior and posterior fins had evolved, they both came under the influence of a group of *Hox* genes that was initially involved in establishing the sequence of structures at the end of the trunk and tail. Research on the distribution of *Hox* genes in the pelvic fin of bony fish, both pectoral and pelvic fins of Chondrichthyes, lungfish, and coelacanths, as well as gene groups 9–13 in jawless fish, would contribute substantially to determining what changes may have occurred during the origin of paired fins and their transition to the paired limbs of tetrapods.

The origin of tetrapod limbs

Neither the fossil record nor study of development in modern genera yet provides a complete picture of how the paired limbs in tetrapods evolved, but this problem

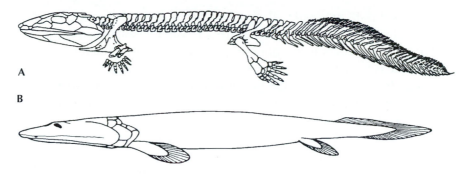

Figure 10.9. **A,** Preliminary reconstruction of the Late Devonian amphibian *Acanthostega*. This is the most primitive tetrapod in which both the fore and hind limbs can be reconstructed. The ribs have been omitted (from Coates and Clack 1995). **B,** The early Upper Devonian osteolepiform fish *Panderichthyes*. The general body form most closely resembles that of tetrapods in the loss of the dorsal and anal fins, but the paired fins retain an external form that is typical of aquatic vertebrates (from Vorobyeva and Schultze 1991).

provides an informative model as to how these two sources of information can be combined to explain how major morphological transitions have occurred.

The closest comparison between the paired fins of obligatorily aquatic fish and animals that were at least facultatively terrestrial is provided by the osteolepiform sarcopterygians *Eusthenopteron* and *Panderichthys* and the stem tetrapods *Acanthostega* and *Ichthyostega* (Figs. 10.9, 10.10). Superficially, the paired fins of the fish appear typical of strictly aquatic vertebrates. They are small relative to the body; they narrow at the base that articulated with the pectoral and pelvic girdles, but broaden distally to form an effective surface for locomotion or directional control in the water. The fin is sheathed with a continuous covering of scales. The proximal scales resemble those on the trunk, but more distally they are narrowed to form jointed dermal fin rays termed **lepidotrichia**.

In contrast, the internal, endochondral bones of the fin are closely comparable to those of terrestrial vertebrates. There is a single proximal humerus and more distal ulna and radius in the forelimb, and the femur, tibia and fibula in the hind limb. They are succeeded distally by bones that are homologous with proximal elements of the wrist (intermedium, ulnare, and centralia) and ankle (fibulare, intermedium, and possibly distal tarsals) of land vertebrates, but they could not have functioned in the manner of these joints in terrestrial vertebrates because they are extensively overlapped by the radius and the tibia. The entire endochondral skeleton is within a functionally continuous fin structure, as seen from its scaly covering. There is no trace of endochondral skeletal elements comparable with the distal carpals or digits of terrestrial vertebrates.

The limbs of *Acanthostega* and *Ichthyostega* have all the major features of later tetrapods. They bear no trace of dermal scales. Much more extensive areas of articulation have evolved between the pectoral and pelvic girdles and the proximal limb bones. The humerus, radius, and ulna, and femur, tibia, and fibula are massive, potentially supporting elements, and the areas of the carpus and tarsus comprise shorter bones that could have served as zones of hinging and/or rotation. The exact patterns of the carpus and tarsus have not yet been determined and are difficult to compare in detail with those of later Paleozoic tetrapods. The carpals are small and poorly ossified, whereas the proximal tarsals are very large. The metacarpals and metatarsals are not clearly distinguishable from the succeeding phalanges. Clearly, these limbs represent a period of transition, but one that has all the potential for evolving into the pattern of typical tetrapods. Most significantly, the elbow, wrist, knee, and ankle joints, while primitive, unquestionably presage those of later land vertebrates.

The digits have the general form of those in fully terrestrial amphibians and reptiles, but their numbers differ significantly. In the most primitive Devonian tetrapods in which the limbs are adequately known, there are eight digits in both the front and rear limb, compared with no more than five in any adequately known post-Devonian tetrapods (Coates 1994, in press; Lebedev and Coates 1995). The retention of no more than five digits in the rear limb and either four or five digits in the forelimb of Carboniferous and later tetrapods was almost certainly the result of subsequent reduction.

In contrast with the clear homology of the more proximal limb bones in osteolepiform fish and early tetrapods, no obvious homologues of the digits is evident in any sarcopterygian. These bones appear de novo in the Upper Devonian tetrapods. How can this be explained?

The structural similarity of the endochondral bones of the upper limb in osteolepiform fish and all tetrapods suggests a similar mode of genetic control during development. This is supported by the expression of comparable *Hox* genes in forms as phylogenetically distant as the zebra fish *Danio,* chickens, mice, and modern amphibians. Not surprisingly, the gene expression in the distal extremities of fish is clearly different from that of tetrapods. In tetrapods, *Hoxa* and *Hoxd* genes within groups 9–13 are expressed to the very extremity of the limb, and the most distal genes are active throughout development. The early expression of *Hoxa* and *Hoxd* in the zebra fish is similar to that of tetrapods, but their later expression in the more distal portion of the fin differs significantly. *Hoxd-11, Hoxd-12,* and *Hoxd-13* are not detected in cells in the anterior half of the fin. Instead, these expression domains are restricted to the posterior margin of the fins early in development and subsequently disappear (Sordino et al. 1995) (Fig. 10.11).

The loss of expression of these genes can be associated both spatially and temporally with the proliferation of cells that form the small, jointed dermal fin rays making up the distal portion of the fin. In early stages of limb formation in bony fish, the fin bud is relatively thick and filled with mesenchyme that differentiates to form the endochondral bones of the girdle and base of the fin (Smith and Hall 1990; Thorogood 1991). At this stage, the distal margin of the fin is formed by a thickened, pseudostratified epidermal ridge, broadly resembling the apical ectodermal ridge (see the following section) of tetrapod limbs (Fig. 10.12). In ray-finned fish, the configuration of this ridge then changes to form the **apical ectodermal fold**, which encloses a very narrow internal space. The mesenchyme at the base of the apical ectodermal fold no longer proliferates to form endochondral bone, but produces two parallel arrays of collagenous fibrils termed **actinotrichia**. The mesenchymal cells then migrate into the distal portion of the limb and generate cells that form the dermal tissue of the lepidotrichia.

Thorogood (1991) argued that formation of the apical ectodermal fold in some

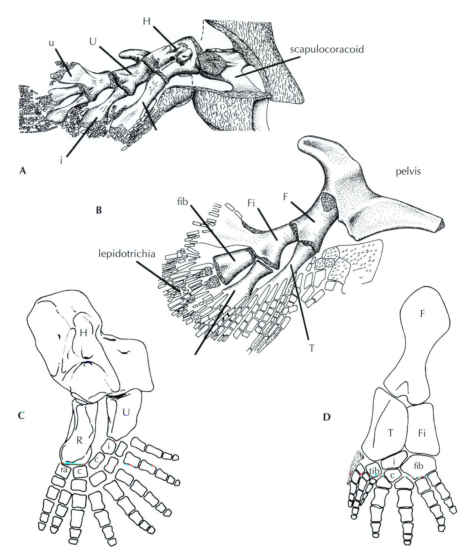

Figure 10.10. **A, B,** Fore and hind fins of the osteolepiform fish *Eusthenopteron* show-ing the endochondral bones of the proximal portion of the fin and the covering of small, jointed, lepidotrichial scales more distally (reproduced by permission of the Roy-al Society of Edinburgh and of Andrews and Westoll from *Transactions of the Royal So-ciety of Edinburgh*, volume 68 [1970], pp. 207–329). **C,** Forelimb of the Upper Devonian amphibian *Acanthostega;* **D,** Hind limb of the Upper Devonian amphibian *Ichthyostega* (C and D with the permission of Michael Coates). Abbreviations as in Figure 10.8.

way alters or blocks signals that previously reached the mesenchyme, so that its function in generating the distal portion of the limb changes. He suggested that dif-ferences in the timing of formation of this fold could be responsible for the differ-ences in the nature of the fins of actinopterygians and sarcopterygians, as well as the formation of the tetrapod limb. Later formation of the fold in sarcopterygians could account for the much greater extent of endochondral ossification in their fins, compared with most actinopterygians. Developmental changes leading to the complete elimination of the fold could provide the mechanism by which ancestral tetrapods were able to continue elaboration of the endochondral skeleton to the very end of the limbs. This is correlated with the complete absence of lepidotrichia in association with the limbs in Devonian tetrapods (Coates 1994). Carboniferous tetrapods do have dermal scales covering their limbs, but these resemble the trunk scales rather than the distal fin scales of bony fish, and were presumably elaborated after the loss of the lepidotrichia from the paired limbs.

The hands and feet of land vertebrates were formed not by a symmetrical extension of endochondral ossification from the end of the osteolepiform fin, but rather by sustained proliferation of the posterior portion. In sarcopterygian fish, including the immediate sister-group of tetrapods, the endoskeleton of both pectoral and pelvic fins are essentially linear structures, with a major proximal-to-distal axis. The *Hox* genes in modern ray-finned fish show a comparable pattern of linear expression. Modern tetrapods show a similarly linear pattern of gene expression for

Figure 10.11. Change in the axis of development and the expression of *Hox* genes between bony fish and tetrapods. **A,** Endochondral bones of the limb of a modern lungfish, showing both preaxial (anterior) and postaxial (posterior) radials, extending from the main axis of development (from Coates 1994). **B,** Endochondral bones of the osteolepiform fish *Eusthenopteron,* in which all radials are preaxial. **C,** Hind limb of the early tetrapod *Ichthyostega,* in which the axis of development angles forward and the distal tarsals and digits develop in a posterior to anterior sequence (B,C modified from Sordino et al. 1995). **D,** Area of expression of *Hoxd* genes in the living zebra fish. **E,** Area of expression of *Hoxd* genes in modern tetrapods (D,E reprinted from *Nature* [Nelson and Tabin, vol. 375]. Copyright © 1995, Macmillan Magazines Limited). Abbreviations as in Figure 10.8.

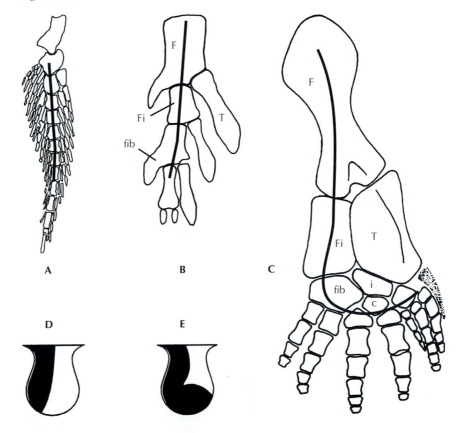

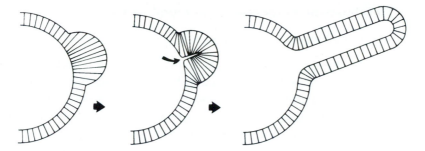

Figure 10.12. Changes from the pattern of an apical ectodermal ridge to an ectodermal fold in modern bony fish. This is accompanied by a shift in developmental processes from formation of the endodermal bones of the proximal radials of the fin to formation of dermal lepidotrichia. The origin of the digits of tetrapods may be attributed to the prolongation of the apical ectodermal ridge throughout limb development. From Thorogood (1991).

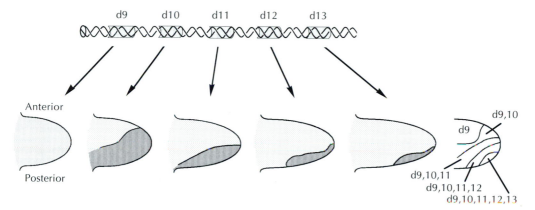

Figure 10.13. Expression of the *Hoxd* gene series in the developing chick limb bud. In the most proximal portion of the limb, which develops first, only *Hoxd-9* is expressed. More distally, successive genes are expressed until the last areas to be formed show overlapping areas of expression of all five *Hoxd* genes.

most of the length of the limb, but in the area of the wrist and ankle, the axis of development angles anteriorly, so that development of the distal carpals, tarsals, and digits proceeds in a posterior-to-anterior direction. This is reflected in the distribution of *Hox* genes, specifically *Hoxd-9–13*, which are expressed primarily in the posterior distal portion of the limb (Fig. 10.13).

Osteolepiform sarcopterygians exhibit branching on the anterior but not the posterior surface of the limb axis. Lungfish, in contrast, have branching radials from both sides of the limb axis. The way in which the digits develop in modern tetrapods suggests that they evolved in the manner of posterior radials, but the anterior bending of the developmental axis would result in their extending laterally; they would hence develop in a posterior to anterior sequence. Coates (1994) suggested that the number of postaxial radials may not initially have been fixed, so that there was considerable latitude in the number of digits in the Devonian tetrapods. Only subsequently were they restricted in all groups to no more than five.

Developmental processes of tetrapod limbs

The patterning of Hox genes

The pattern of the *Hox* genes in the limbs involves only two *Hox* clusters, A and D, rather than all four as in the trunk. Front and hind limbs are patterned in an equivalent manner. The similarity in other aspects of development is sufficiently great that developmental biologists have coined special terms to refer simultaneously to the major elements of both limbs. **Stylopod** applies to the most proximate element of both the forelimbs and hind limbs (the humerus and the femur), **zeugopod** to the units formed by either the ulna and the radius or the tibia and fibula, and **autopod** to the more distal part of the limbs, including the carpals or tarsals (collectively termed the **mesopodials**), the metacarpals or metatarsals (together termed **metapodials**), and the digits and their included phalanges.

Morgan and Tabin (1994) showed that individual genes have different areas of overlap and may influence tissue development in different ways at different times during ontogeny. Both *Hoxa* and *Hoxd* genes are expressed in an anterior to posterior sequence within the limb, but more attention has been focused on *Hoxd* because of its anterior to posterior expression in the most distal portion of the limb, where the digits will develop. *Hoxd-9* is activated first and is expressed throughout the limb; *Hoxd-10* follows but is expressed only in the posterior half of the limb bud; and *Hoxd-11–13* are expressed in successively smaller areas of the posterodistal portion of the limb bud. These expression domains overlap one another, somewhat in the manner of Russian dolls, so that all five *Hoxd* genes are expressed in the posterodistal portion of the limb bud.

Targeted disruptions of *Hoxa* and *Hoxd* in the mouse indicate that development of the major limb elements in successively more distal portions of the limb bud is controlled by successive *Hox* groups (Davis et al. 1995): the shoulder girdle and pelvis by *Hox-9*, the humerus and femur by *Hox-10*, the ulna and radius, tibia and fibula, and the proximal carpals and tarsals by *Hox-11*, the distal tarsals and carpals by *Hox-12*, and the hands and feet by *Hox-13* (Fig. 10.14).

There are five overlapping areas of expression determined by the *Hox-D* series, extending from anterior to posterior: (1) d9 alone, (2) d9–10, (3) d9–11, (4) d9–12, (5) d9–13. The distribution of *Hox* genes in the distal portion of the limb has been used to suggest that they may be instrumental in establishing the number and identity of digits (Tabin 1992). Subsequently, Morgan and Tabin (1994) recognized that the changing domains of the various *Hox* genes during development made it impossible for them to code directly for specific digit number or morphology. They noted that in early development *Hoxd* genes are involved in regulating the growth of the undifferentiated limb mesenchyme, and suggested that the number of digits might be regulated by the amount of tissue produced at this stage. Later these genes regulate the maturation of the nascent skeletal elements, at which time the distinct morphology of the individual digits is determined. They concluded:

> *Hoxd* genes do not act in a simple combinatorial code for "digit identity." While they do contribute to the regulation of digit morphology, our current understanding of their action does not provide an indication of a constraint on potential morphologies. Either such a constraint remains to be discovered in the subtler aspects of *Hox* gene action or else one will have to look elsewhere for it. (pp. 185–6)

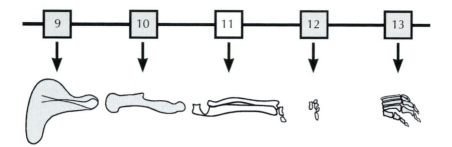

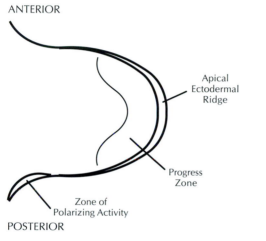

Figure 10.14. Specificity of the *Hox* groups responsible for formation of forelimb elements in the mouse. Homozygous null mutations of *Hoxa-11* and *Hoxd-11* result in the loss of the radius, ulna, and proximal carpals. Reprinted from *Nature* (Davis et al., vol. 375). Copyright © 1995, Macmillan Magazines Limited.

ANTERIOR

Apical
Ectodermal
Ridge

Progress
Zone

Zone of
Polarizing Activity

POSTERIOR

Figure 10.15. Embryological structures associated with development of the limbs.

Even more difficult to explain is the fact that although the areas of expression of *Hoxd* genes in the front and hind limbs are nearly identical, the number and shape of the digits in some groups, particularly birds, are strikingly different (Fig. 10.15). Study of additional paralogous genes may answer this problem (Nelson et al. 1996).

Other factors controlling the development of limbs

While study of the distribution of the *Hox* genes presents a simple, diagrammatic way of looking at the establishment of the pattern of limb development, earlier studies showed that the formation of limb structures is also under the control of several other factors residing within the limbs themselves, including specific areas of differentiation that are present during limb development, molecules termed **morphogens** that diffuse through the developing limb, and other genes that are not part of the *Hox* clusters. The areas of differentiation include the limb fields, the apical ectodermal ridge, the progress zone, and the zone of polarizing activity (Gilbert 1994).

It has long been recognized that the position where the limb buds would develop was established well before they became apparent as extensions from the trunk. These areas were termed the **limb fields** and have since been established as areas of early *Hox* gene expression. *Hoxc-6* has been identified in the area where the forelimb will appear in the zebra fish, the frog *Xenopus,* and the mouse (Oliver et al. 1988; Molven et al. 1990; De Robertis, Morita, and Cho 1991).

The **apical ectodermal ridge (AER)** develops at the tip of the limb bud as it emerges from the flank (see Fig. 10.15). It is formed from superficial ectoderm that is influenced by factors from the underlying lateral plate mesoderm. The ridge is essential to limb growth and differentiation; if it is removed, the limb bud regresses. Under normal conditions, the apical ectodermal ridge maintains the directly underlying mesenchyme in a state of cell proliferation and prevents the cells from forming cartilage. This region of cell division, the **progress zone**, determines the proximal–distal limb axis. The first cells to divide and leave the progress zone form proximal structures; those cells that have undergone numerous divisions within the progress zone become more distal structures. One may imagine the expression of each of the sequential *Hox* genes during the time of maximal cell proliferation: 10 A and D (i.e., *Hoxa-10* and *Hoxd-10*) as the cells that will form the humerus appear; 11 A and D with the precursors of the ulna, radius, tibia, fibula, and proximal carpals and tarsals; 12 A and D with the distal carpals and tarsals; and 13 A and D with the metapodials and digits (Davis et al. 1995).

The anterior–posterior axis of the limb is determined by the **zone of polarizing activity (ZPA)**, which is a group of cells on the posterior margin of the limb whose activity is initiated and maintained by the presence of the apical ectodermal ridge. The ZPA is apparent as early as the 16-somite stage, before the limb bud is even visible. The effect of this zone has been suggested as a result of the diffusion of some substance from the cells in this area, or as a result of successive cell–cell interactions.

Detailed analysis of the functional sequence of gene and structural activation in the ZPA was undertaken by Charité et al. (1994) and Niswander et al. (1994). In the mouse, *Hoxb-8* as well as *Hoxc-6* is expressed in the forelimb field. It is present for only a short period, but this may be sufficient to induce formation of the zone of polarizing activity. Three genes are activated in succession in response to the presence of the ZPA: *Sonic hedgehog (SHH), Fibroblast growth factor 4 (Fgf-4),* and *Hoxd-11. SHH* and *Fgf-4* maintain the activity of the ZPA as *Hoxb-8* expression is lost. *Sonic hedgehog* encodes a protein that is thought to be an intercellular signal molecule controlling anteroposterior patterning in the limb. This molecule may activate the bone morphogenetic proteins. The number of digits that result from transplants within the limb is proportional to the number of posterior cells implanted at the margin of the ZPA (Charité et al. 1994).

Transplantation or removal of the AER or ZPA, disruption or misexpression of *Hox* and other genes, and treatment of developing limbs with the putative morphogen retinoic acid all influence the development of digits, which may be reduced or increased in number, duplicated, or their orientation altered. An especially informative disruption is achieved by removing the cells from the limb bud, disaggregating them, and replacing them in the bud: Digits are formed, but unlike the results of manipulation of the AER and ZPA, the digits do not conform with the specific pattern of any particular, normal digits. This suggests that the capacity to form digits is determined separately from the capacity to establish specific digit identity.

Mutations that affect the presence or extent of the apical ectodermal ridge result in malformation of the limb. In the chick, the mutation *polydactylous,* in which the ridge is longer than normal, results in extra digits. The mutation *eudiplopodia* results in two complete rows of digits on each hind limb. *Limbless,* in which the ridge fails to form, results in complete limb loss. Although these are naturally occurring mutations, the alterations they produce are unlikely to be comparable with those seen in vertebrate evolution. The multiplications of digits that are seen in vertebrate groups can usually be attributed either to the reappearance of previously lost digits, as in polydactyly in dogs and horses (Hall 1984) or can be seen as the splitting of digits distal to the level of the carpals or tarsals, as in ichthyosaurs, rather than the formation of entirely new structures. In snakes and whales, fossil evidence demonstrates that limb loss is progressive, from distal to proximal, rather than occurring in a single step (Haas 1980; Gingerich, Smith, and Simons 1990). Double rows of digits have never been reported from the fossil record.

The configuration of particular elements of the hands and feet may be controlled by specific genes that influence the structure of small groups of bones. Each of several bone- and cartilage-inducing molecules, the **bone morphogenetic proteins (BMPs)**, has different genetic control. For example, the mutations *brachypodism (bp)* and *short ear* in the mouse both disrupt the condensation of mesenchyme cells into outlines of particular skeletal elements, but at different sites (Storm et al. 1994). *Short ear* mutations alter the size and shape of the ears, sternum, ribs, and vertebral processes but do not affect other skeletal elements. *Brachypodism* mutations alter the length of the long bones, change the length of the metapodials, slightly disrupt the organization of the carpals and tarsals, and reduce the number of phalanges in the digits of all four limbs by the fusion of proximal and medial segments. In mice not affected by the mutation for brachypodism, the specific **growth/differentiation factors (GDF)** produced by the wild-type gene is expressed in both distal precartilaginous mesenchymal condensations and in the perichondrium of more proximal skeletal structures. Other mutations differentially affect the forelimbs and hind limbs of mice, but most result in such severe abnormalities that it is unlikely that they provide plausible analogies for changes that might have occurred during their evolutionary history (Lyon and Searle 1989).

Presumably there are other genes that regulate the formation of individual bones, particular aspects of bones, or a simple bone complex such as the assemblage of phalanges that make up a single digit, but these are unlikely to be recognized in screens of populations. It would also be difficult to establish whether such minor differences were under genetic control, appeared randomly, or as the result of environmental factors, but this might be determined in specific cases such as that of the extensive variability that occurs in the hands and feet of very small salamanders (Hanken 1982; Alberch and Blanco 1994; Shubin 1995).

If such genes, which bridge the gap between *Hox* activation and the structure of individual bones, are common, they would answer the question as to how selection on minor anatomical variants could produce significant changes in developmental patterns within the major vertebrate groups. Quantitative genes with such limited effects would provide an explanation for continuous, incremental change over long periods of time, as argued by Darwin and as seen in the fossil record.

The ever-increasing capacity to establish the specific loci and manner of gene expression will make it possible to determine the mechanisms by which individual bones are formed. However, we still seem far from being able to understand how the development of a complex structure such as the vertebrate hand or foot, fin or wing, is controlled – or why, once they have evolved, these structures may remain nearly constant for hundreds of millions of years. We still lack understanding of the molecular blueprint or template that maintains exactly five digits and a phalangeal count of 2,3,4,5,3 in the hands of lizards and their ancestors for a period of 340 million years. Neither can we explain how similar expression of identical *Hox* genes can control development of structures as distinct as the forelimbs and hind limbs of birds (Fig. 10.16).

Although the processes controlling development are still incompletely known, we can look further at the specific patterns of development and evolution in both modern and fossil groups to evaluate just how strictly the rules of developmental constraints have been followed during the history of vertebrates.

Morphogenesis and evolution of tetrapod limbs

Patterns of chondrification

Developmental biology textbooks deal with the differentiation of cells and the elaboration of particular tissues, down to the level of the establishment of the general configuration of the limbs; yet they are not concerned with details of the formation of the individual bones whose evolutionary changes enabled vertebrates to adapt to such diverse activities as swimming, running, digging, and flying. Instead, these subjects have the interest of comparative anatomists and vertebrate paleontologists. The gap between the studies of limb development typically pursued by developmental biologists and those conducted by vertebrate paleontologists has been bridged by a series of very important papers by Alberch and his colleagues (Alberch 1985; Alberch and Gale 1985; Shubin and Alberch 1986) documenting the patterns of mesenchymal tissue condensation and the formation of centers of chondrification that can be seen in the limbs of a wide range of living vertebrates. These studies illustrate a basic pattern common to all major groups of tetrapods, one that serves as a starting point for determining what changes have occurred in more specialized lineages, as well as the degree to which these changes may be constrained by underlying rules of development.

At the cellular level, the formation of the limb skeleton in tetrapods involves three basic processes: de novo condensation of undifferentiated mesenchymal cells, and either branching or segmentation of the condensations as they form. Morphogenesis begins with a single proximal condensation that will eventually become the humerus in the forelimb and the femur in the hind limb (Fig. 10.17). The presence of only a single ossification proximally has been postulated as resulting from the small size of the limb bud when it begins to develop. However, the fact that the base of the paired fin in many groups of primitive fish is very wide and is supported by many parallel radials suggests that some additional factor must be active in sarcopterygians and tetrapods to restrict the base of the limb to a single element. More distally, tissue condensations at the distal ends of the presumptive

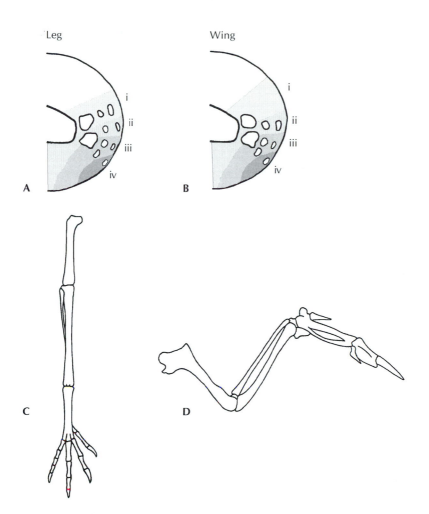

Figure 10.16. Gene expression and limb development. **A, B,** Expression of *Hoxd* genes in the limb buds of the leg and wing of a bird. White patches indicate where digits are forming (reprinted from *Nature* [Morgan et al., vol. 358]. Copyright © 1992, Macmillan Magazines Limited). **C, D,** Adult structure of the foot and wing. The very similar distribution of the areas of *Hox* gene expression is in striking contrast with the very different adult morphology (from Carroll 1987).

humerus and femur begin to bifurcate. As these areas extend distally, they separate at the base to form the more distal bones of the limb. The pattern of condensation of the cells that will form the radius and ulna and the tibia and fibula illustrate three "rules" of limb development in tetrapods:

1. Bones may originate by bifurcation, but never by trifurcation.
2. Bifurcation is always succeeded by segmentation (no limb bones are forked). Segmentation may, however, occur without bifurcation, as in the bones of the fingers and toes.
3. These bones, and all others in the limbs, are initially arrayed in essentially a single plane; none diverges at a significant angle from this plane.

The bifurcation of the radius–ulna or tibia–fibula establishes a pre- and post-axial series (Fig. 10.18). The preaxial series (beginning with the radius or tibia) never undergoes branching. The radius and radiale, as well as the tibia and tibiale, are segmentally divided from one another. Most of the bones that form the carpus, tarsus, and digits are elaborated from the posterior portion of the limb; this is referred to as *postaxial dominance*. Development proceeds by bifurcation distal to the ulna that gives rise to the ulnare and intermedium in the carpus, and of the fibula to give rise to the calcaneum and the intermedium or astragalus of the rear limb. The intermedium produces either a single centrale by segmentation or two centralia by branching.

Rather than development continuing by progressive bifurcation and segmentation across the carpus and tarsus into the digits, what appears as a new axis of development extends across the distal carpals or tarsals. This **digital arch** is essentially a continuation of the proximal–distal axis that extends down the limb, but it is recognized as a distinct entity because of the sharp angle the axis makes as it extends anteriorly from the ulnare or calcaneum. The angled extension into the hands and feet reflects the changes, discussed earlier, that occurred in the origin of the tetrapod limb (see the section "The origin of tetrapod limbs," esp. Fig. 10.11). The digital arch initially appears as a band of tissue in the area where the precursors of the distal carpals or tarsals will differentiate. Condensation of this area proceeds by segmentation from the ulnare, giving rise to the fourth distal carpal, or from the calcaneum to give rise to the fourth distal tarsal. The fourth distal carpal or tarsal then bifurcates, giving rise to the metacarpal of the fourth digit and the third distal carpal, which in turn bifurcates to yield the third metacarpal and the second distal car-

Figure 10.17. Transverse sections through the two-digit stage of *Ambystoma mexicanum* hindlimb. **A,** Lower-power view showing the entire limb field depicting the development of the proximal carpal region and digits one, two, and three. **B,** Close-up of A, showing the cell orientations of the developing precartilage condensations. The perichondral cells (P) are elongate and flattened. The inner zone of cells (S) are rounded in cross section. Other abbreviations: c, centrale; F, Femur; F, fibula; f, fibulare; i, intermedium. From Shubin and Alberch (1986).

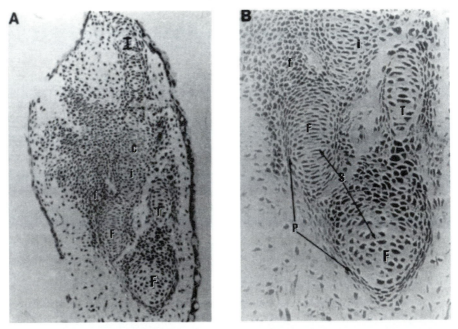

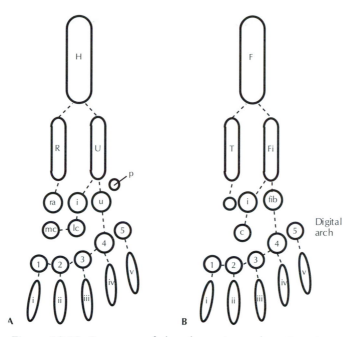

Figure 10.18. Sequence of chondrogenic condensations in a generalized tetrapod limb. The primary axis passes through the humerus–ulna or femur–fibula, meeting the digital arch at the ulnare or calcaneum. The digital arch turns preaxially producing the meso- and metapodials by mesenchymal bifurcation. **A,** Front limb. **B,** Rear limb. Abbreviations as in Figure 10.8. From Caldwell (1994).

pal, and so on. The phalanges of each of the digits arise by segmentation from the appropriate metacarpal. The same process occurs in the foot. The fifth digit develops somewhat independently of the rest of the hand or foot, typically appearing after the fourth. The sequence of digit formation is 4-(5 or 3)-2-1.

Aside from the fifth digit, the digital arch in amniotes always develops in a posterior to anterior direction. This is also true of frogs but, surprisingly, is not the case in the salamanders that have been studied: Their digital arch develops from anterior to posterior, with the sequence of digit formation 2-1-3-4-5. Different patterns of limb development led Holmgren (1933) to suggest that salamanders evolved from a different lineage of fish than did all other tetrapods (i.e., the tetrapod limb arose separately in two lineages). Most paleontologists now think that salamanders shared a common ancestry with other living amphibian groups, but there is no answer as to why their limb development differs from that of all other tetrapods, or at what time in their evolution the change occurred.

Shubin and Alberch (1986) documented the basic consistency in the pattern and sequence of chondrification of limb elements. Where differences are observed among living vertebrates, they nearly always follow a consistent pattern. When limb elements are reduced or lost, their reduction typically reverses the sequence of bone formation. The proximal elements of the limb are highly conserved, but distal elements are more variable in expression and most commonly subject to loss. Progressive limb loss from distal to proximal is well documented among lizards, in which scores of lineages have reduced their limbs (Greer 1991), and frogs and salamanders typically lose digits in the reverse order to the sequence of their formation (Alberch and Gale 1985). Experimental work involving termination of limb growth at various stages by removal of the apical ectodermal ridge shows that the distal and anterior limb elements are always the first to be affected by termination of growth. These examples clearly demonstrate how a simple change in the timing of developmental processes can alter the expression of the adult structure according to a particular pattern.

To what extent do these patterns and processes of development constrain or otherwise influence the direction of evolution? To evaluate this, it is necessary to examine one more level of development.

Ossification sequences in primitive diapsid reptiles

Alberch and his colleagues concentrated their studies on the pattern and sequence of formation of the many centers of *chondrification* in the limb. However, they did not continue their studies to the level of *ossification* of the cartilaginous elements. Although it might be assumed that the sequence of chondrification would be followed nearly exactly by that of ossification, the specific degree of congruence between these two processes must be established in order to study patterns of development from the fossil record, in which only ossified bones are likely to be preserved. In fact, there is one major exception to the similarities in the sequence of these two processes.

In both major tetrapod groups, the amphibians and the amniotes, chondrification of the limbs proceeds continuously from proximal to distal; ossification does not. Although it begins proximally and extends through the humerus or femur to the ends of the ulna and radius or tibia and fibula, the ossification sequence does not continue through the carpals or tarsals but jumps directly to the metacarpals or metatarsals and digits, leaving the areas of the wrist and ankle without ossification until significantly later. Thousands of fossils of immature amphibians and reptiles are known in which the proximal and distal bones of the limbs are ossified, but not the carpals and tarsals (see, e.g., Fig. 10.21A,F).

If the carpus and tarsus ossify out of sequence with the rest of the limb, does ossification of the individual carpal and tarsal bones follow the sequence of formation of their cartilaginous precursors? This gap is now being filled through the work of Rieppel (1992, 1993), who has studied members of many reptilian groups to establish the sequence of ossification of carpals and tarsals following their chondrification. Although there is some coossification of originally separate sites of chondrification and some loss of cartilaginous element, the pattern of ossification among the carpals and tarsals does follow the same sequence as the pattern of chondrification (Table 10.1). Based on this information, it is now possible to make direct comparison between sequences of ossification that are known in fossils and those of living vertebrates to determine the degree of constraint on this aspect of development over the past 300 million years.

Emphasis is being placed on the carpals and tarsals specifically because of the delay in their ossification. Ossification of the rest of the skeleton proceeds so rapidly that it is only rarely possible to determine the sequence of ossification of the individual bones in fossil species. In contrast, the development of the carpus and tarsus is significantly delayed, and the rate of ossification is sufficiently slow that one can observe the sequence of appearance of each individual unit in extensive growth series that are occasionally seen in the fossil record. This provides an ex-

Table 10.1. *The sequence of ossification of carpal and tarsal elements in the lizard* Cyrtodactylus pubisulcus

ulnare	dc4	radiale	dc3	dc1	dc5	dc2	centrale	pisiform	astragalus	calcaneum	dt4	dt3
								+				
								+	+			
								+	+	+		
+								+	+	+		
+	+							+	+	+		
+	+	+[a]						+	+	+		
+	+	+	+					+	+	+		
+	+	+	+	+	+			+	+	+	+	+
+	+	+	+	+	+	+		+	+	+	+	+
+	+	+	+	+	+	+	+	+	+	+	+	+

[a]The only exception is FMNH 1249299, where ulnare and distal carpals 4 and 3 are ossified, but not the radiale.
Source: Reprinted by permission of Oxford University Press from Rieppel (1992), *J. Zool. Lond.*, vol. 227.

cellent model for following the temporal sequence in the expression of a developmental pattern. There are few, if any, other parts of the skeleton that are so well suited to determining the importance of developmental constraints in influencing the patterns of evolutionary change.

The first publication specifically on developmental sequences of the carpals and tarsals of fossil vertebrates was by Caldwell (1994), who studied growth series of early amniotes close to the base of the radiation that eventually led to the origin of all the living diapsid groups (lizards, snakes, *Sphenodon,* crocodilians, and birds) as well as diverse fossil groups including dinosaurs and many lineages of Mesozoic marine reptiles (Fig. 10.19). Caldwell determined that the sequence of ossification of both the carpals and tarsals in Upper Permian diapsids was essentially the same as in modern lizards and *Sphenodon,* taking into consideration the loss of some of the distal carpals and tarsals in the modern genera. This showed, without question, that the digital arch ossified in the same sequence 250 million years ago as it does today. This constancy is matched by the identical number of digits and of phalanges within each digit as those in the majority of living lizards.

Caldwell studied three genera from the Upper Permian of Madagascar (Fig. 10.20). One, *Thadeosaurus,* had body and limb proportions very similar to those of modern terrestrial lizards. *Hovasaurus* differed in the lateral compression of the tail that clearly indicated aquatic locomotion. The third genus, *Claudiosaurus,* is thought to be related to the Triassic and Jurassic marine reptiles – nothosaurs, placodonts, and plesiosaurs – on the basis of the small skull with upper but not lower temporal openings, and a long neck.

These genera all had the same complement of carpals and tarsals as adults. As in modern reptiles, the ossification of the mesopodials was much delayed relative to the more proximal and distal limb bones. The area of the carpus and tarsus is completely lacking in ossification in the smallest individuals, in which the rest of the limb is fully formed. The sequence of ossification of the carpals and tarsals is

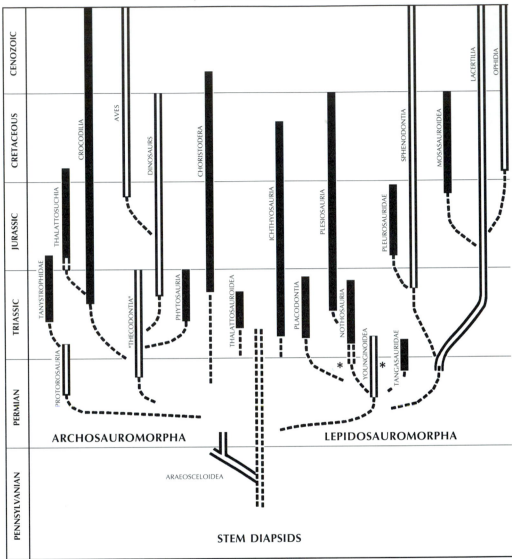

Figure 10.19. Radiation of diapsid reptiles. Open lines indicate primitively terrestrial lineages. Solid lines are aquatic derivatives. Dashed lines indicate probable relationships or range extensions. Asterisks show approximate taxonomic and temporal positions of genera studied by Caldwell (1995). Adapted from Carroll (1985).

best known in *Hovasaurus,* for which the largest number of specimens was available (Fig. 10.21). Although sixteen specimens were available showing the hands or feet, this was not enough to sample every stage of carpal and tarsal ossification. Two or three carpals or tarsals would ossify from one growth sample to the next, giving the impression that they ossified simultaneously. The order of appearance of the carpals was as follows (simultaneous appearance of elements is shown by parentheses): ulnare, (distal carpal 4, intermedium), (lateral centrale, distal carpal 3, distal carpal 1), (radiale, medial centrale, distal carpals 5 and 2). The pisiform is the last to appear; this bone is not part of the basic carpal series, but develops as a

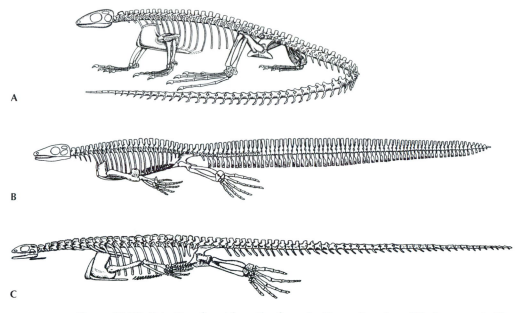

Figure 10.20. Primitive diapsid reptiles from the Upper Permian of Madagascar. **A,** The terrestrial genus *Thadeosaurus.* **B, C,** The aquatic genera *Hovasaurus* and *Claudiosaurus.* From Carroll (1987).

sesamoid element in the tendons of the muscles that extend along the posterior margin of the wrist. In the tarsus, the sequence is as follows: astragalus and calcaneum appear simultaneously, followed by distal tarsal 4, (distal tarsal 3, centrale, distal tarsal 2), then by distal tarsal 1 or 5. Distal tarsal 5 later fuses to distal tarsal 4.

Modern lizards also show irregularity in the sequence of ossification of the distal tarsals. Among the Madagascar genera, the degree of irregularity is greatest in *Claudiosaurus,* in which the order of ossification of the first, second, and fifth distal tarsals varies from specimen to specimen. This degree of irregularity in the sequence of ossification presumably reflects less stringent selection for both the pattern and extent of ossification associated with an aquatic way of life. Ossification of the carpus and tarsus is conspicuously reduced in many, more specialized marine reptiles, for which selection may be acting to produce a more flexible, paddle-like limb that did not have as great a need for the load-bearing capacity of bone.

The most striking difference in limb development between *Claudiosaurus* and both the other Upper Permian genera and most modern reptiles is the relative timing of ossification of the front and rear limbs. In nearly all terrestrial reptiles, the hind limb initiates and completes ossification well ahead of the forelimb. Although the complete ossification sequence is not clear in *Claudiosaurus,* what is known indicates that the carpals reach an advanced state of ossification well ahead of the tarsals. This presumably reflects different use of the limbs in aquatic and terrestrial locomotion. Many more highly derived aquatic reptiles and mammals (e.g., ichthyosaurs, whales, and sirenians) greatly reduce the rear limbs while maintaining the front limbs.

These fossils from the Upper Permian demonstrate that the pattern and sequence of ossification of the carpals and tarsals have remained nearly constant for approximately 250 million years among animals that have retained conservative body

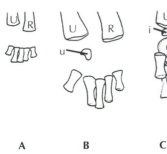

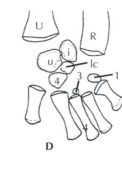

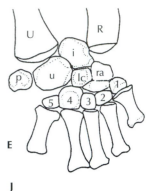

A   B   C   D   E

F   G   H   I   J

1 cm

K

Figure 10.21. Growth series of the carpus and tarsus of the primitive diapsid *Hovasaurus* showing the sequence of ossification. This follows the orientation of the digital arch, seen in Figure 10.18. **A–E,** Carpals. **F–K,** Tarsals. Abbreviations as in Figure 10.8. From Caldwell (1995).

plans and ways of life, such as living lizards and *Sphenodon,* but vary to a limited degree in one lineage that had recently adapted to an aquatic way of life. In the following Triassic period, there was a major radiation of both terrestrial and aquatic diapsids. Small lizardlike forms retained the pattern of the carpals, tarsals, and digits seen in *Thadeosaurus.* This pattern continues to the present in many lizard groups but has been conspicuously altered in the arboreal chameleons, many lin-

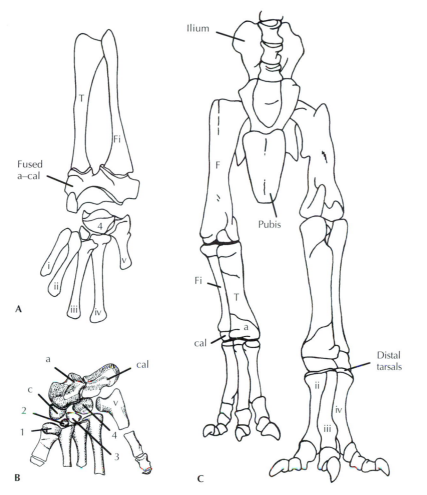

Figure 10.22. Foot structure in advanced terrestrial diapsids. **A,** A modern varanid lizard. In contrast with the primitive diapsid condition illustrated by *Hovasaurus* (Figure 10.21K), the astragalus and calcaneum fuse early in ontogeny and are closely integrated with the tibia and fibula. The main joint in the foot is between the astragalocalcaneum and the fourth distal tarsal. The sequence of ossification of these elements retains the pattern of early diapsids. **B,** The Upper Permian archosaur *Proterosuchus.* The astragalus and calcaneum articulate with one another, somewhat in the manner of crocodiles. The fifth distal tarsal is either lost or incorporated into the head of the fifth distal tarsal. **C,** The dinosaur *Tyrannosaurus.* As in birds, the astragalus and calcaneum have become integrated with the tibia and fibula. The proximal tarsals form a hinge joint with the distal tarsals. Movement of the hind limbs is in a parasaggital plane, in contrast with the sprawling gait of lizards and crocodiles. Abbreviations as in Figure 10.8. From Carroll (1987).

eages that have greatly reduced or lost their limbs, and in highly derived aquatic families.

Other diapsid lineages, leading to larger forms such as crocodiles and dinosaurs, began to modify their limb structure in the Late Permian and Triassic. This is particularly evident in the dinosaurs, which switched from a sprawling to an erect posture. Members of this group show considerable bone loss, fusion, and change in functional relationships. In the rear limb, the proximal tarsals are incorporated with the tibia and fibula and form a simple hinge joint with the distal tarsals (Fig. 10.22).

Development and evolution in Mesozoic marine reptiles

Although many important osteological changes occurred in the limbs of terrestrial vertebrates throughout the Mesozoic and Cenozoic, few if any species are known from extensive growth series that enable developmental studies to be conducted. Aquatic reptiles, on the other hand, provide an excellent model for such work. Plesiosaurs, ichthyosaurs, and mosasaurs were the dominant marine predators of the Mesozoic. They were extremely common throughout the oceans of the world and left a great many fossils that frequently preserve the entire skeleton in natural articulation. The scope of limb evolution in these groups was as great as that which occurred in the origin of tetrapods from fish, although it proceeded in the opposite direction. All of these groups were very diverse and show many different patterns of limb structure, so that even the adults provide information regarding the direction of evolution in the pattern and sequence of ossification of the limb elements. This is further documented through studies of growth series that are particularly well known among the ichthyosaurs. Hence this assemblage provides a singularly informative system for studying the manner in which developmental processes may influence the course of evolutionary change. All of these marine taxa shared a common ancestry with primitive terrestrial diapsids. The structural and developmental patterns of the limbs studied in Late Permian diapsids thus serve as a basis for detailed comparison of the nature and direction of changes observed in the aquatic groups.

The aquatic nothosaurs of the Triassic definitely show evidence of developmental constraints in the manner of carpal and tarsal reduction (Fig. 10.23). Although no informative growth series have yet been described, the pattern of ossification of the adult carpus and tarsus clearly reflects the sequence of ossification both in their Permian ancestors and in living lepidosaurs (Carroll 1985). A very similar pattern is shown among the Upper Cretaceous mosasaurs (Caldwell 1996). Both nothosaurs and mosasaurs show variable reduction of mesopodial ossifications, but the bones that remain are always those that were first ossified in primitive terrestrial diapsids. Bone loss clearly began at the anterior margin and proceeded posteriorly, as would be expected by the general phenomenon of the early loss of the last bones to develop.

This phenomenon also occurs in the early plesiosaurs, which are thought to have evolved from the base of the nothosaur assemblage (Fig. 10.24). They too show reduced ossification of the tarsus, with absence of elements at the anterior margin. Later plesiosaurs, in contrast, show reelaboration of bones in the carpus and tarsus, but much modified so as to form an integrated link between the reduced ulna, radius, tibia, and fibula and the distal portions of the limb (Caldwell in press-c). The number of digits remained constrained to the primitive number, but the number of phalanges was greatly increased. The original constraints were by then considerably stretched; they show further extension in the most highly derived group of Mesozoic marine reptiles, the ichthyosaurs (Figs. 10.25–10.27).

The ichthyosaurs show the highest degree of specialization to an aquatic way of life of any reptiles. In the most derived genera, the body is carangiform, with a high-aspect, lunate tail. The forelimb is a large, paddle-shaped structure, but the hind limb is much reduced. The fossil record of ichthyosaurs begins in the Early

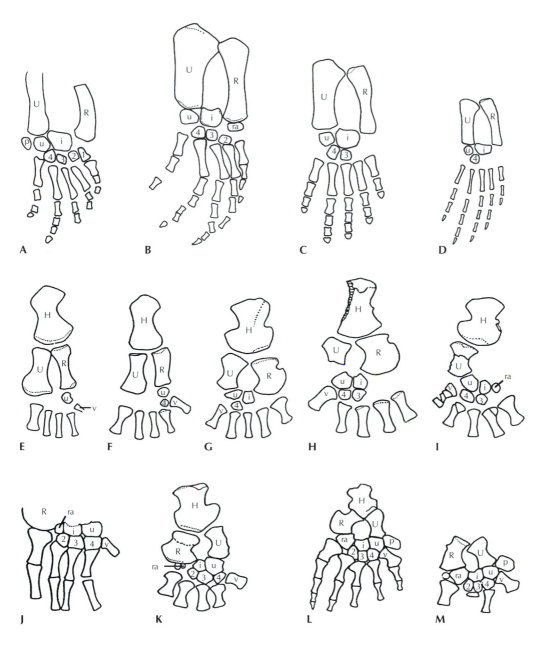

Figure 10.23. Patterns of carpal reduction in the aquatic nothosaurs and mosasaurs. Neither of these groups shows extensive growth series, but the patterns of mesopodial ossification in the juveniles that are known, as well as in the adults, reflect the sequence of ossification in primitive diapsids. The last elements to ossify in ontogeny are the first to be lost phylogenetically. Aquatic nothosaurs (A–D): **A,** The largest complement of carpals among nothosaurs is seen in *Proneusticosaurus,* with the loss of only the radiale, distal carpal five, and the centralia, compared with primitive diapsids. **B,** *Cereiosaurus.* **C,** *Paranothosaurus.* **D,** *Nothosaurus,* showing progressive carpal loss approaching the condition of hatchling early diapsids (A–D from Carroll 1985). Mosasaurs (E–M): **E, F,** Growth stages of *Tylosaurus.* **G, H, I,** Growth stages of *Platecarpus.* **J,** *Ectenosaurus.* **K,** *Plioplatecarpus.* **L, M,** *Clidastes.* The number of phalanges is somewhat reduced in nothosaurs, but may be greatly increased in mosasaurs (E–M from Caldwell 1996). Abbreviations as in Figure 10.8.

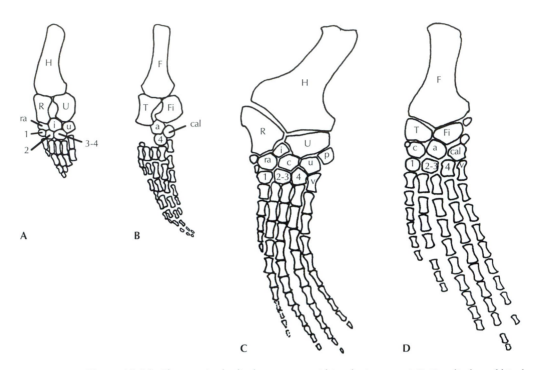

Figure 10.24. Changes in the limb structure within plesiosaurs. **A,B,** Forelimb and hind limb, respectively, of the Lower Jurassic genus *Plesiosaurus*. The hind limb shows the same reduction of ossification of the anterior portion of the tarsus observed in some nothosaurs. **C, D,** Forelimb and hind limb, respectively, of the Upper Jurassic genus *Cryptocleidus*. Several additional bones are added to the carpus and tarsus and the number of phalanges is greatly increased, but five digits are retained. Abbreviations as in Figure 10.8. From Caldwell (1995, in press-a).

Triassic with animals that are clearly aquatic in habitat but whose limbs retain many features of their terrestrial antecedents (Fig. 10.25). The primitive distinction between the bones of the upper and lower arm, wrist, and hand is retained, and there are still five digits. Some genera appear to show a reduction in the number of phalanges; others exhibit a significant increase. By the Middle Triassic, the limbs had attained a more paddlelike shape, and the carpals, metacarpals, and phalanges had formed a continuous series. Five digits were retained, but the number of phalanges was much increased.

Some Late Triassic ichthyosaurs had reduced the number of digits to three, whereas others retained five. All further increased the number of phalanges. In some genera the radius and ulna were much shortened and began to approach the configuration of the carpals, metacarpals, and phalanges. Within the Jurassic and Cretaceous, the radius and ulna came to resemble even more closely the more distal bones of the fin. Some genera exhibited a narrow fin with no more than three digits; others had up to eight rows of phalanges. In some species it is clear that digits have divided to form two rows of phalanges. The most lateral ossifications appear to lie beyond the normal digits, and probably ossified separately within tendons running along the margins of the fins.

Many of the confining parameters of primitive terrestrial reptiles have been exceeded by the advanced ichthyosaurs. The primitive phalangeal count is far sur-

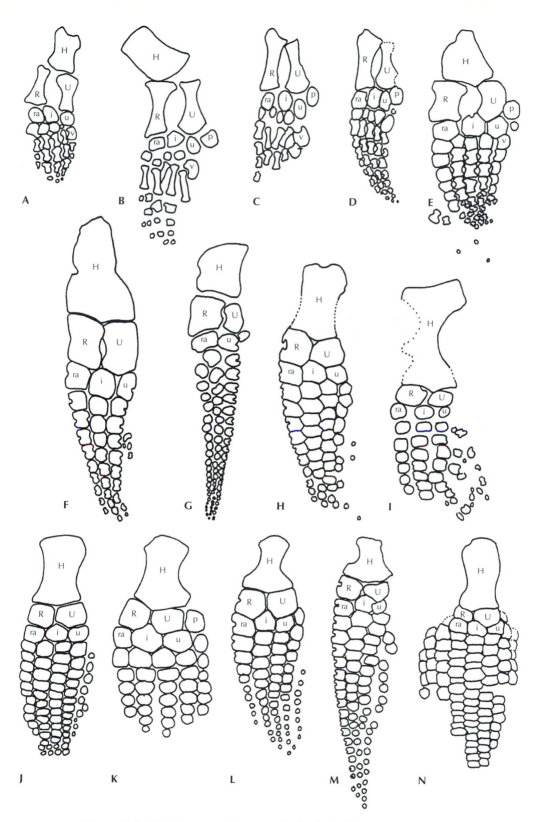

Figure 10.25. Phylogenetic change in the forelimb of ichthyosaurs. Lower Triassic genera: **A,** *Chaohusaurus,* **B,** *Utatsusaurus,* **C,** *Grippia,* and **D,** *Parvinator.* Middle Triassic genus: **E,** *Mixosaurus.* Upper Triassic genera: **F,** *Merriama,* **G,** *Shonisaurus,* **H,** specimen from Williston Lake, and **I,** *Hudsonelpidia.* Jurassic genera: **J,** *Ichtyosaurus,* **K,** *Opthalmosaurus,* **L,** *Stenopterygius,* and **M,** *Leptopterygius.* Mid-Cretaceous genus: **N,** *Platypterygius.* Abbreviations as in Figure 10.8. From Caldwell (1995).

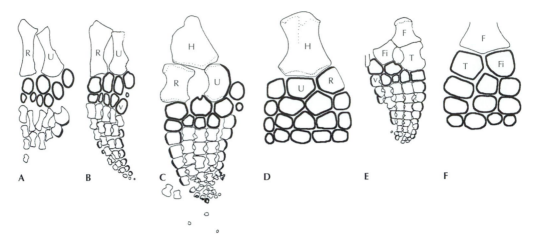

Figure 10.26. Adult condition in progressively more derived ichthyosaurs in which loss of perichondral ossification extends to elements proximal and distal to the carpals. A–D, Forelimbs. E, F, Hind limbs. **A, B,** *Grippia* and *Parvinator* (Lower Triassic). **C,** *Mixosaurus,* Middle Triassic. **D,** *Ophthalmosaurus,* Jurassic. **E,** *Mixosaurus.* **F,** *Ophthalmosaurus.* Thick outlines indicate areas in which perichondral ossification has been much delayed or eliminated. Abbreviations as in Figure 10.8. From Caldwell (in press-b).

passed, the number of rows of phalanges is well above the primitive number of digits, and the morphological distinction among the carpals, metacarpals, and phalanges is almost completely eliminated. More important, a basic feature of the development of terrestrial vertebrates – the lag in ossification of the carpals and tarsals – has been lost (Caldwell 1995). Tiny ichthyosaur limbs preserved in utero show complete ossification of the entire limb at a very early stage, with no gap in the position occupied by the mesopodials.

The change in the sequence of mesopodial ossification is accompanied by changes in the ossification of the ulna, radius, tibia, and fibula as well as the more distal limb elements. In terrestrial reptiles and more primitive aquatic genera, the upper limb bones as well as the metapodials and phalanges are clearly distinct from the carpals and tarsals in their manner of ossification and its timing. The long bones, metapodials, and phalanges ossify in two stages: The first is the rapid ossification of a perichondral sheath, outlining the shaft of the bone; the second is the endochondral ossification of the articulating surfaces at both ends. Ossification of the mesopodials, in contrast, begins with endochondral tissue, which underlies the cartilage of the joint surfaces. Only later, and in areas that do not form articulating surfaces, is a layer of smooth, perichondral bone formed. The fact that the endochondral bone of both the long bones and the mesopodials forms after perichondral bone explains why the carpals and tarsals ossify later. Throughout the Triassic and Jurassic, a period of nearly 100 million years, the ichthyosaurs show gradual and progressive modification in the pattern of ossification of the long bones, metapodials, and digits. Areas of perichondral ossification are gradually reduced so that more and more endochondral bone appears at the margins of the elements. Finally, the long bones, metapodials, and phalanges come to resemble the mesopodials in their reduced perichondral ossification, as well as their overall shape (Fig. 10.26). With the lost of perichondral bone, they also lose the cause for the dis-

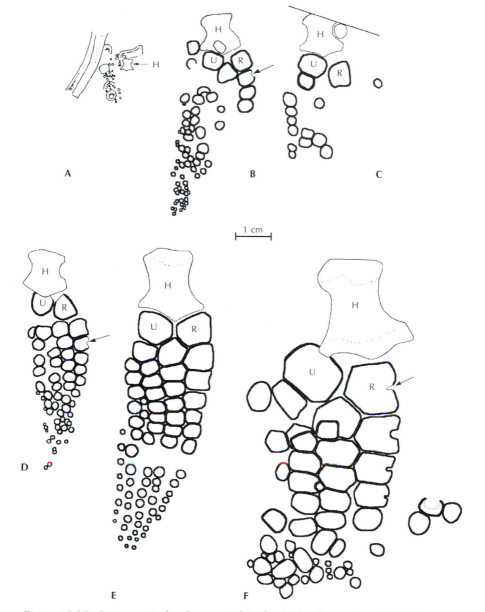

Figure 10.27. Ontogenetic development of the forelimb of Lower Jurassic ichthyosaurs.
A, The smallest known ichthyosaur forelimb, *Ichthyosaurus communis*. **B–F,** Progressively larger specimens of *Stenopterygius quadricissus*. A–E are embryonic. Areas in
which perichondral ossification is lost are indicated by heavy lines. Abbreviations as in
Figure 10.8. From Caldwell (in press-b).

tinction in the timing of their ossification. With the exception of the humerus,
which may ossify slightly earlier, the forelimb ossifies in rapid sequence, with the
mesopodials fully integrated with the rest of the fin developmentally as well as
functionally. We thus see a direct relationship between changes in developmental
processes and the structure and function of the limb in both neonates and adults
(Fig. 10.27).

These developmental changes have occurred in relationship with the progressive unification of function throughout the limb. This can be seen as a complete reversal of the changes that occurred in the evolution of early tetrapods from osteolepiform fish. In that case, a fin, which functioned as a unitary structure throughout its length, evolved toward a limb with a complex joint in the area of the wrist and ankle. This was accompanied by changes in the anatomy of the bones that were to become mesopodials by growing distinct in shape from the more proximal bones of the limb and by forming extensive articulating surfaces on all the bone margins. These articulating surfaces were underlain by endochondral bone, which, like that of the long bones, ossified more slowly than the perichondral covering. Devonian as well as many Carboniferous tetrapods show large unossified areas of the carpus, even in adults. By the mid-Carboniferous, the lag in ossification of both the carpus and tarsus seen in reptiles from the Permian to the Recent is clearly evident. The origin of land vertebrates and their return to the water were both accompanied by heterochronic changes in the sequence of ossification of the mesopodials. These were brought about by changes in the relative extent of perichondral and endochondral ossification, which occur in a sequential pattern.

Returning to aquatic reptiles, plesiosaurs show a similar reduction of perichondral bone on the elements proximal and distal to the mesopodials, but it is not as extensive as that seen in ichthyosaurs. Mosasaurs, gigantic marine lizards of the Upper Cretaceous, show changes in the overall morphology of their limbs comparable to those of ichthyosaurs and plesiosaurs, but without modifying the extent of perichondral ossification seen in those groups. Some turtles became highly adapted to marine locomotion, but their paddles, rather than showing multiplication of phalanges, retain close to the primitive chelonian count of 2,3,3,3,3; each element, however, is greatly elongated.

The variation among marine reptiles suggests that the capacity for long-term change in the configuration of the limb is not strictly constrained by developmental factors. Rather it may be molded by different selective forces acting on each group, or possibly by chance factors involving what mutational changes happened to occur early in their period of aquatic adaptation. The specific nature of these early mutations may have then influenced the course of later evolution. The patterns of phylogenetic loss of carpal and tarsal elements in nothosaurs and mosasaurs directly reflect the sequence of their development in ancestral diapsids, but the modifications in patterns and processes of ossification of the long bones, metapodials, and digits in ichthyosaurs and plesiosaurs do not reflect any developmental processes or patterns exhibited in earlier amniotes.

Summary of patterns and rates of limb evolution

The course of evolution among aquatic diapsid reptiles provides a broad time scale for comparing patterns and rates of morphological change and evaluating the degree to which they may be influenced by factors of development. Most important, the rates, amounts, and directions of change differ markedly from group to group and within groups over time. As a basis for comparison, changes are minimal in the limbs of small, terrestrial diapsids with a lizardlike body form. The basic numbers and patterns of the carpals, tarsals, and digits have remained little modified from the early amniotes of the Upper Carboniferous to conservative lizards and the tuatara, *Sphenodon,* living today. Extremely conservative developmental processes may be at the heart of the long-term constancy of the limb skeleton of these groups, which seem to be inherently constrained beyond the level expected by stabilizing selection and well below the level expected by random variation. On the other hand, the general form of the limbs is certainly associated with the similarity of habitat, behavior, and diet during the past 250–300 million years. During the same period, the skulls of these forms have changed significantly, suggesting that the extreme stasis of the limbs is peculiar to that portion of the skeleton.

In contrast, lizards that adapted to an aquatic or burrowing ways of life, as well as most other diapsid lineages, specialized their limbs in a multitude of different ways. The most extreme examples are seen in the total absence of limbs in snakes and their transformation into wings in birds and Mesozoic pterosaurs. In these cases, natural selection and environmental differences together have governed the course of evolution and led to the escape from nearly all of the specific developmental constraints that may have existed among primitive diapsids. Unfortunately, the rates of change in most of these lineages are difficult to quantify since few continuous evolutionary sequences are yet known from the fossil record.

Among Mesozoic aquatic reptiles, change was gradual and continuous for the ichthyosaurs to the degree that this can be judged by the fossil record. The early stages in their evolution, from some time in the Permian to the Early Triassic, is not yet known; however, the limbs of the oldest known Triassic ichthyosaurs still retain most of the features of their terrestrial ancestors, although their general body form was clearly modified toward an obligatorily aquatic way of life. The forelimb had assumed a paddle shape, with shortening of the digits and changes in their proportions, and the medial row of the carpals had been lost, but these changes do not suggest a major modification of developmental processes (see Fig. 10.25). From this point until the extinction of ichthyosaurs in the Late Cretaceous, there is continuous change in the configuration of the limbs. There is divergent evolution in the proportions of the paddles, from short and wide to long and narrow, and in the number of supporting digits, from three to eight; but there is also progressive change in the integration of all the elements distal to the humerus into a mosaic of small polygonal elements. This is accompanied by progressive loss of perichondral ossification on the elements proximal and distal to the mesopodials. There is no period in this sequence that can be specified as a time of especially rapid or large-scale structural change, nor that reflects the occurrence of a dramatic change in developmental patterns or processes over a short period of time. This spectrum of change is documented over a period of well over 100 million years. No doubt, more complex details of change within this assemblage will be discovered as our knowledge of the fossil record improves, but what we see now suggests progressive change in both morphological and developmental patterns, essentially as predicted by Darwin.

It is generally accepted that plesiosaurs diverged from the base of the nothosaur radiation by the Middle Triassic, but few fossils are known that document a transition between these two groups (Storrs 1993). A single, possibly intermediate lineage is represented by *Pistosaurus* (Sues 1987). Its limbs, although poorly known, broadly resemble those of nothosaurs, not plesiosaurs. The limb structure of all adequately known nothosaurs is similar. There is variable reduction in the number of carpals and tarsals, but the number of digits and even the number of phalanges remain conservative, and the clear distinction between the major portions of the limb is maintained. The main propulsive force for swimming was provided by lateral undulation of the trunk. The latest Triassic and earliest Jurassic plesiosaurs differ radically in their general body form, with the girdles greatly expanded and the limbs highly specialized as paddles that produced the major force in swimming. The body form indicates that selection for limb shape differed greatly between nothosaurs and plesiosaurs (Carroll and Gaskill 1985). However, we lack knowledge of the fossil record both of the early stages in the origin of nothosaurs and of the presumed links between nothosaurs and plesiosaurs. It is conceivable that limb evolution within this assemblage went on in a nearly continuous manner from the Late Permian into the Jurassic and Cretaceous, as it did in ichthyosaurs, but a period of more rapid change between the two groups cannot be precluded.

Current knowledge of the fossil record of mosasaurs (discussed in more detail in Chapter 12) indicates a period of rapid change in limb morphology at the time of their origin from the aigialosaurs. Over a period of approximately 3 million years, the basic limb structure changed from the pattern of obligatorily terrestrial lizards to that of obligatorily aquatic forms. These modifications involved primarily the reduction in the number and degree of ossification of the mesopodials and changes in proportions of the long bones. They can be explained largely by alternations in the timing of chondrification and ossification (heterochrony) rather than by fundamental modifications in developmental processes.

Integration of developmental biology with the evolutionary synthesis

With the great resurgence of developmental biology brought about by the capacity to identify, sequence, and manipulate genes that are associated with developmental processes, it is finally possible to understand the way and degree to which mutational changes and natural selection can influence changes in the development of major structural features. Gilbert (1994, p. 855) and Gilbert et al. (1996) argued that the current explosion of knowledge of developmental biology enabled this discipline to take its place in the further elaboration of the evolutionary synthesis. Similarly, Sean Carroll (1995) called for additional efforts to be made in integrating the new perspective gained from developmental biology within the evolutionary framework of paleontology and population biology. The subject is further pursued by Raff (1996).

One may divide changes in the control of developmental processes among chordates into two major episodes: one specifically associated with duplication of genes of the *Hox* cluster and another involving other genes, especially those that regulate the expression of the *Hox* cluster genes.

Homeobox genes and chordate evolution

The evolution of genes that regulate the expression of other genes in particular parts of the body was presumably the key to the origin of complex metazoans with multiple body parts. Increase in the number and complexity of expression of homeobox genes was certainly one of the prime factors in the differentiation and radiation of the metazoan phyla in the Early Cambrian (Philippe et al. 1994). Once established, the complement of *Hox* and other homeobox genes has apparently remained relatively constant among nonchordate metazoans. In contrast, chordates have undergone at least two episodes of major increase in the number of *Hox* genes. Following initial tandem duplication of *Hox* genes, primitive chordates had approximately the same number and arrangement as do other complex metazoans, including arthropods, annelids, and echinoderms. More advanced chordates are unique in having undergone at least one and more probably two periods of cluster duplication, bringing their number of homeobox genes to roughly fourfold that of other metazoan groups.

The duplication of homeobox genes would not itself have led directly to structural changes. Initially, the new copy, whether produced by tandem duplication within a single chromosome or by duplication of an entire chromosome leading to a new cluster, would simply result in a copy of the existing gene(s) that would presumably have the same function as the original. Subsequent point mutations might gradually alter the DNA of one or both copies so that they would produce proteins with functionally different homeodomains, leading to the activation or deactivation of different genes. Alternatively, mutations might alter the timing of their activation or the position of their expression. The potential for subsequent change would be enormous but would probably have been slowly realized, since most alterations would have to be relatively small if the progeny were to survive.

Natural selection would act on alternative alleles of the new homeobox genes just as it would on any other genetic variability. There may have been a time when organization of both the genome and the organism was sufficiently simple that considerable variability in the number of segments, appendages, or other structures could be tolerated, as is suggested by the great variety of arthropods seen in the fauna of the Burgess Shale and other Cambrian localities (Conway Morris 1994).

Craniates are unique in the achievement of a stable pattern of multiple *Hox* clusters. The duplication of *Hox* clusters made it possible for developmental systems to be regulated by one or more paralogues of each gene group, while others could be modified to serve altered functions or to change their area of expression. This is clearly seen in simple systems such as the linear sequence of vertebral regions. The specific number of vertebrae in each region appears to be altered in relationship to the areas of expression of individual genes, while their paralogues maintain the basic structure of each vertebral pattern. The presence of multiple *Hox* clusters in vertebrates provides an enormous degree of evolutionary flexibility not available in other metazoan groups.

The period of initial tandem duplication of *Hox* genes in chordates began at some undetermined time in the Late Precambrian or Early Cambrian and had proceeded to the level of cephalochordates by 525 mya (Shu, Conway Morris, and Zhang 1996), when they occur in the Chengjiang deposit of China. A hemichordate is also reported from this horizon. Fragmentary remains attributed to craniates are known as early as the Upper Cambrian, but little is known of the general anatomy of these animals. The first unquestioned craniates are known from the

Middle Ordovician, by which time they had presumably achieved a *Hox* complement similar to that of the modern lamprey, with which they share a basically similar configuration of the brain and paired cranial sense organs. The fossil record of craniates in the Ordovician is confined to three principal horizons, represented primarily by jawless fish of broadly similar morphology. All adequately known species are covered with a bony shield over the entire trunk region, precluding the presence of paired fins. Vertebrates with paired fins are not definitely known until the Silurian, by which time jaws had evolved as well. By the end of the Silurian, the lineages leading to the major groups of bony fish and tetrapods had diverged, indicating the end of the known period of *Hox* gene duplication.

For almost a hundred million years advances in craniate structure may have been associated with *Hox* gene duplication. Unfortunately, the fossil record during this period is still so incomplete that specific rates and patterns of evolution cannot be established. All of the major craniate lineages had diverged by the end of the Silurian, but it is impossible to know whether new structures appeared rapidly or arose gradually over tens of millions of years. Unlike the origin of craniates, there are no structures or tissues that can be tied directly to *Hox* gene duplication during this time.

The evolution of regulatory genes

Although the origin of all advanced metazoan phyla may be attributed to the duplication and differentiation of *Hox* genes, the duplication of genes within clusters appears to have ceased soon after the Cambrian radiation. Presumably, a level of complexity was reached in all metazoan groups beyond which very few major changes in the arrangement of the body parts could be accommodated, and selection acted to limit the further increase in *Hox* genes. Nevertheless, evolution among vertebrates has continued at what appears to be an ever-increasing pace since that time. As argued by Ruddle et al. (1994a,b) and Sean Carroll (1995), most of the great range of structural and developmental changes that have occurred over the past 400 million years of vertebrate history have probably resulted from changes in genes that either regulate or are regulated by the expression of the *Hox* genes, rather than in the *Hox* genes themselves. The same is true for all the other major metazoan phyla.

Within the constant framework maintained by the *Hox* genes, the system of regulatory genes provides an enormous potential for change in individual parts of all organ systems. The entire regulatory network certainly contains many more genes than does the *Hox* cluster itself, but the nature, number, and distribution of these genes are only beginning to be determined. Unlike the *Hox* cluster genes, they are not localized to any particular area of the chromosomes, although some are adjacent to the clusters. It has long been assumed that regulatory genes played an important role in evolution, but it was difficult to recognize them or to understand how they operated until the discovery of the significance of the *Hox* genes. Even now, their specific numbers and functions are too poorly known to develop a quantitative model of their effect on development and evolution. Dickinson (1991) has shown the potential scope for such interactions based on knowledge of the most thoroughly known genus, *Drosophila*.

According to Sean Carroll (1995):

> The creative potential of regulatory evolution lies in the hierarchical and combinatorial nature of the regulatory networks. Variation in the morphogenetic output of such a network can arise at many levels simply by altering the relative timing of developmental gene expression or the interactions between members of regulatory network. . . . In this manner, one aspect of gene function can evolve without altering others. Single genes are often regulated by arrays of discrete regulatory elements that control the pattern, position, timing and level of gene expression, and these features can differ between species. (p. 484)

Knowledge of *Hox* genes provides the opportunity to determine how this system has evolved in relationship to changes observed in vertebrate history. There are at least four major levels at which change could have occurred:

1. in the specific base sequence of the *Hox* genes themselves;
2. in the position, timing, or level of *Hox* gene expression;
3. in the regulatory interactions between *Hox* proteins and their targets; and
4. in the target genes.

Although the basic structure, number, and function of the thirty-eight *Hox* genes have apparently not altered over the past 400 million years, there is some degree of capacity for the base sequence within *Hox* genes to vary, as indicated by polymorphisms evident in modern populations (Pendleton et al. 1993). Changes in the homeobox portion of the *Hox* genes must, however, be stringently constrained, since the proteins transcribed by a single *Hox* gene may regulate hundreds of other genes. This is demonstrated by *Drosophila*, in which it has been estimated that 85–170 genes are regulated by the product of a single homeobox gene, *Ultrabithorax* (Mastick et al. 1995). Change or loss of control over such a large number of genes through mutation in the *Hox* gene would affect a very wide range of tissues and almost certainly result in drastic and probably fatal alteration of development. On the other hand, changes in individual proteins regulated by the *Hox* genes, especially those that control timing and area of expression, can be much more readily accommodated. Changes to or control of target genes by different proteins may also be subject to considerable variation, especially if the area and tissue of expression is relatively limited, as in the case of *brachypodia* and *short ear*.

Dickinson (1991) discussed at length the difficulties of studying the evolution of gene regulation in *Drosophila*. There is clearly a great deal of variability in the non-*Hox* genes associated with regulation of protein formation, both within and between species. At least 30 percent of individual "traits" are variable within one species group. This variability provides the potential for the frequencies of these alleles to be acted upon by selection if they influence the viability and fecundity of the organism. Unfortunately, it remains difficult to establish whether these protein polymorphisms have different selective values, since the proteins are typically expressed in several tissues and at different times during development. Although it is difficult to demonstrate at present, it is probably at this level of developmental regulation that we can expect to see results of Darwinian selection leading to gradual and progressive changes in major vertebrate structures, such as the changes in the limb structures discussed in this chapter.

Development and macroevolution

Major changes in processes of development have long been assumed to be basic to large-scale or macroevolutionary changes in morphology. This was most strongly advocated by Goldschmidt (1940), but more recent support has been provided by Gould (1977) and Thomson (1988). It is now possible to identify specific genes associated with a great many aspects of development, ranging from specification of the regions where limbs and other structures will form to control of the specific timing and pattern of cartilage condensation and ossification in individual bones. Although the specific pathways of gene regulation have only begun to be worked out, it is clear that hundreds of separate genes as well as numerous signaling and structural proteins are involved in forming structures. The great diversity of elements in developmental systems points to a much different mode of evolution than was envisioned by Goldschmidt. Instead of single homeotic genes appearing or changing their function as a result of major mutations, we find that the number, position, and basic function of *Hox* genes are extremely stable, but that the position and nature of their expression are regulated by a host of other genes. These other genes exhibit considerable variability both within and between closely related species, suggesting that they have adaptive significance and are subject to natural selection in the same way as genes controlling less vital aspects of morphology, in common with the genes that have been more thoroughly studied by geneticists.

Given our present knowledge of the genetics of development, it seems very unlikely that major changes in vertebrate anatomy have resulted from sudden, rapid changes in developmental processes, at least since the first duplication of the *Hox* clusters associated with the origin of craniates. Major changes in the pattern of expression of *Hox* genes may have occurred fairly rapidly in the origin of the tetrapod limb, but this transition may have required nearly 15 million years, which is not shorter than the time required for a comparable degree of change in other aspects of the anatomy in the evolution of rodents, ichthyosaurs, horses, or elephants, for which the fossil record documents more or less continuous change via numerous transitional stages.

From what we know of the multiplicity of genes on which selection can act within the regulatory system controlling development, there is no obvious point at which sudden, major changes are likely to occur. Rather, this system has the same potential for many different rates of change, as we have seen in other aspects of the morphology and physiology of vertebrates.

Gilbert (1994) argued for a distinct role for developmental biology in explaining macroevolutionary change:

> A new developmental synthesis is emerging that retains the best of the microevolution-yields-macroevolution model and the macroevolution-as-separate-phenomenon model. From the latter it derives the concept that mutations in regulatory genes can create "jumps" from one phenotype to another without necessary intermediate steps. From the former, it derives the notion that genetic

mutations can account for such variants and that selection acts upon them to delete them or retain them in populations. (pp. 855–7)

Certainly, changes in processes and patterns of development are at the base of any large-scale change in morphology. On the other hand, this does not in itself demonstrate or require that change is rapid or proceeds by "jumps." Unfortunately, the fossil record is only rarely sufficiently well known to establish the precise time over which change occurs. It is nearly impossible to determine a minimum period. Nevertheless, there are numerous cases, well documented in the marine reptiles, in which change has proceeded in an apparently continuous manner for as long as 100 million years. There are no cases in which major changes have been substantiated as occurring over less than about 3 million years.

Since the evolutionary explosion among metazoans is shown to have occurred over a period of only about 5 million years during the Early Cambrian (Bowring et al. 1993) this must have been a time of extraordinarily rapid change. This episode is exceptional in two respects: It was unique in being accompanied by the duplication of homeotic genes in all metazoan groups, and it was never again equaled for the origin of diverse body plans. Unfortunately, the fossil record of this time is not sufficiently well known to determine whether there were significant morphological gaps between ancestral and descendant lineages, or if evolution was extremely rapid but essentially continuous. Many lineages of mammals and cichlid fishes show very rapid changes in the late Cenozoic that would appear to be saltatory were the fossil record as incomplete as it is for most intervals in the Paleozoic.

It is not possible to demonstrate that changes brought about by mutations in developmental systems have never produced significant jumps from one phenotype to another. Many large gaps are known between the morphology of ancestral and descendant lineages, but most can be attributed to the absence of appropriate fossil-bearing beds in the intervening period. From what we do know of well-documented lineages, there are no good examples from the post-Cambrian history of vertebrates that require such an explanation.

On the other hand, changes in developmental processes have been instrumental in many of the major changes among early craniates, including the origin of neural crest and sensory placodes that were responsible for the key features of the vertebrate head region. Unfortunately, nothing is known of how this came about, so that no selective scenario can be suggested. The origin of jaws, paired fins, and elaboration of the forebrain occurred roughly synchronously, which may tie these changes to a major increase in developmental capabilities associated with the final stages in duplication of the *Hox* genes, but this requires confirmation through additional knowledge of *Hox* genes in the Agnatha and primitive jawed vertebrates.

The development of paired fins and, later, paired limbs is specifically associated with the elaboration of new embryological features, the apical ectodermal ridge and the zone of polarizing activity. Burke (1989) demonstrated that the development of the turtle shell is also associated with the formation of a structure extremely similar to the AER that initiates comparable epithelial–mesenchymal interactions. Her work illustrates how slight changes in the timing and direction of the proliferation of cell populations can result in major structural changes. In the case of turtles, this involves the elaboration of the carapace and lateral growth of the un-

derlying ribs so that they extend dorsal and lateral to the pectoral and pelvic girdles, in contrast to their medial position in all other tetrapods.

Lee (1993, 1996) argued that turtles evolved from pareiasaurs, in which a carapace had already begun to form. Unfortunately, it is not yet possible to correlate
developmental and fossil evidence to establish how the processes described by
Burke were reflected in phylogenetic change.

Summary

This chapter began with a quotation from Gilbert referring to the split, early in this
century, of the disciplines of genetics, embryology, and evolution from what had
previously been considered as a single field: development. In retrospect, it was almost inevitable that Mendelian and population genetics and evolution would initially concentrate on the study of traits that showed simple patterns of inheritance
and expression. The greatest triumph of Mendelian genetics was to establish the
particulate nature of the hereditary material.

The evolutionary synthesis was based on comparatively simple models of the
relative importance of mutation, selection, and random processes acting on alternative alleles at a single locus and affecting individual traits. Pleiotropy was recognized as a phenomenon that would alter the rate of evolution relative to that which
would result from selection for a single trait, but it was not viewed as a common
property of the genetic material. Evolution has generally been viewed as affecting
each organ system separately; embryology, in contrast, has concentrated on the
integrated nature of tissues, organs, and organ systems – the entire organism must
develop and operate as a unit to survive at all. Somites, for example, are associated with the production of vertebrae, muscles, and the dermis. The lens of the eye
forms in the ectoderm of the skin as a result of induction by the optic cup, an out-
growth of the brain. From this perspective, it was difficult to associate development with what was being learned from Mendelian genetics, or to understand how
selection could act on single traits to modify complex systems.

Knowledge of homeobox genes demonstrates that major aspects of development result from genes that act in a very different way than had been assumed under Mendelian inheritance. The function of homeobox genes is not limited to particular tissues or systems but rather to spatial regions of the body. Expression of
specific *Hox* genes in the trunk influences tissues as disparate as surface ectoderm,
mesoderm, neural tube, neural crest, gut, and presumptive gonads. For *Hox* genes,
pleiotropy is the rule rather than the exception.

The other major difference is that genes controlling development function in a
hierarchical pattern. Only thirty-eight *Hox* genes are known in higher vertebrates,
but each may regulate the expression of scores or even hundreds of other genes.
Some of the genes that are regulated by the *Hox* genes may in turn influence a
wide range of cellular processes, such as cell proliferation, differentiation, adhesion, movement, and cell death, in a host of different tissues. In contrast, other
genes may affect primarily a single tissue (such as bone or cartilage) either throughout the area of *Hox* gene expression or in a more circumscribed space.

The stability and integrative capacity of *Hox* genes ensure that individual organisms develop as a functional whole. The absolute constancy of *Hox* gene number and position throughout higher vertebrates indicates that alterations at this level are very unlikely to be incorporated into long-term evolutionary change. This is reflected in the basic constancy of most of the organ systems throughout vertebrate history.

On the other hand, changes in some specific areas of *Hox* gene expression may occur with relative ease, at least among vertebrates, in which each *Hox* group may have several paralogues. This is most clearly seen in the case of variation in the number of vertebrae in the different regions of the vertebral column. In this example, there is a clear-cut difference in the degree of flexibility in different taxonomic groups. In birds, there is considerable variability in the number of cervical vertebrae, with a range of twelve to eighteen; in mammals, the range is only from six to eight, and nearly all species, from shrews to giraffes, have seven. The length of the neck is controlled by changing the length of the individual vertebrae rather than by increasing or decreasing vertebral number. Among aquatic reptiles, nothosaurs and plesiosaurs show a wide range of variation in the total of cervical and trunk vertebrae, but the primitive genus *Claudiosaurus* (see Fig. 10.20C) achieved a long neck by shifting the pectoral girdle posteriorly while retaining the same total number of trunk and cervical vertebrae as in primitive diapsids.

The amount of change that can be tolerated in individual genes regulated by the *Hox* genes depends on their degree of pleiotropy and area of expression. Those that control only the size and shape of cartilage and bone and are expressed in a limited area may be as capable of change through mutation and selection as any of the quantitative traits studied through Mendelian genetics.

In summary, genes that govern changes in patterns and processes of development vary greatly in the scope of their effects and so in their probability of perpetuation following mutational change. At the level of genes influencing details of the shape of individual bones, there is probably a great deal of variability that can be acted upon by natural selection. As in the case of general genetic variability discussed in Chapter 9, there may be much more variability among populations then is typically expressed in long-term evolution, but detailed examples are limited to only a few cases. It is almost certain that the degree of variability in regulatory genes known from modern populations is sufficient to account for much greater rates of evolution than are ever reported from the fossil record.

Natural selection of alternative alleles presumably provides the same general manner of evolutionary control over developmental patterns and processes as it does over other traits described by Mendelian genetics. It is expressed primarily as stabilizing selection in relationship to the structure and position of the *Hox* genes themselves, but it can act in a continual, directional manner on genes that more directly control the configuration of individual elements.

As yet, knowledge of developmental biology does not require a major modification of the Darwinian paradigm of natural selection. What it does provide is an understanding of how development occurs, what changes there have been in the patterns and processes of development during vertebrate history, and how they have influenced the patterns and rates of evolution.

11 Physical constraints

Extreme conservatism in body form evidenced by the fossil record cannot be attributed specifically to developmental constraints, but in many cases it can be understood in terms of physical properties of the environment that have remained constant throughout the history of vertebrates. These properties function as absolute limits to the extent of organismal adaptation. Natural selection can act to optimize body form and structure in relationship to these limits, but they cannot be exceeded. There are a host of physical properties that limit different aspects of vertebrate form and function, of which a few conspicuous examples will be discussed in this chapter: (1) properties of water and air; (2) gravity; (3) transfer of substances across membranes; and (4) properties of body dimensions.

Constraints on body form in fast-swimming vertebrates

The properties of water are associated with all aspects of life. All organisms require water in the fluid state for all biochemical processes. Life is ultimately constrained to environments where cells can operate between the extremes of water in the solid or gaseous states – roughly 0–100 °C. The movements of all aquatic vertebrates are subject to constraints of the density and viscosity of water. These vary slightly depending on the temperature and salinity, but for practical purposes are constants to which organisms must adapt.

In Chapter 10, Mesozoic marine reptiles were used as examples of the degree to which attributes of development influenced changes in the number and morphology of individual limb bones during transitions between fully terrestrial animals and obligatorily aquatic forms. Evolution proceeded at a nearly uniformly slow rate over a period of more than 100 million years in ichthyosaurs, but it occurred more rapidly during the origin of mosasaurs. Although differing in their rate and specific nature, changes in the limbs appear to have been continuous throughout the evolutionary history of these groups.

The general body form of these and other moderate- to fast-swimming vertebrates shows a very different pattern of evolution. In most aquatic groups, locomotion at slow to moderate speeds is achieved by lateral undulation of the entire trunk and tail in animals with long, fusiform bodies. More advanced forms have a more rigid trunk region, with the force of propulsion concentrated in the tail or occasionally the limbs. In the fastest-swimming vertebrates, the body becomes spindle-shaped rather than fusiform, and the tail has a high lunate shape. In contrast with changes that we have seen in the limbs, the pattern of evolution in body form is not continuous throughout the history of these groups; rather, it proceeds rapidly until an optimal general shape is reached, after which it remains nearly constant for tens of millions of years. This pattern of evolution results from fixed properties of the water dictating a particular body form that cannot be improved upon within the biological constraints of the tissues and metabolic properties of the vertebrate body.

Aquatic locomotion in vertebrates is produced by forces generated by the organism that are opposed by the water's viscosity and density. Most aquatic vertebrates propel themselves primarily by undulation of the trunk and tail. This is based on a neuromuscular system that was established within the cephalochordates, with alternating rhythmic contraction of the myotomes on either side of the notochord. The amplitude of lateral undulation increases from the head to the tip of the tail. The effective force is proportional to the angle that the moving part of the body makes with the water, with the maximum force when it is at right angles to the direction of motion: Force at any other angle is wasted in moving the body laterally. The force against the water is also proportional to the area of the surface in motion and the distance and speed of its movement. In a particular organism, speed is generally controlled by increasing the rate of undulation.

The forces generated by the animal are opposed by drag resulting from the viscosity and density of the water. Two types of drag are recognized: friction drag and pressure drag. **Friction drag** results from the resistance that occurs between the *boundary layer* – a thin layer of water that adheres to the surface of the animal – and the surrounding fluid. At slow speeds, flow of water over the animal is laminar, and frictional drag is proportional to velocity $V^{1.5}$. At higher speeds, water flow is turbulent, and the exponent increases to $V^{1.8}$. **Pressure drag** results from displacement of the water in front of and behind the animal and around any irregularities in its surface; it is proportional to V^2. The combined drag is proportional to $V^{2.5}$ to V^3. Friction drag is lower for a short, plump body than for a long, slender one of the same volume; pressure drag is lower for a slender body. As a compromise between these two factors, total drag is minimized in a streamlined animal that is about 4.5 times as long as its maximum diameter. This ratio is not critical, however, for total drag is only 10 percent higher when it is 3 or 7 (Alexander 1974).

In fast-swimming vertebrates, the combination of these opposing forces acts through natural selection to produce a body shape that is optimal for rapid locomotion. Among vertebrates propelled by lateral undulation, optimizing the angle at which the force is transmitted to the water is achieved by concentrating lateral bending in the posterior portion of the body and ultimately at the base of the tail. The trunk becomes more rigid to limit the amount of wasted lateral movement. This is accompanied by a body profile that is high and narrow with maximum thickness anteriorly. This results in a spindle-shaped body rather than the elongate, fusiform body of slower-swimming forms. The base of the tail, termed the *peduncle,* is narrow and highly flexible to permit maximum lateral bending (Fig. 11.1).

Drag increases in proportion to the linear distance over which the water moves. This influences the shape of the tail. For a given surface area, a very high, narrow tail has less drag than does a broadly triangular or rectangular tail. One can recognize fast-swimming vertebrates by this character alone. Tail dimensions are measured in terms of the **aspect ratio**, which is the **span** (the maximum dorsoventral

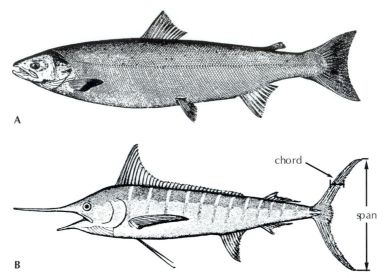

Figure 11.1. Body profiles of moderate- and fast-swimming bony fish showing changes in configuration to increase the forces of locomotion while reducing the forces of drag. The salmon (**A**) swims at moderate speeds, up to about 5 km/h. The body is moderately long and fusiform. Propulsion is achieved by lateral undulation of the entire trunk and tail. The most effective force is provided by the tail when it acts at right angles to the direction of movement. The blue marlin (**B**) swims at speeds up to 40 km/h. The trunk is high and narrow anteriorly and stiffened so as to resist lateral undulation. Most of the force is concentrated in the tail. The caudal peduncle is narrow. Small lateral keels reduce turbulence. The high, lunate shape of the tail maximizes its area while reducing drag. From Bigelow and Schroeder (1953).

extent of the tail) divided by the **chord** (the average of its anteroposterior extent) (Fig. 11.1B). Fast-swimming vertebrates such as dolphins and tuna, with speeds up to 40 km/h, have an aspect ratio in the range 4–6, whereas slow but steady swimmers such as salmon (5 km/h) have an aspect ratio of 1–2.

Primitively aquatic vertebrates

Vertebrates evolved from obligatorily aquatic primitive chordates and inherited the pattern of swimming seen in living cephalochordates. The earliest adequately known craniates were covered from the head to the base of the tail with heavy bony plates/or scales. The speed of their swimming was constrained by the great weight of their body and, in many groups, by the close overlap of bony scales; thus none achieved the body form of modern fast-swimming vertebrates. The extent and thickness of bony covering was reduced in several lineages of placoderms, the most primitive group of jawed vertebrates to diversify widely, but even those genera most highly specialized for pelagic life must have remained slow swimmers.

Sharks

The early history of more advanced jawed vertebrates is poorly represented in the fossil record. Remains of scales and denticles that can be attributed to cartilaginous and bony fish have been reported from the Silurian, and even suggested from the Ordovician (Sansom, Smith, and Smith 1996), but remains showing the form of the body and the configuration of the fins are not known until the Devonian. Sharks were the first craniates to have a body form indicative of rapid swimming. This is evident in Upper Devonian deposits that preserved the first complete body remains of this assemblage. Although primitive in other respects, the Devonian genus *Cladoselache* (Fig. 11.2A) has a superficially symmetrical caudal fin with a high aspect ratio. The peduncle, while not strongly constricted, bears lateral keels, as in the most rapid-swimming modern sharks and bony fish. We do not know how long it took for this body form to evolve, but it was certainly short compared with the subsequent history of sharks, and the outline shows little improvement even among the fastest-swimming modern sharks, more than 350 million years later.

Sharks remained the dominant large marine pelagic carnivores well into the Mesozoic. During the Paleozoic they exhibited extensive radiation into a variety of other marine and also freshwater habitats. Most of these lineages became extinct near the end of the Permian, to be succeeded in the Triassic and Jurassic by a second radiation leading to the modern orders. Concentrating on the pelagic families, the living shark groups show a remarkable consistency in their body form from that time to the present. This may be attributed to the historical constraints of buoyancy control in this group, coupled with the physical constraints of aquatic locomotion.

Sharks are exceptions among jawed vertebrates in the absence of bone as a skeletal material. Shark teeth, denticles, and spines are all related to bone, structurally and developmentally, but no sharks ossify their internal skeleton or have had external dermal plates. Either the ancestors of sharks never had the capacity to ossify their endochondral skeleton, or this capacity was reduced and lost very early in their evolution. The absence of endochondral ossification and dermal plates resulted in a much lower specific gravity among Chondrichthyes than in primitive jawless fish, placoderms, or early osteichthyans. Cartilage has a specific gravity of approximately 1.1, compared with twice that in bone. Modern sharks, and presumably their early ancestors, further reduced their density by the presence of very large livers filled with the oil squalene, which has a specific gravity of only 0.86. Sharks never evolved the capacity to control their specific gravity afforded by the swim bladder of bony fish but retained a nearly constant buoyancy, slightly greater than that of water, that requires continuous swimming to maintain their position in the water column.

Another historical constraint on sharks and other chondrichthyans involves the structure of the fins. Bony fish have both endochondral and dermal fin supports; the latter provide a great deal of functional and evolutionary flexibility, which is totally missing in sharks. Sharks do undergo considerable change in the endochondral fin supports, especially during the Paleozoic, but there are few, if any significant changes among pelagic forms during the Mesozoic and Cenozoic. In contrast, the fin structure changes drastically among the skates and rays during this same period of time.

Like early bony fish, sharks typically have a conspicuously heterocercal tail, with the notochordal axis sharply angled into the dorsal lobe of the caudal fin. It has long been argued that the heterocercal tail of primitive jawed fish acted to lift the posterior portion of the body as it drove it forward. This would occur because the

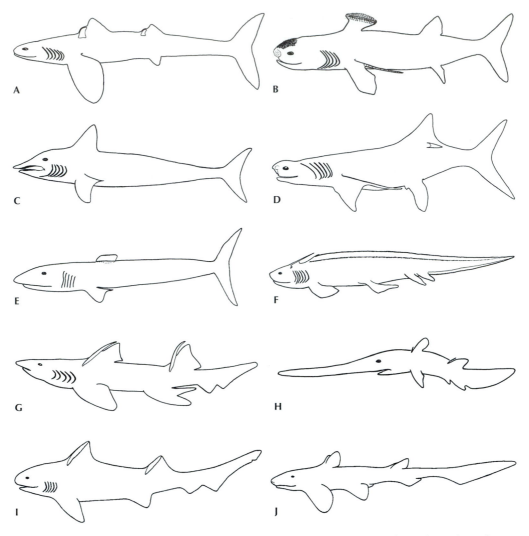

Figure 11.2 (above and facing). Diversity of body form in pelagic elasmobranchs. Paleozoic pelagic elasmobranchs (A–H): From the Late Devonian until the mid-Mesozoic, elasmobranchs were the dominant predators of the open ocean. In the Paleozoic, they showed a great variety in the anatomy and distribution of the median and paired fins while retaining a generally fusiform body plan. Numerous benthonic lineages evolved during this time, but they are omitted from this illustration. **A,** *Cladoselache* from the Upper Devonian, one of the earliest sharks in which the body form is known. The tail is symmetrical and has a high aspect ratio indicative of rapid swimming. This trait was achieved very rapidly after the initial radiation of sharks. **B–E** are primitive sharks that lack the anal fins, but otherwise have a body form suggestive of moderate to rapid swimming. **F,** A pleurocanth, which is highly specialized in its elongate body and coalescence of the dorsal fins into a single, low structure that extends the length of the trunk. Pleurocanths were common in shallow freshwater. **G, H,** Ctenoacanthoidea sharks, the sister-group of modern elasmobranchs (A–H redrawn from Zangerl 1981). Mesozoic sharks (I, J): **I,** *Hybodus,* a common Mesozoic shark derived from primitive ctenoacanths, but not part of the radiation leading to modern sharks. **J,** The Lower Jurassic neoselachian *Palaeospinax,* showing the ancestral pattern for all the modern pelagic elasmobranchs (I, J, redrawn from Schaeffer and Williams 1977). Representatives of modern shark families (K–U), showing the basic conservatism of body form: **K,** *Rhincodon,* the whale shark, Suborder Orectoloboidea (among the most primitive of living shark groups). This huge shark reaches lengths of 15–18 m. Although very slow swim-

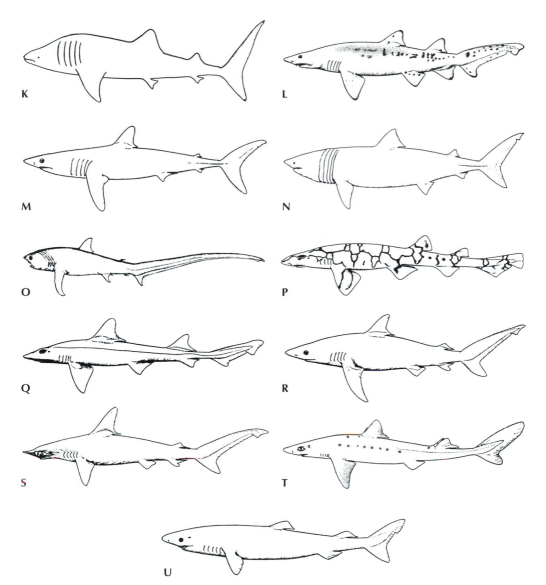

Caption to Fig. 11.2 (cont.)
ming, it has a high-aspect, lunate tail (K from Bigelow and Schroeder 1948). Suborder
Lamnioidea (L–O). **L,** *Carcharias,* Family Carchariidae. **M,** *Isurus,* Family Lamnidae.
N, *Cetorhinus,* Family Cetorhinidae. **O,** *Alopias,* Family Alopiidae. Suborder Carcha-
rhinoidea (P–S): **P,** *Scyliorhinus,* Family Scyliorhinidae. **Q,** *Mustelus* Family Triakidae.
R, *Carcharhinus* Family Carcharhinidae. **S,** *Sphyrna,* Family Sphyrnidae. Order Squa-
lomorpha, suborder Squaloidea (T, U): **T,** *Squalus,* Family Squalidae. **U,** *Somniosus,*
Family Dalatiidae (L–U from Bigelow and Schroeder 1953).

flexible ventral portion of the tail lagged behind the movement of the stiff notochordal axis. The tail as a whole thus exerted a force that acted ventrally as well
as transversely and posteriorly, so that the tail was forced dorsally, around the center of gravity, thereby lowering the more anterior portion of the body. This was
compensated by a planing effect of the pectoral fins that raised the head region.
These forces would have balanced one another to keep the body on an even keel
as long as the body was in motion.

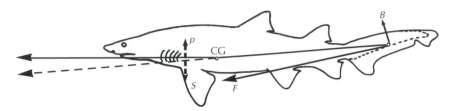

Figure 11.3. Profile of a shark showing the forces generated by the tail in swimming. The line of net thrust from the tail is directed through the center of gravity (CG). This produces a net sinking effect *S* that is offset by a lifting effect *P* from the planing action of the pectoral fins. *B* and *F,* resolution of force from the tail into lift and a forward force, respectively. From Thomson (1976).

To judge by similarity of body shape and distribution of fins in sharks as early as the Upper Devonian, the specific gravity and hence the mechanics of swimming in pelagic genera must have been broadly similar since that time. There is considerable variety in the number and position of the dorsal fins and the shape of the paired fins in the Paleozoic, but a much more conservative pattern is evident in the Mesozoic and Cenozoic. The conservation of body form is clearly demonstrated by the similarity of diverse families of sharks with modern representatives (Fig. 11.2K–U). Genera within some of these families are known as early as the Lower Cretaceous, and others from the Upper Cretaceous or early Cenozoic. Many of these lineages have retained a nearly constant body form for more than 100 million years.

Most sharks swim constantly at a low to medium speed, but some are extremely rapid swimmers. According to studies by Thomson (1976) and Thomson and Simanek (1977), all modern sharks swim in much the same manner, which is associated with a very stereotyped body form. All retain the heterocercal tail of their Paleozoic ancestors, long thought to function in the way previously described for all primitive fish with heterocercal tails; but observations of living fish, as well as analysis of tail form and body proportions in a host of other genera, demonstrate that the heterocercal tail in modern sharks has the capacity to move the caudal portion of the body either dorsally or ventrally by controlled changes in the movement of both the dorsal and ventral lobes. Force can be directed at a slightly downward angle toward the center of balance, with a net sinking effect. This is offset by a planing effect of the pectoral fins and the ventral surface of the head (Fig. 11.3). Alternatively, the tail can be rotated so that the shark either dives or swims upward at a high angle. The latter is important in feeding, which is frequently directed toward the ventral surface of the prey.

Thomson and Simanek recognized four basic modes of swimming within this overall stereotyped pattern, linking specific angles and proportions of the tail with different speeds. The earliest adequately known representative of the modern shark radiation is *Palaeospinax* from the Lower Jurassic (Fig. 11.2J). The elongate body, low angle of the axis of the heterocercal caudal fin, and absence of a distinct ventral caudal lobe indicate that this fish was a slow swimmer; but by the end of the Jurassic it had given rise to a number of lineages that include extremely fast and powerful swimmers. This is shown by the shorter and stouter trunk, high symmetrical tail, and a more rounded ventral surface of the head, which indicates that it does not form a planing surface.

The basic body plan of all pelagic sharks has remained broadly similar since the Late Jurassic. The fastest swimmers, regardless of the specific family to which they belong, have all evolved the same details of caudal fin construction. This implies a fixed constraint, established by the physics of the water and the constituents of their own bodies, which they cannot exceed.

Within the limits of their buoyancy control and mode of swimming, subtle changes in the configuration of the tail and position of the paired and dorsal fins enabled sharks to achieve a fairly wide range of diet, feeding behavior, and size, but they never achieved the wide adaptive radiation of bony fish.

Bony fish

Complete skeletons of bony fish are known as early as the Lower Devonian, but these were very heavy and show no evidence of rapid swimming. A heterocercal tail is retained in the most primitive living bony fish, the chondrosteans, including the sturgeon and the gar, and in an abbreviated form in *Amia,* but the caudal fin becomes superficially symmetrical in the vast assemblage of teleost fish. The change in the shape of the caudal fin in advanced actinopterygians can be associated with a lightening of the body through the loss of the heavy bony layer of the scales and the greater effectiveness of the swim bladder in reducing the specific gravity. Once the body was lightened in teleost fish, the function and configuration of the pectoral and pelvic fins changed as well, resulting in a great variety of different shapes among advanced ray-finned fish (Nelson 1984).

The evolution of ray-finned fish was characterized by a series of radiations, with that of the perciform fish in the Late Cretaceous and early Cenozoic leading to the majority of living species. From this radiation evolved a number of very fast-swimming lineages. The Family Scombridae (including the tunas and mackerels) is known from the Lower Paleocene. The Istiophoridae (the marlins; see Fig. 11.1B) are represented by modern genera as early as the Upper Cretaceous. *Xiphius* (the swordfish) is known from the Eocene, and other members of the Xiphiidae from the Upper Cretaceous. Since the initial radiation of all perciform fish took place in the Late Cretaceous, the achievement of the morphology of modern genera must have occurred very quickly thereafter, relative to the long period of their subsequent duration. The basic body form of these fish must have been nearly static for most of the Cenozoic, as was the case for most families with an adequate fossil record (Carroll 1987).

This section has emphasized rapidly swimming fish because they operate at the limits imposed by the physics of fluids, for which a single body form is optimal for all groups. Several other individually stereotyped body patterns are recognized among aquatic vertebrates that inhabit particular environments and/or have relatively constant locomotor and feeding habits (Webb 1994). Among vertebrates in which great acceleration rather than sustained moderate to rapid swimming is selected for – such as the pike among ray-finned fish and *Eusthenopteron* among the sarcopterygians – the dorsal and paired fins are concentrated posteriorly, where they form the largest surface area that can act against the medium at the maximum angle. The body shape of eels is optimal for life in very restricted spaces, where fast locomotion is of no advantage. Fish with nearly round profiles are common in the still waters of reef environments. There are also many fish whose body forms are

selected for primarily by factors other than those of the fluid dynamics of locomotion, and these show a remarkable variety of different shapes: for example, fish
that rely on the body form to provide camouflage, and deep-sea fish with enormous mouths and tiny bodies that wait for the occasional meal of any size.

Secondary aquatic adaptation among groups with terrestrial ancestors

During the evolution of fast-swimming fish, one sees relatively rapid change in
body proportions and tail dimensions early in their history, resulting in achievement of an optimal configuration. Changes in body form are illustrated even more
dramatically among groups such as ichthyosaurs, mosasaurs, and whales that
evolved from initially terrestrial animals with no specializations for aquatic locomotion. All began with quadrupedal ancestors with a short vertebral column and a
narrow, elongate tail. After initial adaptation to an aquatic way of life, most multiplied the number of trunk vertebrae to increase the capacity for lateral undulation.
Whales and other groups of aquatic mammals are exceptional in practicing dorsoventral undulation in the water as the result of ancestral constraints that emphasized dorsoventral rather than lateral bending of the vertebral column. Subsequent
evolution resulted in either dorsoventral or lateral (in mammals) expansion of the
tail, and ultimately in the formation of a lunate structure with a high aspect ratio.

The pattern of evolution in body and tail form is most clearly seen in fossil ichthyosaurs because the tail is partially supported by bone, and in many specimens
the outline of the tail and dorsal fin are preserved. The earliest known ichthyosaurs
had an essentially straight, narrowly cylindrical tail (Motani, You, and McGowan
1996). The Middle Triassic genus *Mixosaurus* had a caudal fin supported by elongate neural spines, but the tail was essentially straight, or no more than gently
curved ventrally in even the Upper Triassic genera. In contrast, the vast assemblage
of ichthyosaurs that appeared in the latest Triassic and Early Jurassic have a sharp
ventral bend in the tail, formed by several wedge-shaped centra, and the more posterior portion supports a high-aspect lunate caudal fin. The general body form attained the pattern of the fastest-swimming modern fish, with a ratio of length to
maximum diameter ranging from 3 to 7, with many in the most effective range of
4–5 (Fig. 11.4).

Unfortunately, it is not possible to determine the exact time frame over which
this change occurred, although it is unlikely that it was initiated earlier than the
Upper Triassic, in the absence of any Middle Triassic specimens showing a sharp
tail bend. Once established, this body form persisted until the extinction of ichthyosaurs late in the Cretaceous. This suggests a period of rapid evolution in this trait,
no more than 10 million years in duration, followed by 120 million years of essential stasis. A similar pattern of change is evidenced by whales (see Chapter 12).

Flight

In terms of evolutionary progression, it would appear natural to discuss terrestrial constraints between those inherent to aquatic and aerial ways of life, but the

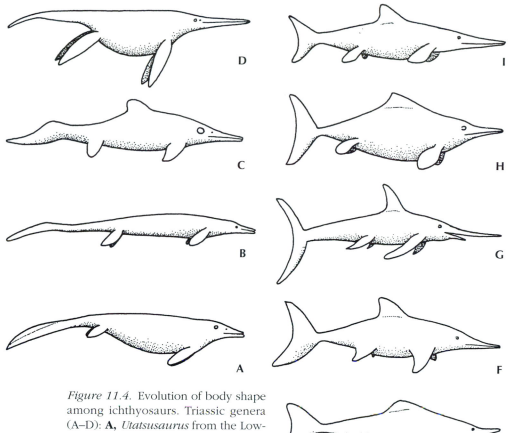

Figure 11.4. Evolution of body shape among ichthyosaurs. Triassic genera (A–D): **A,** *Utatsusaurus* from the Lower Triassic of Japan. **B,** *Cymbospondylus* from the Middle Triassic of North America. **C,** *Mixosaurus* from the Middle Triassic of Spitzbergen, central Europe, Canada, and the East Indies. **D,** *Shonisaurus* from the Upper Triassic of western North America. Jurassic and Cretaceous ichthyosaurs (E–I), all of which have a similar body form indicative of fast swimming. **E,** *Ichthyosaurus,* lower Liassic, Lower Jurassic. **F,** *Stenopterygius,* upper Liassic, Lower Jurassic. **G,** *Eurhinosaurus,* upper Liassic. **H,** *Ophthalmosaurus,* Upper Jurassic and Lower Cretaceous. **I,** *Platypterigius,* Upper Cretaceous. From *Paleontology: The Record of Life* by Stearn and Carroll (after McGowan 1983). Copyright © 1989. Reprinted by permission of John Wiley & Sons, Inc.

laws of fluid dynamics for rapidly swimming fish and all flying vertebrates are more similar to one another than are either to the constraints of terrestrial locomotion.

Actively flying vertebrates encounter the same problems of propulsion and drag in a fluid medium as do aquatic forms. Pterosaurs, birds, and bats responded to these factors by evolving similar wing shapes that are extremely highly constrained for most of their evolutionary history. The mechanical and physiological constraints of flight in these groups are explained in considerable detail by Norberg (1990). Because each group evolved from different ancestors, their ways of achieving a flight surface differ significantly, but the shape and size of the wings follow the same rules. Each group shows a great variety of feeding habits and resting environments, but the necessities of flight alone are the primarily factors in constraining their skeletal anatomy over tens of millions of years.

Hildebrand (1995) lists four major requirements of flight:

1. sufficient upward force, either from muscles or from the environment, to counter the pull of gravity;
2. the limitation of drag, particularly if flights are long or fast;
3. the capacity for propulsion at various speeds; and
4. the ability to retain stability, maneuver, brake, and land.

In all flying vertebrates these requirements depend on a very light but strong and rigid body. Generation of lift requires a large, controllable flight surface, or wing.

As in aquatic locomotion, flight is a balance between power for forward movement and drag, but with the added necessity of countering the force of gravity. The forces acting on the wing and derived from its motion can be divided into drag – by definition in line with the airstream and opposite to the direction of flight – and lift, which is at a right angle to drag. Lift depends on the angle of the wing or airfoil to the airstream; this is limited to less than 15°, beyond which turbulence is produced that reduces the lift and results in stalling. The anterior margin of the wing's upper surface is convex so that the airstream passes more rapidly over this surface than the lower one. Following Bernoulli's theorem, the faster flow of the airstream over the upper surface creates a vacuum, so that the wing rises. The amount of lift is proportional to the square of the speed times the area of the wing. Lift is also influenced by the angle of attack and the camber of the wings. Fast flyers can generate the same amount of lift with relatively smaller wings, less camber, and a lower angle of attack than slow flyers. In flapping flight, forward movement is generated by anteroventral movement of the wings, which results in lift acting in a forward direction.

As in aquatic vertebrates, there are two components of drag, but their definitions differ. **Profile drag** results from the friction of the air against the body, the displacement of air, the formation of pressure gradients in the air, and the creation of eddies. **Induced drag** is produced by air that is flowing around the wing tip and inward over the wing. It is much reduced by narrow wings with pointed tips.

In common with the tail of fast-swimming fish, the geometry of the wings is expressed as the aspect ratio, which is equal to the span of the wings divided by the average width or chord, or span²/area. The relative effect of the wings, referred to as **wing loading**, is a ratio of the weight of the body to the area of the wings. Since gravity is acting on the mass or weight of the flyer, and lift is related to the area of the wing, the relative area of the wing must be larger in heavier flyers. This is most clearly shown in a measure of wing span (Fig. 11.5). This and many other aspects of the size and shape of the wings retain a close functional relationship among each of the three groups of actively flying vertebrates.

The relationship between the force of gravity and the capacity of muscles to produce power for flight acts as the ultimate constraint to the mass of actively flying vertebrates. The largest birds with continuous-powered (flapping) flight weigh 12–15 kg (Norberg 1990). The largest bat weighs only 1.5 kg. Larger animals do not have the power to beat their wings fast enough to achieve the lift needed for horizontal flight.

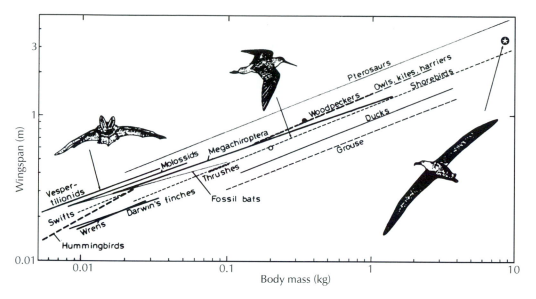

Figure 11.5. Constraints on the geometry of flight illustrated by the correlation between wingspan and body mass in birds, bats, and pterosaurs. Filled circle shows the span of the fossil bat *Archaeopteropus,* the open circle that of *Archaeopteryx,* and the star that of *Diomedea.* From Norberg (1990), *Vertebrate Flight,* Springer–Verlag, Inc.

Larger size is reached by birds that rely on power from the wind for takeoff and use air currents for gliding flight. The largest size for a gliding bird was reached by the Pleistocene condor *Teratornis,* which had an estimated mass of 40 kg. Among the pterosaurs, *Pteranodon* is estimated as having a mass of 17 kg and wingspan of 7 m. The largest of all vertebrate gliders was *Quetzalcoatlus,* with an estimated wing span of 11–15 m and a mass of 86 kg (Langston 1981).

There is also a minimum size for vertebrate flight. This is set by a limitation for maximum wing-beat frequency, resulting from the time that vertebrate muscles require to reset the contractile mechanism after each beat. For the hummingbirds, the smallest size is 1.5 g. The smallest bat weighs 1.9 g. Insects can be much smaller because of fundamental differences in the structure of their flight muscles.

Among vertebrates, powered flight has evolved only three times: among pterosaurs, birds, and bats. The mechanical requirements of flight muscles have resulted in fairly similar patterns of the shoulder girdle and sternum that remained stereotyped in most members of each group (Fig. 11.6). All require a large surface area for the attachment of flight muscles and maximum buttressing by the bony skeleton. Bats evolved from small placental mammals, birds from bipedal theropod dinosaurs, and pterosaurs from a distinct clade closely related to the dinosaurs. The fossil record does not provide evidence of the transition toward either pterosaurs or bats: The earliest known members of these groups had already evolved an advanced flight apparatus (Figs. 11.7A, 11.8A). Plausible ancestors of bats are known from approximately 10 million years before their first appearance. The time of divergence of the lineage leading to pterosaurs may have been as great as 30 million years prior to the first appearance of this group, although this cannot be established with any assurance. In the case of birds, *Archaeopteryx* provides an al-

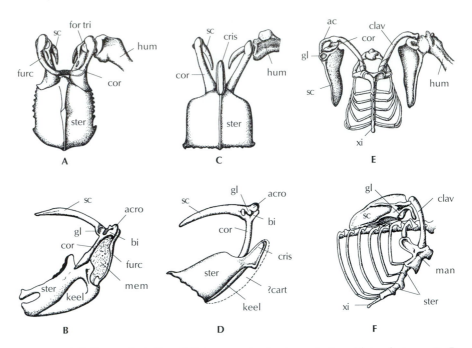

Figure 11.6. Pectoral girdles of flying vertebrates in ventral and lateral views. **A, B,** birds. **C, D,** pterosaurs. **E, F,** bat. Flight evolved separately in these three groups, but a similar pattern of the bones buttresses the sternum to resist the force of the flight muscles. In birds and pterosaurs this is provided by the coracoid, but in bats by the clavicle. Abbreviations: ac, acromion process; acro, acrocoracoid process; bi, biceps tubercle; ?cart, possible cartilagineous extension of sternal keel; clav, clavicle; cor, coracoid; cris, cristospine; for tri, foramen triossium (junction of clavicle, coracoid, and scapula for passage of tendon from supracoracoideus muscle that raises the wing); furc, furcula; gl, glenoid fossa; hum, humerus; keel, sternal keel; man, manubrium; mem, membrane connecting sternum, coracoid, and clavicle; sc, scapula; ster, sternum; xi, xiphoid process of sternum. Drawings are not to scale. From Padian (1983).

most ideal intermediate between theropod dinosaurs and skeletally more modern birds. The precise time at which the line leading to *Archaeopteryx* diverged from the most closely related dinosaurs cannot be accurately specified, but was at least 15 million years earlier.

Although the length of their previous history is difficult to determine, it is obvious that, once established, the basic anatomy of the flight structures remained extremely conservative for the remainder of their history in all actively flying vertebrates. This is strikingly well documented by bats; The skeleton of an early Eocene species is almost indistinguishable from modern forms (Fig. 11.7). Despite the great conservatism of their flight mechanism, bats are among the more speciose and diverse of all placental orders. They have a very wide range of diets, from insects and other arthropods to fish, blood, nectar, fruit, and pollen, with accompanying specialization of the cranial morphology and dentition. One can readily separate the stasis associated with the requirements for flight from the extensive adaptive radiation reflected in other structures.

Pterosaurs show a similarly conservative pattern in the bones supporting the wing, from the Upper Triassic until the end of the Cretaceous, compared with an

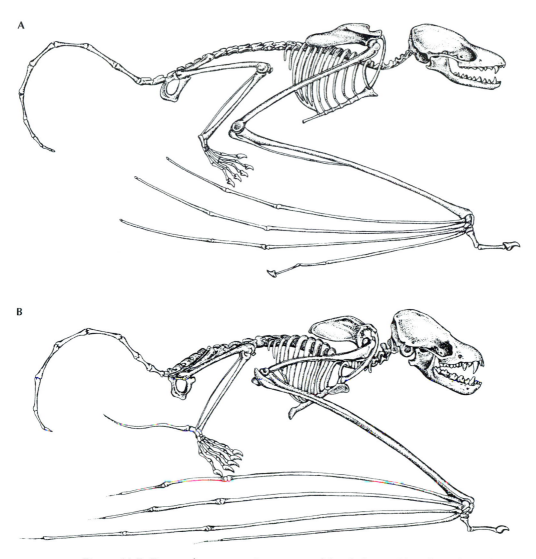

Figure 11.7. Extremely conservative pattern of the skeleton of bats from the early Eocene *Icaronycteris* (**A**) to the modern *Myotis* (**B**). Drawn to the same scale. From Jepsen (1970).

extremely diverse pattern of cranial structure and a wide size range (Wellnhofer 1978, 1991).

The wings of birds differ significantly from those of pterosaurs and bats in that the flight surface is formed by individual feathers rather than a membranous structure. In contrast with pterosaurs and bats, the bony skeleton of the wing changes significantly between *Archaeopteryx* and all more advanced birds. On the other hand, the structure of the individual feathers and their arrangement on the wing retain an extremely conservative pattern in all flying birds. The feathers of *Archaeopteryx* are preserved as impressions in the very fine-grained lithographic limestone from the Upper Jurassic of southern Germany. They show details of structure down to the level of the interlocking barbs that unite the elements that make up

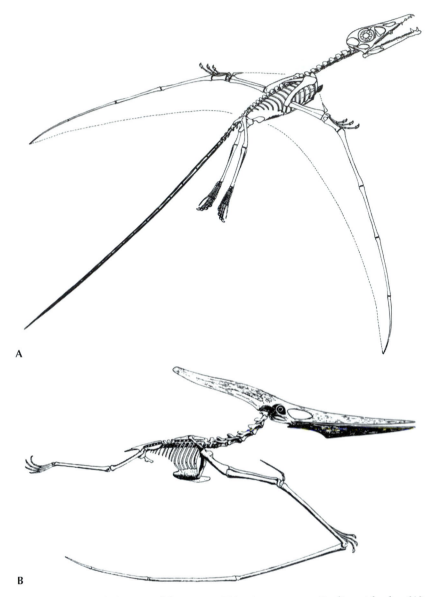

Figure 11.8. Skeletons of the Upper Triassic pterosaur *Eudimorphodon* (**A**) and the Upper Cretaceous genus *Pteranodon* (**B**) to show the conservative pattern of wing support, in contrast with the very different structure of the skull and reduction of the tail between the two genera. From Wild (1978) and Eaton (1910), respectively.

the vane on either side of the shaft (Fig. 11.9). Of particular importance is that the feather is asymmetrical, with the shaft nearer the anterior margin. This gives each individual flight feather the characteristic of an airfoil to produce lift. The geometry of the flight feathers of *Archaeopteryx* is identical with that of modern flying birds, whereas nonflying birds have symmetrical feathers. The way in which the feathers are arranged on the wing also falls within the range of modern birds. In the Berlin specimen of *Archaeopteryx* there are nine primary feathers, compared with nine to twelve in modern birds, and the first three are progressively reduced

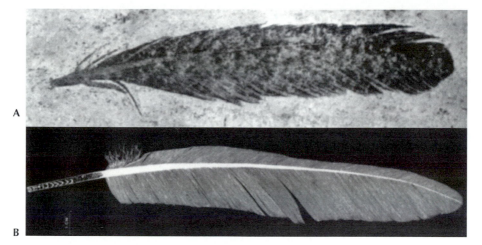

A

B

Figure 11.9. Conservative pattern of flight feathers over 150 million years. **A,** A flight feather of *Archaeopteryx* from the Lithographic limestone of Solnhofen, Germany (courtesy of John H. Ostrom). **B,** A flight feather of a domestic chicken (from Lucas and Stettenheim 1972). **C,** Distal ends of the sixth primary feathers of *(left to right)* a flying rail, *Archaeopteryx,* and a flightless rail showing the loss of asymmetry (reprinted with permission from *Science,* vol. 203, Feduccia and Tordoff; copyright © 1983, American Association for the Advancement of Science).

C

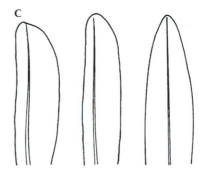

in length inward. *Archaeopteryx* has fourteen secondaries; modern birds have seven to thirty-two. The only feature of modern birds missing from the wing of *Archaeopteryx* is the midwing slot (Feduccia 1980). According to Van Tyne and Berger (1976), the relative size and shape of the wing of *Archaeopteryx* are similar to that of birds that move through restricted openings in vegetation, such as gallinaceous birds, doves, woodcocks, woodpeckers, and most passerine birds.

As is discussed in Chapter 12, many aspects of the bony skeleton and almost certainly the soft anatomy and physiology change significantly between *Archaeopteryx* and modern birds of the early Cenozoic, but the basic flight apparatus does not. The flight feathers have been in stasis for at least 150 million years as a result of the unchanging physical constraints of active flight.

The flight feathers in birds provide an informative example of how a particular factor may be identified as the primary cause of long-term stasis. Changes in both genetic and developmental factors were certainly necessary for the transformation of scales into feathers in the ancestors of birds. On the other hand, one cannot attribute the maintenance of the feather structure to either genetic or developmental constraints because wing feathers have lost the characteristics necessary for flight in a great number of lineages of secondarily flightless birds. In many cases, this must have occurred over relatively short periods of time, as in the case of the numerous flightless rails known from recent populations on islands. As long as selection acted to maintain flight, physical constraints on the flight apparatus would

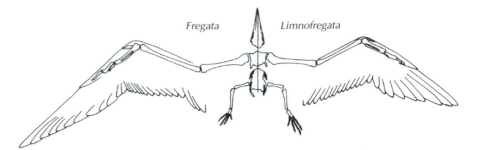

Figure 11.10. Outline comparing the overall skeletal proportions of modern and fossil frigate birds. *Fregata* is depicted on the left half of this composite drawing and *Limnofregata* from the lower Eocene on the right. The body size of the two genera is virtually the same. From Olson (1977).

have maintained the most effective feather configuration possible within the limits of its organic constituents. On the other hand, in situations such as isolated islands where flight may be disadvantageous, variations already present in the genes and developmental programs for the formation of flight feathers would be selected to reduce and finally eliminate characteristics necessary for flight. At least in this case, stasis is clearly maintained by physical constraints.

The extreme conservatism of the shells of ostracods, foraminifera, and radiolarians may also be maintained by consistent physical constraints associated with buoyancy or other aspects of life in the water.

According to Feduccia (1995, 1996), most of the modern bird orders diverged at the beginning of the Cenozoic, subsequent to the extinction of dinosaurs and most archaic avian groups at the end of the Cretaceous. Except for the passerines, nearly all avian orders had achieved an almost modern morphology by the early Eocene. Where adequately known, their basic anatomy at their first appearance in the fossil record was very similar to that of modern genera. This is graphically shown in the frigate bird, which has hardly changed since the lower Eocene (Fig. 11.10).

Terrestrial constraints

In contrast with aquatic vertebrates, the primary physical constraint on terrestrial locomotion, gravity, is expressed not in general body form but in the configuration of the limbs and girdles. The optimal shape and proportions are related to both support and locomotion.

Historical constraints of aquatic locomotion dominated the patterns of evolution in Paleozoic tetrapods. Many features of the skeletal anatomy that change between aquatic and terrestrial ways of life can be attributed to the necessity of resisting the force of gravity. These involve aspects of the dermal skull, the braincase, axial skeleton, and limbs. As is discussed in Chapter 12, these changes occurred in a mosaic fashion over time and among different groups.

The oldest known tetrapods inherited their pattern of locomotion from their aquatic ancestors, substituting their limbs for the paired fins and tail of fish in pushing against the substrate, but retaining sinusoidal lateral undulation of the trunk for the main propulsive force. The function of the limbs changed in that they came to be used to pull the body forward and lift it above the ground at each stride.

Since the limbs were initially extended from the side of the body, all of the body weight was originally supported by the limb muscles. The limbs in Devonian tetrapods, especially the more vertically held distal elements, were originally very short, indicating that the body was not raised far above the substrate, nor moved very far at each stride. Most Carboniferous tetrapods had a very thick covering of dermal scales between the pectoral and pelvic girdles that may have served for protection as the animals dragged themselves across the ground. The earliest known tetrapods were more than a meter in length, and later animals with very primitive limb and girdle structures attained a length of several meters; the largest animals, such as the embolomeres, may have been commonly, if not obligatorily, aquatic.

Even the most derived of the pelycosaurs, which were eventually to give rise to mammals, retained a sprawling posture, with the humerus and femur extending nearly horizontally from the trunk and the body only barely raised above the ground. This was definitely a physical constraint that resulted in very similar girdle and limb structure in most amphibians and reptiles within particular size classes during the Paleozoic. Among primitive land vertebrates, long slender limbs suggestive of agile locomotion were only achieved among animals less than about 15–20 cm in length from the head to the base of the tail. Animals of this size can readily support the body by muscles linking the girdles and limbs, even when the proximal limb elements are held at right angles to the substrate.

From large early tetrapods, evolution took several divergent paths. Several lineages returned to a habitual or obligatorily aquatic way of life, thus obviating the problem. Other lineages lost their limbs and look up a snakelike mode of locomotion, either on land or in the water. Medium-sized to large animals – including members of the lineage leading to mammals and that leading to crocodiles and, ultimately, to dinosaurs and birds – drew their limbs closer to the midline, so that the movement of the limbs was closer to the sagittal plane, and the bone of the limbs could assist in support of the body.

Other lineages attained smaller size and became particularly effective in the use of a sprawling gait; one such lineage eventually gave rise to lizards. Animals clearly related to the modern lizard infraorders are not known until the Jurassic, but more primitive animals with some characteristics of lizards are known from the Triassic, and some key attributes of modern lizards were attained by the Late Permian. These included a mechanism for significantly extending the range of the forelimbs. A new bone evolved along the ventral midline, the sternum, which serves as a surface for rotation of the base of the scapulocoracoid in a horizontal plane. Thus, the forelimb could have a much greater range of anterior–posterior movement than was possible through its articulation with the fixed glenoid of more primitive tetrapods (Fig. 11.11). Cursorial mammals and quadrupedal cursorial dinosaurs evolved analogous ways of extending the stride of the forelimb but in the vertical plane. The key difference of the specialization in lizards and their ancestors was that it involved movements of the limb in the horizontal plane, which perpetuated locomotion based on the sinusoidal bending of the trunk of their aquatic ancestors. This type of coracosternal articulation was present in two of the three genera whose patterns of carpal and tarsal development were discussed in Chapter 10. The extensive reliance on sinusoidal movement of the trunk and the sprawling posture of the limbs may be associated with the extreme constraint in carpal, tarsal, and digital structure seen throughout typical terrestrial members of this assemblage since the Permian.

Another group that is striking in the stereotypy of basic locomotor patterns is the anurans (frogs and toads). As adults, most members of this assemblage use primarily symmetrical movements of the elongate rear limbs in both terrestrial and aquatic locomotion. This pattern has not been observed in any Paleozoic amphibians. An initial stage is illustrated by the Lower Triassic genus *Triadobatrachus,* in which the vertebral column is shortened and the blade of the ilium is elongated. A nearly modern configuration of the jumping apparatus is achieved in the Lower Jurassic genus *Prosalirus,* described by Shubin and Jenkins (1995) (Fig. 11.12). By the Upper Jurassic, the structure of the rear limb and girdle of frogs is almost identical with that of modern genera. For the past 180 million years, this particular pattern of specialization of the rear limbs has dominated the locomotion of anurans, be they primarily aquatic, terrestrial, or arboreal.

The basic adaptive radiation of the modern mammalian orders occurred during a relatively short period of the Late Cretaceous and early Cenozoic. The common ancestors of the dominant groups of living mammals, the placentals and marsupials, were small animals, perhaps less than 100 g in weight. The ancestors of all the modern placental orders retained a relatively small size and a semisprawling stance into the early Cenozoic. It was only with the achievement of a considerably greater size and/or a cursorial gait that the limbs were brought under the body in a fully erect stance.

All elephantine mammals have similar, although certainly not identical, limb structure throughout their evolutionary history. All have massive limbs, held as vertical as possible, and take short strides. Most show little reduction in the primitive mammalian digital number.

Moderate-sized to large cursorial mammals, whether carnivores or herbivores, have long slender limbs, with elaboration of the distal elements, reduction of the medial and lateral digits, and specialization of the shoulder girdle and spinal column to allow the greatest possible anteroposterior extent of limb movement. The mechanical and geometrical constraints of these and other specializations of limb structure are thoroughly discussed by Hildebrand (1995).

These physical constraints, like those of flight for bats and aquatic locomotion in cetaceans and sirenians, acted quickly to establish the basic form of the locomotor apparatus in the various orders of placental mammals, which has been refined in detail throughout the latter part of the Cenozoic.

On the other hand, the stasis in general body form and structures associated with locomotion did not preclude significant diversity in feeding habits, dentition, and so on seen in all groups, clearly demonstrating that stasis is not a general characteristic of the body, but one that is specific to particular systems. The obvious exception lies among the so-called living fossils, but these constitute a small minority of vertebrates and show only one extreme within the general pattern of vertebrate evolution.

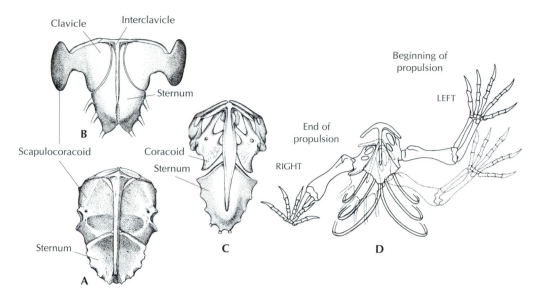

Figure 11.11. Conservative structure and function of the shoulder girdle in lepidosaur-ian reptiles. **A,** Ventral view of the shoulder girdle of a primitive diapsid from the Up-per Permian. The anterior margin of the sternum is groved for the coracoids, which ro-tate around it in a horizontal plane to increase the stride of the forelimb. **B, C,** The same view of the shoulder girdle in (respectively) a modern sphenodontid and an iguanid lizard, which have the same functional relationships. **D,** The effect of movement be-tween the coracoids and sternum in a varanid lizard, revealed by X-ray cinematography (from "The functional anatomy of the shoulder of the Savannah monitor lizard [*Varanus exanthematics*]" by Jenkins, Jr., and Goslow, Jr., *Journal of Morphology;* copyright © 1983; reprinted by permission of John Wiley & Sons, Inc.). The right forelimb is depict-ed in a posture typical of the end of propulsion, the left forelimb in a posture typical of the beginning of propulsion. Translation at the coracosternal joint advances the trunk relative to the glenoid and simultaneously rotates the contralateral shoulder ahead of the propulsive phase shoulder (dark lines). With a fixed coracosternal joint (dashed lines), the same limb excursion results in a shorter step length. The conservative func-tion of the shoulder girdle in these diapsids may be correlated with the extremely con-servative structure of the manus throughout terrestrial lepidosauromorphs.

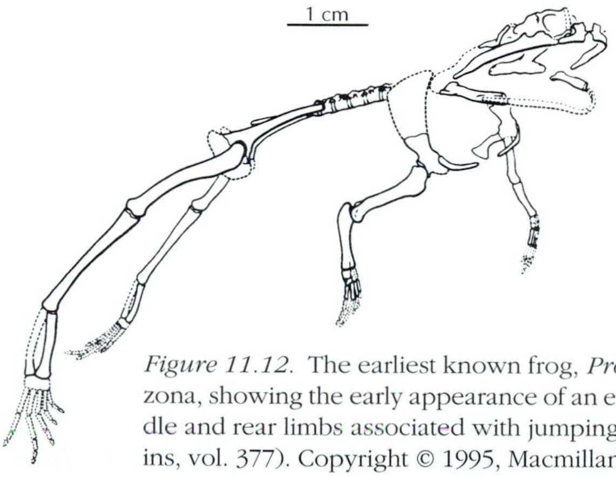

Figure 11.12. The earliest known frog, *Prosalirus bitis* from the Lower Jurassic of Ari-zona, showing the early appearance of an essentially modern structure of the pelvic gir-dle and rear limbs associated with jumping. Reprinted from *Nature* (Shubin and Jenk-ins, vol. 377). Copyright © 1995, Macmillan Magazines Limited.

Discussion

Constraints of the physical environment on general body form and locomotor structures act in a similar manner and over a similar time span in many different lineages. They are among the most clearly understood factors in constraining the course of evolution to particular pathways and provide a general model to explain long-term stasis in major groups. Animals as phylogenetically distinct as Mesozoic sharks and ichthyosaurs, and Cenozoic bony fish and whales, have achieved functionally similar body forms despite the radically different skeletal structure and metabolism of their immediate ancestors and details of their genetic and developmental systems. In all cases, factors of their physical environment and way of life have produced a similar pattern of evolution. In all of these lineages for which the fossil record is adequate, their history begins with a period of relatively rapid change, followed by slower refinement of details, leading eventually to much longer-term stasis (Fig. 11.13). The long-term pattern of the evolution of functional units appears very different from that hypothesized by Darwin, who envisioned continuous progressive change in most attributes influenced by selection, to judge by the one diagram in the first edition of *On the Origin of Species*. On the other hand, the period of change at the beginning of all these lineages occurred over several million years, and during this time it may have transpired in a more or less continuous and progressive manner, as implied by Darwin. As we saw in Chapter 10, details of structure, such as the specific patterns of the bony elements of the fins of aquatic lizards, turtles, ichthyosaurs, and plesiosaurs, do show what appear as gradual and progressive change over tens and even hundreds of millions of years. Similar progressive changes in the proportions, fusion, and loss of individual bones can also be seen in the evolution of the limbs in terrestrial vertebrates (e.g., horses; MacFadden 1992).

Transfer of substances across membranes

All of life is dependent on the transfer of substances and properties into and out of cells and the larger bodies that they may constitute. Movement across cell membranes follows the general formula

$$R = \frac{K \cdot A(C_{ex} - C_{in})}{D}$$

where R is the rate of diffusion; K is a constant specific to the substance crossing the membrane; A is the area of the membrane; C is the concentration of the substance on the two sides of the membrane; and D is the distance across the membrane. If the membrane is the cellular membrane, the distance is essentially constant. For any given substance, the most important variable is the area: Whatever the substance being considered, the rate of its diffusion across the membrane and throughout the cell or organism is proportional to the surface area to volume ratio. The larger the body or the cell, the smaller is the amount of diffusion of any substance, relative to the requirements of the organism.

This general formula applies to the transport of respiratory gases, water, heat, nutrients in solution, and so on. A simple example of constraints associated with

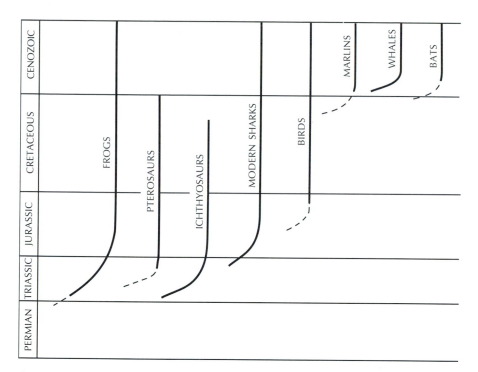

Figure 11.13. Diagrammatic representation of the patterns of evolution of lineages that are strongly influenced by physical constraints. Changes in patterns of locomotion tend to occur relatively rapidly, compared with much longer periods of subsequent stasis. It is not practical to indicate the scale of morphological change. Data are from studies discussed in the text.

transfer of substances across membranes is evident in vertebrate respiration. In contrast with the cephalochordates, the body volume in all vertebrate groups is too great, and the metabolic rate too high, for respiratory gases to be exchanged passively through the general body surface. Nearly all vertebrates have either gills or lungs, with their expanded surface area and extensive vascularization for exchange of oxygen and carbon dioxide. The modern amphibian orders, however, have a sufficiently low metabolic rate and large surface-to-volume ratio that exchange of gas through the skin provides a significant supplement to the lungs. Cutaneous respiration, augmented by a system of blood vessels just beneath the skin, is practiced throughout this group. Plethodontid salamanders are so small that they have been able to lose their lungs entirely. Cutaneous respiration brings its own constraints, however. It is only possible if the skin is kept moist, which precludes these groups from evolving a more impervious skin, such as that in amniotes. Hence, all modern amphibian groups are more susceptible to water loss and are generally precluded from dry environments. In addition, as long as they are reliant on cutaneous respiration, the modern amphibians are severely constrained in their body size, except for obligatorily aquatic genera such as *Cryptobranchus* and *Andrias*.

Passive diffusion of respiratory gases also limits the size of tetrapod eggs, unless they are provided with supplementary respiratory structures. The origin of the pattern of reproduction that distinguishes amniotes can be associated with constraints involving gas exchange. Primitive tetrapods, in common with many modern frogs and salamanders, laid small eggs in the water, where they hatched as tiny, limb-less larvae that were dependent on an aquatic medium for support, gas exchange, and food. The immaturity of the hatchlings results from the small size of the egg, which in modern frogs and salamanders is less than 10 mm in diameter. The small size of most amphibian eggs can be attributed to the need for a high surface-to-volume ratio so that respiratory gases can be exchanged with the developing embryo by passive diffusion.

In contrast, all amniotes are characterized by the possession of specialized extra-embryonic membranes within the egg – the allantois, chorion, and amnion – that provide protection, water retention, and extensive surface area and vascularization for gas exchange. Without these membranes or other accessory structures for gas exchange (e.g., the embryonic gills in caecilians), amphibian eggs apparently cannot exceed a size of 10 mm in diameter. The nutrients that can be carried in eggs of this size are sufficient only for a limited period of embryonic growth. In order for the young to grow to the size of many early land vertebrates (a meter or more in length), most development must take place in the water through a protracted larval stage. This is essentially a physical constraint on the mode of development.

The necessity for a long aquatic larval stage in amphibians can be evaded, however, by reducing the size of the adult. If the adult is sufficiently small, enough food can be provided in an egg less than 10 mm in diameter for the hatchling to reach a much more advanced developmental stage, essentially equivalent to that of modern lizards, which hatch as miniatures of the adults. Plethodontid salamanders and small lizards require approximately the same volume of nutrients for the development of the young to this stage, as indicated by the overlap in egg size among small members of both groups (Fig. 11.14). Although tiny plethodontids can reach a terrestrial stage of development without extraembryonic membranes, eggs less than 10 mm in diameter presumably cannot contain enough nutrients to support the growth of terrestrial young that achieve larger adult size than about 100 mm in

Figure 11.14. Correlation between mean diameter of eggs and adult body size in geckos (open circles) and plethodontid salamanders (closed circles). The adult size of direct developing plethodontids is presumably constrained by the size of their eggs, which do not exceed 10 mm in diameter. Gecko eggs have extraembryonic membranes that enable gas to be exchanged in much larger eggs, from which can hatch larger but fully developed young. From Carroll (1970).

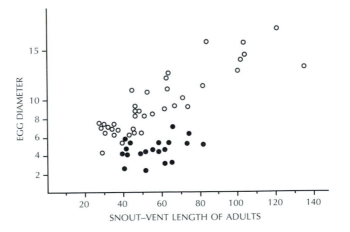

Figure 11.15. **A,** The plethodontid salamander *Batrachoseps* with eggs in the process of hatching. Although the eggs lack extraembryonic membranes, the relative size of the young is so great that they can assume terrestrial habits immediately following hatching. The length of the adult, minus the tail, is approximately 40 mm (from Stebbins, *Amphibians of Western North America;* copyright © 1951, University of California Press). **B,** A small but highly ossified specimen of the oldest known amniote *Hylonomus*. The adult is approximately 100 mm in snout–vent length but may have hatched from an egg no larger than the 10-mm diameter of the aspirin tablet. The constraint of a highly developed, terrestrial young hatching from a nonamniotic egg restricted animals undergoing the transition between aquatic and terrestrial development to roughly this body size.

snout–vent length. For larger animals to hatch as well-developed terrestrial young, they require larger eggs, which in turn depend on the evolution of extraembryonic membranes to supplement the passage of respiratory gases by passive diffusion.

The initial stage in the evolution of the pattern of reproduction common to amniotes, the hatching of well-developed young on land, would have been constrained to animals with an adult size of less than approximately 100 mm in snout–vent length. This corresponds well with the size of the first animals identified as amniotes on the basis of their skeletal anatomy (Carroll 1970). These animals, from the Upper Carboniferous of Joggins, Nova Scotia, are nearly as small as plethodontid salamanders that reproduce via nonamniotic eggs laid on land (Fig. 11.15B). Most amniote lineages known from later in the Paleozoic show progressive size increase, which would only have been possible for animals reproducing on land if they had evolved extraembryonic membranes.

Heat absorption and transfer

Control of body temperature is an important factor in the life of all organisms. As in the case of the exchange of respiratory gases, loss and gain of body heat occur at rates dependent on the surface-to-volume ratio. In small, noninsulated animals with a low metabolic rate, metabolic heat is lost nearly immediately into the surrounding water or air. These animals maintain the temperature of their surroundings unless they absorb radiant heat from the sun. Most terrestrial reptiles are able to raise their body temperature well above the ambient level through basking. Lizards can maintain a temperature near 30 °C even when the ambient temperature is near zero, as long as the sun is shining (Greenberg 1980).

Large ectotherms, such as crocodiles and the largest lizards and turtles, retain heat much longer but cannot maintain a constant temperature overnight or in cold climates. On the other hand, Spotila (1980) and his colleagues have shown that even larger animals, such as most dinosaurs, could have maintained a nearly constant body temperature approaching that of birds and mammals in the equable climate of the Mesozoic, even with very low metabolic rates. From their size alone, almost all dinosaurs would have been homeothermic.

The evolution of the reproductive pattern of early mammals was also influenced by the problem of heat transfer, but in this case in very small animals. The Permian and Early Triassic ancestors of mammals were relatively large animals that show evidence of high metabolic rates. The body size in the lineage leading to Upper Triassic mammals became progressively smaller throughout the Triassic, probably, as in the origin of amniotes, through selection for more effective feeding on insects and other small arthropods. By the Late Triassic and Early Jurassic, the ancestral mammals had decreased to about 20 g. This approaches the theoretical size limit for maintaining a high, internally produced body temperature, calculated by the increasing metabolic rate of the smallest living mammals (Fig. 11.16). Smaller animals would need a food intake approaching infinity in order to survive. How, then, do the young of these animals survive, since they must be smaller than this limit? By not being endothermic. They are naked, helpless, and essentially embryonic when born. They could only survive through being fed, warmed, and protected by their parents. This is the most primitive condition for newborn mammals, prior to their increase in size in the Late Cretaceous. By that time, they had diverged into two major lineages: the marsupials, which retained a tiny body size at birth, and the placentals, which evolved an entirely new method of development within the body of the mother.

Miniaturization

In adaptation to specific physical environments, we have seen how structures approach an optimum against the limits of physical constraints. Vertebrates may also be constrained to large or small size or to particular patterns of development by physical limits to biological processes. Other constraints produce not constancy but a sharp break in morphology in response to physical limits. These are clearly seen in the skull in relationship to the size of sense organs and the brain in organisms that are much smaller than their immediate ancestors. Small size is a highly significant feature in the origin of several major vertebrate groups, notably amniotes (Carroll 1991), mammals (Lillegraven 1979), and a number of amphibian groups, as well as lineages within the lizards (Rieppel 1984). A more general review is provided by Hanken and Wake (1993).

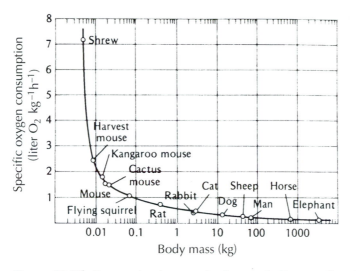

Figure 11.16. Progressive increase in the metabolic rate of small mammals. From Schmidt-Nielsen (1990).

Small skulls are dominated by their sense organs and the brain. Neither eyes nor the middle ear can be reduced to the same degree as the outline of the entire skull in small animals. For the eyes, the wavelength of visible light requires a certain minimum size of receptor cells, below which they would not function in image formation. For the inner ear, size is limited by the response time for fluid movements in the semicircular canals (Jones and Spells 1963) (Fig. 11.17). Hence, these organs, as well as the brain, come to overshadow or crowd out other parts of the skull when it reaches a size of less than about 2 cm in width.

The results of miniaturization are clearly seen in salamanders studied by Hanken (1984) (Fig. 11.18). Although closely related, they differ markedly in their size and in the configuration and number of the dermal bones of the skull. If size were not considered, it might appear that the smaller species had undergone a host of genetic changes to account for all these differences. In fact, there was probably only a single genetic change – that resulting in smaller adult body size. In the context of small size, the sense organs must retain their primitive dimensions while the configuration of the remainder of the skull simply adjusts to the smaller area within which it develops.

Physical constraints on size reduction played a major role in the origin of the novel cranial morphology of the modern amphibian orders. One of the most serious problems in vertebrate phylogeny concerns the interrelationships of frogs, salamanders, and caecilians (long-bodied, limbless amphibians of the wet tropics) and the nature of their origin from Paleozoic tetrapods. All three orders differ significantly in their structure and way of life from all of the vast assemblage of Paleozoic tetrapods (Duellman and Trueb 1986; Carroll 1992). Frogs and salamanders are notable for their very open skulls, with large orbits and extensive loss of bones from the temporal region. In contrast, most modern caecilians have solidly roofed skulls, retaining most of the bones common to small Paleozoic amphibians. No members of this group have large eyes, and in many species they are much re-

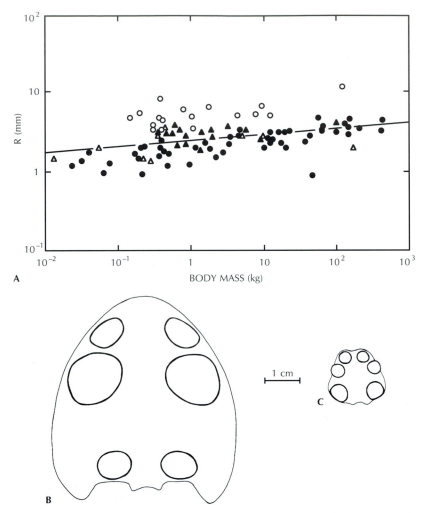

A

BODY MASS (kg)

B

C

Figure 11.17. **A,** Relationship between radius of curvature of the semicircular canals of the inner ear and body mass for all animals. Solid circles, mammals; open circles fishes; solid triangles, birds; open triangles, reptiles (from Jones and Spells, "A theoretical and comparative study of the functional dependence of the semicircular canal upon its physical dimensions" in *Proc. of the Royal Society B* 157: 403–419; copyright © 1963, Royal Society of London). **B,C** Skull of the Paleozoic amphibian *Tersomius* (B), which is thought to belong to the sister-group of some, if not all, of the modern amphibian orders, and of a much smaller amphibian, *Pseudocaecilia* (C). The heavy circles indicate the size of the sensory organs. The otic capsules, which house the semicircular canals, are nearly the same absolute size in both genera, although they appear much larger in the smaller genus. Note in particular how much closer they are to the eye socket. The distance between the otic capsules is similar in both genera because it is constrained by the width of the back of the brain (from Carroll 1990).

duced and covered by the bone of the skull roof (see Fig. 11.18). Both the reduction and loss of vision and the extensive covering of bone over the entire head region are associated with the burrowing habit of many members of this group. This accounts for their loss of limbs as well.

The earliest fossils of frogs, caecilians, and salamanders all appear in the Early to Middle Jurassic. All show most of the important attributes of their living descen-

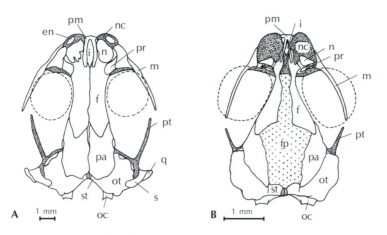

Figure 11.18. Skulls of two closely related salamanders showing striking differences in the pattern of the dermal bones to accommodate the great relative increase in size of the sense organs as a result of the smaller absolute size of the skull. **A,** *Pseudoeurycae goebeli.* **B,** The much smaller animal, *Thorius narisovalis.* Scale bars = 1 mm. Abbreviations: en, external naris; f, frontal; fp, frontoparietal fontanelle; i, internasal fontanelle; m, maxilla; n, nasal; nc, nasal capsule; oc, occipital condyle; ot, otic capsule; pa, parietal; pm, premaxilla; pr, prefrontal; pt, pterygoid process; q, quadrate; s, squamosal; st, synotic tectum. The coarsely stippled area is cartilaginous. From Hanken (1984).

dants. Only a single specimen of a single species, *Triadobatrachus massinoti,* provides a link between any of these groups and any Paleozoic amphibians. *Triadobatrachus,* from the Lower Triassic of Madagascar, shows several characters in common with frogs, including a fused frontoparietal bone, an embayment in the squamosal that may have held a tympanic ring, a short vertebral column, and long ilia. It shows no specific features that support affinities with either salamanders or caecilians. The presence of a large area of support for a tympanum is a feature in common with a particular assemblage of Paleozoic tetrapods, the temnospondyls, including *Tersomius* (Fig. 11.19A), which suggests that frogs may have evolved from that group. Many recent authors argue that salamanders and caecilians also share a common ancestry among the temnospondyls, although this requires that the latter groups have lost the tympanum and an impedance-matching middle ear (Bolt 1991; Trueb and Cloutier 1991; Milner 1993).

Whether all three modern amphibian orders evolved from temnospondyls, or only the frogs, their origins all involved significant reduction in the size of the skull and corresponding increase in the size of the sense organs and brain relative to other structures. In all three groups the skull is supported by a massive braincase.

In primitive frogs, salamanders, and caecilians the area primitively occupied by the major jaw-closing muscles is much reduced by the relatively wider back of the braincase and "expanded" otic capsules. This problem is further exacerbated in the frogs and salamanders, but not caecilians, by the large size of the eyes. Both frogs and salamanders compensate for this by the reduction or loss of most of the bones that originally surrounded the adductor chamber. The muscles expand out of this chamber and take origin from the lateral wall of the braincase and otic capsule, and may extend out over the neck region.

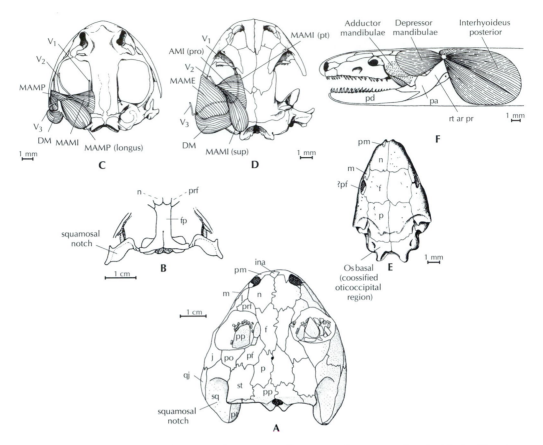

Figure 11.19. Changes in the pattern of jaw musculature associated with significant reduction of skull size among amphibians. **A,** Skull of the Lower Permian dissorophid
Tersomius (from Carroll 1964, Museum of Comparative Zoology, Harvard University).
B, The Lower Triassic frog relative *Triadobatrachus* (from Rage and Roček 1989). **C, D,**
The primitive living frog *Ascaphus* and the salamander *Ambystoma,* respectively. In
frogs and salamanders, the adductor jaw musculature has expanded extensively outside the primitive adductor chamber (from Carroll and Holmes 1980). **E,** The modern
caecilian *Grandisonia.* Despite the small size of the skull, the jaw musculature does not
extend out of the primitive adductor chamber (from Carroll and Currie 1975). **F,** Lateral
view of the skull of the caecilian *Ichthyopis* showing the elaboration of the interhyoideus posterior as a jaw-closing muscle to supplement the very small area of the adductor
chamber (Nussbaum 1983). Abbreviations of muscles and nerves: V_1, V_2, V_3, branches
of the fifth nerve; DM, depressor mandibulae; IHP, interhyoideus posterior; MAME, adductor mandibulae external (large in salamanders and missing in most frogs); MAMI,
adductor mandibulae internus; MAMI (pro), profundus head of MAMI; MAMI (pt),
pterygoideus head of MAMI; MAMI (sup), superficialis head of MAMI (large in salamanders, but missing in frogs); MAMP, adductor mandibulae posterior (elaborated in frogs,
but not in salamanders); MAMP (longus), longus head of MAMP. Abbreviations of bone
names: f, frontal; fp, frontoparietal; ina, internasal; j, jugal; l, lacrimal; m, maxilla; n, nasal; p, parietal; pa, pseudoangular; pd, pseudodentary; pf, postfrontal; pm, premaxilla;
po, postorbital; pp, postparietal; ppc, palpebral cup; pt, pterygoid; qj, quadratojugal; sm,
septomaxilla; sq, squamosal; st, supratemporal.

The evolution of the jaw adductors took a very different course among the caecilians. By far the oldest and most primitive caecilian, *Eocaecilia,* from the Lower Jurassic, had a solidly roofed adductor chamber, retaining most of the bones common to Paleozoic tetrapods (Jenkins and Walsh 1993). As in most living caecilians, the adductor muscles were confined to an extremely small chamber surrounded by the bones of the cheek and palate.

Early in their evolution, caecilians evolved an alternative method of closing their jaws (Fig. 11.19F). *Eocaecilia* already shows the full elaboration of a structure common to all later caecilians: a greatly elongated retroarticular process of the lower jaw that serves as a lever for the attachment of the interhyoideus posterior. This muscle, common to early tetrapods, constricts the ventral portion of the throat region and assists in swallowing. The anterior extent of this muscle reaches the level of the back of the jaws in larval frogs and adult salamanders, and presumably in their Paleozoic antecedents.

In caecilians, this muscle is attached to the retroarticular process, a structure not present in either frogs or salamanders, and pulls down on the jaw posterior to the jaw articulation so that the toothed portion of the jaw is lifted up. The interhyoideus posterior is much larger than the normal jaw adductors in all caecilians and has a much more effective leverage. It has essentially taken over the role of jaw closing from the normal adductors throughout the entire group. Caecilians solved the problem of a restricted adductor chamber in a different manner than frogs or salamanders, but in all three groups remodeling of this area of the skull resulted ultimately from constraints on the size of sensory structures and the brain.

Summary

Many attributes of vertebrates are constrained by physical properties of the environment. This is especially evident in aquatic and flying animals, whose locomotor structures are governed by laws of fluid mechanics. The laws of transport of gases and fluids across membranes influence the physiology of animals with different surface-to-volume ratios, elucidating why all modern amphibians are small and why the earliest amniotes were tiny. The laws of heat transfer demonstrate that dinosaurs maintained a high and nearly constant body temperature and explain why the young of small mammals and birds are altricial. Physical constraints on size reduction of sensory structures can lead to radical changes in cranial anatomy resulting from general miniaturization of the body. This is well exemplified by changes in the jaw-closing apparatus of ancestral frogs, salamanders, and caecilians.

12 Major evolutionary transitions

When a major group of organisms arises and first appears in the record, it seems to come fully equipped with a suite of new characters not seen in related, putatively ancestral groups. These radical changes in morphology and function appear to arise very quickly, especially in comparison with the normal pace of evolutionary change within a given lineage. If real, how do such changes occur? One must find out as much as possible about the actual pattern of change, and then try and fit a mechanistic explanation to it. – Thomson (1988, p. 98)

Introduction

The most exciting events of evolution are major transitions from one type of organism to another and between major habitats and ways of life. How can we explain the evolution from primitive chordates to craniates, from fish to amphibians, and from dinosaurs to birds, or the origin of marine reptiles, bats, turtles, and whales?

At the time of Simpson's major books on this subject (1944, 1953), the fossil record was still too poorly known for any of the major transitions among vertebrates to be well understood, either phylogenetically or anatomically. Given the impetus of Simpson's own work on Mesozoic mammals (1928, 1929), the transition from primitive amniotes to early mammals was the first of importance to be studied in detail (Kemp 1982; Hopson 1994). The transitions between osteolepiform fish and early tetrapods (Clack and Coates 1995; Coates and Clack 1995; Ahlberg, Clack, and Lukševičs 1996), theropod dinosaurs and birds (Ostrom 1994; Chiappe 1995a), and terrestrial mammals and whales (Gingerich et al. 1994; Thewissen 1994) are now the focus of attention as a result of many new fossil discoveries.

Simpson emphasized the problems then shrouding major transitions, including the extreme rarity of intermediate forms and the apparent speed of anatomical and habitat changes. So struck was he with the apparent rapidity that he coined a new term, **quantum evolution**, to designate such transitions. He assumed that all transitions occurred through continuous change, with no actual breaks or saltation, but at a rate that would make it difficult to find an uninterrupted sequence of intermediate fossils because of the inherently incomplete nature of the stratigraphic record.

Although several of the major transitions have been extensively studied during the past fifty years, others remain nearly as mysterious as they were when Simpson wrote. We still have no fossil evidence of the nature of the transition between cephalochordates and craniates. The earliest adequately known vertebrates already exhibit all of the definitive features of craniates that we can expect to have preserved in fossils. No fossils are known that document the origin of jawed vertebrates. Only fragmentary remains are known of animals that may represent early stages in the differentiation of placoderms, chondrichthyes, and osteichthyes. The specific lineage that links primitive chondrosteans with advanced neopterygian osteichthyes has yet to be recognized with assurance. Other important transitions and radiations still poorly known include the radiation of advanced tetrapod groups in the Lower Carboniferous, the origin of the modern amphibian orders, the transition between early tetrapods and the earliest definitive amniotes, early stages in the radiation of diapsids, and the origin of pterosaurs, ichthyosaurs, nothosaurs, turtles, and bats.

The first question that we must investigate is whether transitions between major groups are exceptionally poorly represented, compared with the fossil record of well-established groups, or if the gaps between major adaptive zones are simply a further example of the general incompleteness of the fossil record.

Diagrammatic phylogenies of large groups such as those illustrated in this book (see Figs. 1.2–1.5, 10.19) conventionally represent all major lineages with continuous lines, drawn from the horizon in which the oldest remains are found to the time of their last occurrence. This is also the convention for the record of all families illustrated in *The Fossil Record 2* (Benton 1993). In reality, there are long periods within nearly every lineage for which the fossil record remains unknown (Raup 1995). For example, no fossils of the Latimeriidae have been found between the Late Cretaceous and the Holocene, and no members of the Sphenodontidae are known in the fossil record since the Lower Cretaceous. Other examples are provided by the nothosaurs and the Choristodira, in which many genera and species have been described from only a single horizon (Figs. 12.1, 12.2).

These examples demonstrate that individual lineages are typically very poorly represented in the fossil record, whether they are part of well-established groups or represent transitions between groups. There is no adequate evidence to demonstrate that intermediate lineages are particularly poorly represented by fossils, as would be expected if they were evolving extremely rapidly or consisted of very small or localized populations. On the other hand, periods of transition do appear to be characterized by relatively few lineages, compared with those known from well-established groups, which implies a limited capacity for adaptive radiation in intermediate habitats or for animals with intermediate ways of life. The rarity of fossils documenting major transitions may simply attract more attention because of our desire to know the patterns and rates of major changes in structure and habitat, and the specific phylogenetic affinities of the derived groups.

We shall now examine a few of the better-known transitions, from which we can judge whether or not it is necessary to postulate special macroevolutionary factors or processes in addition to those that are known to occur at the level of populations, species, and genera. The following evidence is necessary to establish patterns and rates of change between major adaptive groups:

1. knowledge of the anatomy and approximate time of occurrence of immediately antecedent forms;
2. knowledge of the anatomy and approximate time of first appearance of the descendants; and
3. knowledge of intermediate forms, from intervals of 10 million years or less.

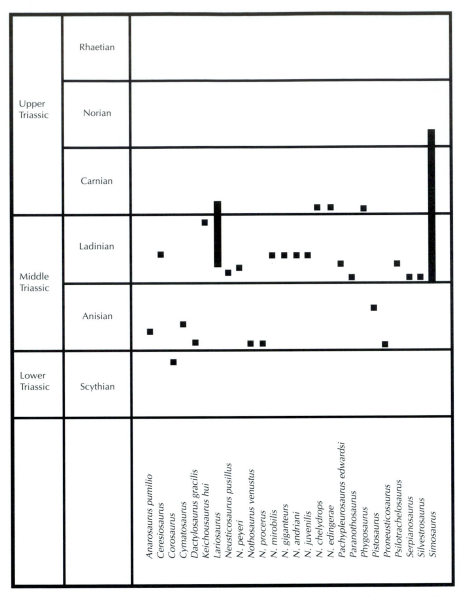

Figure 12.1. Stratigraphic record of nothosaurs showing gaps in the fossil record during the period of their origin from primitive terrestrial diapsids, during the early radiation within the group, and within individual lineages. Systematics of *Lariosaurus* and *Simosaurus* are not sufficiently well known to establish ranges of included species. Drafted by Lin Kebang.

It is never possible to establish the exact period over which transitions occur. It is always possible that later fossils of the antecedent group will be discovered, or earlier fossils of the descendants. In any case, the actual transition may have occurred a significant length of time before the last appearance of the antecedent assemblage. As an extreme example, the living cephalochordates provide a good model for the ancestors of craniates, although the transition between these groups must have occurred more than 500 million years ago.

The time within which a transition occurs is also difficult to establish because it

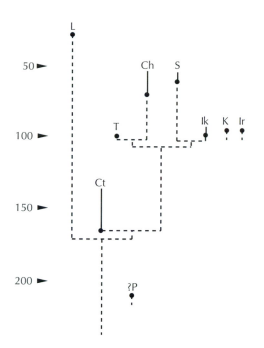

Figure 12.2. Fossil record of the Choristodira. Abbreviations: Ct, *Ctenoiogenys;* Ch, *Champsosaurus;* Ik, *Ikechosaurus;* Ir, *Irenosaurus;* K, *Khurendukhosaurus;* L, *Lazarus-suchus;* S, *Simoedosaurus;* T, *Tchioria;* P, *Pachystropheus.* From Evans and Hecht (1993).

is inherently part of a considerably longer continuum. Study of the origin of mammals once concentrated on early Mesozoic fossils that showed stages in the evolution of the skull and dentition that approached those of the living marsupials and placentals. In contrast, the initiation of more fundamental aspects of mammalian anatomy and physiology can be traced well back in the Carboniferous, to the time of the initial radiation of amniotes (Carroll 1986). Specification of particular anatomical or physiological changes as indicative of the period of transition is, to varying degrees, arbitrary. One may view the transition between fish and amphibians as the period in the Late Devonian during which a fishy fin evolved into a tetrapod limb, but this process was dependent on other changes in the fin structure that may have been initiated in the Late Silurian.

This chapter concentrates on transitions involving marked changes in physical habitats – between aquatic and terrestrial ways of life and from terrestrial to aerial. These transitions involved the greatest number of structural, physiological, and behavioral changes, many of which can be associated with specific physical constraints. The key periods of these environmental shifts apparently all took place within a relatively short time frame, on the order of 10–20 million years, with a minimum of intermediate radiations.

The major question to be investigated is the degree to which changes that occurred during the limited period of transition *between* major adaptive groups differ from those that occurred *within* the evolution of the ancestral and descendant segments of the larger monophyletic assemblage. Three attributes may be considered: the rate of change of individual traits, the relative number of traits that change significantly, and the nature of the traits that change.

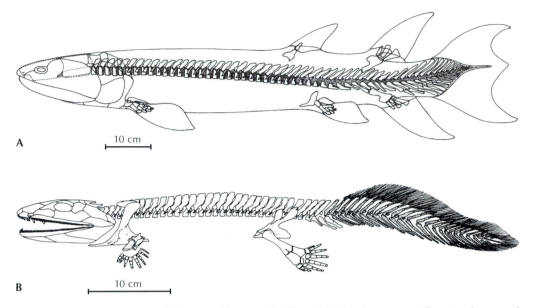

A

B

Figure 12.3. **A,** Skeleton of the osteolepiform fish *Eusthenopteron* (from Andrews and Westoll 1970). **B,** Skeleton of the uppermost Devonian amphibian *Acanthostega* (from Coates and Clack 1995).

The origin of terrestrial vertebrates

The most significant structural and adaptive change among vertebrates was the transformation of obligatorily aquatic fish to terrestrial amphibians. The descendants of this transition have dominated the land and air for the past 360 million years. The magnitude of this change can be seen by comparing the skeleton and body form of an early Upper Devonian sarcopterygian and a primitive tetrapod from the latest Devonian (Fig. 12.3). The primary focus of this transition has been on the evolution of the paired fins of fish into the digitate limbs of tetrapods. This can be conservatively dated as occurring between the early Frasnian and late Famennian stages of the Upper Devonian, a period of approximately 15 million years (Fig. 12.4).

This transition occurred between a particular group of lobe-finned fish, the osteolepiforms, which uniquely share with land vertebrates the possession of internal nostrils surrounded by the dermal bones of the palate and the marginal tooth-bearing bones of the skull. The general pattern of the dermal bones of the skull and palate are comparable in the two groups, as is the configuration of the proximal bones of the paired limbs. Osteolepiform fish are first known in the early Middle Devonian and continue into the Permian. The best-known genus, and the one that is typically used as a basis for establishing the ancestry of tetrapods, is *Eusthenopteron* from the lower beds of the Upper Devonian.

Five Late Devonian tetrapods can now be recognized: *Acanthostega* and *Ichthyostega* from East Greenland (Clack and Coates 1995; Jarvik 1996; Coates in press), *Tulerpeton* from Russia (Lebedev and Coates 1995), *Ventastega* (Ahlberg, Lukše-vičs, and Lebedev 1994) from Latvia, and *Hynerpeton* from eastern North America

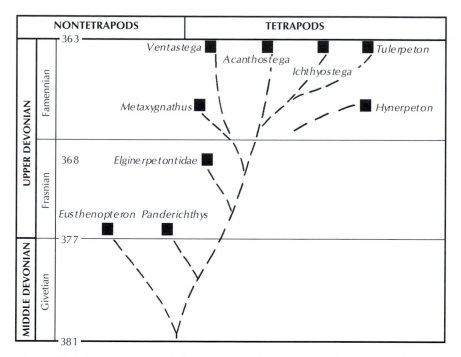

Figure 12.4. Time scale and phylogeny of Late Devonian osteolepiforms and early amphibians.

(Daeschler et al. 1994). *Acanthostega,* the best known, provides the main basis for establishing the structure of early tetrapods.

Eusthenopteron and *Acanthostega* may be taken as the end points in the transition between fish and amphibians. Of 145 anatomical features that could be compared between these two genera, 91 showed changes associated with adaptation to life on land (Carroll 1995). The distribution of these changes within the skeleton is as follows: dermal bones of the skull, palate, and operculum, 21; braincase, 5; lower jaw, 2; visceral arches, 9; vertebrae and ribs, 6; median fins, 3; girdles and limbs, 43; scales, 2. This is far more than the number of changes that occurred in any one of the transitions involving the origin of the fifteen major groups of Paleozoic tetrapods.

On the other hand, many of the major aspects of the skeleton of the ancestral and descendant groups remained similar. Changes in the dermal bones of the skull were primarily associated with consolidation of the areas of the snout, cheek, and skull table that articulated with one another in primitive osteolepiforms, and the loss of a number of small bones that had been undergoing reduction since earlier in the Devonian (Fig. 12.5). A major change was the loss of most of the opercular bones, thus freeing the skull from the bony attachment with the shoulder girdle. The basic elements of the vertebrae were present in *Eusthenopteron,* but specialization of the cervical vertebrae to articulate with the back of the skull is not known to have occurred prior to *Acanthostega.* The elements of the vertebrae became progressively better integrated with one another among early amphibians, but the notochord remained the major supporting element of the column. Small ribs were present in osteolepiform fish linking the neural arch and the centrum, but the main shaft of the rib only became elaborated in tetrapods to support the lungs and viscera against the force of gravity.

By far the most modifications occurred in the structure of the girdles and limbs. All the bones of the amphibian shoulder girdle were present in osteolepiforms, although details of their structure and proportions changed during the transition. The pelvic girdle became much larger and separated into three areas of ossification. The ilia became attached to the vertebral column by the sacral ribs.

Two very important features of terrestrial vertebrates had already evolved much earlier in the history of sarcopterygian fish: internal nares and the pattern of the proximal bones of the paired fins. Internal nostrils, present in all adequately known osteolepiform fish, are presumed to have evolved to facilitate smelling food within the mouth cavity. In tetrapods they permit aerial respiration while the mouth is closed and enable the mouth to be used as a pump to force air into the lungs.

Large endochondral bones forming the axis of the paired fins appeared even earlier, at the base of sarcopterygian radiation in the Lower Devonian. It is assumed that they evolved in relationship to the near-bottom habitat of most early members of this group: In contrast with the many smaller, parallel radials of ray-finned fish, they would have been more effective in pushing the body along the substrate. The specifically tetrapod pattern of a single proximal element followed by two more distal bones was established by the early Middle Devonian in the osteolepiforms.

Structures that have a particular adaptive value in one group but serve a different function in its descendants – such as the internal naris, with its olfactory role in fish and its respiratory role in tetrapods – have long been termed *preadaptations*. The teleological connotations of that term led Gould and Vrba (1982) to suggest the alternative *exaptations*. Regardless of the word, these are important phenomena in the evolution of all groups, but are particularly significant in major transitions, where many structures may alter their function from one environment to another. Without the presence of stout endochondral supports in the paired fins of demersal osteolepiforms, the history of land vertebrates would have been very different – or they might never have arisen.

While a great number of characters changed between early Upper Devonian osteolepiforms and latest Devonian tetrapods, others that have been thought of as typical of early amphibians only evolved subsequent to this transition. Several bones once thought typical of fish are retained in the skull and shoulder girdle of *Acanthostega*. These include a persistent rostral (also termed an *internasal*), the anterior tectal, the preopercular, and the anocleithrum, which had linked the shoulder girdle with the operculum. The cleithrum remains a much larger bone in the shoulder girdle than does the scapula, and a chamber is still present for the internal gills. Both *Acanthostega* and *Ichthyostega* retain a fishlike tail.

The transition between aquatic and terrestrial life also involved modifications in several sensory structures. The focal length of the eye would have had to change because of the different optical properties of water and air, and the receptors responsible for smell would have had to change to detect airborne molecules rather than those carried in the water. It is unlikely that modifications in either of these sensory structures are recorded in the bony anatomy that is capable of fossilization.

The other major sensory structures in fish are the lateral-line canals of the head and trunk that detect vibrations in the water. These are vital for both prey capture

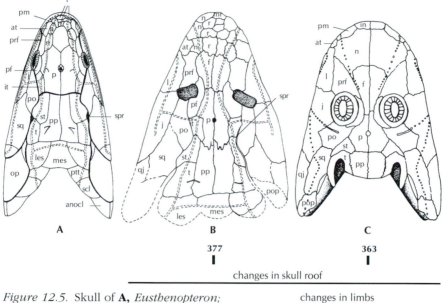

A   **B**   **C**

377   363

changes in skull roof

Figure 12.5. Skull of **A**, *Eusthenopteron;*   changes in limbs
B, *Panderichthys;* and **C**, *Acanthostega.*
Abbreviations: anocl, anocleithrum; at, an-   changes in brain case
terior tectal; f, frontal; in, internasal; op,
opercular; p, parietal; pf, postfrontal; pm, premaxilla; po, postorbital; pop, preopercular; pp, postparietal; prf, prefrontal; ptt, posttemporal; qu, quadratojugal; r, rostral; scl,
supracleithrum; spr, spiracle; sq, squamosal; st, supratemporal; ta, tabular. Numbers indicate age of specimens in millions of years. Redrawn from *Nature* (Carroll, vol. 381).
Copyright © 1996, Macmillan Magazines Limited.

and avoidance of predation. Vibrations in the water are passed via small pores to
fluid within the tubes, where they are detected by the displacement of sensory
hairs and transmitted to the central nervous system. The operation of this system
depends on the similarity of the density of the fluid external to the fish and that
within the lateral-line canals. This system cannot function in terrestrial animals because the density of the fluid within the body remains the same, whereas that of
the surrounding air is only about 0.1 percent that of water and can exert only about
one-sixtieth of the pressure per unit area on the surface of the body. In addition,
the lateral-line canals would be a source of water loss from the body. Lateral-line
canals are lost in the adult stage of all terrestrial amphibians, although they are retained in the larvae and in species that remain aquatic as adults.

Most lineages of Paleozoic tetrapods are not known to have possessed structures
capable of detecting vibrations in the air; presumably, sight and smell were sufficient for locating prey and avoiding predators. Only two groups evolved structures
comparable to those of living tetrapods: the temnospondyls and the seymouriamorphs. In both of these groups, the squamosal bone at the back of the cheek is
deeply recessed as it is in modern frogs for support of the tympanic annulus. As in
frogs, a slender stapes is directed toward the center of this recess. In frogs, the large
tympanum and the slender stapes are part of an impedance-matching system that
magnifies the force of vibrations carried in the air so that they can be detected by

the movement of fluid in the inner ear. This is accomplished by the large surface area of the tympanum compared with the much smaller area of the footplate of the stapes that fits into the fenestra ovalis, plus a lever system formed by the stapes and an extrastapedial process, which are free to vibrate in an air-filled chamber (Robinson 1973).

The similarity of the impedance-matching system in temnospondyls and frogs is one of the main reasons why they are thought to be closely related. The seymouriamorphs, which are thought to have close affinities with amniotes, evolved this system independently. An impedance-matching system also evolved separately in the origin of mammals, turtles, lepidosaurs, and archosaurs (Lombard and Bolt 1988).

Several of the early tetrapod lineages are known to have had a massive stapes that formed a structural link between the braincase and the cheek. Given this role, it would have been physically impossible to serve as an effective unit in an impedance-matching system. The oldest fossil in which the stapes is sufficiently small to have served in transmitting high-frequency airborne vibrations, as it does in frogs and lizards, is *Balanerpeton* – a small temnospondyl from the Lower Carboniferous, about 25 million years after the appearance of tetrapod limbs (Milner and Sequeira 1993).

Another system that did not change fundamentally until long after the critical period of change in the locomotor pattern was that of reproduction. There is no fossil evidence of the reproductive pattern in osteolepiform sarcopterygians, but neither is there reason to think that they reproduced differently than most bony fish, via a large number of tiny eggs laid in the water. This is certainly the case in early amphibians, for which a great number of fossils of small aquatic larvae are known. As in most modern amphibians, they must have hatched from tiny eggs laid in the water. In common with many modern amphibian families in which the adults are primarily terrestrial, most Paleozoic amphibians must have undergone a period of metamorphosis from obligatorily aquatic larvae to habitually terrestrial adults. A fully terrestrial pattern of reproduction was achieved only among the amniotes, in which large eggs could be laid on land, and the young hatch out as miniature replicas of the adults. This is not known to have occurred until the Upper Carboniferous. The changes in sound detection and reproduction demonstrate that some major aspects of adaptation to the new physical environment occurred only tens of millions of years after the initial shift in the patterns of behavior and locomotion.

Nevertheless, a very large suite of characters did change significantly over the 15 million years between *Eusthenopteron* and *Acanthostega*. What is the pattern of this change, and did it require special processes that are not evident in preceding and succeeding periods in the evolution of choanates? Within the past few years, several animals that are intermediate in time and morphology have been described: *Panderichthys* and *Elpistostega* from the upper Middle Devonian and lower Upper Devonian (Vorobyeva and Schultze 1991; Ahlberg et al. 1996), and *Elginerpeton* and *Obruchevichthys* from the middle Upper Devonian (Ahlberg 1995). These help to establish the pattern of evolution of three structural complexes across the fish–amphibian transition: the median fins, the skull roof, and the braincase.

Eusthenopteron shows no features that are specifically associated with life on land, and few that differ from typical osteolepiforms 10 million years earlier. *Pan-derichthyes* shows the first appearance of characteristics associated with life in es-pecially shallow water, indicative of an early stage in the transition toward land. This genus has lost the dorsal and anal fins, and the caudal fin is much reduced. The skull roof is better consolidated, with a smaller number of individual bones and sutural connections between areas of bone that articulated with one another in *Eusthenopteron*.

In most osteolepiforms, both the braincase and skull roof were separated into several units that contributed to a complex pattern of cranial kinesis that facilitated both feeding and respiration in the water (Thomson 1967). The operation of this complex depended on specific skull proportions, including a relatively short snout. The longer snout in *Panderichthys* together with the sutural integration of the bones in this area and at the back of the skull indicate that this system was no long-er operative. Change in skull proportions in *Panderichthys* may have been asso-ciated with a shift of feeding habits in the water, but the consolidation of the skull roof would also have facilitated adaptation to land, where a solidly integrated skull and braincase would have been necessary to resist the force of gravity in a head that was freely movable relative to the trunk.

Although the mobility of the elements of the skull roof and braincase were close-ly linked in *Eusthenopteron,* the consolidation of the skull roof in *Panderichthys* was not accompanied by any change in the structure of the braincase, which is nearly identical in these two genera (Ahlberg et al. 1996). All changes in the brain-case must have occurred more rapidly, subsequent to *Panderichthys*.

The primary focus of the fish–amphibian transition has been on changes in the paired appendages from a functional fin to a limb with digits. *Panderichthys* shows no advances toward the tetrapod condition. Ahlberg (1991) provided evidence for a more derived structure of the proximal limb elements in *Elginerpeton,* which is intermediate in age between *Panderichthys* and the Upper Devonian genera with terrestrial limbs; but, unfortunately, there is still no evidence at all of an intermedi-ate stage in the evolution of the critical distal elements of the limb – the wrist and ankle joints and the digits.

Evidence discussed in Chapter 10 indicates that formation of endochondral bone in the area of the digits in tetrapods arose de novo through extension of the zone of mesenchymal condensation that was confined to the more proximal portion of the limb in bony fish. In addition, the linear orientation of the main axis of devel-opment angles anteriorly, and elements that may be homologous with posterior ra-dials are generated and assume the segmental pattern of digits. This was clearly a local revolution in developmental processes, but we have no idea as to the pattern or rate of its evolution. Did all these processes change simultaneously, or in very rapid succession, as the result of one or more major changes in genes regulating development? Or did these changes occur in a more protracted, stepwise manner as a result of selection for minor variants in developmental processes, leading to the progressive elongation of the area of presumptive carpal and tarsal ossification, and continuing with the elaboration of segmented posterior radials and the reori-entation of the distal portion of the limb axis? This can be determined only by the discovery of a succession of fossils bridging the gap between lower Upper De-vonian osteolepiform fish and the tetrapods of the uppermost Devonian. The maximum time involved is less than 15 million years. The presence of fully developed hands and feet in members of three clearly distinct lineages in the uppermost Devonian suggests that these features must already have evolved in their immediate common ancestor some undetermined but not trivial period of time earlier.

Of the four genera known from the time interval between *Eusthenopteron* and the uppermost Devonian tetrapods, three are known primarily from skulls and lower jaws, while the fourth is known only from the shoulder girdle; however, other material from these localities may yet answer the question of the pattern and rate of evolution of the manus and pes during this period. Clearly, this transformation was much more rapid and extensive than other changes in the fin or limb structure in osteolepiforms or Early Carboniferous labyrinthodont tetrapods, but there is yet no evidence as to whether it can better be attributed to continuous, stringent, directional selection of minor genetic variants, or to major genetic-developmental changes that provided a new opportunity for rapid adaptive change. Patterns of the evolution of the limbs in transitions for which there is a more complete fossil record are discussed throughout the remainder of this chapter.

The origin of birds

One of the most dramatic transitions in the history of vertebrates was the origin of birds. Birds are the most clearly distinct of all vertebrate classes, with a unique manner of locomotion and way of life as well as a unique tissue, feathers. Their metabolic rate and sustained body temperature are higher than in all other groups, and the geometry and mechanics of their respiratory system are unparalleled.

Archaeopteryx *and dinosaurs*

Despite the enormous gap in anatomy, physiology, and way of life between modern birds and the other long-recognized vertebrate classes, the fossil record provided singularly informative evidence of the origin of birds long before we understood the ancestry of tetrapods, amniotes, or mammals. Historically, the question of the origin of birds has concentrated on a single genus, *Archaeopteryx* from the Upper Jurassic, which appears as an almost ideal intermediate between "reptiles" (specifically dinosaurs) and birds (Hecht et al. 1985). The first evidence of this genus was provided by a single, clearly avian feather reported in 1861, just two years after the publication of *The Origin of Species* (see Fig. 11.9A). This was followed shortly by the discovery of two nearly complete skeletons, the first of which was acquired by the British Museum (Natural History) and the second by the Humboldt Museum, Berlin. Together these specimens include nearly every bone in the body, as well as impressions of the feathers forming the wings and extending the length of the long tail (Fig. 12.6). Thomas Huxley (1868) soon proclaimed *Archaeopteryx* as an intermediate between reptiles and birds, and as the strongest evidence for the truth of Darwin's theory of evolution.

In contrast, until recently little was known of either the ancestry of *Archaeopteryx* or of animals intermediate between this genus and essentially modern birds of the later Mesozoic. Within the past twenty years, a host of new discoveries have

Figure 12.6. Photograph of the Berlin specimen of *Archaeopteryx lithographica,* courtesy of Dr. Hans-Peter Schultze, Humboldt Museum, Berlin.

begun to fill both these gaps, outlining the accumulative evolution of avian characters over a period that spans approximately 40 million years, from obligatorily terrestrial dinosaurs to an essentially modern avian anatomy.

Were it not for the presence of feathers, *Archaeopteryx* would almost certainly have been identified as a dinosaur; in fact, one of the subsequently recognized specimens *was* originally identified as a member of the dinosaur genus *Compsognathus,* and only later were faint feather impressions observed. Among the original specimens, only a single bone, the furcula or wishbone, has a specifically avian rather than dinosaurian configuration. Huxley argued strongly for a specific link between *Archaeopteryx* and dinosaurs, but Broom (1913) and later Heilmann (1926) maintained that all known dinosaurs were too specialized to have given rise to birds, specifically because they had lost the clavicles, the bones that in birds form the furcula. Broom and Heilmann argued that the origin of both dinosaurs and birds were to be found among a more primitive ancestral assemblage, the Triassic thecodonts (primitive archosaurs thought to include also the ancestors of crocodiles and pterosaurs). This view dominated thinking regarding the origin of birds until the 1970s, when Ostrom (1974, 1975, 1976a,b) provided the first thorough documentation of the many derived similarities between *Archaeopteryx* and theropod dinosaurs. By this time, several of these bipedal carnivorous dinosaurs had been shown to possess clavicles, thus eliminating the one specific reason that Broom and Heilmann used to dismiss relationships between these groups. On the basis of meticulous redescription of all of the specimens of *Archaeopteryx* and extensive knowledge of advanced theropod dinosaurs, Ostrom demonstrated the near identity of almost every bone in the skeleton of these two groups. Of particular importance are the structural and functional similarities of the bones of the front and rear limbs (Figs. 12.7, 12.8). What had previously been overlooked in establishing the relationship of *Archaeopteryx* was the fact that birds, like advanced theropod dinosaurs, have a uniquely specialized rear limb for effective bipedal locomotion. In contrast, bats certainly evolved flight from a quadrupedal ancestor, and whether or not pterosaurs were effectively bipedal continues to be argued (Wellnhofer 1991; Padian 1983, 1996). The anatomy of the rear limb is one of the primary features for determining the specific relationship of avian ancestors among the dinosaurs. Equally important is the fact that the forelimb of *Archaeopteryx* is typical of small, agile theropods and much more primitive than that of later birds.

Similarities of specific aspects of the skull between birds and primitive Triassic archosaurs, although still discussed (Feduccia and Wild 1993; Welman 1995), provide little evidence of their relationship to the avian lineage and ignore all the shared derived features of the postcranial skeleton uniting theropod dinosaurs, *Archaeopteryx,* and later birds. One particular Triassic genus, *Protoavis* (Chatterjee 1991), has been proposed as an immediate ancestor of birds, even more advanced than *Archaeopteryx,* but its anatomy is not sufficiently well known to demonstrate this relationship (Ostrom 1994; Wellnhofer 1994; Chiappe 1995a).

The temporal distribution of fossils documenting the evolution between dinosaurs and modern birds is shown in Figure 12.9. It is important to recognize that there are still very significant gaps in time, which give the record of morphological change a somewhat punctuated appearance. This is accentuated by the still very incomplete knowledge of the anatomy of the early members of several key lineages.

Dinosaurs have long been divided into two major groups, the saurischians and the ornithischians. Although the Ornithischia were named on the basis of the bird-like configuration of the pelvis, the relationship of birds certainly lies with the saurischians and particularly the carnivorous theropods, which include such well-known forms as *Allosaurus* and *Tyrannosaurus.* Avian relationships lie more specifically with smaller, more gracile animals that were long grouped as coelurosaurs, such as the previously cited genus *Compsognathus* (Fig. 12.10A). Within this assemblage, the Dromaeosauridae show the greatest number of derived characters in common with *Archaeopteryx* and later birds (Fig. 12.10B) (Holtz 1994).

What emerges from phylogenetic analysis of the position of *Archaeopteryx* among the dinosaurs is that this genus, as well as later birds, evolved through a protracted step-by-step accumulation of derived characters, rather than through a sudden appearance of an entire suite of new, specifically avian features.

There is no nonarbitrary point that can be identified as the beginning of the origin of birds. Bipedality, which is a necessary step prior to the elaboration of the forelimbs for flight, had already been established in the ancestors of all dinosaurs

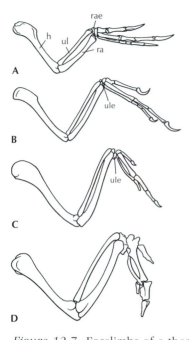

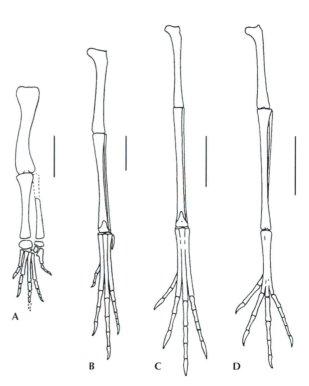

Figure 12.7. Forelimbs of a theropod dinosaur and birds. **A,** *Deinonychus.* **B,** *Archaeopteryx.* **C,** *Sinornis,* from the Lower Cretaceous. **D,** A chicken. Abbreviations: h, humerus; ra, radius; rae, radiale; ul, ulna; ulc, ulnare. Reprinted with permission from *Science,* vol. 255, Sereno and Chenggang. Copyright © 1992, American Association for the Advancement of Science.

Figure 12.8. Hind limbs of archosaurs and birds. **A,** The thecodont *Euparkeria.* **B,** The small bipedal theropod *Compsognathus.* **C,** *Archaeopteryx.* **D,** A modern pigeon. From Ostrom (1975). Reproduced, with permission, from the *Annual Review of Earth and Planetary Sciences,* vol. 3. Copyright © 1975, Annual Reviews Inc.

by the Middle Triassic. Gatesy and Dial (1996) proposed a model of sequential changes in locomotor modules between primitive theropods and birds. They pointed out that primitive tetrapods used the entire trunk and tail as a single module for terrestrial locomotion. Theropod dinosaurs completely freed the forelimbs from a role in terrestrial locomotion, but in primitive genera the rear limbs were still strongly coupled with the tail (Fig. 12.11). As in earlier tetrapods, the caudofemoralis, a large muscle extending from the tail and inserting on the fourth trochanter of the femur, is the major retractor of the rear limb. The great bulk of muscles in the tail also served to counterbalance the weight of the anterior part of the body. In the more gracile coelurosaurs, including the immediate sister-group of birds, the size of the tail, as reflected by the number of caudal vertebrae, is progressively reduced (Fig. 12.12). This suggests that the coelurosaurs may have begun to develop the hamstring as a retractor of the rear limb, as in modern birds. Ostrom (1969) postulated that the tail in the dromaeosaur *Deinonychus* was used as a dynamic stabilizer, a function that would be compromised if the tail were strongly attached to the rear limb. The functional separation of the two units is indicated by the loss of the fourth trochanter in all dromaeosaurs.

Gatesy and Dial argued that freeing of the tail from a functional connection with the rear limb was also a prerequisite for the origin of flight. The tail is not absolute-

Figure 12.9. Temporal distribution of fossils documenting the early evolution of birds, indicated on a cladogram showing their interrelationships. Nodes on cladogram: (1) Highly modified proximal caudal vertebrae; pubis extends almost directly ventrally; pubic foot projects only posteriorly; loss of fourth trochanter; lesser trochanter nearly confluent with proximal head of femur. (2) Unserrated crown and crown–base constriction of teeth; short prezygopophyses on caudal vertebrae; caudal vertebrae fewer than twenty-six; sternum; furcula; great proportional increase in length of forelimb; opistopubic pelvis; partial fusion of distal tarsals to metatarsals; fusion of metatarsals at their proximal ends; completely reversed hallux; flight feathers. (3) Broad furcular arms that are grooved posterodorsally (noncompressible); outer condyle of femur with a dorsal crest; anterodorsal ischial process; unique anteromedially directed scapular process. (4) Elongated coracoid with a raised scapular facet; triangular depression on the dorsal surface of the coracoid; greatly elongated hypocleidium on furcula; sternal keel (if present) much posterior to the anterior margin; posterior margin of sternum much emarginated; outer metacarpal distal to the others. (5) Laterally flexible furcula; rounded concave scapular facet on coracoid; distinct straplike procoracoidal process; keel almost reaching anterior margin of sternum; sternum elongated and reaching the abdominal region; uncinate processes on ribs; absence of pubic foot; distal to proximal metatarsal fusion;

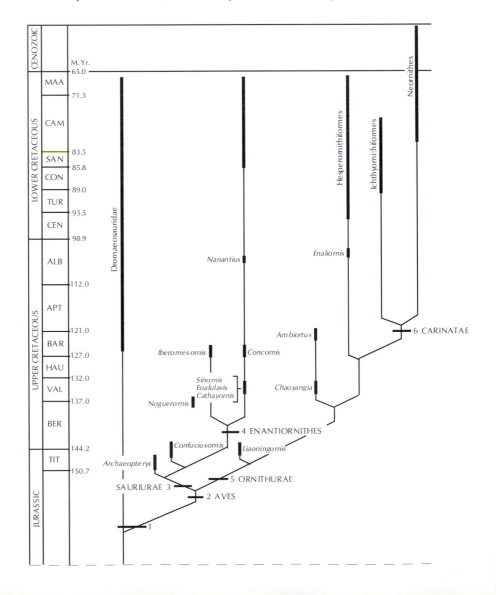

Caption to Figure 12.9 (cont.)
fusion of the distal ends of the outer two metacarpals; tarsal cap on the tarsometatarsus. (6) Supratendonal bridge on the tibiotarsus. Length of heavy lines accompanying the names of the Lower Cretaceous genera indicate not their stratigraphic range but rather the degree of uncertainty regarding their age. All but *Archaeopteryx* are restricted to a single horizon. Abbreviations (geological stages): Tit, Tithonian; Ber, Berriasian; Val, Valanginian; Hau, Hauterivian; Bar, Barremian; Apt, Aptian; Alb, Albian; Cen, Cenomanian; Tur, Turonian; Con, Coniancian; San, Santonian; Cam, Campanian; Maa, Maastrichtian. Reprinted (modified) from *Nature* (Chiappe, vol. 378). Copyright © 1995, Macmillan Magazines Limited. Additional data from Hou et al. (1996).

Figure 12.10. **A,** *Deinonychus,* one of the best known of the dromaeosaurs, the family that shares the greatest number of synapomorphies with *Archaeopteryx* and later birds of any group customarily recognized as dinosaurs. **B,** The tiny coelurosaur *Compsognathus,* from the same horizon as *Archaeopteryx.* It differs in having only two fingers and a much more primitive tail and foot structure. **C,** *Archaeopteryx* from the Tithonian near Solnhofen in southern Germany. From Ostrom (1969).

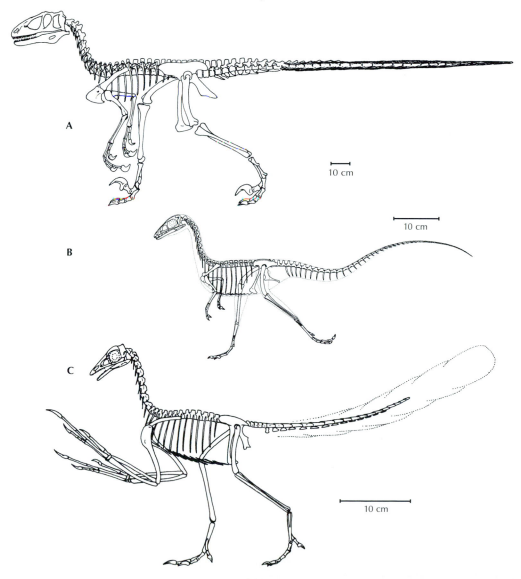

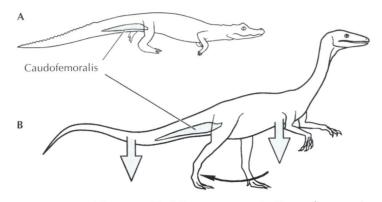

Figure 12.11. The crocodile (**A**) represents primitive archosaurs, in which the entire trunk and tail are involved in terrestrial locomotion. *Syntarsus* (**B**) is a primitive thero- pod dinosaur in which the forelimb is freed from terrestrial locomotion, but the tail and the rear limb are functionally integrated by the caudofemoralis muscle, which retracts the femur. The tail also serves to counterbalance the anterior half of the body above the hip joint. From Gatesy and Dial (1996).

ly necessary for flight in modern birds, but it may have been in *Archaeopteryx* be- cause of the relatively small size of the wings and the configuration of the shoul- der girdle, which would have provided little area for muscle attachment and little buttressing to resist strong muscle contraction. Freeing the tail from the rear limbs would have provided greater maneuverability, so that the tail feathers could have been used more effectively in providing lift.

The following shared derived characters unite *Archaeopteryx* with dromaeo- saurs as a monophyletic assemblage, distinct from all other animals customarily re- ferred to as dinosaurs (from Benton 1990; Holtz 1994):

1. highly modified proximal caudal vertebrae (neural spines limited to caudals I–IX, boxlike centra in caudals I–V, vertically oriented zygapo- physeal facets);
2. chevrons longer than deep;
3. pubis extending almost directly ventrally, rather than anteroventrally;
4. pubic foot projecting only posteriorly;
5. loss of fourth trochanter; and
6. lesser trochanter nearly confluent with proximal head of femur.

Other derived characters are shared with other coelurosaurs: reduction of pre- frontal, subrectangular coracoid, elongate forelimb, ulna bowed posteriorly, semi- lunate carpals, very thin metacarpal II, digit IV of foot longer than digit II.

Dromaeosaurs are known no earlier than about 20 million years *after* the ap- pearance of *Archaeopteryx*. The better-known genera may hence show some fea- tures (e.g., the extremely elongate caudal prezygapophyses) that evolved after the two lineages diverged. However, characters of dromaeosaurs that are more primi- tive than those of *Archaeopteryx* may be accepted as representing the pattern from which this genus evolved. On this basis, the following characters are considered to be synapomorphies that distinguish *Archaeopteryx* and later birds from primitive dromaeosaurs:

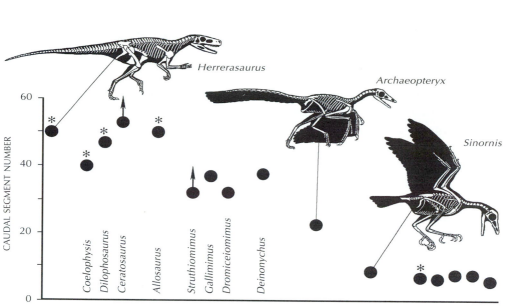

Figure 12.12. Plot of caudal segment number through theropod phylogeny. *Archaeopteryx* has fewer caudal segments than dinosaurs, but a much longer tail than in Lower Cretaceous birds, which have a pygostyle. Arrows indicate an actual number of caudal segments greater than those preserved. Asterisks designate estimates of total segment number. The more advanced birds, represented by closed circles, are *Columba, Apterxy, Dimedea,* and *Chordeiles*. From Gatesy and Dial (1996); body outlines courtesy Sereno.

1. unserrated crown and crown–base constriction of teeth,
2. caudal vertebrae fewer than 26,
3. sternum (at least in one specimen),
4. furcula,
5. great proportional increase in total length of forelimb,
6. partial fusion of distal tarsals to metatarsals,
7. fusion of metatarsals at their proximal ends,
8. completely reversed hallux,
9. flight feathers, and
10. reduced body size.

Several of the derived characters of *Archaeopteryx* are specifically associated with flight. These include the much larger relative size of the forelimb (although the anatomy of the individual bones is not altered), the reduction in the number of caudal vertebrae, and the formation of the furcula and sternum. It may not be possible to demonstrate that feathers were not present in dromaeosaurs or other, earlier theropods: They are known in *Archaeopteryx* only because of the unique quality of preservation in the lithographic limestone from the Solenhofen quarries in Germany. *Compsognathus* from that locality definitely lacks feathers (Ostrom 1985), but a specimen that resembles that genus from the Lower Cretaceous of China is feathered (Browne 1996). It is highly improbable that flight feathers were present in dromaeosaurs, whose weight (estimated at 30–80 kg) and relatively small forelimbs would have made flight physically impossible (Ostrom 1990).

The fossil record provides no direct evidence of the origin of feathers, but they are known to develop from the same tissue as scales. In some modern birds there is a continuous transition from typical scales to typical feathers along the length of the legs, but the two tissues never develop in the same place, further demonstrating their homology. In addition to the origin of the feathers themselves, their particular distribution on the wings and bodies would have had to be specified. This is accomplished in living birds through the formation of feather tracts early in development and through inductive relationship between the ectoderm and underlying mesoderm.

It is difficult to account for the initial evolution of feathers as elements in the flight apparatus, since it is hard to see how they could function until they reached the large size seen in *Archaeopteryx*. In contrast, feathers may have evolved initially to provide insulation. This could have been achieved by gradual increase in the size and "feathering" of scales. Insulation would have been necessary in dinosaurs as small as *Archaeopteryx,* and especially in other early birds that were much smaller, were they to retain the high body temperature necessary for sustained activity.

Although it seems probable that the origin of feathers was associated with relatively small size, it is certain that small size was necessary for powered flight. As we saw in Chapter 11, horizontal powered flight is not possible for animals weighing more than about 12 kg, and takeoff, without assistance from the wind or gravity, is difficult even for animals of this weight. *Archaeopteryx* is estimated as weighing less than 0.5 kg. As in the origin of amniotes in the mid-Carboniferous, small body size in the lineage leading to birds may have been selected for in animals that fed on insects or other small prey. One may conceive of the differentiation of this lineage beginning with a shift in behavior to smaller prey than that commonly taken by dromaeosaurs.

Two theories have been advanced for the way in which avian flight originated. One is that flight arose in arboreal animals that initially used the feathers to break their fall or assist in gliding; their first flights would have been gravity assisted. Because of the clear evidence for their ancestry among bipedal cursorial dinosaurs, Ostrom (1994 and references cited) has argued that flight began instead from the ground up. The feet of *Archaeopteryx* are not specialized for perching, as are those of slightly later birds; but a great range of animals can climb trees, including goats, so *Archaeopteryx* too may have had this capability. Many papers discussing the flight-origination issue were published in *The Beginnings of Birds* (Hecht et al. 1985). Unfortunately, neither structural nor physiological arguments have yet settled this controversy conclusively.

Equally important is the question of the origin of the power stroke in flight. Lift and forward propulsion in birds are produced by the forward, downward, and inward movement of the forearm. Although the bones of the wing in *Archaeopteryx* differ significantly from those in modern birds, the similarity of the pattern of the feathers and the aerodynamic requirements of flight indicate that they must have moved in a very similar manner. To judge by the similarity of the anatomy and proportions of the front limb in dromaeosaurs and *Archaeopteryx,* the immediate theropod ancestors of birds were capable of comparable movements. Ostrom (1974) argued that dromaeosaurs used movements very similar to those of avian flight in the capture of prey. He suggested that they may have used their forelimbs, with evolving feathers, like nets to capture insect prey. This specific hypothesis is difficult to test, but whatever they fed upon, the major muscles to move the forelimbs must have been located ventrally, medially, and anteriorly if the hands were used together in the capture of prey. Hence, the movement of the forelimbs in any bipedal dinosaur in which they were used to capture prey would have approximated the flight stroke of birds.

Since *Archaeopteryx* was almost certainly capable of powered flight, and the dromaeosaurs were not, it seems logical to use the point of divergence between these lineages as the base of the taxon Aves, as does Chiappe (1995a), although Gauthier, Cannatella, de Queiroz (1989) and Gatesy and Dial (1996) chose to limit Aves to the immediate common ancestry of the living avian orders.

While the structure of the individual feathers and their distribution on the forelimb of *Archaeopteryx* are nearly identical with those of modern birds, the skeleton shows only the very minimum of changes necessary for flight. The increased relative size of the forelimb is only just enough to provide the wing loading necessary for weak flight. The reduction in tail size may have given the flexibility needed for control of an additional area of lift provided by the tail feathers. The fusion of the clavicles into the median furcula would have increased the area of fixed attachment for the flight muscles. A sternum is not present in any of the earlier discovered specimens of *Archaeopteryx,* but one has been identified in the last individual to be found. This specimen, one of the smallest, comes from several meters above those previously found in the main quarry – a stratigraphic occurrence that might be interpreted as evidence that this bone evolved, or gained the capacity for early ossification, within the genus. Wellnhofer (1993) used this character, together with the relatively longer legs, to designate this specimen as a separate species, *A. bavarica,* with all other specimens in the species *A. lithographica.* However, what little evidence is provided by the fact that this specimen comes from the same quarry as other members of the genus does not support a speciation event; rather, these differences might be attributed to phyletic evolution within a single lineage.

Aside from these features, *Archaeopteryx* is far more primitive than all other adequately known birds and is typically classified as the only member of a major subdivision of the Class Aves, the Subclass Archaeornithes. It remains the only certainly documented Jurassic bird. Only within the last five years have specimens from the Lower Cretaceous begun to bridge the gap between *Archaeopteryx* and the essentially modern birds of the Upper Cretaceous. These discoveries have recently been reviewed by Wellnhofer (1994), Chiappe (1995a), and Hou et al. (1996).

Early Cretaceous birds

The most important discoveries have come from a series of freshwater lake deposits in China. Most are from the Valanginian stage, near the base of the Cretaceous, but one may be as early as the Tithonian, in the latest Jurassic (see the geological time scale in Fig. 12.9). Others come from the Valanginian and Barremian of Spain. Although the earliest of these fossils range from only 5 to 10 million years after *Archaeopteryx,* they demonstrate that birds had already divided into three major lineages, which were much advanced in the structure of either the skull or the flight apparatus.

The earliest deposits are in the Yixian Formation of northeastern China. Biostratigraphic information has been interpreted as indicating either a latest Jurassic age

(Hou et al. 1995, 1996), or a very Early Cretaceous age (Chiappe 1995a). Whichever of these dates is accepted, the fossils from these deposits demonstrate either extremely rapid evolutionary change in the latest Jurassic or earliest Cretaceous, or the presence of more advanced birds significantly earlier than *Archaeopteryx*.

The most primitive genus recognized from the Jurassic–Cretaceous transition is *Confuciusornis* (Hou et al. 1995, 1996) (Fig. 12.13A). This genus and an undescribed animal of apparently Jurassic age from the People's Republic of Korea retain the relative large size and primitive postcranial features of *Archaeopteryx*, in contrast with other, much smaller Early Cretaceous birds. *Confuciusornis* is represented by many specimens from one small area, showing nearly all aspects of the skeletal anatomy and much of the plumage. In most characteristics, *Confuciusornis* resembles *Archaeopteryx*. There is no fusion of the skull bones. The humerus is longer than the ulna and radius and retains a primitive configuration, in contrast with all assuredly Early Cretaceous birds. Three long, clawed digits remain in the hand, as does the first metacarpal. The hallux is slightly elevated, and the fifth metatarsal is present, as in *Archaeopteryx*, whereas it is lost in other birds. A small, nonkeeled sternum is present. The only features of the postcranial skeleton in which *Confuciusornis* is significantly advanced over *Archaeopteryx* are the shortening of the tail and the fusion of the distal caudal vertebrae. In strong contrast with the primitive postcranial skeleton, both the upper and lower jaws have numerous sensory and nutrient foramina that in modern birds are associated with the presence of a horny beak. Unlike any other early birds, the teeth were completely lost.

Until recently, most authors have assumed that all more advanced birds evolved from an ancestor closely resembling *Archaeopteryx*. Hou et al. (1996), on the other hand, argued that all birds should be divided into two subclasses, Sauriurae and Ornithurae, which probably diverged prior to the occurrence of the known specimens of *Archaeopteryx*. The Ornithurae include all living birds and their immediate Mesozoic ancestors; the Sauriurae include *Archaeopteryx* and *Confuciusornis* and a wide variety of primitive Cretaceous birds, the Enantiornithes. The Enantiornithes and Ornithurae both show rapid evolution at the base of the Cretaceous. We shall first discuss the pattern of evolution in the early members of the Enantiornithes, since they are currently better known.

The term Enantiornithes ("opposite birds") refers to the fact that ossification of the tarsometatarsus proceeds from proximal to distal, whereas it occurs in the opposite direction (distal to proximal) in the modern, ornithurine birds. The oldest known enantiornithine bird, *Noguerornis*, comes from the end of the Berriasian or the beginning of the Valanginian in Spain, about 11 million years after *Archaeopteryx* (Lacasa Ruiz 1989). It is much smaller, with the humerus only 22 mm long. It is incomplete and disarticulated but shows some advanced characters indicative of improved flight capabilities, including interlocking of manual elements into a rigid structure and enlargement of the wing surface. A laterally compressed extension below the rami of the furcula, termed the *hypocleidium*, is large, in common with enantiornithine as opposed to ornithurine birds.

Putting aside the poorly known *Noguerornis*, the most primitive bird clade above the level of *Archaeopteryx* is represented by *Iberomesornis* from the Barremian of Spain (Sanz and Bonaparte 1992; Fig. 12.13B). It shows a host of derived

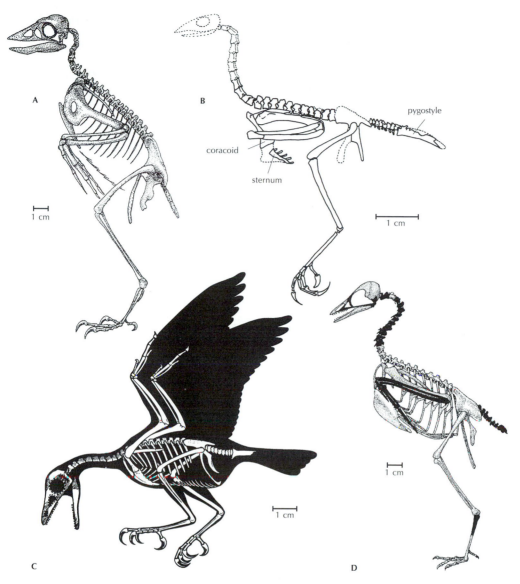

Figure 12.13. Lower Cretaceous birds. **A,** *Confuciusornis* from the lowest bird-bearing beds in China. It is about the same size as *Archaeopteryx* (reprinted from *Nature* [Lian-hai Hou et al., vol. 377]; copyright © 1995, Macmillan Magazines Limited). **B,** *Iberome-sornis* from the Barremian of Spain (from Sanz and Bonaparte 1992). **C,** *Sinornis* from the Valanginian of China (reprinted with permission from *Science*, vol. 255, Sereno and Chenggang; copyright © 1992, American Association for the Advancement of Science). **D,** *Chaoyangia,* from the Early Cretaceous of China, the oldest known ornithurine bird for which a reconstruction is possible (from Hou et al., 1996).

characters in common with more advanced birds. The most conspicuous features are the great reduction of the tail to eight free vertebrae and the formation of a slender pygostyle that would have supported a fan of feathers to provide lift, enhance aerial maneuverability, and function as a brake during landing. There are only eleven dorsal vertebrae, down from thirteen or fourteen in *Archaeopteryx*. The sternum is much increased in size, although the portion that might support a keel is missing. The coracoid, which had a platelike appearance in *Archaeopteryx,* now forms a long strut that articulates with the sternum via a broadly expanded base to buttress it against the force of the flight muscles. As in *Noguerornis,* the furcula bears a large hypocleidium. The scapula comes to a sharp point posteriorly; the ulna is longer than the humerus and thicker than the radius, in common with modern birds, where it provides the surface for attachment of the flight feathers. The foot is specialized for perching, suggestive of arboreal habits. Unfortunately, the one known specimen of this clade lacks the head and the hand.

Iberomesornis is known from beds about 125 million years old, but a more advanced genus, *Sinornis* (Figure 12.13C), is known approximately 9 million years earlier, from the early Valanginian lake beds of China (Sereno and Chenggang 1992). This and all subsequent birds are advanced over *Iberomesornis* in having a synsacrum with more than eight vertebrae (rather than five), heterocoelous cervical vertebrae, complete calcaneal–astragalar–tibial fusion, and distal tarsals completely fused to the metatarsals. *Sinornis* itself was more primitive in retaining *gastralia,* or ventral scales, between the sternum and the pelvic girdle – a character apparently lost independently in *Iberomesornis* and more advanced birds.

Sinornis is more completely known and shows more clearly how rapid the advance was in avian characters in no more than 15 million years after *Archaeopteryx.* Except for retaining gastralia, *Sinornis* has all the derived characters seen in *Iberomesornis* and in addition shows several advanced characters of the forelimb. The glenoid socket faces laterally, rather than posteroventrally, permitting excursion of the humerus above the vertebral column. The wrist joint is modified for hyperflexion of the manus against the forearm during the recovery phase of the flight stroke and for folding the wing against the trunk during rest. The second metacarpal is beginning to predominate, indicating an early stage in the specialization of the carpometacarpus, although there is no fusion of these bones, nor evidence that the locking mechanism described by Vazquez (1992) as necessary for effective flight in modern birds had yet evolved (Fig. 12.14). The hallux is lowered, making the foot better suited for perching. In terms of its skeleton, *Sinornis* differs more from *Archaeopteryx* than the latter genus differs from the most birdlike dinosaurs.

Sinornis is primitive in retaining teeth and three fingers, two of which bear claws; the sternum lacks a keel, and the pubis retains an expanded foot. However, a very low keel is reported in *Cathayornis* from the same horizon, which also shows fusion of the carpals with individual metacarpals, although the metacarpals themselves are not fused (Zhou, Jin, and Zhang 1992). Also within the enantiornithines, *Concornis,* from the same locality as *Iberomesornis,* is more advanced than *Sinornis* in the apparent distal fusion of metacarpals II and III. It also has a carinate sternum, although the keel occupies only the caudal half of the sternum (Sanz, Chiappe, and Buscalioni 1995).

Among the most tantalizing specimens from the Barremian of Spain is the thoracic region of a goldfinch-sized bird, *Eoalulavis,* in which the first digit, like that of modern birds, bears a single feather on the proximal phalanx that formed an *alula,* or bastard wing, necessary for maintaining lift at slow speed (Sanz et al. 1996). This structure is not reported in any other Early Cretaceous bird. Surprisingly, this genus has a very narrow sternum with a low keel and no articulating surfaces for the clavicles.

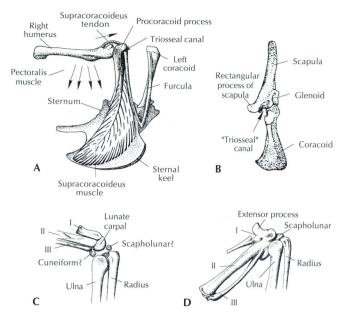

Figure 12.14. **A,** Anterolateral view of the pectoral girdle and sternum of a modern ornithurine bird, the pigeon, in which the triosseal canal is formed by the procoracoid process. The upper arrow indicates the course and action of the supracoracoideus tendon from the insertion toward the triosseal canal. The lower arrows indicate the location of the pectoralis, which is not illustrated (from Ostrom 1976b). **B,** Articulated scapula and coracoid of the enantiornithine bird *Enantiornis* in anterior view, showing the rectangular process of the scapula that forms the margin of the canal for passage of the tendon of the supracoracoideus tendon (from L. D. Martin 1995). **C, D,** Comparison of the wrist area between *Archaeopteryx* and a modern bird, represented by *Cathartes,* showing homology of elements. The pattern in *Archaeopteryx* retains all the features of coelurosaur dinosaurs. Abbreviations: I, II, III, metacarpals. This complex is one of the slowest to evolve in Cretaceous birds. It is not known for certain until the Upper Cretaceous (from Ostrom 1976b).

In addition to these genera from the Early Cretaceous of China and Europe and other specimens from Australia, the Enantiornithes were also the main component of the Late Cretaceous fauna of continental birds in North and South America and Asia (Chiappe 1992, 1995b). The Asian genus *Gobipteryx* was advanced in the loss of teeth, and *Enantiornis,* from South America, had a wingspan of over 1 m. The carpometacarpus is extensively fused in the Coniacian–Santonian *Neuquenornis,* but its metatarsals never achieved the complete distal fusion of more advanced birds.

According to Chiappe (1995a), ornithurine birds, which constitute the entire Cenozoic avian fauna, diverged from among the early enantiornithines and first appeared in the fossil record some 25 million years after *Archaeopteryx*. In contrast, new evidence provided by Hou et al. (1996) shows that they had diverged by the time of the Jurassic–Cretaceous transition, perhaps as early as 5 million years after the oldest known Jurassic birds.

Ornithurine birds are distinguished by the much greater posterior extension of the sternum, which is fully keeled among the first members of this group, the presence of uncinate processes on the ribs, and an elongate procoracoid that forms an important part of the triosseal canal. In modern birds, the triosseal canal forms a pulley over which passes the tendon of the supracoracoideus muscle to raise the wing (Figure 12.14A,B). A passage for this tendon also evolved among the enantiornithines, but in that assemblage a rectangular extension of the scapula rather than a process of the coracoid formed the major portion of the canal.

Hou et al. (1996) emphasized that the large sternum and ribs with uncinate processes form the respiratory pumping apparatus that is necessary for the extremely high metabolic rate in modern birds. The great posterior extent of the sternum also serves to support the major air sacs. They argue that the initiation of these bony structures in Early Cretaceous ornithurine birds indicates a great physiological advance over the contemporary enantiornithines. The bone histology of Late Cretaceous enantiornithine birds supports the argument that they retained a much lower metabolic rate than ornithurine birds (Chinsamy, Chiappe, and Dodson 1995).

The first evidence of ornithurine birds is provided by a partial skeleton of *Liaoningornis* from the same beds as *Confuciusornis* (Hou et al. 1996). It has a deeply keeled sternum that extends much further posteriorly than that of enantiornithine birds and is not deeply emarginated along the posterior margin. The tarsometatarsus is fully fused distally but not proximally. Ribs are not known, so that the presence of uncinate processes cannot be established. The sharp, recurved pedal claws and short broad tarsometatarsus suggest that *Liaoningornis* spent much of its time in trees.

Much more complete material is available of *Chaoyangia* from the Valanginian, approximately 10 million years later than *Liaoningornis,* permitting reconstruction of nearly the entire skeleton (see Fig. 12.13D). The configuration of the sternum and the uncinate processes are comparable with those of modern birds, as is the flexible nature of the arms of the furcula, giving it the character of a spring (Jenkins, Dial, and Goslow 1988). The coracoid has a long, straplike procoracoid process, indicating the typical ornithurine configuration of the triosseal canal. The pubis has lost its pubic foot. On the other hand, the premaxilla and dentary remained toothed. The small pedal claws suggest adaptation for wading. Although the wings are not known in *Chaoyangia,* they were probably not more advanced than those of the next younger ornithurine bird, the Barremian/Aptian genus *Ambiortus,* which retains two well-developed wing phalanges and a claw on the second digit (Kurochkin 1985; Olson 1985; Feduccia 1996).

Comparison between *Sinornis* and *Chaoyangia,* both from the Valanginian, and earlier Cretaceous birds demonstrates that the pattern of ossification of the tarsometatarsus, changes in the configuration of the sternum and its keel, and the formation of the triosseal canal all evolved separately in the enantiornithines and ornithurines.

It is surprising that the ornithurines, although they appear to have evolved further and faster in the earliest Cretaceous, show only a very limited radiation prior to the Cenozoic. They are represented by only two genera later in the Early Cretaceous: the very incompletely known *Ambiortus* and *Enaliornis* from the Albian. Four major adaptive morphotypes are known from the Upper Cretaceous: *Patagopteryx,* a terrestrial flightless form from South America (Chiappe 1996); flightless diving birds (Hesperornithiformes); the ternlike Ichthyornithiformes; and the ex-

Figure 12.15. **A,** *Ichthyornis* from the Upper Cretaceous; this advanced ornithurine bird still retains teeth (from Marsh 1880). **B,** *Lithornis,* a primitive neornithine bird from the Paleocene of Montana, described by Houde (1988) (reprinted with permission from *Science*, vol. 214, Houde and Olson; copyright © 1981, American Association for the Advancement of Science).

tremely rare Neornithes, including the ancestors of all the modern orders (Fig. 12.15). However, it is possible that there were many other ornithurine birds living in the Late Cretaceous, but that their fossil remains have gone undiscovered, as was the case for nearly all Lower Cretaceous birds until about ten years ago.

Cracraft (1986) cited the following features as distinctive of the Neornithes: loss of teeth, loss of coronoid bone, a supratendinal bridge of the tibiotarsus, a bony mandibular symphysis, fused uncinate processes, obturator foramen, quadrate articulating with the prootic, bony eustachian tubes, and dentary forked posteriorly. Unfortunately, the line leading to the Neornithes is separated from the Early Cretaceous ornithurines by more than 50 million years, precluding any estimates of the rate of acquisition of these modern anatomical features.

Numerous remains attributed to the Neornithes have been identified from the last 15 million years of the Cretaceous, but there is very little evidence for the origin of the modern avian orders prior to the beginning of the Cenozoic, when they underwent a major radiation (Feduccia 1995, 1996), discussed in Chapter 13.

Rates and patterns of evolution

The nature of the changes at different stages in the origin of birds is shown in Figure 12.16. Unfortunately, it is very difficult to determine actual rates of change because of long gaps in the fossil record and the incomplete preservation of most specimens.

The transition from obligatorily terrestrial dinosaurs to birds possessing most osteological attributes of the modern flight apparatus may have occurred over fewer than 20 million years. Compared with the great longevity of the modern avian orders, this is a relatively short period of time, but it is apparently longer than that required for the Cambrian explosion of metazoan body plans, or the differentiation of the placental orders in the early Cenozoic. It certainly took place over a sufficiently long period of geological time that we can expect to find many more intermediate fossils documenting more specifically the rates and patterns of evolution.

We may consider the beginning of the dinosaur–avian transition as the time of divergence of the avian lineage from dromaeosaurids. This can be only roughly established, by the prior appearance of other divergent dinosaur lineages, as Middle Jurassic, about 165 million years ago. This provides at least 15 million years for the evolution of the derived anatomical, size, and behavioral characters of *Archaeopteryx,* although these changes could have taken place over a much shorter period. By the Valanginian, no more than 15 million years after *Archaeopteryx,* most of the osteological refinements of the flight apparatus had already occurred in both the Enantiornithes and the Ornithurae. The only area of the wing that had been little modified was the manus. The carpometacarpus is beginning to consolidate in the Barremian enantiornithine *Concornis.* The final step in the modernization of the wing skeleton and the loss of teeth was probably achieved separately in the enantiornithines and the ornithurines, early in the Late Cretaceous. Although none of these dates is firmly established, the progressive addition of derived characters does suggest that this is a fairly reliable model.

Some 20–25 million years passed between *Archaeopteryx* and essentially modern ornithurine birds but with teeth. The greatest amount of change, occurring over perhaps the most accurately established time span, was the modernization of the flight apparatus between the level of *Archaeopteryx* and that of *Chaoyangia,* which required no more than 15 million years. The few characters seen in *Liaoningornis* had achieved a modern appearance after only 5 million years. These changes could, of course, have taken place in even less time, but there are no fossils to document this time interval.

Although the Late Jurassic and Early Cretaceous were not times of extensive and long-lasting radiation of major avian taxa, the later stages of the dinosaur–avian transition did involve the proliferation of a large number of minor lineages that show a strikingly mosaic pattern of evolution, especially within the flight apparatus. The rate of evolution differs from element to element and from taxon to taxon. The furcula and coracoids evolved the most rapidly, with no intermediates known between *Archaeopteryx* and an essentially modern pattern in the oldest known ornithurine birds in the Lower Cretaceous. The sternum had begun to ossify in *Archaeopteryx,* but it was still without a keel in some Early Cretaceous birds. The keel

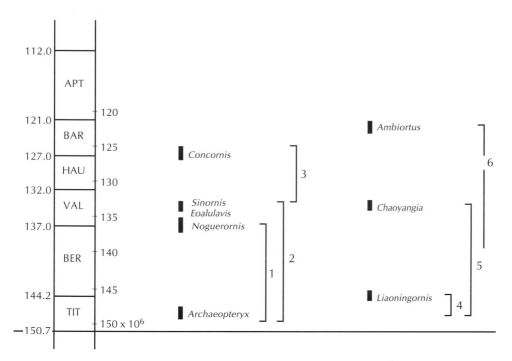

Figure 12.16. Sequence of skeletal modernization among Lower Cretaceous birds. (1) Great reduction in size; formation of hypocleidium; ulna longer than humerus and thicker than radius; interlocking of manual elements into a rigid structure. (2) Reduced cervical neurapophyses; fused posterior cervical ribs; eleven dorsal vertebrae; only eight free caudals; pygostyle; strutlike coracoid having broad contract with sternum; sharp caudal end of scapula; increase in synsacrum from five to eight vertebrae; complete fusion of astragalus and calcaneum to tibia; fusion of distal tarsals to metatarsals; hallux lowered; glenoid socket faces laterally; wrist joint modified for hyperflexion of manus; second metacarpal enlarged; appearance of alula. (3) Heightening of keel of sternum; apparent distal fusion of metacarpals II and III. (4) Sternum extending far posterior; posterior edge of sternum not emarginated; long deep keel of sternum; distal end of tarsometatarsus fused (distal to proximal tarsal fusion). (5) Uncinate processes; internally flexible furcula; small hypocleidium; strap-shaped procoracoid process (modern configuration of triosseal canal); pubic foot lost. (6) Fusion of distal ends of outer two metacarpals.

increased progressively in depth and anteroposterior extent over the next 10 million years within the enantiornithines, but was fully formed by the earliest Cretaceous in the ornithurines. Coossification of the carpometacarpus was not even initiated until nearly 25 million years after the time of *Archaeopteryx*.

Chiappe (1995a) pointed out that changes in the rear limb structure, which had begun even earlier than the divergence of the dromaeosaurs from more primitive theropods, continued on at a slower pace well into the Lower Cretaceous. Proximal fusion of the metatarsals was achieved between dromaeosaurs and *Archaeopteryx,* but distal fusion had still not been achieved in the Late Cretaceous Enantiornithes. On the other hand, *Gansus,* a possible ornithurine known from little more than the foot, exhibited complete fusion of these bones in the Lower Cretaceous (Hou and Liu 1984).

The origin of mosasaurs

Although this transition is of less significance phylogenetically than the origin of amphibians or birds, it is well documented in the fossil record, and its relative simplicity provides a good model for analysis. The skeletons of ancestral, intermediate, and descendant forms are sufficiently well known that tabulations of nearly all character changes can be made over several periods of the transition (DeBraga and Carroll 1993).

Mosasaurs were gigantic lizards that dominated the world's oceans for the last 24 million years of the Mesozoic. Their ancestry can be traced to primitive anguimorph lizards of the Late Jurassic or Early Cretaceous. The aigialosaurs, from the mid-Cretaceous of the Adriatic, provide an informative example of a transitional stage between habitually terrestrial lizards and the obligatorily aquatic mosasaurs (Fig. 12.17). Their skulls are almost indistinguishable from mosasaurs. The tail is laterally compressed and the pelvic girdle resembles that of another group of aquatic reptiles, the pleurosaurs, but the limbs were unchanged from those of terrestrial lizards.

Six stages in the evolution of mosasaurs may be recognized, beginning soon after the initial radiation of the modern lizard groups:

1. divergence of a distinct lineage from ancestral anguimorphs (this lineage might not be recognizable initially by major anatomical differences, but may be identified as a uniquely important biological entity on the basis of subsequent habitat and structural changes);
2. the transition between the most primitive members of the aigialosaur clade and those known from the Cenomanian–Turonian boundary (during this period aigialosaurs became behaviorally and structurally adapted to an aquatic way of life);
3. the transition between aigialosaurs and the earliest known members of the Mosasauridae;
4. an initial dichotomy within the Mosasauridae resulting in a derived pattern of cranial kinesis;
5. divergence of the three mosasaur subfamilies; and
6. progressive changes within each of the mosasaur subfamilies.

The transition between primitive anguimorphs and aigialosaurs involved at least forty-two character transformations, which were concentrated in the skull and mandibles. There is no way to judge when these changes occurred, within a time span of nearly 65 million years. Between twenty-three and thirty-three changes happened between the mid-Cretaceous aigialosaurs and the most primitive mosasaurs. The appearance of the earliest mosasaur may precede the occurrence of the last aigialosaurs. Since the aigialosaurs exhibit no characters that are more derived than those of mosasaurs, it is unlikely that the two lineages diverged more than 1–3 million years before the appearance of the oldest mosasaurs. The initial radiation of the four major mosasaur lineages is estimated as having occupied about 3 million years, but it involved sixty-three character transformations. After their initial divergence, the three major mosasaur subfamilies underwent a total of 153 character transformations over a 23-million-year period.

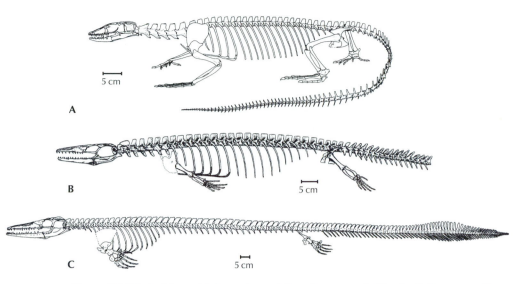

Figure 12.17. **A,** A terrestrial anguimorph lizard, represented by the living genus *Varanus.* **B,** An aigialosaur from the mid-Cretaceous. **C,** An Upper Cretaceous mosasaur, *Clidastes.* From Caldwell, Carroll, and Kaiser (1995).

The total number of character transformations during the transition between aigialosaurs and mosasaurs is lower than that which occurred in the prior evolution of the aigialosaur lineage or the subsequent history of mosasaurs, although the latter changes took place over much longer periods. None of these time intervals is sufficiently accurately established to determine actual rates of evolution. Most of the changes in the skull occurred within the aigialosaur lineage, as did the achievement of a vertical angle between the centra of the caudal vertebrae. The angle of the trunk centra changed during the period of transition.

The most significant change that occurred during the transition was the transformation of the limbs from a pattern typical of terrestrial lizards to that of fishlike fins (Fig. 12.18). This involved primarily a significant shortening of the proximal portion of the limb and a reduction in the ossification of the carpals and tarsals. Little change is apparent in the metapodials or digits. The nature of the changes in the limbs closely resemble those leading to the earliest ichthyosaurs and nothosaurs, discussed in Chapter 10. While the general outline of the limb and the nature of the wrist and ankle joints changed significantly in the formation of the fin, there was actually less quantitative and qualitative change in the individual bones than occurs in the subsequent history of mosasaurs, during which the appearance of the proximal bones was dramatically altered and the number of phalanges greatly increased (Fig. 12.18C–E).

Thus, there appears to be no significant difference in either the number or nature of changes that occur between aigialosaurs and mosasaurs as compared with those that occurred in either the ancestral or the descendant groups. The changes that did occur in the proportions and degree of ossification of the limb were key to the subsequent success of mosasaurs as open-water swimmers, but they required no special evolutionary processes. Once these changes occurred, the group underwent a rapid and extensive radiation.

Modes of evolution in the origin of mosasaurs

The history of the mosasaur clade provides a fairly simple pattern (inasmuch as it is known) from which to consider the tempo and mode of evolution. Although the specific ancestors of the clade are not known, mosasaurs are certainly closely related to anguimorph lizards, represented in the modern fauna by the varanoids and having a fossil record going back to the Middle Jurassic. In common with other groups of lizards, terrestrial anguimorphs have a sufficiently conservative skeletal anatomy that well-known Late Cretaceous varanoid lizards may be used as a reliable pattern for the origin of the characters we see in aigialosaurs.

It is not known when the lineage leading to mosasaurs diverged from primitive varanoids. It may have taken place at any time from the Upper Jurassic until shortly before the appearance of aigialosaurs. Regardless of the duration of this phase of evolution, some processes can be reconstructed even without specific fossil evidence. This clade must have begun, as do all other clades, with a speciation event. We have no knowledge of the specific structure or adaptive niche of the early members of the mosasaur clade, but there is no reason to think they differed significantly from other early varanoids, which were small, terrestrial insectivores or carnivores (Evans 1994). Several major changes occurred between these forms and the mid-Cretaceous aigialosaurs: a substantial size increase from a skull length in Middle and Upper Jurassic anguimorphs of approximately 5 cm to 16 cm in aigialosaurs, presumably a concomitant shift to feeding on much larger prey, and a shift in habitat. The fossil record provides no direct evidence of which of these changes took place first.

All of these changes, although they are evidenced by structural modifications, were almost certainly initiated by behavioral changes. Many predators are capable of making use of a wide variety of diets; what they eat depends primarily on local availability. The capacity for opportunistic feeding is a particularly important aspect of the continuing evolution of the cichlid fishes in the East African Great Lakes. As demonstrated among the finches of the Galápagos by Peter and Rosemary Grant (see Chapter 3), selection acts very stringently on the behavior of feeding. One need only postulate a situation in which the ancestors of aigialosaurs could take advantage of a succession of larger prey species over the period of time required to select for the genetic characters necessary to increase body size substantially.

The rate of size increase cannot be established in the lineage leading to aigialosaurs, but an informative analogy is provided by size increase in later species within the Varanidae. The largest living varanid is *Varanus komodoensis,* in which the skull–trunk length reaches at least 1.5 m. It was greatly exceeded in size by *Megalania prisca* from the late Pleistocene, which was twice the length of the komodo dragon and approximately eight times the mass. Its skull alone reached 50 cm. According to Hecht (1975), it was an integral element of a fauna dominated by giant herbivores in the Pleistocene of Australia; with their extinction at the end of the Pleistocene, *Megalania* also passed from the scene. Smaller, related species are known in the Pliocene, suggesting that it took no more than 3–5 million years for the achievement of a size increase that was even greater than that which occurred between primitive Mesozoic anguimorphs and early mosasaurs.

In Chapters 10 and 11, we considered several groups of diapsid reptiles that

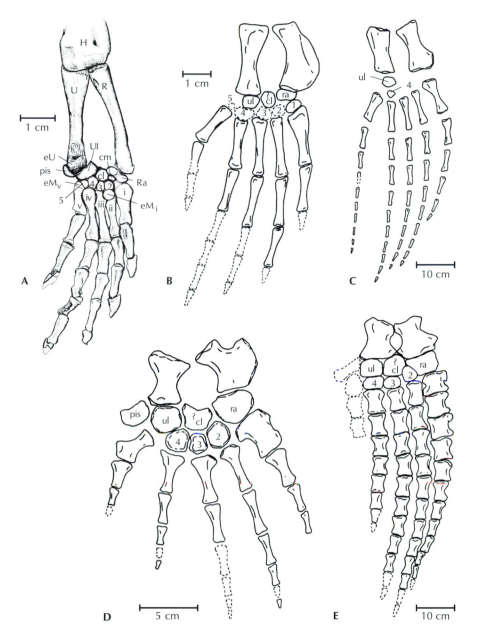

Figure 12.18. Lower forelimb of mosasauroids. **A,** An aigialosaur, showing the typical pattern of terrestrial lizards. **B,** The primitive mosasaur *Halisaurus sternbergi.* More advanced mosasaurs (C–E) show divergent specializations of the carpus and phalanges: **C,** *Tylosaurus proriger,* in which the carpus is very incompletely ossified, but the number of phalanges is greatly increased; **D,** *Clidastes liodontus,* in which the carpus is fully ossified, and the manus is greatly expanded laterally, but with limited hyperphalangy. **E,** *Plotosaurus bennisoni,* exhibiting a well-ossified carpus and hyperphalangy. Abbreviations: cl, lateral centrale; cm, medial centrale; eU, eM$_i$, eM$_v$, epiphyses of ulna and of first and fifth metacarpals, respectively; H, humerus; pis, pisiform; R, radius; ra, radiale; U, ulna; ul, ulnare; 1–5, distal carpals and tarsals; i–v, metacarpals and metatarsals. From Carroll (in press).

adapted to life in the water. At least sixteen major lineages within this assemblage became aquatic. Aquatic adaptation is relatively easy among primitive diapsids for both anatomical and physiological reasons. Terrestrial locomotion among quad-rupedal diapsids retains the basic pattern of their fish ancestors. The primary pro-pulsive force is provided by alternating lateral contractions of the trunk and tail. Crocodiles and habitually terrestrial lizards accommodate to aquatic locomotion behaviorally by drawing the limbs toward the trunk to reduce drag and lateral movements of the front part of the body. Aquatic locomotion is energetically cheaper for lizards, which may swim vigorously for much longer periods of time than they can move actively on land, where they are subject to rapid muscle fatigue from buildup of lactic acid. The low metabolic rate and associated tolerance to cold and lack of oxygen make it possible for lizards and turtles to spend long stretches in the water without returning to the surface.

The modern marine iguanid *Amblyrhynchus* provides a model for adaptation of lizards to an aquatic way of life. This genus depends entirely on aquatic vegeta-tion on which it feeds below waterline around the islands of the Galàgagos archi-pelago. Detailed comparisons of this genus with its closest terrestrial relative, *Cono-lophus,* by Dawson, Bartholomew, and Bennet (1977) demonstrated the absence of either structural or physiological specializations for an aquatic way of life. Its adaptation to this environment appears to be entirely behavioral. Immunological data suggest that these genera diverged from one another as long as 15–20 mya (Wiles and Sarich 1983), but as much as 25 million years may have been available for their evolution according to recent geological evidence of the age of the Galá-pagos (Christie et al. 1992). Unfortunately these figures provide no information re-garding the actual time during which aquatic adaptation occurred, which may have been over a much shorter period.

Amblorhynchus provides an informative example of an early stage in the achievement of an aquatic way of life such as may have occurred in the lineage leading to mosasaurs. In contrast, aigialosaurs also show significant skeletal mod-ification indicative of aquatic locomotion: a laterally compressed tail with elongate haemal and neural spines and modification of the pelvis. More important, changes in the middle ear, including ossification of the tympanum and formation of a mas-sive quadrate, suggest adaptation to feeding at depth. Comparable changes are seen in early whales, which are discussed in the next section.

Although the limbs of aigialosaurs were shorter relative to trunk length than those of active terrestrial varanoids, the detailed structure of the individual bones, their proportions, and degree of ossification remain unaltered (see Fig. 12.18). Ai-gialosaurs may have been habitually aquatic but were not obligatorily so. It is pos-sible that significant changes in limb structure were delayed because of conflicting selective pressure to maintain the primitive pattern of terrestrial reproduction. A recently discovered mosasaur with skeletons of the young in the body cavity strongly suggests that they gave birth to live young (Bell 1996). This does not prove that they gave birth in the water, as do ichthyosaurs and whales, but it is sugges-tive. It is possible that a switch in reproductive behavior resulted in a sudden change in selective forces, leading to the rapid evolution of fins and loss of the at-tachment between the vertebral column and the pelvic girdle that would have made terrestrial locomotion nearly impossible.

Commitment to an aquatic way of life was also associated with a final increase in size from aigialosaurs to mosasaurs, making it possible for them to compete actively in a fully marine environment, as the group quickly expanded from local shallow water habitats to the oceans of the world. This appears to have occurred very soon after the attainment of a fully mosasaurian structural grade.

Comparable changes in skull configuration, increase in body size, assumption of an aquatic way of life, and the origin of viviparity were achieved separately by other groups of lizards. Individually, none of these changes requires explanations beyond those provided by Darwin. What sets mosasaurs apart was the combination of these changes and their persistent evolution over tens of millions of years. The significance of these advances is further emphasized by the great radiation of mosasaurs and their dominance in marine environments for almost 25 million years.

The origin of whales

In common with diapsid reptiles, several mammalian lineages have returned to an aquatic way of life, including the sirenians and the pinnipeds. The most extensive marine adaptation and the largest adaptive radiation was achieved among the whales, Order Cetacea. Most modern families are known from the Miocene, and one first appears in the mid-Oligocene. Much more archaic whales, the archaeocetes, are known from the Eocene, including genera that illustrate the transition from primitive terrestrial mammals.

Many similarities of the skull and dentition demonstrate that whales are closely related to a distinctive group of Paleocene and Eocene mammals, the mesonychids (Fig. 12.19, and see Fig. 12.22). The mesonychids, common from the late Paleocene to the end of the Eocene, exhibit an odd combination of characters. Many aspects of the postcranial skeleton, including hoofs, are indicative of primitive herbivores, whereas features of the skull and dentition are suggestive of carnivores. Most anatomical and molecular data indicate that their closest relatives among living mammals are the artiodactyls, but other molecular evidence implies closer affinities with perissodactyls (Thewissen 1994). In fact, neither of these orders is known as early as the earliest mesonychids. It may be more realistic to think of the ancestors of horses, cows, and whales as all having diverged from near the base of the radiation of large herbivores.

Mesonychids and primitive whales were unique among early mammals in the structure of their molar teeth. The lower molars were very narrow; their upper surface ground against a basin formed by the medial portion of the upper molars. Shearing occurred between the medial side of the lateral cusps of the upper molars and the lateral surface of the lower molars.

It is not possible to identify a sequence of mesonychids leading directly to whales, although some teeth now recognized as belonging to primitive whales were originally described as from mesonychids. All adequately known mesonychids were terrestrial in most aspects of the skeleton, and some show specializations for cursorial locomotion. However, O'Leary and Rose (1995) noted that the Lower Eocene *Pachyaena* possessed features analogous to those of semiaquatic tapirs, suids, and capybaras.

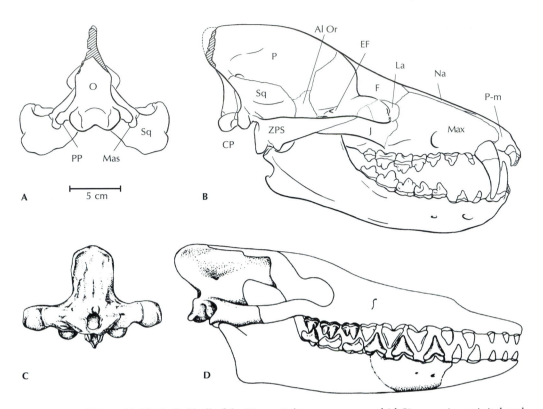

Figure 12.19. **A, B,** Skull of the Upper Paleocene mesonychid *Sinonyx,* in occipital and lateral views, from Zhou et al. (1995). **C, D,** Skull of the whale *Pakicetus* from the latest lower Eocene, in occipital and lateral views (reprinted with permission from *Science,* vol. 220, Gingerich et al.; copyright © 1983, American Association for the Advancement of Science). Abbreviations: Al Or, alisphenoid and orbitosphenoid; CP, condyloid process; EF, ethmoid foramen; F, frontal; J, jugal; La, lacrimal; Mas, mastoid part of petrosal; Max, maxilla; Na, nasal; O, occipital; P, parietal; P-m, premaxilla; PP, paroccipital process; Sq, squamosal; ZPS, zygomatic process of squamosal.

The transition between mesonychids and primitive but obligatorily aquatic whales is represented by a sequence of intermediate animals from the upper portion of the lower Eocene and the lower half of the middle Eocene of Pakistan, continuing into the later middle and upper Eocene of Egypt and southeastern United States (Fig. 12.20). This sequence extends over a period of 10–12 million years, beginning with riverine sediments, including primarily fossils of terrestrial mammals, through shallow coastal marine, to deep neritic deposits at the edge of the continental shelf. Several genera are recognized, showing progressive reduction in the size of the appendicular skeleton, freeing of the tail for aquatic locomotion, and a succession of modifications in the structure of the middle ear.

The oldest fossil now recognized as a whale is *Pakicetus* from the latest early Eocene (see Fig. 12.19C,D). It is known only from cranial remains. The front of the skull is elongate, as in most early whales, and the otic auditory bulla is only weakly attached to the base of the skull. Although the skull and dentition are similar to those of later whales, these remains were found in a riverine deposit together with clearly terrestrial mammals, including rodents, primates, creodonts, artiodactyls,

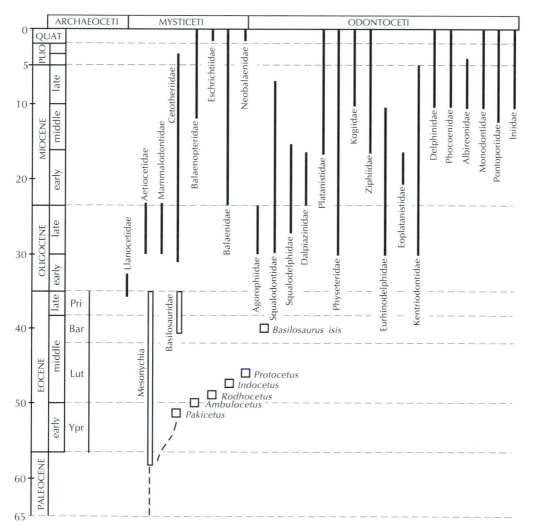

Figure 12.20. Fossil record of whales. Stratigraphic ranges of Oligocene and modern whales from Fordyce & Barnes (1994). Stratigraphic position of Eocene whales from Gingerich et al. (1994). Geological time scale from Harland et al. (1990). Abbreviations of Eocene stage names: Bar, Bartonian; Lut, Lutetian; Pri, Priabonian; Ypr, Ypresian.

perissodactyls, and proboscideans. Gingerich et al. (1983) argued that *Pakicetus* was an amphibious animal that fed in the water. An elongate skull and simple teeth are associated with feeding on fish in many other vertebrate groups. Deciduous as well as adult teeth are found in this deposit, implying that the young were born on land.

Hearing in whales

In the absence of postcranial remains, the strongest evidence for the aquatic nature of *Pakicetus* is the structure of the middle ear. Whales and other vertebrates that have returned to an aquatic way of life face problems of hearing comparable to those of amphibians emerging from the water. Having evolved an air-filled middle ear chamber, a large tympanum, and a series of ossicles to amplify the pressure exerted in the fluid-filled inner ear, they must readjust to the much stronger pressure received from the aquatic medium. The middle ear must now be protected, not only from extremely great external pressures while submerged but also from the radical changes in pressure every time they emerge from the water to breathe air.

One may assume from the rate of modification of the middle ear in early whales – much more rapid than that which occurred in the origin of amphibians – that the necessity to evolve an effective means of hearing underwater in previously terrestrial vertebrates was much greater than the need for Carboniferous tetrapods to detect airborne vibrations. The shape of the skull and the nature of the dentition in early whales indicate feeding on fish. Without a sensory system to supplement vision, this would only be possible in shallow, clear water. Both to protect the middle ear structures and to develop an effective means of directional hearing, many changes occurred.

In primitive placentals, including early ungulates, the tympanic bulla was largely cartilaginous, with only a thin ring of bone, the tympanic, which supported the tympanum. In at least some mesonychids, including *Sinonyx* (Zhou et al. 1995), the entire bulla had become ossified, as it is in all whales. In *Pakicetus* the bulla is much thicker and the bone far denser than that of the rest of the skull. This greater density contributed in three ways to improve hearing in cetaceans: It strengthened the bulla to resist compression more effectively, insulated it acoustically from the rest of the skull better than ordinary bone, and raised the frequency of sound that could be detected.

Detection of the directionality of sound requires that the ears be acoustically isolated from each other so that they can register the slight difference in the time at which sound reaches the two sides of the head. In modern whales, the tympanic bulla is firmly attached to the periotic bone that supports the inner ear. This bony complex is separated from the rest of the skull by extensively vascularized sinuses that also protect the ear from the changes in external pressure.

Except for its greater thickness and density, the auditory bulla in *Pakicetus* does not show evidence for the acoustic isolation necessary for directional detection of sound (Thewissen and Hussain 1993). The bulla is only loosely attached to the surrounding bones of the skull, but there are still four points of articulation, two to the occipital and one each to the squamosal and periotic. There are no openings for the sinuses present in later whales. On the other hand, the ear ossicles were intermediate between those of terrestrial mammals (specifically artiodactyls, which may be the sister-group of whales) and later whales. The chain of ossicles rotated in the origin of whales and *Pakicetus* exhibits an intermediate orientation (Lancaster 1990) (Fig. 12.21). Unfortunately, the ear ossicles are not known in any mesonychids, so it is not possible to determine when this transformation was initiated.

Among later Eocene whales, *Rodhocetus* possessed larger auditory bullae and had evolved a large mandibular foramen and mandibular canal that were absent in *Pakicetus*. In modern odontocetes the mandibular foramen is occupied by a fat pad that transmits vibrations to the middle ear. However, *Rodhocetus* remains primitive in lacking pterygoid fossae that house the accessory air sinuses in more advanced whales. In the upper Eocene *Protocetus* and later whales, connections with the squamosal and occipital are lost.

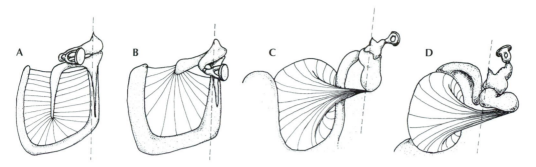

Figure 12.21. Rotation of the ear ossicles from primitive placental mammals to whales. Left side of middle ear, viewed from a posteromedial perspective. The vertical dotted line is the axis of rotation of the malleus–incus system. **A,** Adaptation of the hypothetical ancestral therian middle ear. **B,** the manubrium of the malleus and long process of the incus have shortened and the evolutionary rotation has begun. **C,** The anterior limb of the tympanic bone has become broader and the posterior limb has become less distinct as the tympanic bulla has begun to form. The axis of rotation has been displaced out of the plane of the tympanic membrane. **D,** The sigmoid process is in place as a buttress to the malleus. The intermediate condition is illustrated by *Pakicetus.* The modern condition is achieved in upper middle Eocene whales. From Lancaster (1990).

Locomotion

The earliest evidence of the transformation of the appendicular skeleton for aquatic locomotion is provided by *Ambulocetus,* about two million years later than *Pakicetus* at the boundary between the lower and middle Eocene (Fig. 12.22). Found in a shallow near-shore deposit together with marine molluscs, *Ambulocetus* was approximately the size of a sea lion, with a weight of about 300 kg. The limbs are substantially shorter than those of mesonychids but retain many features of primitive terrestrial ungulates. Overall, the limb structure and posture suggest a pinniped or otterlike mode of locomotion in the water, assisted by dorsoventral flexion of the trunk, with the capacity for limited movement on land.

The triangular shape of the head of the radius indicates that the forearm was fixed in a semipronated posture; otherwise, the elbow, wrists, and digital joints were flexible rather than fixed as in modern whales. The five digits were long and divergent, forming a paddle that was probably used in steering, as in modern whales, rather than propulsion. Thewissen, Hussain, and Arif (1994) suggested that the forelimb had a posture like that seen in sea lions. The lever arms provided by the large olecranon of the ulna and the pisiform of the wrist would have enabled strong locomotion on land, assisted by humeral retraction.

The femur was short, compared with the very long foot that may have contributed substantially to locomotion in the manner of fur seals, with considerable extension and retraction of the vertebral column in the vertical plane. Reflecting their cursorial ancestry, the knee and ankle joints limited movements primarily to the sagittal plane. As in mesonychids, the terminal digits bore hoofs. A single long caudal vertebra suggests that the tail was long and not used in the manner of more advanced whales. The cervical vertebrae were also long. This genus demonstrates that the origin of whale locomotion passed through a stage resembling that of living pinnipeds, with dorsoventral flexion of the trunk to power the rear limbs, before the origin of a tail fluke.

Rodhocetus from the early Lutetian, approximately 2 million years later, represents a slightly more advanced stage in the origin of whales, although with less evidence of the limb skeleton (Gingerich et al. 1994). The femur is only slightly shorter than in *Ambulocetus,* but the discovery of these remains in deep neritic waters suggests that this genus was further committed to a marine way of life. This is supported by the structure of the sacrum, unknown in *Ambulocetus,* in which flexibility is achieved by the loss of fusion of the four sacral vertebrae. This would have permitted greater dorsoventral excursion of the tail, indirectly suggesting that a fluke may have evolved. Nevertheless, the presence of a strong articulation between the last sacral rib and the pelvis and the large size of the pelvic girdle indicates that *Rhodocetus* could still have supported itself on land. The cervical vertebrae are shortened, as in later whales, to restrict the movement of the extremely large skull, but the presence of long spinal processes on the thoracic vertebrae suggests that the head could be lifted above the level of the trunk, as in terrestrial vertebrates. Although it may have been capable of swimming further out to sea than *Ambulocetus,* it probably also remained capable of terrestrial locomotion.

The slightly later *Indocetus,* from shallow marine water in Pakistan, and *Protocetus,* from deep neritic water in Egypt, are less well known. *Indocetus* had long hind limbs, and the sacral centra were solidly fused to one another, but the large size of the proximal end of the tail suggests that it was propelled caudally. *Protocetus* had only a single sacral vertebra, and the sacrum did not articulate directly with the pelvis, indicating that it could not have supported itself on land; the tail, meanwhile, was approaching the configuration of advanced whales.

A gap of 4–5 million years separates the poorly known *Protocetus* from the well-known basilosaurid whales of the later Eocene, common in Egypt and the southeastern United States. The best known of the archaeocete whales, these may be considered as marking the end of the transition from their terrestrial ancestors. The body is extremely elongate, and the nature of the terminal caudal vertebrae indicates the presence of a horizontal fluke; yet Gingerich et al. (1990) have shown that they retain substantial remnants of the hind limbs (Fig. 12.23). The pelvic girdle, although no longer attached to the vertebral column, remains large, as do the femur and patella. The tibia and fibula are shorter and fused to each other, as are the bones of the ankle. Three metatarsals and accompanying phalanges are also retained, some 10 million years after the beginning of the transition. Modern whales have only rodlike vestiges of pelvic bones, femora, and rarely tibiae embedded in the musculature of the body wall. Gingerich and his colleagues suggested that the hind limbs in *Basilosaurus isis* may have served as copulatory guides, which might have led to their retention long after their function in propulsion was lost.

Although the period of transition leading to the emergence of fully aquatic whales may have lasted no more than one-fourth of the time of their subsequent evolution, it was still a substantial period of geological time, during which a host of sequential morphological changes occurred. While the upper Eocene marked the end of this early stage in whale evolution, it was only the beginning of the divergence of the modern whale groups, the odontocetes and mysticetes, which show continuous, significant changes in skull structure throughout the remainder of the Cenozoic (Fordyce and Barnes 1994).

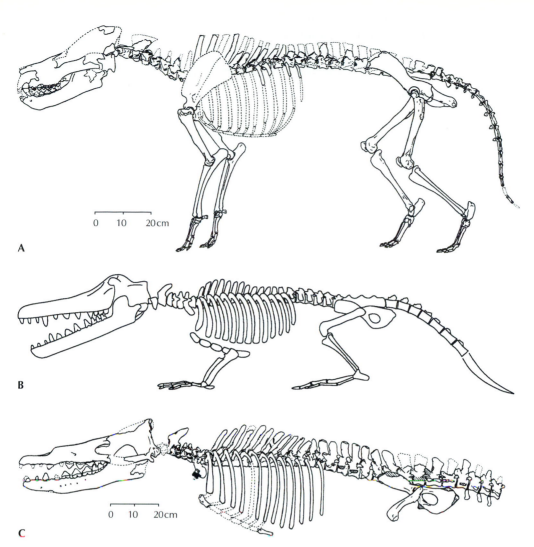

Figure 12.22. **A,** The early Eocene mesonychid *Pachyaena ossifraga* (reprinted from *Nature* [Gingerich et al., vol. 368]; copyright © 1994, Macmillan Magazines Limited). **B,** The earliest middle Eocene whale *Ambulocetus* (from Thewissen et al. 1994). **C,** The early middle Eocene whale *Rodhocetus* (reprinted from *Nature* [Gingerich et al., vol. 368]; copyright © 1994, Macmillan Magazines Limited.

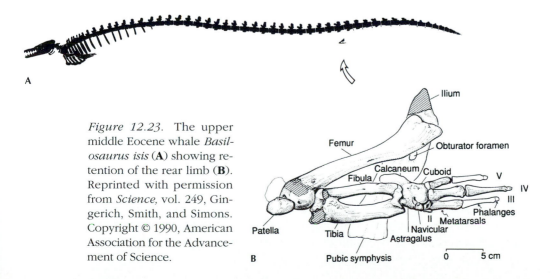

Figure 12.23. The upper middle Eocene whale *Basilosaurus isis* (**A**) showing retention of the rear limb (**B**). Reprinted with permission from *Science,* vol. 249, Gingerich, Smith, and Simons. Copyright © 1990, American Association for the Advancement of Science.

Other major transitions

Two other major transitions – that between primitive amniotes and the ancestors of modern mammals (Kemp 1982; Hopson 1994) and the origin of teleosts from primitive chondrostean fishes (Patterson 1994) – are well documented in the fossil record but more difficult to characterize in a simple manner, since they involved many different adaptive shifts and intervening radiations. Both occurred within a single broad environment, and so lack the obvious physical parameters that governed the direction of evolution seen in the examples chosen. Identification of any particular period as key to these transitions is even more difficult. Other forces that influenced the evolution of these transitions are discussed in Chapter 14.

General features of major transitions

The examples emphasized in this chapter were selected because they are among the best-documented transitions from one major adaptive zone to another. Concentration on changes from one physical environment to another simplified determination of the polarity of character change and the primary direction of selection. It also made it possible to narrow the most critical portion of the transition to a specific period of time.

Gaps in time and the absence of information regarding the transformation of particular structures continue to limit our knowledge of these transitions, but it is no longer necessary to assume that the origin of major new structures and ways of life occurred so rapidly that they are unlikely to be documented in the fossil record. Although the number of divergent lineages recognized during these transitions is certainly less than in either the ancestral or descendant groups, the fossil record of the individual lineages is not necessarily less complete than is that within particular lineages among well-established groups.

Time and rate of change

Focusing on changes in the skeleton associated with locomotion, the critical period of these transformations may have lasted from less than 1 million to between 15 and 20 million years. The shortest period for a major change in limb structure and function may have occurred in the transition between aigialosaurs and mosasaurs, in which there is a short overlap in the range of the ancestral and descendant groups; however, the transition may have begun significantly earlier. Changes in limb structure in the origin of tetrapods may have occurred any time during an approximately 12–15-million-year time span. The major period of limb change in the origin of whales is documented over about 12 million years.

Changes in the wing structure of birds occupied two different periods. There is no specific evidence of the time during which the flight feathers evolved, except that it must have followed the divergence of the lineage leading to *Archaeopteryx* from its closest sister-group among the theropod dinosaurs. This divergence probably occurred at least 15 million years prior to the appearance of *Archaeopteryx*.

Subsequent to the appearance of *Archaeopteryx,* elements of the pectoral girdle and the forelimb underwent a period of essentially continuous modification lasting 9–13 million years.

In all cases, the critical period of these transitions was much shorter than the subsequent history of the descendant groups. The rate of change of key characters during this period is generally more rapid than over equivalent periods in the ancestral or descendant groups, but there is no evidence that it approached the rate of morphological change over short time spans observed in living and late Cenozoic species (Gingerich 1983; Endler 1986).

More important than a higher than average *rate* of evolution during these transitions was the maintenance of a nearly constant *direction* of change. This can be attributed to the strong, unidirectional selective forces involved in changes between major physical environments. Gould (1990) argued that selection pressures are rarely sufficiently constant for long periods to account for long-term directional change in morphological characters; however, this condition was certainly met in all transitions between environments dominated by major differences in physical parameters.

These transitions probably all manifest a threshold effect, but they may also have been influenced by critical changes in particular aspects of the organisms' biology. For example, early in the period of aquatic adaptation in the ancestors of mosasaurs and whales, the limbs maintained most aspects of their forerunners, presumably because the capacity for terrestrial locomotion was necessary for reproduction on land. Subsequently, descendants became so dependent on feeding in the water that selection for effective aquatic locomotion became dominant. The specific time at which the balance shifted may have depended on changes in reproductive behavior.

For some features, the rate of change cannot be judged, either because it happened over a very short period of time or because the particular interval has not been discovered in the fossil record. The structure and arrangement of flight feathers are modern when they first appear in the fossil record, and as yet no intermediates are known between fish fins and tetrapod limbs. Other features can be documented as showing irregular but more or less additive change for tens of millions of years – best seen in the progressive changes in the pectoral girdle and forelimb among Lower Cretaceous birds. This transition is particularly informative in showing the variable rates and times of change in different elements of a single functional complex.

The structure of the furcula, coracoid, and sternum had achieved a nearly modern configuration in the ornithurine birds within less than 15 million years after *Archaeopteryx.* In contrast, among the enantiornithines, the sternum had not reached a modern configuration even after 25 million years. Changes in the carpus and manus did not begin until 12 million years after *Archaeopteryx* and were not complete until at least 10 million years later.

From the standpoint of evolutionary constraints, it is striking that changes in the flight apparatus occurred most rapidly in the proximal bones (the furcula, coracoid, and sternum). In contrast, the carpometacarpus was not significantly altered until tens of millions of years later. There is no reason to think that there would have been less genetic variability in the carpus and manus than in the shoulder girdle, and developmental constraints can probably be ruled out as well. One would expect that developmental changes in the distal portion of the limb could be more readily accommodated than changes in the shoulder girdle since the structure of the limb is established in a proximal–distal sequence. Differential selective pressures provide a much more probable explanation for the different temporal patterns. If there are limits to the capacity of organisms to accommodate a large number of changes in structural and behavioral complexes, it is logical to assume that differing selection pressures will govern the sequence of change.

Amount of change

In all these transitions, we can recognize structural and/or behavioral characteristics in the ancestral lineage that distinguished them from other members of the larger monophyletic group to which they belong. These features evolved, commonly in a stepwise fashion, toward a way of life that enabled their descendants to enter a new environment.

   As transitions have become better known, the relative number of character changes associated with the crucial period are frequently reduced. For example, the early tetrapod *Acanthostega* is now known to retain many characters that were once thought to be restricted to fish, whereas the sarcopterygian *Panderichthys* had already achieved a number of characters previously thought unique to tetrapods. Although a great many changes occurred in the girdles and limbs in the transition between advanced osteolepiform fish and early tetrapods, many features of the skull, vertebral column, and the proximal limb elements did not change significantly during this transition but were significantly modified in either the ancestral or the descendant groups. As the specific relationship of *Archaeopteryx* to theropod dinosaurs has become better established, only a very few skeletal features associated with flight are known to have changed between the two groups. Many more modifications occurred between *Archaeopteryx* and more advanced birds within a short period of time in the Early Cretaceous.

Nature of change

Many of the individual changes that occurred in these transitions are matched quantitatively and/or qualitatively by changes that are known to occur elsewhere in the same monophyletic assemblage. This is especially evident in the origin of mosasaurs, which encompassed many structural complexes that evolved in a similar manner among other anguimorph lizards or other diapsid groups. In some transitions, as between aigialosaurs and mosasaurs and between mesonychids and whales, changes in habitat preceded major changes in limb structure, and did not involve novel developmental or structural changes. Many changes in limb and girdle structure could be attributed to the heterochronic retention of juvenile characters such as small size and incomplete ossification. In contrast, both the origin of feathers in the lineage leading to *Archaeopteryx* and the changes in the distal limb structure of early tetrapods were unique among vertebrates and required major changes in developmental processes and patterns, without which the transitions would not have occurred.

Our knowledge of these transitions remains very limited. An average of approximately 5 million years separates the horizons from which have been discovered intermediates between obligatorily aquatic osteolepiform fish and amphibians. Equivalent intervals separate localities from which have come the succession of Lower Cretaceous birds. A period of 2–3 million years separates the known specimens illustrating the transition between terrestrial mesonychids and whales with extremely reduced rear limbs.

In nearly all cases, individual genera are known from only a single horizon. *Archaeopteryx* is exceptional in being preserved at more than one level in the quarries around Solnhofen. *A. bavarica* may either represent phyletic evolution from *A. lithographica* or a distinct clade within the genus. It provides the only example of evolution at the species level recognized so far in these transitions. There are relatively few examples from the Paleozoic or Mesozoic of successive horizons yielding fossil vertebrates that are less than 2 or 3 million years apart. It may be some time before it is possible to study the patterns and rates of evolution in these transitions consistently at the species level. At present, there is no reason to think that species-level phenomena occurred any differently in these transitions than they do in Late Tertiary and Quaternary populations. The known rates of mutation, the amount of variability in natural populations, and their documented response to selection are sufficient to account for the amount of morphological change that occurred over the time intervals available in these transitions. Initiation of habitat shifts can be attributed to the population-level phenomenon of behavioral change.

Factors unique to these transitions

On the other hand, the evolution of some structures – specifically, feathers in birds and digits in tetrapods – certainly required major changes in the patterns and processes of development that are unique to these transitions and could never have been hypothesized on the basis of the study of modern populations. In the case of the origin of whales, long-term geological changes may have played an important role in altering the environment in which the whales originated. During the Eocene, the Indian subcontinent was approaching Asia. The area in which the early whales lived became an enclosed saline epicontinental sea that may have had unusually high productivity, providing new opportunities for feeding on shallow-water coastal fish, and later in deeper waters enriched by oceanic upwelling (Gingerich et al. 1983). This is another situation in which a major transition was contingent upon factors that would be impossible to extrapolate from those that govern living species.

13 Patterns of radiation

Introduction

Large-scale, long-term radiations are the most conspicuous of all macroevolution-ary phenomena: the radiation of metazoan phyla that began in the Cambrian, that of primitive fish and early vascular plants in the later Paleozoic, of dinosaurs and flowering plants in the Mesozoic, and of placental mammals and birds in the Ceno-zoic (see Figs. 1.2–1.5). The patterns exhibited by all major groups of multicellu-lar organisms show similar features. Vertebrates, the major invertebrate phyla, and vascular plants have all radiated into a comparative small number of major groups, each of which retains a fundamentally similar body plan that is clearly different from that of all other groups. The historical, genetic, and developmental constraints resulting from the common ancestry of each of the groups maintain the integrity of the basic body plan and organ systems regardless of subsequent adaptive change, no matter how major.

Within each physical environment, adaptive modes are reinforced by constraints on anatomy and physiology that are intensified over time. Taxa in aquatic groups become increasingly well adapted for aquatic feeding, buoyancy control, and loco-motion, while nearly all terrestrial lineages have intensified adaptation to that envi-ronment. This type of canalization is clearly evident down to the level of families.

There are also biological constraints that maintain a significant degree of adapta-tional difference between major groups. As long as there are a multiplicity of taxa inhabiting a particular environment, competition at all taxonomic levels will tend to reinforce anatomical differences that facilitate adaptation to different diets, modes of feeding and locomotion, and other aspects of habitat utilization. Togeth-er, these factors maintain the integrity of the anatomy and general adaptive patterns of families and orders for tens and even hundreds of millions of years.

Although it is fairly easy to explain the maintenance of the morphological dis-tinction between clades within major groups once they are established, it is much harder to determine the manner in which these patterns were initially determined. As in the case of major transitions, the first problem is inadequate knowledge of the fossil record. Once established, most vertebrate families and orders are sufficient-ly common and long-lived that they are almost certain to leave enough of a fossil record for the general pattern of their history to be adequately known. In contrast, cladogenesis begins with the subdivision of a single species into two lineages whose chances of preservation are much smaller, as indicated by the plethora of dashed lines or gaps at the base of nearly all taxa at the time of their divergence. Establishing the nature of the initial radiation of major clades is even more difficult than that of major transitions, since one must know the time of origin of not one but a host of descendant groups. It is also more critical to establish the existence of a strictly monophyletic assemblage if a large number of highly divergent lineages are involved. For example, the pattern of radiation of early craniates would be very different depending on whether all known agnathans belong to a single monophyletic group distinct from the gnathostomes, or if known agnathans are a paraphyletic assemblage (Forey and Janvier 1993).

In common with major transitions, the processes of radiation have long been the subject of controversy. Some of the largest-scale radiations appear to have been initiated over particularly short periods of time and involved major adaptive changes, arguing strongly for the influence of phenomena that are not evident among modern populations and species.

To judge from the figures in Chapter 1, nearly all the major groups of multicellular organisms are presumed to have begun with an episode of radiation that occurred over a period significantly shorter than the subsequent history of the group. Unfortunately, this can only rarely be adequately documented, as one must know the time range of both the ancestral and most of the descendant groups within a span of 10–20 million years. If there are larger gaps in the fossil record, it is possible that groups may have undergone not one but a series of successive radiations, or that major lineages have diverged individually over long time intervals, so that each divergence requires a separate explanation. Only a few large-scale radiations are sufficiently well known to distinguish which of these patterns they exhibit.

During the past decade, a great deal of research has been concentrated on establishing phylogenetic relationships among the various vertebrate groups (Prothero and Schoch 1994). Unfortunately, this work does not directly contribute to our understanding of the temporal patterns of major radiations, since it establishes only the relative sequence of divergence rather than the period over which a series of divergences occurred. On the other hand, it does provide a means of assessing the relative degree of completeness of the fossil record. Norell and Novacek (1992a,b) demonstrated that this can be estimated by comparing the degree of congruence between the time of first occurrence of taxa established from the fossil record and the time of divergence expected on the basis of phylogenetic analysis (Fig. 13.1). Comparison of the times of first appearance of all families of vertebrates compiled by Benton (1993) with recent phylogenetic analysis at the level of orders and classes shows that the temporal pattern of radiation is poorly known in most major taxa of Paleozoic and Mesozoic vertebrates.

A few better-known radiations are reviewed in the following sections to determine what temporal patterns they exhibit, and to consider the degree to which they may have been influenced by phenomena that are not evident at the level of populations and species. We begin with the radiation of metazoans in the Early Cambrian, from which evolved the earliest chordates.

The Cambrian explosion

The greatest radiation in the history of multicellular organisms occurred during the Early Cambrian. The first metazoans appeared during the last major subdivision of

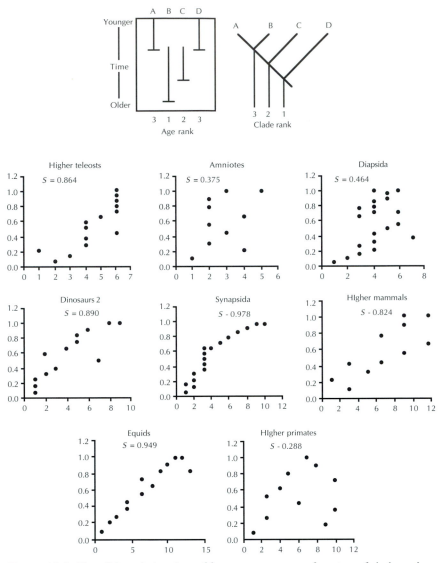

Figure 13.1. Plot of the relative time of first occurrence as a function of clade rank provides a measure of the relative accuracy of the fossil record in determining the pattern of radiation. *Top,* age rank and clade rank are determined from fossil records *(left)* and cladistic phylogenies *(right).* The oldest taxon, B, has the age rank of 1. Taxa A and D appear at the same time and, hence, have equal age ranks. The cladogram *(right)* shows the phylogenetic relationships of these four taxa. Clade ranks are determined by order of branching from the base of the main cladogram axis. *Bottom,* plots for selected groups analyzed by Norell and Novacek (1992a). Age rank is plotted on the *x* axis, and clade rank on *y.* A diagonal distribution of points indicates complete concordance. Values of S = Spearman rank correlation are statistically significant at $P < 0.01$ for higher teleosts, synapsids, higher mammals, and equids; at $P < 0.05$ for diapsids and dinosaurs. Correlations are not statistically significant at these levels for amniotes or higher primates. Reprinted with permission from *Science,* vol. 255, Norell and Novacek. Copyright © 1992, American Association for the Advancement of Science.

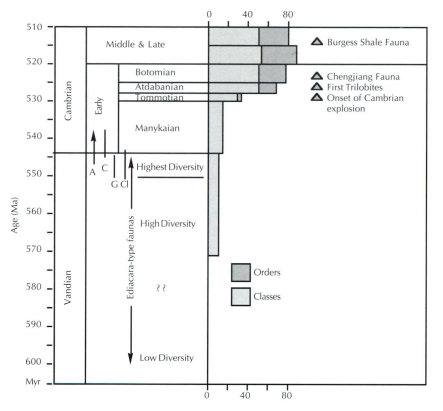

Figure 13.2. Revised time scale for the Vendian and Cambrian periods. A, C, G, & Cl, indicate the ranges of Vendian age skeletalized fossils. Data for this chart are combined from Grotzinger et al. (reprinted with permission from *Science,* vol. 270, Grotzinger et al. Copyright © 1995, American Association for the Advancement of Science) and Valentine, Erwin, and Jablonski (1996).

the Precambrian, the Vendian (Fig. 13.2). Their remains are common and easily recognized in deposits 565 million years old, but more primitive forms appeared as long as 600 mya (Grotzinger et al. 1995). By 525 mya, all of the modern phyla had differentiated.

Most animals from the Vendian were without mineralized skeletons and were preserved as impressions, or are known from tracks and trails. These are collectively referred to as the Ediacaran Faunas. Many of the impressions represent animals with a medusoid or frondlike body form, which are assumed to have had only two principal tissue layers, in common with modern corals, hydras, and jellyfish; they are commonly attributed to the Cnidaria. Others superficially resemble arthropods, annelids, and echinoderms, although they lack specific synapomorphies of these phyla.

Tracks and trails from the Late Vendian indicate the presence of bilaterally symmetrical, wormlike animals that were round in cross section and large enough to require a circulatory system. The discovery of fecal pellets reflects a digestive system with both mouth and anus. The complexity of the tracks and burrows indicate that they had well-developed body muscles and a fairly sophisticated nervous system to integrate sensory input and control motor reactions. These characteristics indicate the presence of animals late in the Precambrian that were more advanced than modern flatworms.

By the very end of the Vendian, a few small animals had evolved the capacity to deposit mineralized skeletons. These cannot be attributed to any of the modern metazoan phyla but are collectively referred to as the *small shelly fauna*. They became much more common and varied in the Manykaian, at the base of the Cambrian, 544 mya. The first fossils unequivocally exhibiting characteristics of modern protostome and deuterostome phyla, including molluscs, brachiopods, and echinoderms, appeared in the Tommotian, approximately 530 mya. Within another 5 million years, all of the major modern phyla, with the exception of the bryozoans, are recognized in the fossil record. Primitive relatives of the vertebrates, the hemichordate *Yunnanozoon* and the cephalochordate *Cathaymyris,* have been described from the end of the Atdabanian, 525 mya (Shu, Conway Morris, and Zhang 1996; Shu, Zhang, and Chen 1996). Both come from the Chengjiang fauna, which preserves a host of soft-bodied animals.

The term *Cambrian explosion* has been used in reference to the extremely rapid appearance of nearly all the metazoan phyla between the beginning of the Tommotian, 530 mya, and the end of the Atdabanian, 5 million years later (Bowring et al. 1993). By the end of this time there were more than forty metazoan classes and approximately seventy orders. However, Fortey, Briggs, and Wills (1996) argued that the extremely rapid achievement of this level of taxonomic diversity may be exaggerated by the incompleteness of the fossil record. They pointed out that the first appearance of several phyla – arthropods, molluscs, brachiopods, and echinoderms – include representatives of several evolutionary levels that must have diverged from one another over an appreciable period of time. Earlier members of these groups may not have been found either because they lacked preservable hard parts or because they were so small that they are unlikely to be discovered by normal collecting techniques. Evidence from later horizons indicate that the primitive sister-groups of several phyla were extremely small animals, living between the grains of the sediments. Primitive trilobites and brachiopods are known to have lacked mineralized exoskeletons. Among the earliest known trilobites, particular genera are known from different regions, indicating that their ancestors had undergone extensive geographical distribution and significant morphological change prior to their appearance in the fossil record.

Until earlier fossils are found, it will not be possible to determine the actual period of time during which the modern phyla diverged from one another. Valentine, Erwin, and Jablonski (1996) and Knoll (1996b) argued that divergence probably began approximately 565 mya and lasted at least 40 million years. Fortey et al. (1996) emphasized that the Cambrian explosion actually involved a succession of events. The radiation of the metazoan phyla is typically associated with the appearance of many divergent body plans, but this must have been preceded by cladogenetic events separating lineages that were initially very similar to one another and lacking any of the attributes that distinguish the early fossil representatives of the different phyla.

Knowledge of the fossil record is currently inadequate to establish either the time of cladogenesis or the length of time involved in the establishment of the derived body plans. Valentine et al. (1996) provided three models of how these pro-

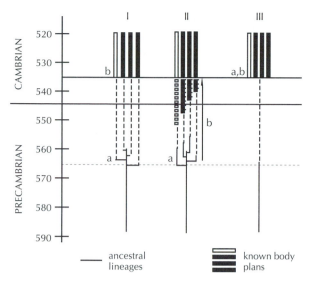

Figure 13.3. Models of the sequence of evolutionary events surrounding the Cambrian explosion. The events include: (a) the protostome–deuterostome split and divergence of the major lineages and (b) the origin of body plans. *Model I:* The lineages that eventually lead to higher metazoan phyla branch early, but a burst of body-plan diversification occurs near 530 mya. Genomes achieve complexity early, but undergo possible secondary expansion near 530 mya. *Model II:* Metazoan lineages, genomes, and body plans branch, diversify, and increase in complexity progressively during the Vendian and Manykaian, achieving a threshold in developmental controls that permits a final burst of advanced body-plan diversification near 530 mya. *Model III:* Higher metazoan lineages, genomes, and body plans all diversify explosively near 530 mya. From Valentine et al. (1996).

cesses may have been related to one another in time (Fig. 13.3). In the first model, cladogenesis began approximately 565 mya, but the burst of body-plan diversification did not occur until just prior to 530 mya, when the first body fossils of the modern phyla appear in the fossil record. In the second model, cladogenesis also began 565 mya, but body-plan diversification occurred at different times in the various phyla between the Late Vendian and their appearance in the fossil record. The third model shows nearly simultaneous and instantaneous occurrence of cladogenesis and evolution of diverse body plans in all phyla 530 mya. The authors strongly favored the second model.

The presence of sophisticated tracks and burrows by the end of the Vendian indicates that at least some advanced phyla had appeared by that time, and hence that the divergence of the major phyla had begun. The Manykaian may have been a time of progressive attainment of the basic body plans of all the modern phyla, although none yet produced mineralized skeletons that can be identified as characteristic of their later Cambrian descendants. The Tommotian and Atdabanian marked the final attainment of the characteristics of the diverse phyla, through the capacity to form a mineralized skeleton and achieve much larger body sizes than their tiny sister-groups that lived among the sediment particles.

The fact that major changes were occurring during the Tommotian and the Atdabanian is demonstrated by the stepwise rather than instantaneous appearance of the major phyla. While the molluscs, brachiopods, and echinoderms all appeared in the Tommotian, the far more diverse arthropods are not known until the following Atdabanian. The absence of a fossil record of arthropods during the Tommotian can be explained by the delay in attainment of larger size and the capacity to mineralize their skeletons until the end of this stage.

By analogy with the patterns of major transitions discussed in Chapter 12, the critical time of metazoan radiation may have been preceded by a long period during which the definitive features of each phylum evolved. This was followed by a much shorter episode during which these features were suddenly manifest in what appears as an explosive radiation. Although the initial stages in the evolution of the metazoan level of organization and the preliminary phylogenetic divergence of the major taxa may have taken place gradually over nearly 60 million years of the Vendian, no more than 20 million years of the early Cambrian were required for the formation of essentially modern body plans and the radiation of most of the classes and orders that were to dominate the remainder of the Paleozoic and give rise to all phyla present in the modern biota. By any measure this was the most rapid and large-scale radiation and differentiation of unique body plans in the history of metazoans.

Despite the absence of fossil evidence of the time of lineage divergence and the mode of attainment of the individual body plans, the relationships of the many phyla, and hence the sequence of their divergence, can be established from the anatomy and molecular systematics of their modern representatives (Fig. 13.4). These show a succession of progressively more complex phyla, beginning with sponges and proceeding through Cnidaria and Platyhelminthes (flatworms) to a major divergence between protostomes and deuterostomes. The lack of resolution among most of the protostome phyla may reflect their original divergence over a very short interval of time.

What factors may have been responsible for this unique radiation? Knoll (1996a,b) emphasized the importance of the increase in atmospheric oxygen toward the end of the Precambrian. Until this time, there may not have been enough oxygen available for the metabolic requirements of complex metazoans, or for the deposition of mineralized skeletons. These factors may have limited Ediacaran genera to very simple body plans and high surface-to-volume ratios, and restricted the size of animals with more complex body plans until the Tommotian. Two other aspects of the physical environment that may have influenced the origin of the metazoan phyla were the end of an ice age that had affected the world's climate in the Early Vendian and a major change in the configuration of the continents, which had been massed together in the late Precambrian but began to split apart in the early Phanerozoic.

However, the major factor enabling the metazoan phyla to radiate was the evolution of the capacity for cell differentiation and control of development. The most primitive cellular aggregates include only a single cell type; what distinguishes true metazoans is the differentiation of cells and tissues with a variety of structural and functional properties, and their assembly into a functional organism. Genetic control of specific aspects of cell differentiation, growth, movement, adhesion, and assembly into particular structures must have begun to evolve prior to the radiation of the individual phyla, early in the Vendian.

The temporal and anatomical integration of all these processes is governed by

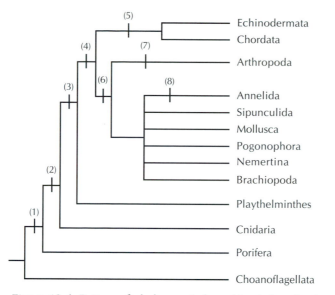

Figure 13.4. Pattern of phylogenetic branching inferred to have led to selected metazoan phyla, based on 18S rRNA trees. Branch lengths are not scaled to molecular distance measures or to geological time. From Valentine (1995). Reprinted with permission, *Tempo and Mode in Evolution.* Copyright © 1995 by the National Academy of Sciences. Courtesy of the National Academy Press, Washington, D.C.

another set of genes: the *Hox* genes emphasized in Chapter 10. It was the evolution of the *Hox* genes that provided the level of developmental organization necessary for the evolution of all the advanced metazoan phyla. *Hox* genes are first recognized in the cnidarians, which are among the earliest of true metazoans to appear in the fossil record. Development in more complex animals is controlled by a larger number of *Hox* genes, which evolved by the duplication and subsequent modification of genes present in the early cnidarians. *Hox* genes themselves are not responsible for forming the specific structures that distinguish more complex phyla, but there is a close correlation between the number of *Hox* genes and the degree of complexity, since their presence is necessary to control where and when tissues and structures develop.

The next level of tissue complexity above the cnidarians was achieved by the flatworms. Since the tracks and burrows seen in the Upper Vendian were made by animals more complex than flatworms, we can assume that some of the late Precambrian metazoans were already approaching the level of higher phyla such as arthropods and chordates, which are characterized by an increased number of *Hox* genes. Valentine et al. (1996) argued that the immediate common ancestors of the protostomes and deuterostomes had at least six *Hox* genes, in contrast with only three in cnidarians and flatworms.

The number and nature of the *Hox* genes did not themselves govern the nature of the changes that occurred in each phylum, but provided the capacity to reach a certain level of complexity. Early chordates and early arthropods had approximately the same number of *Hox* genes but different patterns of development and very distinct adult structure. Both groups evolved paired limbs that are controlled by similar genes, but the limbs evolved separately in both groups from limbless ancestors. The conservative nature of the *Hox* genes themselves is demonstrated by the fact that genes from insects can control comparable aspects of development in mice or chicken, and vice versa.

However, the capacity to evolve a great variety of complex body plans was only one side of the story involving the massive radiation of metazoans in the early Cambrian. The other was the nature of the biological environment. In contrast with the modern world, in which every new species must compete with many of the thousands of previously evolved species with which it interacts, the world of the Lower Cambrian was initially almost unoccupied by other animal life. Multicellular plants may have evolved more than a billion years earlier, and unicellular organisms 3 billion years before; but the entire spectrum of adaptive niches that could be occupied by metazoans was empty. The sponges and cnidarians were already engaged in sessile filter feeding, but the entire range of feeding habits for which mobility was necessary was open for adaptation.

Briefly, it was a non-Malthusian world where population growth had extremely wide limits and neither predation nor competition was a serious problem at the individual level. Structural and behavioral variability among individuals and species could express their full potential in adapting to the widest possible adaptive niches. Initially such variability may have extended to major structural elements and even body plans, to judge by the range of anatomical patterns seen in the Burgess Shale fauna from the Middle Cambrian.

Conditions rapidly changed, however, as the metazoans themselves quickly occupied nearly all of the adaptive zones available to marine organisms. By the Middle Cambrian, marine arthropods already exhibited essentially the same degree of anatomical and adaptive diversity attained by their living counterparts (Briggs, Fortey, and Wills 1992, 1993; Foote and Gould 1992). This level was achieved extremely quickly, to judge by the similar diversity of forms in the extensive assemblages of soft-bodied organisms present in the Lower Cambrian Chengjiang fauna (525 mya) and the Middle Cambrian Burgess Shale fauna, only 10 million years later. This can be attributed to both an increase in developmental control, as suggested by McNamara (1986) to explain progressive decline in variability in Late Cambrian trilobites, and as a direct result of increased specificity of habitat occupation.

By the end of the Lower Cambrian, all the major phyla had appeared, as had most classes among the marine groups. No new phyla have appeared in the succeeding 500 million years. (Groups as distinct as flying insects and terrestrial vertebrates evolved later, but these are not designated as new phyla since they retained the basic body plans of phyla that had diverged in the Cambrian.) Never again did life radiate into as many different adaptive patterns, even after the most catastrophic extinctions.

The Cambrian radiation was clearly a unique event in the history of life. It can be attributed to the combination of at least three major phenomena that were themselves unique: a substantial increase in the amount of atmospheric oxygen, the elaboration of *Hox* and other genes that enabled the development of complex organisms, and an Earth nearly devoid of other organisms with a comparable level of complexity. Knoll (1996b, p. 6) summarized the Ediacaran–Cambrian diversification of animals as reflecting "the interaction of genetic possibility with environmental opportunity."

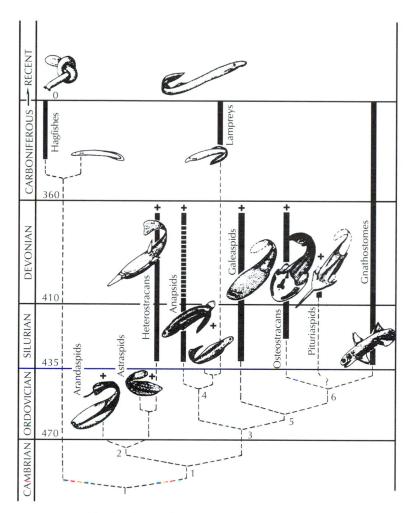

Figure 13.5. The fossil record of primitive craniates superposed on a putative phylogeny, indicating the inadequacy of the fossil record to establish the temporal pattern of their radiation. Reprinted from *Nature* (Forey and Janvier, vol. 361). Copyright © 1993, Macmillan Magazines Limited.

Radiations among primitively aquatic vertebrates

Radiations appear to have followed the achievement of each of the major anatomical advances in vertebrate evolution, but few are known to have occurred as quickly, or can be so clearly characterized from the fossil record, as the Lower Cambrian radiation of metazoans (Carroll 1987). Detailed studies of many such radiations are hampered by the incompleteness of the fossil record – in most cases because of gaps in the sedimentary record, but in others because of the inherently limited capacity for preserving relevant fossils.

Early stages in the radiation of craniates are obscured by the fact that many major lineages apparently went through a long period in which they were unable to form a bony skeleton. When they first appeared in the fossil record, each of these lineages was very distinct from the other, particularly in the form of the exoskeleton, and there is little evidence of the time at which they diverged from one another (Fig. 13.5). The absence of consolidated skeletons may also obscure our knowledge of the time at which jaws evolved and the major groups of jawed vertebrates first radiated.

The absence of ossified tissue or even calcified cartilage in the early history of Chondrichthyes precludes analysis of the pattern of their initial radiation from perhaps as early as the Late Ordovician until the Late Devonian, when a plethora of highly distinct lineages appears in the fossil record. Inadequacies of the fossil record also preclude specifying the pattern or precise timing of the radiation of the modern neoselachians between their first occurrence in the fossil record in the Late Triassic and the appearance of most of the modern taxa in the Late Jurassic, an interval of approximately 70 million years.

The early radiation of ray-finned fish is obscured by the almost total absence of fossils until the end of the Devonian. A number of divergent lineages appear throughout the Carboniferous and Permian, but their specific interrelationships and actual time of divergence have not yet been determined. Other late Paleozoic and Mesozoic groups can be recognized on the basis of progressive achievement of more derived features of the feeding and locomotor apparatus in the lineage leading to advanced teleosts.

It is only among teleosts that a very extensive radiation from a particular anatomical pattern can be adequately documented as occurring over a relatively short period of time: This was the radiation of the percomorphs and, more specifically, the beryciforms and their derivatives. Within this assemblage, the Perciformes alone include nearly eight thousand species grouped in 150 families of the modern fauna. This assemblage is first known in the Cenomanian, approximately 97 mya, represented by three families. Six families appeared later in the Cretaceous and eleven in the Paleocene. The bulk of the new families, eighty, appeared in the first 14.5 million years of the Eocene; a smaller number of families appeared at irregular intervals during the remainder of the Cenozoic. Although the early history of this assemblage extended over more than 50 million years, most of the new appearances of families occurred during a 15-million-year interval.

Paleozoic and Mesozoic tetrapods

Sarcopterygian fish diverged, in an as-yet-undetermined pattern, into several major lineages during the Devonian, with the choanates appearing by the base of the Middle Devonian. The initial radiation of tetrapods occurred between the Upper Devonian and the middle of the Carboniferous. Very few fossils, representing only two or three lineages, are known from the first 25 million years of the Lower Carboniferous. Nine major lineages appear between the later Lower Carboniferous and the end of the Upper Carboniferous. Their first appearances are scattered over a period of approximately 50 million years, in a pattern that does not accord with the probable sequence of phylogenetic divergence (Carroll 1992).

No subsequent stage in the differentiation of nonamniote tetrapods is sufficiently well known to describe the temporal pattern of radiation within intervals of less than 20 million years. There seems to have been a succession of radiations among assemblages with common anatomical patterns, including the Paleozoic "labyrinthodonts" and "lepospondyls," the early Mesozoic "stereospondyls," and the Mesozoic and Cenozoic "lissamphibians," but none of these groups can be established as being monophyletic. In all cases, there is a broad range of times during which individual lineages differentiated.

The earliest known amniotes appeared approximately 311 mya. Two major lineages are represented at this time, one of which eventually gave rise to mammals and the other to all groups commonly referred to as reptiles as well as birds. Current evidence suggests that amniotes did not radiate rapidly or extensively early in their history; rather, they gradually differentiated and progressively replaced the archaic amphibian groups, first on land and then in the water, during the Upper Carboniferous, Permian, and Triassic. By the end of the Triassic, the diapsids dominated nearly all terrestrial habitats, and many in the marine environment.

Several of the major diapsid clades appear to have diverged in a series of simple dichotomies or stepwise radiations, none of which can be closely constrained in time (see Fig. 10.19). Fossil records of crocodiles and dinosaurs appear to show a fairly strong concordance with their postulated phylogenetic relationships (Figs. 13.6, 13.7). Crocodiles show a continuous proliferation of new families throughout their history, including fairly highly specialized marine forms, clearly terrestrial predators, and a host of amphibious groups. Within the Upper Triassic, dinosaurs diverged into three major groups: the bipedal carnivorous theropods, the larger, primarily quadrupedal sauropodomorphs, and the primitively bipedal, herbivorous ornithischians. Each of these major groups gave rise to a succession of divergent families throughout the Jurassic and Cretaceous, within which smaller-scale radiations occurred.

There is some evidence for more rapid radiation among secondarily aquatic groups, including nothosaurs, ichthyosaurs, and mosasaurs, all of which are represented by many lineages soon after their first occurrence. In the former groups, the time of their divergence from terrestrial ancestors is not known, and so the actual rate of radiation cannot be established. The transition between aigialosaurs and mosasaurs and the initial radiation of mosasaurs may have occupied no more than 3 million years, by which time some sixteen genera in three families had diverged. These lineages were to dominate the marine environment for the next 23 million years. The fossil record of pterosaurs does not reveal a period of extensive early radiation, but probably more closely reflects the irregular occurrence of sediments appropriate to their preservation than it does their actual history.

The history of synapsids, leading toward mammals, appears to show a pattern of successive periods of major divergence, following a series of structural and physiological transitions – but again, few are sufficiently well represented in the fossil record for detailed temporal analysis. The largest number of first occurrences of families is associated with the radiation of therapsids in the Late Permian and Early Triassic. One specimen of a therapsid of uncertain phylogenetic position, *Tetraceratops* (Laurin and Reisz 1996), appears in the Lower Permian, approximately 265 mya. During a 15-million-year period in the Upper Permian and Lower Triassic, fifty-one therapsid families appear; but these may have begun to differentiate at any time in the preceding 35 million years, from putative ancestors in the Upper Carboniferous.

For most groups of terrestrial vertebrates, from the Late Paleozoic through the Mesozoic, the pattern of their first occurrences is more a reflection of our incomplete knowledge of the fossil record than a reliable indication of the temporal pattern of their radiations.

Early Cenozoic mammals

Placentals

The most dramatic radiations that can be accurately documented among terrestrial vertebrates involve replacement after mass extinction. The best known is the radiation of placental mammals in the early Cenozoic, following the sudden extinction of dinosaurs and the decimation of marine species. Placentals are first recognized in the Early Cretaceous, but only at the very end of the Mesozoic are any of

Figure 13.6. Ranges of crocodylomorph families suggestive of repeated divergence throughout the Mesozoic and Cenozoic. Family names: 1, Saltoposuchidae; 2, Sphenosuchidae; 3, Protosuchidae; 4, Orthosuchidae; 5, Teleosauridae; 6, Metriorhynchidae; 7, Crocodileimidae; 8, Notosuchidae; 9, Comahuesuchidae; 10, Baurosuchidae; 11, Libycosuchidae; 12, Sebecidae; 13, Peirosauridae; 14, Atopsauridae; 15, Dyrosauridae; 16, Trematochampsidae; 17, Hsisosuchidae; 18, Goniopholididae; 19, Pholidosauridae; 20, Paralligatoridae; 21, Bernissartiidae; 22, Brillanceausuchidae; 23, Hylaeochampsidae; 24, Stomatosuchidae; 25, Gavialidae; 26, Thoracosauridae; 27, Dolichochampsidae; 28, unnamed; 29, Crocodylidae; 30, Pristichampsidae; 31, Alligatoridae; 32, Nettosuchidae. Modified from Fig. 39.4 in *The Fossil Record 2* by Benton, M. J., Chapman & Hall, 1993.

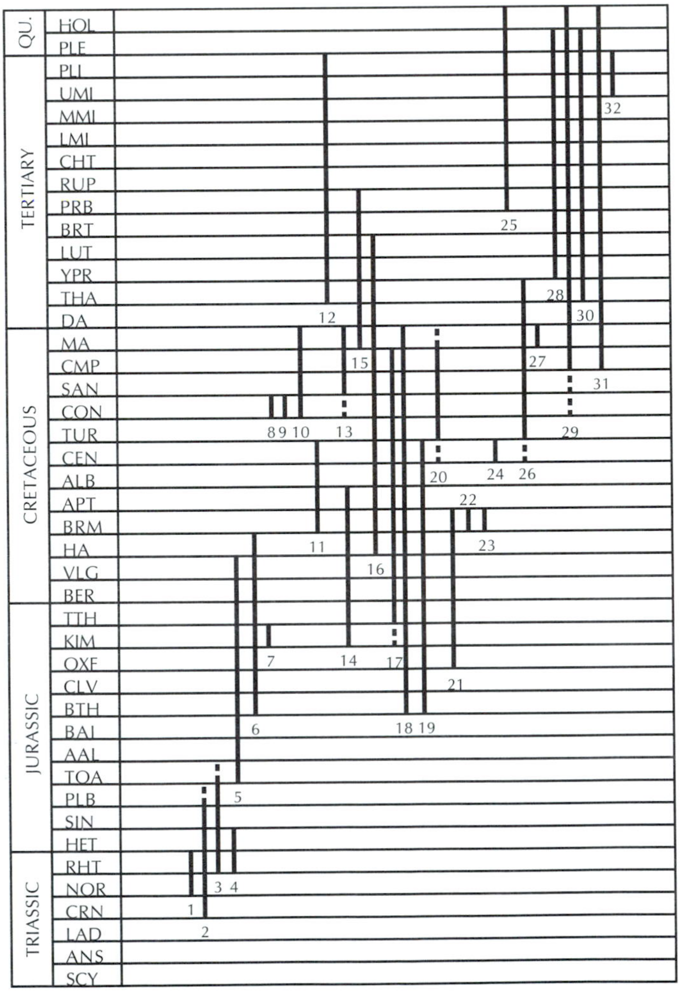

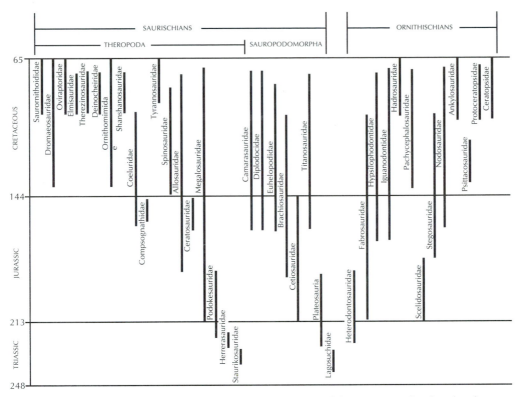

Figure 13.7. Stratigraphic ranges of the families of dinosaurs. Each suborder shows a pattern of sequential radiation at the family level. From Carroll (1987).

the modern orders known from the fossil record. By the beginning of the Eocene, only 10 million years later, the ancestral lineages of approximately twenty-five orders had diverged (Fig. 13.8).

The early radiation of placental mammals may have begun nearly simultaneously throughout the continents of the Northern Hemisphere. However, the only place where a continuous, well-studied stratigraphic sequence spans the uppermost Cretaceous and the lowermost Tertiary (referred to as the *K–T boundary*) is in western North America. This sequence was studied by Archibald (1983) and further discussed in several subsequent papers (Archibald et al. 1987; Archibald and Lofgren 1990; Archibald 1993, 1994).

Archibald investigated a series of faunas from northeastern Montana covering an approximately half-million-year period. The earliest of them, Hell Creek, is latest Cretaceous (Lancian); the remainder are from the earliest Paleocene (Puercan). Later faunas from higher in the Puercan and in the succeeding land mammal age, the Torrejonian, were used as a basis for comparison. The fossils from the earliest horizons consist almost entirely of tiny, disarticulated bones and teeth, but they document an exceptionally rapid increase in the number of species and the amount of adaptational diversity.

Unfortunately, no more recent analysis of the rates of evolution has been published since Archibald's 1983 paper, although many additional species have been recognized, and the age of the individual beds has been extensively revised. Dan

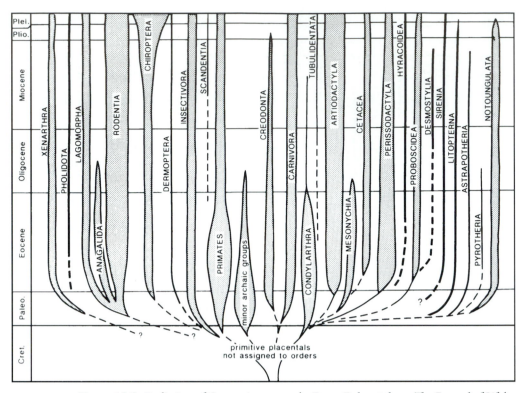

Figure 13.8. Radiation of Cenozoic mammals. From *Paleontology: The Record of Life* by Stearn and Carroll. Copyright © 1989. Reprinted by permission of John Wiley & Sons, Inc.

Riskin's (pers. commun. [1996]) unpublished review of subsequent literature incorporates this new data to provide a revised analysis of the first half a million years of placental evolution in the early Paleocene of North America.

Based on recent correlations, the Hell Creek Formation lies within about 0.1 million years of the K–T boundary. It is followed by three horizons in the earliest Paleocene, termed bk (for Bug Creek) 1, 2, and 3, which occur at approximately 0.1-million-year intervals. All the Bug Creek localities are included in the lower portion of the Puercan Land Mammal Age, designated Pu0. The next biostratigraphic interval, which lasted about 0.2 million years, is termed Pu1; Pu2 ends at about 1.1 million years into the Paleocene, and Pu3 at 1.3 million years. Only the horizons from the Hell Creek through Pu1 are sufficiently limited in their duration and close enough to the K–T boundary to document the patterns and rates of evolution at the very base of the placental radiation.

In contrast with the marsupials, which show a significant drop in diversity across the K–T boundary, the species richness of placental mammals more than doubles between the Hell Creek and bk1, from six to fourteen. The number remains constant from bk1 to bk2, but then increases at a progressively more rapid rate to nineteen in bk3, forty-one in Pu1, and forty-eight in Pu2. By the end of the Paleocene, the number of genera reached a plateau that is maintained for the remainder of the Cenozoic (Tables 13.1, 13.2).

The rate of appearance of new species increased steadily from the K–T bound-

Table 13.1. *Species number and rate of turnover during the latest Cretaceous and early Paleocene*

	HCF	bk1	bk2	bk3	Pu1	Pu2	Pu3	To1
Duration of stratigraphic intervals (in 10^6 yr)	0.1	0.1	0.1	0.1	0.2	0.6	0.2	1.5
Total no. of species	6	14	14	19	41	48	32	30
No. of first appearances (per 10^4 yr)		0.1	0.3	0.7	1.5	0.6	1.3	0.1
No. of disappearances (per 10^4 yr)		0.3	0.2	0.9	1.7	0.6	1.2	0.1

Abbreviations: HCF, Hell Creek Formation; bk, Bug Creek; Pu, Puercan Land Mammal Age; To, Torrejonian.
Source: Dan Riskin (pers. commun.).

Table 13.2. *Number of species and percentage (in parentheses) of total fauna for major groups of placental mammals in the early Paleocene*

	Lan	bk1	bk2	bk3	Pu1	Pu2	Pu3	To1
Arctocyonidae	4 (29)	4 (29)	6 (43)	11 (58)	21 (51)	18 (38)	12 (38)	9 (30)
Carnivora						1 (2)		1 (3)
Cimolesta	7 (50)	6 (43)	4 (29)	4 (21)	4 (10)	3 (6)	5 (16)	3 (10)
Dermoptera							1 (3)	1 (3)
Hyopsodontidae					1 (2)	1 (2)	1 (3)	3 (10)
Insectivora							1 (3)	1 (3)
Leptictida	2 (14)	3 (21)	2 (14)	1 (5)	2 (5)	1 (2)	2 (6)	
Mioclaenidae					1 (2)	7 (15)	1 (3)	2 (7)
Pantolesta							1 (3)	2 (7)
Periptychidae	1 (7)	1 (7)	2 (14)	2 (11)	11 (27)	14 (29)	7 (22)	5 (17)
Primates					1 (5)	1 (2)		1 (3)
Taeniodontia						3 (6)		2 (7)
Total	14	14	14	19	41	48	32	30

Abbreviations: Lan, Lancian; remainder as in Table 13.1.
Source: Dan Riskin (pers. commun.).

ary through Pu1, from 0.1 to 1.5 new species per ten thousand years. With only very incomplete fossil remains, it is not possible to differentiate whether individual "new" species resulted from anagenetic change within lineages or from cladogenesis; but the progressive increase in the total number of morphologically defined species demonstrates that cladogenesis and low extinction rate must have been important factors throughout this sequence. In contrast, immigration does not appear to have been significant in western North America during this time. According to Woodburne and Swisher (1995), no intercontinental immigrations occurred from the K–T event to the end of the Torrejonian. They argued that only "background" migration, involving four or fewer species, occurred during this interval. This would not have had a significant effect on the amount of species increase in the areas represented by these deposits; rather, the total increase in species may be attributed almost entirely to local adaptive radiation.

Stanley (1979) calculated the average rate of species increase over time in seven families of late Cenozoic mammals as 0.22 per million years, while that of early Paleocene placentals was 0.36 per million years – 75 percent more rapid. Archibald (1983) calculated this rate for a single group of early Paleocene mammals, the condylarths that make up 50 percent of the fauna, and found the average rate to be 1.16 per million years – more than four times as fast. Riskin's figures suggest an even more rapid rate over shorter time intervals.

The increase in diversity of the fauna can be associated with shorter species longevity than has been recorded for most intervals in the Tertiary. Stanley had calculated an average species duration of 1 million years for later Cenozoic mammals, with only 1 percent having durations of less than 350,000 years. In contrast, Archibald determined that at least 30 percent of the early Paleocene species had durations shorter than 350,000 years. Riskin calculated a rapid decline in species longevity during the earliest Paleocene, reaching a low in the Pu2, followed by a gradual increase into the Torrejonian.

The great increase in the total number of mammalian species and genera in the lower Paleocene is unique for the Cenozoic. From the end of the Paleocene to the present, the number of genera of North American placental mammals has remained nearly constant, at approximately 110, despite continuing short-term fluctuation (Alroy in press-b). The extremely rapid increase in the number of species during the first million years of the Paleocene can certainly be attributed to the great increase in the number of available adaptive zones following the extinction of the dinosaurs. Dinosaurs had dominated not only a wide spectrum of carnivorous and herbivorous feeding habits but also a broad range of sizes, if juvenile individuals are included.

Even more important than the increase in the number of placental species in the early Cenozoic was their rapid divergence into a diversity of adaptive zones, as indicated by the nature of their dentition and their body size. This is recognized by their inclusion in an increasing number of higher taxonomic groups (Fig. 13.9). Nearly all placentals living during the last stage of the Cretaceous have been included in a single assemblage referred to informally as "proteutherians." Most were of small size and presumedly insectivorous habits. Within 1 million years, their descendants had diverged sufficiently to be recognized as the ancestors of modern carnivores, primates, two archaic herbivore groups (the Taeniodonta and the Pantodonta), as well as a host of more derived herbivores collectively termed the Condylarthra that included the ancestors of horses, cows, elephants, whales, and several extinct orders of South American ungulates. By the end of the Paleocene (about 10 million years after the extinction of the dinosaurs), all the modern placental orders had become clearly differentiated.

Although the number of distinct habitats and ways of life represented by the many orders that evolved by the end of the Paleocene is striking, the early members of these lineages remained small and little differentiated from one another. Their remains are not well enough known for their differences to be quantified in terms of darwins or other rates of morphological change, but their similarity can be judged by the fact that the early members of the lineages that gave rise to orders as

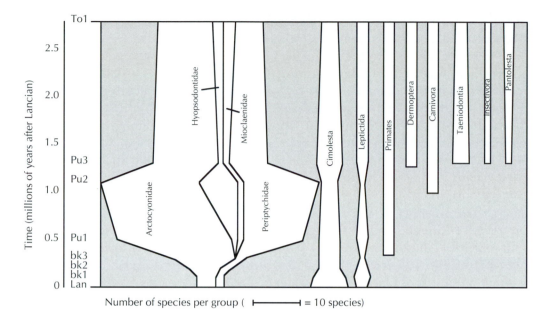

Figure 13.9. Relative species richness over time of major taxa of placental mammals in the lower Paleocene. (From Dan Riskin, pers. commun.)

diverse as the cetaceans, proboscidians, artiodactyls, and perissodactyls, plus many South American ungulates, were once all placed in a single order, the Condylarthra. A close analogy can be made between the very early stages in the differentiation of the placental orders and that of the cichlids in the East African Great Lakes.

Looking back from the perspective of the modern mammalian fauna, the latest Cretaceous and first 1.5 million years of the Cenozoic can be seen as a time of extremely rapid divergence of all the major placental lineages. On the other hand, if we look at the rates of species turnover and amount of morphological change within any particular lineage, their magnitude is not greater than those observed within some individual genera in the later Tertiary and Quaternary. For example, although the overall appearance rate of about 1.4 species per 10^4 years in the Lower Paleocene calculated by Archibald (1983) is faster than the average for Cenozoic mammals, it is equaled during the early Eocene, when there was a large-scale immigration of mammals from Eurasia coincident with the extinction of many archaic local lineages. What *is* unique about the earliest Cenozoic is that a great number of lineages were changing at the same time, and continued to change in the same general direction for millions of years thereafter. The rate of change at any one time can be accounted for by selection coefficients that are no more stringent than in the later Cenozoic, but they were exceptional in being much more nearly constant in their direction and intensity.

Constancy in the direction of selection was also seen in the evolution of particular traits that were modified over spans of several million years in lineages adapting to different physical environments (see Chapter 11). It is more difficult to quantify the requirements of tooth structure necessary for change from insectivory to carnivory, or from frugivory to herbivory, but once adapted to a particular dietary niche, continuous change in the dentition would be expected until geometrical, histological, and tooth-occlusion patterns were achieved that were optimal for whatever food sources were the major dietary components.

In common with the probability of a considerably earlier divergence of metazoan phyla prior to their sudden appearance in the Tommotian and Atdabanian, the initial differentiation of the major placental groups may have occurred long before their rapid radiation in the early Cenozoic. The fossil record of representatives of the modern placental orders from the Upper Cretaceous is certainly very sparse despite extensive collecting wherever appropriate beds are exposed. None of the six placental species from the Hell Creek Formation identified by Archibald (1982) had yet achieved definitive characteristics of any of the modern orders, and only two living groups have been definitely identified from other Upper Cretaceous deposits: the Insectivora (Fox 1984) and the Ungulata (Archibald 1996). However, the presence of several ungulate species from 85-million-year-old deposits in Uzbekistan indicates that this lineage had diverged long before the Cenozoic, although none of these species is directly related to any of the known Cenozoic forms. Phylogenetic analysis of placentals (Novacek 1994) indicates that ungulates were the last of the major groups to evolve, and hence that all other groups must also have diverged prior to 85 mya. On the basis of molecular studies of living representatives of the modern placental orders, Hedges et al. (1996) placed the time of their initial divergence at an even earlier data, more than 100 mya.

Regardless of how long ago the initial differentiation of the major groups may have occurred, the great radiation within each of the modern lineages appears to be confined to a very short period in the latest Cretaceous and early Paleocene, judging by the successive appearance of extremely primitive sister-taxa of each of the modern orders.

Marsupials

Marsupials also radiated rapidly at the beginning of the Cenozoic, but in a pattern strongly influenced by geography as well as by the prior extinction of dinosaurs. Marsupials were moderately diverse in North America at the end of the Cretaceous but were reduced to a single lineage following the end-Cretaceous extinction. They later dispersed to Europe, Asia, and northern Africa, but these lineages all became extinct with little further radiation. Another lineage, which entered South America, radiated extensively in that continent, spread to Antarctica, and from there extended its range into Australia as these continents first came in contact with one another and then drifted apart between the end of the Cretaceous and the beginning of the Eocene. They radiated very rapidly in both South America and Australia (Woodburne and Case 1996).

Later Cenozoic mammals

The evolutionary history of most of the individual orders of mammals is generally very well known during the Cenozoic (Carroll 1987; MacFadden 1992; Szalay, Novacek, and McKenna 1993; Prothero and Schoch 1994). The patterns of radiation within each of the placental orders resemble those seen at higher taxonomic levels. The Perissodactyla underwent a major radiation in the Eocene, giving rise to a number of superfamily-level lineages that persisted throughout much or all of the Cenozoic. The Artiodactyla, in contrast, show a pattern of successive radiations in the Eocene, Oligocene, and Miocene, culminating in the advanced ruminant families. The sirenians and Tubulidentata persisted throughout the Cenozoic without any major subradiations. In these groups, changing features within a single broad adaptive zone appear to follow more closely the general pattern outlined by Darwin (see Fig. 1.1) and do not call for unique macroevolutionary phenomena.

The Carnivora and Cete (including mesonychids and whales) are exceptional among placental orders in having given rise to major secondary radiations in the marine environment. The sea lions, walruses, and seals arose in a series of sequential dichotomous branchings during the Oligocene and Miocene from the lineage including bears. Following the achievement of a fully aquatic body form by the basilosaurids in the early mid-Eocene, little is known of whale radiation until near the end of the Oligocene, when fourteen families appeared within an interval of about 10 million years. Their sudden radiation, some 25 million years after obligatorily aquatic whales had evolved, has been attributed to changes in oceanic circulation that resulted in upwelling in the southern oceans, providing an extremely rich food source for marine vertebrates.

Birds

The modern bird orders are almost entirely restricted to the Cenozoic. Numerous specimens from the late Mesozoic have been described as possible ancestors, but as knowledge of their anatomy has increased, fewer and fewer can be assuredly attributed to living orders. Other than the Hesperonithiformes and Ichthyornithiformes, which are advanced birds but clearly not ancestral to the living orders, most of the Cretaceous genera have been placed in the Enantiornithes (Chiappe 1995a), which are certainly not related to any living birds. Feduccia (1994, 1995) has recently suggested that nearly all of the Cretaceous bird lineages became extinct at the end of the Mesozoic, at the same time as the dinosaurs. The almost complete absence of ancestors of the modern bird orders in the Cretaceous is not the result of an incomplete fossil record, but because they had not differentiated until the early Cenozoic. Feduccia argued that nearly every type of modern bird evolved within about 5–10 million years, most from an assemblage termed the "transitional shore birds."

All modern bird orders had diverged by the Eocene. Even the passerines have now been reported from the Lower Eocene of Australia (Boles 1995). The passerine families began to radiate in the Northern Hemisphere only in the lower Oligocene, but many families of the other orders appeared by the late Eocene. Many modern genera are known by the Miocene.

Discussion

In contrast with Darwin's assumption that large-scale evolutionary change took place gradually and continually over hundreds of millions of generations, the fossil record demonstrates that the history of life is punctuated by a relatively small number of rapid radiations that resulted in the appearance of a wide range of anatomical patterns and adaptive modes.

The fossil record is frequently too incompletely known to establish the specific pattern and duration of radiations, but the very-large-scale radiations of metazoans in the Lower Cambrian and placental mammals and birds in the early Paleocene were concentrated in intervals of 10 million years or less.

All radiations are influenced by a combination of factors, some inherent to the group in question and others external to it. Factors inherent to the group may be referred to as their *capacity* to radiate, whereas external factors provide the *opportunity* to radiate. In cases where major anatomical changes coincide with or just slightly precede a major radiation, it may be assumed that the radiation can be attributed primarily to a newly evolved capacity. This was almost certainly the case for the initial radiation of terrestrial vertebrates, although the fossil record is not well enough known in the early Lower Carboniferous to determine the time span over which this radiation occurred. The timing is better established for the initial radiation of mosasaurs, which appears to have occurred immediately after the change in limb structure and increase in size that differentiate them from aigialosaurs.

On the other hand, there was no major change in the anatomy, nor apparently in the physiology, of either placental or marsupial mammals during the Late Cretaceous or early Cenozoic. These groups show little evidence of significant morphological change until after the initiation of their radiations in the early Cenozoic. Many of the characteristics that distinguish both of these groups from more primitive synapsids had evolved by the end of the Triassic, at which time many of the skeletal, physiological, and behavioral traits present in the most primitive living species had already evolved. They had almost certainly attained a high constant body temperature, high metabolic rate, hair, mammary glands, extended care of young, acute hearing, smell, and sense of touch, and a brain size well above that of comparably sized reptiles.

Although the capacity of mammals to radiate into a broad spectrum of terrestrial habitats may have evolved by the end of the Triassic, or certainly by the beginning of the Cretaceous when marsupials and placentals began to diverge, the remainder of the Mesozoic was spent in the shadow of the dinosaurs. At the close of the Cretaceous, most were still restricted to small, presumably nocturnal climbers, jumpers, and arboreal forms. The dental characteristics of placentals suggest adaptation to insectivory, and some degree of carnivory, omnivory, or frugivory (Kielan-Jaworowska, Bown, and Lillegraven 1979).

The fact that nearly all Mesozoic mammals were small, many comparable in size to living mice and shrews, may be attributed to predation and competition from dinosaurs. The absence of color vision in primitive mammals suggests that the early members of this group were nocturnal and/or adapted to life where they could remain hidden from diurnal predators. The late Brian Patterson referred to the Mesozoic as a "lawn mower ecology," where any animal large enough to protrude from the vegetation would be quickly eliminated. Only after the extinction of dinosaurs did mammals finally have the opportunity to radiate. The diverse fauna of contemporary lizards also had an increased opportunity to radiate, but their failure to do so may have resulted from their lack of capacity to adapt to the great number of habitats that were occupied by early Cenozoic mammals. The great adaptive success of mammals may be attributed to their much more versatile dental structure, feeding mechanics, and behavior.

New opportunities for radiation have also been provided through the appearance of recently emerged land areas, such the Hawaiian archipelago and the Galápagos Islands that formed from undersea volcanoes, and the continental drift of previously inaccessible land masses, such as South America, Antarctica, Australia, Africa, and India, which permitted successive radiations of marsupials and placentals in the early Cenozoic. Opportunities for local radiations of aquatic vertebrates were provided by the formation of new bodies of water, such as the East African Great Lakes in the late Cenozoic and the lakes of the Newark Basin in the Late Triassic and Early Jurassic.

The early metazoans did not face competition or predation from previously existing multicellular organisms of a comparable level of complexity, but they may have been prevented from prior radiation by the low level of oxygen, which precluded the development of large complex bodies and the formation of a mineralized skeleton. It may have been a final stage in the increase of oxygen that provided the opportunity for their explosive radiation.

The most rapid radiations – such as those of Early Cambrian metazoans, early Cenozoic placental, marsupials, and birds, and late Cenozoic cichlids – occurred in less than 10 million years in environments almost totally lacking in competitors or predators. Other radiations – those of the neoselachian Chondricthyes, the percomorph actinopterygians, early amniotes, therapsids, and the succession of dinosaur and mammalian families – occurred in environments already occupied by related forms, with which there was presumably competition for both space and resources. These radiations appear to have required significantly longer periods for their expression, and may have proceeded by a series of divergences rather than a single short period of large-scale radiation. More precise correlation between the temporal pattern of radiations and the prior occupation of the environment in which they occurred depends on additional knowledge of the many radiations for which the fossil record is currently inadequately known.

The largest-scale and most rapid radiations can certainly be attributed to factors that were not considered by Darwin. There is no way that knowledge of modern populations could be extrapolated to encompass the effects of mass extinctions, changes in the distribution of continents, significant modifications in the nature of the genome and developmental processes, or a major increase in the amount of atmospheric oxygen. On the other hand, the changes in anatomy and behavior that must have occurred in the radiation of placental mammals in the early Paleocene were not significantly different than those that occurred during a similar time frame within individual mammalian groups at other times in their history. Changes that occurred in the early Paleocene differed from those observed in modern populations only in continuing in the same general direction for much longer periods and in involving a large number of lineages simultaneously.

14 Forces of evolution

> On the geological timescales on which evolution is played out, the physical development of our planet may be a major engine of evolutionary change.
> – Knoll (1989, p. 287)

The past several chapters have concentrated on aspects of large-scale, long-term evolution – demonstrating the importance of physical constraints, showing the nature and time scales of major adaptive changes within lineages, and investigating the patterns of large-scale radiations. This brings us back to questions raised at the very beginning of this text regarding the degree of applicability of processes of evolution that can be studied at the level of populations and species to changes over much longer time scales.

Forces of evolution evident at the level of populations and species

Although processes of evolution at the level of populations and species are extremely difficult to infer from fossil evidence in deposits earlier than the Cenozoic, there is no reason to think that the potential for population growth, genetic variation, and natural selection were any less effective in changing the expression of characters within species during earlier stages of vertebrate history.

Population growth, dispersal, and behavior

From Malthus, Darwin accepted the potential for exponential population growth as a major force of evolution. This is clearly inherent to all forms of life from as long ago as their first appearance in the fossil record. Large populations of prokaryotes are known from particular localities far back in the Precambrian (Schopf 1995), and literally millions of fragmentary specimens of the primitive vertebrate *Astraspis* are known from localities distributed over hundreds of miles in the Late Ordovician. Since the potential for population growth in all species far exceeds the possibility of survival for more than a small percentage of the progeny, this remains an essentially constant force of evolution throughout the history of life.

Additional important forces are dispersal and other aspects of behavior that enable organisms to spread geographically and into a diversity of potential habitats. Behavior is rarely emphasized in textbooks as a major evolutionary force, but West-Eberhard (1989) cited numerous papers that support this concept in her review of phenotypic plasticity and the origin of diversity. It may have required no more than a change in individual feeding habits toward seeking food at the margins of the sea to initiate the divergence of the modern marine iguana and the earliest whales from their terrestrial ancestors. Differences in feeding behavior enable incipient cichlid species and Galápagos finches to reduce competition with each other. One need not go much further to envisage behavioral differences in geographically isolated populations of early Paleocene mammals to account for the origin of distinct placental orders.

Genetic variability

Mutation rates and the degree of heterozygosity differ from character to character and between species, but they do not vary in a systematic way between major taxa of living vertebrates. Studies of modern populations demonstrate that both the accumulation of inherited variation and the generation of new mutations are sufficiently high for organisms to respond rapidly to selection coefficients far higher than those that would be necessary to account for most changes observed in the fossil record. There is no reason to think that either average mutation rates among vertebrates or the potential for retaining variability differed significantly over the past 500 million years.

Prior to this, morphological change may have occurred at a more rapid rate during early stages in the evolution of chordates and the origin of craniates, when *Hox* and other genes that are responsible for major aspects of development were undergoing duplication and significant functional change. This may have led to significant modifications in body form within only a few million years, but comparable changes in the genetics of development did not occur in later stages of vertebrate evolution.

Speciation

Genetic variability among vertebrates, as in other biparental, diploid organisms, is perpetuated throughout populations for thousands or even millions of generations, providing extensive variation that is continually subject to natural selection. As long as interbreeding is possible, populations within species respond to a single selective regime, without permanent division into separate lineages. However, once physical barriers to reproduction occur within species, the newly isolated populations have the capacity to respond to different forces of selection and can separate into distinct species. The phenomenon of speciation is hence a vital force in vertebrate evolution leading to the subdivision of previously existing niches and the continual expansion of the total adaptive range beyond that of the single ancestral species. Speciation has been sufficiently common in all vertebrate groups and at all times in their history to provide a diversity of different lineages that can explore a great variety of potential adaptive zones. The amount of speciation differs considerably from genus to genus within each of the vertebrate classes, but most classes include some highly speciose taxa (Minelli 1993).

Speciation also provides the most direct link between species level phenomena and the most significant macroevolutionary process, the origin of major taxa. Be-

yond proliferation of new lineages within genera, new species have the potential
to give rise to new clades at all taxonomic levels, including classes or phyla.

Response to selection

Population growth, genetic variation, dispersal, and speciation are all factors inherent to species that have the potential for producing continuous radiations, up to
the carrying capacity of the inhabitable biosphere. Throughout the history of life,
these factors have been counterbalanced by aspects of the physical and biological
environment. All organisms are subject to competition and predation, as well as
physical and biological limits to natural resources. These are the forces that Darwin
(1859, p. 84) argued were ". . . daily and hourly scrutinising, throughout the world,
every variation, even the slightest; rejecting that which is bad, preserving and adding up all that is good. . . ."

Studies of both wild and laboratory populations demonstrate that differing selection pressures can produce a wide range of evolutionary rates that may be several orders of magnitude greater than those associated with the most rapid changes
that have been determined from the fossil record (Gingerich 1983; Endler 1986;
Grant and Grant 1989). Natural selection may act in many different ways on characters within populations and species, resulting in stasis, oscillatory change, or
nearly continuous directional change. Any of these patterns may persist throughout the duration of a species (up to a million years or more) or change over time
and differ from population to population. Over time scales of tens to hundreds of
thousands of years, competition and predation may be relatively constant factors
that perpetuate a nearly balanced equilibrium with the forces of population growth,
mutation, and dispersal. Alternatively, if physical or biological aspects of the environment change in a consistent and directional manner over many generations, the
balance will shift, leading to the expansion of some species and the gradual extinction of others, as argued by Darwin.

Evolutionary forces that can be studied in modern populations are sufficiently
powerful to explain the amount and rate of morphological change throughout the
entire course of vertebrate history, even during transitions between different physical environments and periods of major adaptive radiation. We can think of the
forces of population growth, generation of genetic variability, behavioral change,
speciation, and selection acting together as an engine that has driven vertebrate
evolution for the past half-billion years. It is a powerful engine that is capable of
great speeds when the opportunities or requirements of the environment demand,
or it may idle at rates so slow that they are barely detectable. The way in which
these forces act is essentially constant throughout all vertebrate groups and at all
times in their evolution, from the earliest stages in their radiation to the final process of species and group extinction. They are as regular in their expression as the
genetic code, meiosis, and the cellular constituents of vertebrate bodies.

If these were the only forces acting to control the rate and direction of change,
one might imagine that the overall pattern of evolution would appear similar at all
time scales throughout vertebrate history, as charted by Darwin – a gradual and
continuous unfolding, limited to a narrow range of rates and directions, and with
a similar potential for change at all times. However, we know that life has not fol-

lowed such a uniformly diversifying pattern. The patterns and rates of vertebrate evolution have been irregular at the largest scales, in ways that cannot be directly extrapolated from knowledge of the current world.

Additional factors of long-term evolution

Some aspects of this irregularity, although not directly observable in living populations, can be explained by processes acting in the modern world. These include both physical and historical constraints. Physical constraints help to explain the constancy of body form within major adaptive zones and particular size animals, as well as the rapidity of change that occurs in groups undergoing major transitions.

All evolutionary change, in both modern and extinct groups, has been constrained by the structure, physiology, and behavior of their immediate ancestors. One can hence explain many of the differences in the nature and amount of morphological change and the extent and patterns of radiation observed between clades by the differences in their founding members. Gould (1989) referred to this as **historical contingency**. This gives an overall directionality to evolution, and a degree of control over what is possible in any particular clade. Vertebrate evolution began in an aquatic environment, and the basic attributes of all subsequent vertebrates were initially established in response to the requirements of that way of life. All terrestrial vertebrates owe their basic limb structure to that of the most primitive amphibians, and all amniotes to the reproductive specializations of the early members of that clade. Humans owe their acuity of sight and brain, and flexibility of body, to their primate ancestors of the early Cenozoic.

Evolutionary trends

Each major group of vertebrates has a unique assemblage of capabilities and limitations. Because each clade begins at the species level with few differences from its sister-group, its definitive anatomical characteristics evolve in a stepwise manner, as we saw in the origin of tetrapods and advanced birds. However, behaviorally directed adaptive changes may occur even more quickly, thus setting the pattern for strong, directional selection that rapidly establishes the course of future evolution. In the case of the initial radiation of placental mammals in the early Cenozoic, this may have occurred within the first one or two hundred thousand years of the Paleocene. Trends were conceivably established even more rapidly in amniote groups that gave rise to secondarily aquatic lineages. The habits of aquatic locomotion and feeding might have been initiated in only a few generations in small isolated populations, in which there was strong selection for a change in habitat.

As behavioral traits become genetically fixed and anatomical and physiological changes begin to accumulate, the clade becomes committed to the new diet, environment, or way of life, and will respond primarily to selection pressures that perpetuate this trend. This is most clearly seen in lineages that have adapted to environments that differ dramatically in the nature of their physical constraints, as emphasized in Chapters 11 and 12. Major adaptive characteristics common to all subsequent members of the clade may become established within 5–10 million years, but some long-surviving lineages continue to perfect these features throughout their duration.

Gould (1995a and previous papers) has argued that evolutionary trends established within populations will not be continued across speciation events; instead, they must be perpetuated through species selection. All amniote lineages that returned to an aquatic way of life certainly faced strong selection for more effective locomotion, feeding, sensory input, and modes of reproduction, with no regard to speciation. Pervasive selection must also have acted upon a great number of lineages in diverse orders of placental mammals that were faced with adaptation to changing climatic conditions during the late Cenozoic. Here we have clear fossil evidence of continuous morphological change through millions of years and many speciation events.

Other very-long-term evolutionary trends were expressed within single, relatively constant adaptive zones. Among the most significant in terms of the modern biota were trends in the clades leading to modern teleost fish and mammals. In contrast with the transitions that were discussed in Chapter 12, these trends have continued over hundreds of millions of years and have involved a series of major intermediate radiations prior to those that dominate the world today. They also differ in involving nearly all aspects of the body, so that a complete description of changes would require books as long as this text. (Longer reviews are provided by Carroll 1987.) In fact, it is the complex integration of a number of functional systems that has driven these trends. These clades exemplify yet another force of evolution that is inherent to organisms, but is expressed to significantly different degrees from group to group. The history of teleosts illustrates the evolution of a host of characteristics that are of advantage in nearly all habitats and ways of life within the aquatic realm, whereas mammals achieved features that would enable them not only to dominate the terrestrial environment, but also to give rise to subgroups that radiated in the sea and air.

The origin of teleosts

The ancestors of teleosts diverged from a yet-undetermined lineage of primitive ray-finned fish in the late Paleozoic. The earliest members were small, with a thick covering of scales over the entire trunk and fins. These provided protection and support for the body, but limited the capacity for movement. Teleost ancestors had a swim bladder, but the weight of the scales and the skull made it difficult for them to achieve neutral buoyancy. The jaw muscles were confined by the bones covering the head, and the gape was limited. From the Carboniferous throughout the early Mesozoic, this clade underwent a series of changes in the skull roof, braincase, and respiratory apparatus that greatly increased the gape and the force of the jaw muscles, enabling them to feed on an enormous spectrum of prey. From the Triassic to the Late Cretaceous, changes in the feeding apparatus were accompanied by major improvements in a complex of structures associated with locomotion. The weight of the body was greatly lessened by the reduction in scale thickness, while the swim bladder closed. Together these changes enabled buoyancy to be controlled with little expenditure of energy. The vertebral centra became ossified so that they assumed the primary role in support of the trunk and resistance to the force of the swimming muscles. The fins were altered so that they became more maneuverable, and their location on the body changed as their role in maintaining the position of the fish in the water column was taken over by the swim bladder. The complex of changes involving nearly all aspects of their adaptation to an aquatic way of life enabled the ancestors of teleosts to undergo the greatest radiation in all of vertebrate history in the Late Cretaceous and early Cenozoic, resulting in the proliferation of more than 20,000 living species.

What distinguished major aspects of evolution in the ancestors of teleosts was the close integration of a number of functional complexes. Feeding and respiration are closely integrated in bony fish, while the skull is closely connected with the pectoral girdle and the muscles of the trunk serve to open the mouth. Changes in the capacity for buoyancy control and an increasing diversity of locomotor patterns extended the range of diets and feeding behavior. This positive feedback system has continued to drive their evolution for almost 300 million years.

In contrast, sharks, which have an even longer history, show an extremely constrained pattern of evolution. The absence of bone throughout the known history of this group precludes the complex integration of feeding, respiration, and locomotion that is a hallmark of bony fish. A complex, self-perpetuating feedback system such as that which distinguished the evolution of teleosts never emerged.

The origin of mammals

Kemp (1982) coined the phrase **correlated progression** to describe a mechanism of evolution among the ancestors of mammals that resembles the one just described in teleosts. In common with the predecessors of modern teleosts, the late Paleozoic and early Mesozoic mammalian ancestors show a complex, interrelated series of structural changes, beginning with cranial structures associated with feeding and respiration, and continuing with alterations in the locomotor apparatus. In addition, profound modifications in physiology, the central nervous system, sensory apparatus, and the pattern of reproduction also occurred within this group.

The earliest members of the mammalian clade from the Late Carboniferous were small animals resembling unspecialized living lizards in most aspects of their skeletal structure and physiology (Carroll 1986). They had a low metabolic rate and depended on external sources of heat; they were covered with scales, laid eggs, and showed little care for their young. By the Late Jurassic, they had evolved most of the structural and physiological features of small, primitive living marsupials and placentals.

The key to the origin of mammals lies in the early expression of a trend toward a metabolic rate higher than that of other, contemporary amniotes. By the end of the Permian, the early cynodont ancestors of mammals showed a complex of cranial features including multicusped cheek teeth, advanced jaw structure and mechanics, together with a secondary palate that indicated the capacity for extensive mastication of food while maintaining a constant supply of oxygen, both necessary for a continuously high metabolic rate. Changes in the jaw musculature led to modifications in the bones at the back of the skull that resulted in a restructuring of the braincase. This later enabled the brain to attain the large size that is characteristic of advanced mammals. Changes in the jaws also led to the evolution of the structures of the middle ear that made it possible for mammals to detect airborne vibrations with much greater acuity and over a much greater range of frequencies than their early amniote ancestors (Hopson 1994).

Changes in all these functional complexes interacted and reinforced one another. Modifications of the muscles and bones of the lower jaw enabled more controlled use of the teeth and resulted in the evolution of precise tooth occlusion. Improved mastication permitted increased metabolic rate with a concomitant demand for more food.

By the Middle Triassic, increase in the efficiency of prey capture and mastication made it possible for some lineages to develop a metabolic rate sufficiently high that they could maintain a high, constant body temperature while significantly reducing their body size. Small body size – presumably selected for by the availability of small prey, particularly insects and other arthropods – brought ancestral mammals into an environment close to the ground, and even into cracks and burrows where they pursued their prey. Movement in this environment resulted in selection for modification of tactile structures necessary for sensing the size of these passages, even in the dark. Primitive amniotes presumably already possessed stretch receptors associated with the scales, as do modern lizards. Maderson (1972) hypothesized that such structures could have evolved into tactile hairs, such as the *vibrissae,* or whiskers, of cats and rodents. If such hairs covered a large portion of the body, they would also have an insulatory function. Although this story of the origin of hair is largely hypothetical, it is well established that the presence of hair was closely associated with the origin of both sweat and mammary glands. Hence, we have another complex of reinforcing behaviors and structures leading from more effective teeth to increased metabolic rate, smaller body size, hair, mammary glands, and longer parental care. Increased input of sensory information – from more sophisticated ear structures, tactile hairs, and eyes operating in areas of reduced light – required and benefited from increased integrative capacity of the central nervous system, which could now occupy the large braincase whose form was established by the elaboration of the jaw muscles.

Although there are few if any stages in the origin of mammals in which evolution can be studied at the level of populations and individual species, these basic trends continued without major breaks through a number of adaptive radiations. With this level of complexity, it is hardly surprising that this system required 100 million years to evolve to the level of the most primitive mammals known from the Upper Triassic. What is more striking is that the level of mammalian structure and physiology did not advance appreciably for the next 150 million years, until the extinction of the dinosaurs at the end of the Cretaceous.

Continental drift

Although far beyond the scale of change that can be seen even over tens of millions of years in the evolution of late Cenozoic vertebrates, most aspects of the origin of mammals are readily accommodated within Darwinian theory. In contrast, other features of the history of vertebrates are difficult, if not impossible, to explain on the basis of what we see around us in the modern world. Given the current position of the continents, how can we account for a vast radiation of Paleozoic amphibians, amid a flora of tropically adapted plants, in a belt extending from north central Europe across northeastern North America? How do we explain the presence of tropical forests filled with primates as far north as the Arctic islands in the Paleocene and early Eocene of North America, their disappearance in the Oligocene, and the progressive replacement of the succeeding subtropical and temperate forests with dry savanna, tundra, and finally Arctic conditions? Why did a great many families and even genera of amphibians, reptiles, and primitive mammals have worldwide distributions from the Late Permian into the Early Jurassic, whereas each continent now has its own distinct fauna? These questions require explanations that go beyond the constraints of the climate and geography of the current world, to major changes in the physical nature of Earth that have been occurring ever since its formation 4.6 billion years ago.

Darwin's writings on evolution were strongly influenced by the uniformitarian theory of geology elaborated by Lyell in *Principles of Geology* (1832–3), published during Darwin's voyage on the *Beagle*. Lyell not only felt that all geological processes could be understood on the basis of those acting today, but that they had acted in the same manner and force throughout the Earth's history. As emphasized by Gould (1995b), Darwin accepted Lyell's uniformitarian vision of geology in all its uncompromising intensity and applied the same approach to the history of life.

It is only in the past thirty-five years that geology has broken free of the belief that the general form and position of the continents and oceans have remained nearly constant for much of the Earth's history. We now recognize that the major features of the Earth's crust have changed dramatically and continuously over the past 4.6 billion years. These changes have, in turn affected the climate and other aspects of the environment of all organisms, thus strongly influencing the direction of their evolution and their very survival. The history of the continents and the oceans and their influence on life are discussed in recent publications by Klein (1994), van Andel (1994), and Windley (1995).

A brief history of the continents

Approximately 4 billion years ago, the continental crust consisted of small island continents in an extensive ocean. Movements in the mantle generated by the heat of the Earth's core resulted in accretion of these elements into larger continents and their subsequent breakup (Fig. 14.1). There were at least four major episodes of continental accretion and breakup during the Precambrian, with the last supercontinent beginning to break apart early in the Cambrian (Fig. 14.2). By the Ordovician, the areas that we now recognize as North America, Europe, and Asia were island continents, separated by an extensive ocean from the areas that later formed South America, Africa, Antarctica, Australia, and India. These southern continents were then integrated in a supercontinent called Gondwanaland (Fig. 14.3).

During the later Paleozoic, all the continents came together to form a single supercontinent, Pangaea (Fig. 14.4). A single ocean, Panthalassa, surrounded this continent, except on the east, where there was a huge embayment between the area of Eurasia and the southern continents. This is termed the Tethys Sea, and lat-

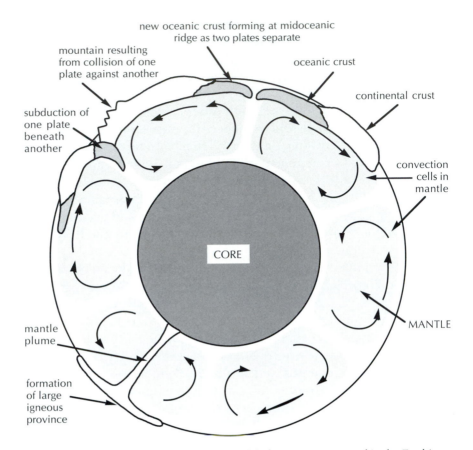

Figure 14.1. The mechanism of continental drift. Heat is generated in the Earth's core, which results in circulation within the mantle. Movement in the upper mantle transports the crustal plates: They move away from one another where heat from the mantle is rising, and converge where the currents are sinking. Mantle plumes, which form large igneous provinces, arise from the base of the mantle independent of the position of the crustal plates.

Figure 14.2. Distribution of continents in the late Precambrian, when they were beginning to disperse. Most are clustered in the Southern Hemisphere. Their proximity to the South Pole resulted in extensive glaciation, indicated by the wavy shading. Abbreviations: Af, Africa; Ant, Antarctica; Au, Australia; Ba, Baltica; N. Am, North America; S. Am, South America; Si, Siberia. From van Andel (1994).

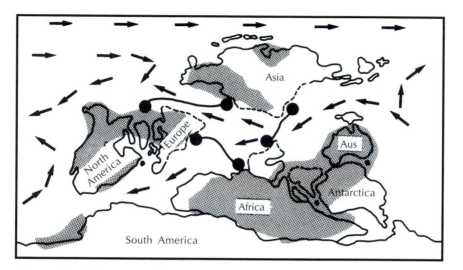

Figure 14.3. The pattern of continents in the Late Silurian, when they were beginning to reassemble. Areas of land are shaded; the arrows indicate possible ocean currents, and the black dots mark opposing areas that, according to fossil evidence, ought to be connected. The odd shape of South America is due to the projection. From van Andel (1994).

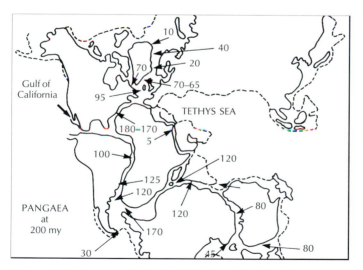

Figure 14.4. Pangaea during the Early Jurassic, when the continents were beginning to disperse. The numbers indicate the time, in millions of years ago, when particular areas began to separate. The earliest new oceans were the southern North Atlantic and the one separating Africa from Antarctica; the South Atlantic followed considerably later. The process is not complete: The Red Sea began to open 5 million years ago, and the Gulf of California is even younger. In 50 million years Baja and Southern California will be a long, narrow continent somewhere in the Gulf of Alaska. From van Andel (1994).

er became confluent with basins that formed between North America and Europe on the north, and South America and Africa.

Pangaea achieved its maximum degree of coalescence approximately 230 mya, in the Late Triassic, and then began to disperse as a result of the force of the heat from the core that built up under the enormous continental mass. Appreciable separation was evident by the Middle Jurassic. The various elements of Pangaea separated at different times and at different rates, providing a changing mosaic for the distribution of both aquatic and terrestrial organisms. One may associate particular episodes of attachment and separation of the continents with almost all stages in the evolution and dispersal of late Mesozoic and Cenozoic vertebrates. As the southern continents separated from one another, Australia, Africa and India began to move north, and the latter two made contact with Eurasia by the Miocene. Movement of plates between North and South America resulted in the connection of these continents approximately 3 mya (Fig. 14.5).

In addition to changes in the relative position of the continents, their orientation with regard to the poles and equator also changed extensively, as did the portions that were above sea level (Fig. 14.6). During much of the Paleozoic and Mesozoic, extensive areas of North America were under water, but the modern coastlines were achieved rapidly by the beginning of the Cenozoic. In general, the sea level was low when the continents were massed together and higher when they were dispersed. Periods during which sea level dropped precipitously, when the water suddenly withdrew from the continental shelves, were associated with the largest-scale extinction events (Fig. 14.7).

Figure 14.5. Changing positions of the continents during the Mesozoic and Cenozoic. **A,** Mid-Jurassic. **B,** Late Cretaceous. **C,** Middle Oligocene. **D,** Miocene–Pliocene. During the Mesozoic, the surface circulation of the oceans evolved from a simple pattern in a single ocean with a single continent to a more complex situation in the new oceans of the Cretaceous. Throughout this time, the open circumequatorial path and the absence of circumpolar currents resulted in a temperature distribution that was more even than today. The Cenozoic history of ocean circulation is dominated by two events: the opening of the Antarctic circumpolar seaway 25–30 million years ago, and the closure of the circumequatorial seaway by the emergence of the Isthmus of Panama. Climatic deterioration during the Cenozoic is commonly attributed to changes in oceanic circulation. From van Andel (1994).

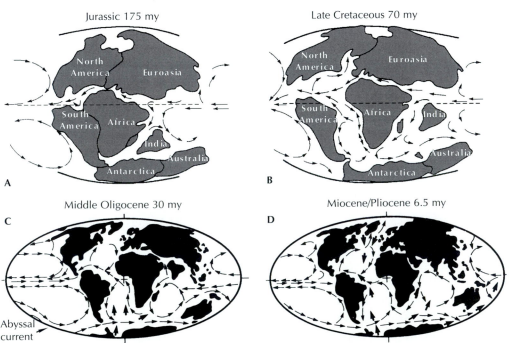

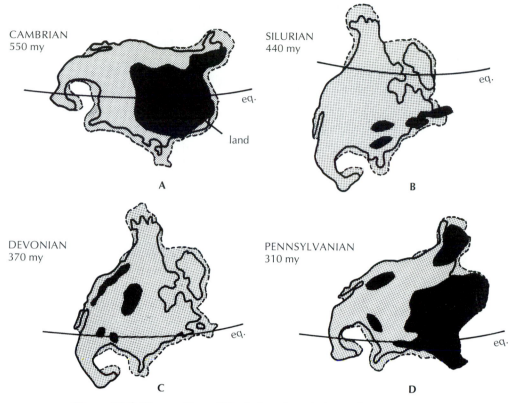

Figure 14.6. The position of North America relative to the equator in successive periods of the Paleozoic. **A,** Cambrian. **B,** Silurian. **C,** Devonian. **D,** Pennsylvanian. Dark shading indicates the extent of land. The lighter shading shows the extent of epicontinental seas and the continental shelves. The present coast of the continent is given for guidance, but does not reflect the geography of Paleozoic North America. From van Andel (1994).

Biological effects of continental drift

The relative position of the continents and the nature of the connections between the ocean basins strongly influenced the climate. In general, when the continents were separated from one another and had relatively low relief, the climate was maritime – moderately warm and moist, and nearly equable from the equator to the poles. Such conditions prevailed during the period of the dominance of dinosaurs, throughout the Jurassic and Cretaceous. The larger the supercontinents, the more continental their climate, with extremes of hot and cold and wet and dry. Throughout the past 800 million years of its history, Earth has oscillated between major ice ages and longer periods of warmer climate. These are referred to as **icehouse** and **greenhouse** conditions (Fig. 14.8).

The largest supercontinents covered the area of the South Pole during the Late Precambrian and from the Carboniferous into the Permian. These times coincided with the most extensive ice ages in the Earth's history. The late Cenozoic glaciation, on the other hand, occurred when the continents were more dispersed, but their particular locations relative to one another led to patterns of oceanic circulation that isolated both the north and south polar regions. The growth of the Antarctic ice sheet in the Oligocene followed the separation of Antarctica from the southern continents and the development of a circumpolar current that isolated the polar waters from the temperate and tropical circulation (see Fig. 14.5C). In contrast, the north polar region was nearly surrounded by land in the Pleistocene, trapping the coldest waters in the north.

We saw in Chapter 5 how gradual cooling of the northern continents in the latter half of the Cenozoic resulted in the spread of grasslands, steppes, and finally tundra and arctic conditions. In fact, episodes of cooling had begun as early as the late Paleocene (Prothero and Berggren 1992; Janis 1993). The changes in climate and vegetation in turn led to evolutionary modifications in the habits and dentition of a great number of mammalian lineages, and eventually to the origin of the tundra and arctic faunas. In the Southern Hemisphere, the origin and radiation of penguins centered about the margins of the ice-covered land masses.

The extensive cooling associated with Permo–Carboniferous glaciation in Gondwanaland may have had even more profound effects on the early stages in mammalian evolution, for it was during the Late Permian that the therapsid ancestors of mammals in southern Africa first showed skeletal features, such as the formation of a secondary palate and complex tooth and jaw structures, that indicate early stages in the evolution of a high, constant metabolic rate. These changes may have begun during adaptation to a cold and strongly seasonal climate. The *Glossopteris* flora that is common in these horizons shows clear evidence of growth rings; these are absent in plants of that time from the then-tropical Northern Hemisphere.

Although the gradual movement of land masses and the changing patterns of the climate resulting from continental drift would have required continuous evolutionary adjustments by organisms both in the water and on land, other phenomena associated with the release of heat from the Earth's core produced gigantic lava flows and released enormous amounts of CO_2 and other gases into the atmosphere over periods that may have been too short for evolutionary accommodation. Such phenomena are obvious candidates as causes for worldwide extinctions.

Large igneous provinces

The material of the Earth's mantle is not restricted to the area beneath the crust, but may burst forth beneath the ocean and onto the land as lava flows and volcanic eruptions. Much of this activity occurs at the margins of the plates, along the mid-oceanic ridges and in island arcs, but other eruptions result from **mantle plumes** that come from areas deep within the mantle and reach the Earth's surface without regard to crustal features (see Fig. 14.1). The linear arrangement of the Hawaiian Islands and the contiguous Emperor chain was due to the movement of oceanic crust across a relatively stationary plume over a period of 70 million years. Mantle plumes have been active in other areas throughout the Earth's history, but their great extent has only recently been recognized (Coffin and Eldholm 1993; Windley 1995). The largest structures formed by mantle plumes, termed **large igneous provinces**, incorporate millions of cubic kilometers of molten lava spread out over millions of square kilometers and accumulating up to several kilometers in thickness (Fig. 14.9). The Ontong Java Plateau contains at least 36 million km^3 of igneous rock and covers nearly 2 million km^2, two-thirds the area of Australia. It formed

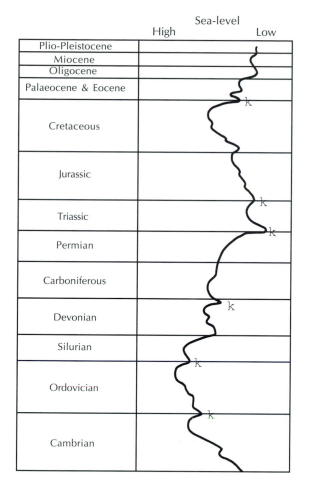

Figure 14.7. Phanerozoic sea-level curve with six main extinction events shown by asterisks. From *Phanerozoic Sea-level Changes* by A. Hallam. Copyright © 1992 by Columbia University Press. Reprinted with permission of the publisher.

Figure 14.8. Alternation of greenhouse and icehouse conditions on Earth over the past 800 million years, with a glance into the future. From van Andel (1994).

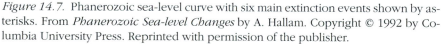
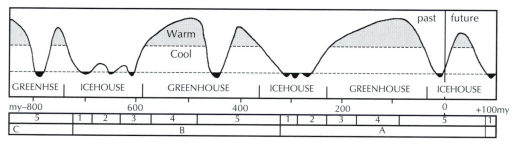

in less than 3 million years, 120–125 mya, compared with 50 million years for the formation of the Rocky Mountains. The CO_2 emitted during this eruption is estimated to have raised worldwide temperatures 7.6–12.5 °C higher than today's mean. Much of the Indian subcontinent was covered by such flows, termed the Deccan traps, formed within a span of less than 1 million years, 65 mya. Others include the Columbia River flood basalts (with 1.3 million km³ of rock) and the Siberian traps, implanted approximately 250 mya, just before the end of the Permian. A single flood basalt event generating 1,000 km³ of lava emits 16 trillion kg of CO_2, 3 trillion kg of sulfur, and 30 billion kg of halogens (Coffin and Eldholm 1993).

Mass extinctions

Although Darwin considered the extinction of individual species to be a very important factor of evolution, he argued that extinction, like other evolutionary processes, was slow and gradual. Major floral and fauna changes at the end of the Permian and Cretaceous had been used as early as 1840 to demonstrate the biological distinction among the Paleozoic, Mesozoic, and Cenozoic eras, but Darwin did not believe that they were the result of rapid, global extinctions. Rather, he attributed the apparently rapid changes to gaps in the fossil record.

As knowledge of the sedimentary and fossil records improved, it became increasingly evident that extremely large numbers of species common to the Upper Permian and Upper Cretaceous did not occur in subsequent horizons and must have become extinct over a geologically short period of time. Since 1980, with the publication of the paper by Alvarez and colleagues hypothesizing that the Age of Dinosaurs came to an end as the result of a collision with a large meteorite or comet, great interest has been focused on this and other periods of mass extinction (e.g., Chaloner and Hallam 1989; Sharpton and Ward 1990; Raup 1991, 1995; Ryder, Fastovsky, and Gartner 1996).

Species extinction has occurred throughout the history of life, but there have been some intervals when a much higher than average number of species and the larger taxa to which they belong became extinct relatively quickly. If such **mass extinctions** occurred throughout the world and involved many different types of organisms, it is logical to assume that they had a common cause that distinguishes them markedly from the type of extinctions discussed by Darwin, in which the loss of each species would have a unique cause.

According to Darwin, extinction was the natural outcome of the relative fitness of species. Species that did not have the capacity to adapt to changes in the biological and physical environment as rapidly as others would contribute fewer progeny to the next generation and would gradually become extinct. This manner of extinction is central to the concept of natural selection as a force for continuing biological change.

In contrast, mass extinctions, such as those at the end of the Cretaceous and the end of the Permian, resulted in the extinction of vast numbers of species without regard to their relative fitness. Whatever their cause, they were catastrophes that wiped out entire environments and nearly all the species they contained. For many major groups, no lineages survived.

The mass extinctions at the end of the Permian and Cretaceous certainly resulted in major changes in the nature of the biota and in the very direction of evolution of many of the surviving groups. They exemplify a force of evolution that differs fundamentally from Darwinian selection, and are frequently spoken of as "resetting the evolutionary clock." One may most readily recognize their significance in terms of the total diversity of life, in which the number of species or families showed a sudden decline and a slow subsequent recovery (Fig. 14.10).

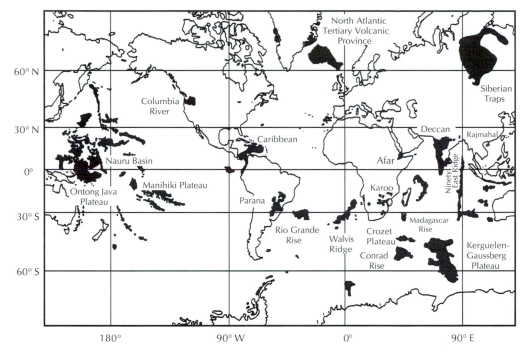

Figure 14.9. Location of large igneous provinces, which include continental flood basalts, plume tracks, extinct ridges, and oceanic plateaus. From Saunders et al. (1992).

Gould (1985) argued that one can think of the forces of evolution as acting at three tiers. The first is at the level of populations, which evolve over limited time spans in the manner hypothesized by Darwin. The second is at the level of species selection, which results in significant morphological change over longer periods of time as well as major adaptive shifts. The third is at the level of mass extinction, which would explain the largest-scale evolutionary patterns. The latter is a completely non-Darwinian process since there is no way in which selection can act on a generation-by-generation basis to improve survival rates during catastrophes of an unpredictable nature that occur at intervals of tens to hundreds of millions of years.

The basic pattern of evolution, involving many groups of both aquatic and terrestrial vertebrates, was indeed altered between the late Mesozoic and early Cenozoic. Mammals became dominant in the early Cenozoic not because of gradual competitive replacement of dinosaurs but because the dinosaurs were wiped out by sudden changes in the environment. Was the end-Cretaceous extinction a unique event in the history of vertebrates, or did large-scale extinctions play an important role in their evolution throughout the Phanerozoic?

Since 1980, studies of the fossil record have led to the recognition of an increasing number of mass extinctions. Raup and Sepkoski (1982) noted five major peaks of extinction in the Late Ordovician, Late Devonian, Late Permian, Late Triassic, and Late Cretaceous (Fig. 14.11). In 1984, using more detailed taxonomic analysis, they recognized ten peaks since the Middle Permian, each separated by approximately 26 million years. They suggested that this degree of regularity indicated some periodically repeating cause. Because no earthbound phenomenon appears to show such striking regularity, they postulated that these extinctions resulted from

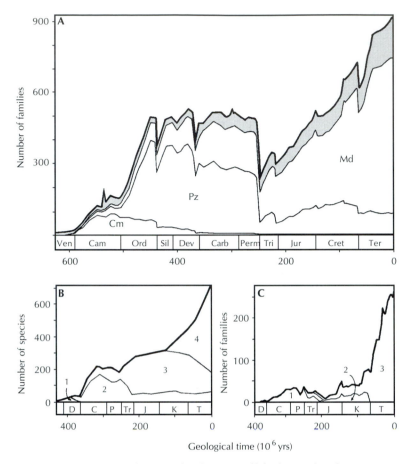

Figure 14.10. Curves showing the diversity of life during the Phanerozoic. Each curve is divided into fields that illustrate the diversity of constituent faunas or floras. **A,** Marine animal families: Cm, Cambrian evolutionary fauna; Pz, Paleozoic fauna; Md, modern fauna; stippled field represents known diversity of families with rarely preserved members that lack heavily mineralized skeletons. Sharp drops in diversity mark the end-Ordovician, Late Devonian, end-Permian, Late Triassic, and end-Cretaceous extinction peaks. **B,** Terrestrial vascular plant species: 1, initial Silurian–Devonian flora; 2, pterido-phyte-dominated flora; 3, gymnosperm-dominated flora of seed plants; 4, angiosperm flora (reprinted from *Nature* [Niklas et al., vol. 303]; copyright © 1983, Macmillan Maga-zines Limited). **C,** Terrestrial tetrapod families: 1, labyrinthodonts, anapsids, and synap-sids; 2, early diapsids, dinosaurs, and pterosaurs, 3, lissamphibians, turtles, crocodiles, lizards, birds, and mammals. A,C from Sepkoski (1990).

astronomical phenomena causing regular, impact-produced extinctions. Sepkoski (1990) extended this analysis by comparing extinction peaks in nine well-known groups of nonvertebrates (foraminifera, bryozoans, corals, brachiopods, marine gastropods, marine bivalves, ammonoids, marine arthropods, and echinoderms) as well as marine vertebrates. He used 10,383 genera, omitting single occurrences. The time of extinction was apportioned to forty-four stratigraphic intervals with an average duration of 6 million years. All groups gave an essentially similar pattern, summarized in Figure 14.12. If there were indeed such short spans of time between episodes of mass extinction, then this phenomenon may be one of the most impor-tant factors determining the course of evolution. At the other extreme, the appar-

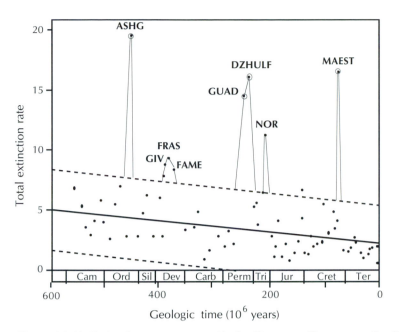

Figure 14.11. Extinction rate, measured in families per million years, of well-preserved marine organisms during the Phanerozoic. Letter codes and peaks indicate intervals with high extinction rates: ASHG, Ashgilian, Late Ordovician; GIV, Givetian; FRAS, Frasnian; FAME, Famennian, all Late Devonian; GUAD, Guadalupian; DZHULF, Dzhulfian, Late Permian; NOR, Norian, Late Triassic; MAEST, Maestrichtian, Late Cretaceous. Reprinted with permission from *Science*, vol. 215, Raup and Sepkoski, Jr. Copyright © 1982, American Association for the Advancement of Science.

ent periodicity of extinction peaks may be little more than another reflection of the incompleteness of the sedimentary and fossil records or an artifact of the way the data were analyzed (Jablonski 1996). To evaluate this problem, it is necessary to examine more closely the nature and evidence of the end-Cretaceous event.

End-Cretaceous extinction

The most thoroughly studied extinction is that at the end of the Cretaceous. In total, 14 percent of families, 38 percent of genera, and 65–70 percent of marine species became extinct (Raup 1991). Its results were particularly conspicuous in the marine environment: Eleven families of ammonites and three of belemnites became extinct at or just before the K–T boundary, bringing these groups to an end. Unicellular organisms with calcareous shells were almost completely wiped out, apparently because of a change in the *p*H of the water that resulted in dissolution of the shells. The extinction of foraminifera is typically described as occurring over as short a time as a hundred thousand years and involving nearly all the species synchronously; only a single species is said to have survived to begin a new radiation in the Cenozoic. On the other hand, more detailed stratigraphic analysis reported by MacLeod (1995) indicates that the pattern of foraminiferan extinction was much more complex and occurred over a period of 1–2 million years. It does not provide evidence of a single catastrophic event, but resembles extinctions at the end of the Eocene that are associated with changes in oceanic circulation.

Larger, benthonic invertebrates passed through the K–T boundary with lower levels of extinctions than those reported among planktonic forms. Among marine vertebrates, 21 percent of chondrichthyan families became extinct, but less than 10 percent of the vast evolving horde of bony fish.

There was considerable local extinction of land plants. In the northern Great Plains and Rocky Mountains, 79 percent of species known from their leaves disappeared at the K–T boundary, including most of the dicotyledonous angiosperms (Johnson and Hickey 1990). This coincided with a major loss of pollen species but an explosive increase in the amount of fern spores. These changes indicate sudden, catastrophic deterioration of the environment, such as results from extensive forest fires. In contrast, plant extinction levels across the K–T boundary in the rest of the world appear low to moderate.

According to Spicer (1989), the end of the Cretaceous was ecologically catastrophic, with some selective extinction of broad-leaved evergreen species and long-term vegetational restructuring. The changes were most strongly expressed in middle and low latitudes in the Northern Hemisphere. Plants have a greater capacity to survive catastrophic events than do most animals. Because their seeds are extremely resistant and may remain dormant for long periods of time, they have a greater ability to recover from events that kill vast numbers of mature organisms. The capacity to remain dormant during periods of darkness and cold would have saved high-latitude species, whereas midlatitude forms would have been more vulnerable to loss of the standing crop.

As outlined by Benton (1989), thirty-six families of land vertebrates became extinct – only 14 percent of those present in the Late Cretaceous. These extinctions, however, were concentrated in particular groups that had previously dominated the land, sea, and air: dinosaurs (19 families), mosasaurs (1), plesiosaurs (3), and pterosaurs (2). In addition, three families of crocodiles were lost, four of birds, two of marsupials, and two among turtles and squamates. The effect of the loss of dinosaurs on the radiation of placental mammals was emphasized in Chapter 13. If mosasaurs and plesiosaurs had not become extinct, they might have delayed or precluded the evolution of sirenians, whales, and pinnipeds. The nature of Late Cretaceous pterosaurs and early Cenozoic neornithine birds appears so different that it is unlikely that these two groups would have interacted in any way, but the extinction of enantiornithine birds may have been significant. Turtles, the rest of the crocodiles, lizards, snakes, and placental mammals do not show evidence of indiscriminate extinction at the K–T boundary.

Causes of the end-Cretaceous extinction

The cause of the end-Cretaceous extinction has long fascinated paleontologists and geologists. The extinction of dinosaurs was attributed to global cooling associated with mountain building, higher continents, and changes in the pattern of oceanic and atmospheric circulation. Marine extinction seemed to be explained by a major regression of the sea at the end of the Cretaceous. Following the hypothesis of Alvarez et al. (1980), attention has been focused on the possibility of the impact of a large meteorite or comet. (The term *bolide* is frequently used for such an object, whose actual identity is unknown.) This appears to be confirmed by the discovery

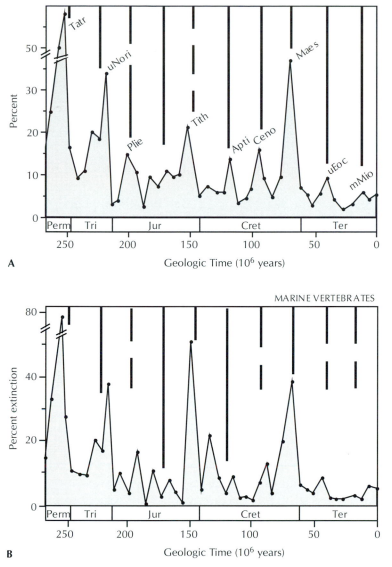

Figure 14.12. **A,** Percentage extinction of marine animal genera from the Middle Permian to Recent, including data from vertebrates, foraminifera, bryozoans, corals, brachiopods, gastropods, arthropods, bivalves, echinoderms, and ammonoids. The dashed vertical line indicates where a periodic peak is expected but not clearly defined in the Middle Jurassic. **B,** Percent extinction of marine vertebrates. Tatr, Tatarian; uNori, upper Norian; Plie, Pliensbachian; Tith, Tithonian; Apt, Aptian; Ceno, Cenomanian; Maes, Maastrichtian; uEoc, upper Eocene; mMio, middle Miocene. From Sepkoski (1990). Reprinted with permission from the Geological Society Publishing House.

in northern Yucatán of a structure 180 km in diameter, with the properties of an impact crater, whose age is coincident with the K–T boundary (Sharpton et al. 1992; Hildebrand et al. 1995). A crater of this size could have been caused by an object approximately 10 km in diameter.

There are a great many ways in which the impact of an object of this size could have affected the Earth's biota (Sharpton and Ward 1990). The heat of its passage through the atmosphere may have started fires across the planet's surface; there is, in fact, evidence of a thick layer of soot at just this horizon in many parts of the world (Wolbach, Gilmour, and Anders 1990). It might have caused acid rain that acidified the oceans. The impact would have thrown up enormous amounts of pulverized and vaporized rock: If large quantities reached the upper atmosphere, they would have circled the Earth, blotting out the light and heat of the sun, thereby shutting down photosynthesis and sending the world into a nuclear winter. On the other hand, the large amount of CO_2 resulting from the wild fires would have produced a sudden greenhouse effect, raising the temperature and further perturbing the climate.

No matter how horrendous the effects of the bolide, there is evidence that many lineages had actually become extinct several million years prior to the impact. The ammonites had been in decline for the last 19 million years of the Cretaceous. House (1989) argued that the pattern of ammonoid diversity throughout the Mesozoic can be explained almost entirely in relationship to the extent of marine transgressions and regressions. Only four families are present in the latest stage of the Cretaceous, compared with twenty-three earlier in the period. Rudists, bivalves that were major reef-building animals in the Late Cretaceous, also became extinct well before the end of the period. Clearly, the decline of these groups had started much earlier. No more than eight families of dinosaurs are known in the latest beds of the Cretaceous. However, dinosaur-bearing beds at the very end of the period have been thoroughly studied in only one part of the world: western North America. We have no knowledge of the pattern or specific time of their extinction elsewhere, although there is no solid evidence of their survival into the Cenozoic. In view of the nearly worldwide radiation of placental mammals at the very base of the Paleocene, it is unlikely that any survived the end of the Cretaceous by more than a few hundred thousand years.

Coincident with the bolide and a major regression at the end of the Cretaceous, were massive outpourings of lava that formed the Deccan traps. The fact that the flora in India was less strongly affected by the end-Cretaceous event than was that in North America (Spicer 1989) implies that large-scale volcanic eruptions were not as disruptive to life on a worldwide scale as has been thought. Nevertheless, the timing argues strongly for a multiplicity of causes for the end-Cretaceous extinction.

From an evolutionary standpoint, one of the most important aspects of this extinction is its apparently selective nature. Large groups that were expanding at the time of the extinction, including angiosperms, teleosts, squamates, and placental mammals, were generally not as severely affected as those, such as ammonites, dinosaurs, and pterosaurs, that were previously in decline. Relatively diverse groups, including chondrichthyans, turtles, and crocodiles, also suffered few extinctions. The total extinction of the moderately diverse plesiosaurs and mosasaurs may be attributed to the severe depletion of many other marine forms; some mosasaurs, for example, unquestionably fed on ammonites.

Permo–Triassic extinction

The extinction at the end of the Permian is accepted as the most catastrophic in the history of life, with a loss of 90 percent of the species and more than half the families. This was particularly striking among marine invertebrates (Erwin 1993, 1994, 1996). Erwin considers the evidence of mass extinction of terrestrial plants to be equivocal. In the 25 million years between the Lower Permian and the Middle Triassic, there was a drop in diversity at the family level of about 50 percent as the moist-adapted paleophytic community was gradually replaced by a mesophytic flora, but there were also marked changes at the Permo–Triassic boundary. Retallack, Veevers, and Morante (1996) emphasized a sudden climatic change that occurred globally at this time, with the almost complete cessation of coal deposition for approximately 20 million years. They attribute this to the extinction of a variety of Paleozoic species that were specifically adapted to the acid conditions of peat-forming environments, and their gradual evolutionary replacement by other species in the Late Triassic.

The number of extinctions appears very high for vertebrates. Maxwell (1992) calculated that 75 percent of the terrestrial vertebrate families became extinct during this interval, a total of thirty-seven families of amphibians and reptiles. On the other hand, forty new families appeared near the base of the Lower Triassic. The figures for jawed fish are so limited that they probably have little significance, but thirteen families continued across the boundary, five were lost, and nine new ones appeared in the Lower Triassic. For neither terrestrial nor aquatic vertebrates is there evidence for an interruption in the general pattern of evolution seen on either side of this boundary.

There is no evidence for an extraterrestrial cause for the extinctions at the end of the Permian. Erwin (1994) identified an interrelated series of earthbound events that contributed in a complex manner to these extinctions. The first was a major marine regression that dried out many marine basins, reduced habitat area, and increased climatic variability. Further regression triggered the release of gas hydrates and the erosion and oxidation of marine carbon. Coincident with the eruption of the Siberian flood basalts, these carbon sources increased climatic instability. Elevated atmospheric CO_2 may have produced oceanic anoxia and global warming. The final phase involved the flooding of near-shore terrestrial habitats during the rapid transgression in the earliest Triassic. The temporal sequence of the main events, shown in Fig. 14.13, demonstrate that the end-Permian extinction was not instantaneous, but may have lasted 1–2 million years. Changes in sea level and volcanic activity may both be associated with the tectonic history of Pangaea. The buildup of heat and pressure caused by the presence of such a widespread, unbroken area of continental crust may have brought about the explosive outpourings of lava and CO_2 at the end of the Permian.

Other times of major extinction

Other times of mass extinction identified by Raup and Sepkoski (1984), Benton (1989, 1995), and Sepkoski (1990) show less significant reduction in overall diversity or obvious effects on the course of vertebrate evolution. These studies all concentrated on the number of taxa that were present in one stratigraphic unit but absent in the next. There are several problems with this approach.

1. *The problem of an incomplete fossil record.* Darwin was certainly correct in implying that an incomplete fossil record would give an erroneous impression of

rapid extinction. For example, until recently it was thought that there was a major extinction late in the Precambrian between the last occurrence of the Ediacaran fauna and the emergence of the small shelly fauna at the beginning of the Cambrian. Discoveries in southern Africa and elsewhere have filled in this gap, showing that the Ediacaran fauna had its greatest diversity in the latest Precambrian and continued on up into the Cambrian (Knoll 1996b). There had been no mass extinction. Another factor that gives the false impression of mass extinction is the occurrence of a particular horizon that has an extremely diverse flora or fauna between long periods with a much less complete fossil record – for example, the temporally scattered horizons in the Carboniferous with a rich fauna of early tetrapods, and the Upper Jurassic Solnhofen beds in southern Germany. The geological stage in which such horizons occur will be tabulated as a time of extraordinary increase in the number of taxa, and the end of the stage will be considered as a time of mass extinction. It is equally probable, however, that the number of species was relatively constant across a longer interval of time.

2. *The problem of pseudoextinction.* Patterson and Smith (1987) specifically criticized Raup and Sepkoski (1984) for failing to differentiate monophyletic and paraphyletic taxa. Paleontologists have long recognized that many apparent extinctions are because of the replacement of one taxon by its own descendants: The ancestral form has not become extinct but undergone a so-called *pseudoextinction,* in which the same lineage continues under another name. Pseudoextinctions are very difficult to recognize in groups that are undergoing rapid morphological change and radiation. In periods when a large number of new species are appearing, as in the radiation of placental mammals in the early Paleocene, an almost equal number of species are disappearing. This may be accounted for by pseudoextinction, or by extinction resulting from competition; but it is biologically entirely different from what happened to dinosaurs, pterosaurs, or ammonites at the end of the Cretaceous.

3. *Intervals reported.* In all these studies, a particular stratigraphic unit – the stage, with an average duration of 6 million years, for Raup and Sepkoski (1984); a combination of stages and epochs, with an average duration of 7.5 million years, for Benton (1995) – is used as the basis for recording occurrences and absences. That is, all the genera or families whose last occurrence was during a particular stage or epoch are lumped together and tabulated as having become extinct at the *end* of that interval. This is necessary because fossils are frequently not dated more precisely than by stage or epoch, but it gives the impression that all extinctions occur at 6- or 7.5-million-year intervals. This procedure imposes periodicity whether it exists or not. A further bias is introduced by the fact that the boundaries of stages and epochs were established on the basis of changes in lithology. Hence the apparent disappearance of genera or families from one stage to the next may result from the improbability of their living in different sedimentary environments rather than from their extinction.

One may use times of extinction and origination of tetrapods during the Permian and Triassic to illustrate these points (Fig. 14.14). The Artinskian was listed as a time of mass extinction by Benton (1989) but in fact reflects a long gap in productive horizons between the extremely rich Lower Permian red beds of the United States and later deposits in Russia and South Africa. The replacement of pelycosaurs by therapsids occurred during this interval, but there are no fossil-bearing

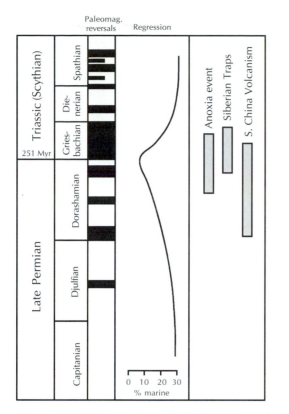

Figure 14.13. Temporal correlation between geological events and the Permo–Triassic boundary. Reprinted (simplified) from *Nature* (Erwin, vol. 367). Copyright © 1994, Macmillan Magazines Limited.

horizons that provide direct evidence of this event. In the Ufimian there is a simultaneous peak in both extinctions and originations that marks the beginning of the extremely rich deposits of the Russian Platform and the Karroo beds of southern Africa. This was a time of rapid turnover and pseudoextinction but overall faunal expansion. The Tatarian, at the end of the Permian, shows a modest extinction, but this cannot be accurately established in the otherwise extremely productive African deposits because of significant gaps in sedimentation at the end of the Permian. The Triassic in southern Africa begins with an entirely different depositional environment and the temporary loss of several lineages, but most reappear later in the Triassic, at which time there is a major radiation of diapsids (Smith 1995). King (1991) concluded that there were two episodes of decrease in the number of genera in the Late Permian but that both were well before the end of the period. When we look at the larger-scale patterns of evolution, rather than simply count the apparent loss of families from one horizon to the next, the end of the Permian appears to represent little more than a hiatus in the preservation of vertebrate fossils.

In the Triassic, the Carnian and Rhaetian are both considered times of major extinction. The Carnian shows more originations than extinctions, whereas the Rhaetian shows a larger number of extinctions. Olsen, Shubin, and Anders (1987) documented a series of stepwise extinctions at the family level in the Triassic, from the end of the Spathian (8), Anisian (5), Ladinian (1), Carnian (10), to the end of the Norian (including the Rhaetian) (14). Although the Carnian ended with a number of extinctions, it also saw the origin of three major tetrapod groups: Sphenodontidae, Pterosauria, and Dinosauria. The Testudines, Crocodylia, and Mammalia all appeared during the time of greatest "extinction," the Norian. This was a period of important turnover, but not of destruction of one fauna and replacement by another. The dominant diapsid reptiles of the Triassic, the thecodonts, overlapped with the dominant diapsids of the Jurassic over a period of 12–27 million years (Benton 1989). Benton (1991) further emphasized the importance of the Carnian extinctions relative to those at the very end of the Triassic, while pointing out the need for more precise study of the relative completeness of the fossil record, the specific timing of extinction events, and their geographic extent. He attributed the loss of nonmarine tetrapod taxa to major phases of turnover (i.e., high extinction and high origination rates) in contrast to extinction events that occurred in a number of nonvertebrate groups (Fig. 14.15).

Further evidence that several of the peaks of mass extinction recognized by Raup, Sepkoski, and Benton were artifacts of the method used to identify and tabulate extinctions is provided by Alroy (in press-a), who surveyed all the ranges of genera and species among Eocene and later mammals from North America. Rather than using previously identified stage or epoch boundaries to separate time intervals, Alroy used radiometrically established 1-million-year divisions. Also, as far as possible, he used monophyletic rather than paraphyletic taxa. He found no significant variation in extinction rate for the period beginning with the lower Eocene and ending just before the Wisconsin glaciation. He was unable to recognize any of the three extinction peaks identified by Benton (1989), nor the two peaks of Raup and Sepkoski. It should be noted, however, that such large-scale statistical studies can obscure significant shorter-term episodes of higher than average extinction within individual taxonomic groups, correlated with short-term climatic change and important periods of migration. For example, during an interval such as encompasses the Eocene–Oligocene boundary (Prothero and Berggren 1992), high extinction rates in one group could be compensated by low rates in others.

Periodicity

Considerable doubt is also cast on Raup and Sepkoski's (1984) hypothesis of a 26-million-year periodicity of mass extinctions by recent studies of bolide impacts (Sharpton and Ward 1990). Although there is mounting support for the correlation between a devastating impact by a large extraterrestrial object and the extinction at the K–T boundary, there is little evidence of large impacts or iridium peaks associated with any of the other major periods of extinction (Jansa, Aubry, and Gradstein 1990; Orth, Attrep, and Quintana 1990). Orth and his colleagues, analyzing 8,000 rock samples in a search for iridium in other horizons where mass extinctions were postulated, concluded that the K–T boundary was unique in its high concentration of this element. They did not find compelling evidence that any horizon older than the terminal Cretaceous event showed evidence of an impact-related cause. Jansa et al. (1990) studied the absence of biological effect from other impacts of objects up to 3 km in diameter and concluded that objects large enough to cause significant disruption of life have probably collided with Earth no more often than once every 100–500 million years. Surprisingly, quite large craters have been formed in horizons that provide no evidence for mass extinction. These in-

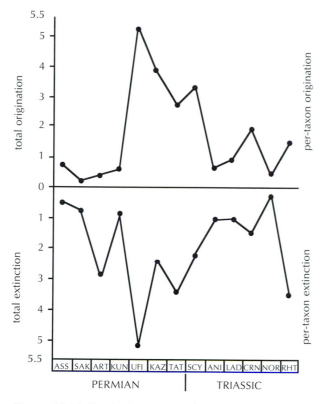

Figure 14.14. Total origination and extinction rates for all tetrapods during the Permian and Triassic. ASS, Asselian; SAK, Sakmarian; ART, Artinskian; KUN, Kungurian; UFI, Ufimian; KAZ, Kazanian; TAT, Tatarian; SCY, Scythian; ANI, Anisian; LAD, Ladinian; CRN, Carnian; NOR, Norian; RHT, Rhaetian. From Maxwell (1993).

Figure 14.15. Overview of extinction events (solid circles) and major phases of turnover (i.e., high extinction and high origination rates; open circles) of various marine and nonmarine groups during the Late Triassic. From Benton (1991).

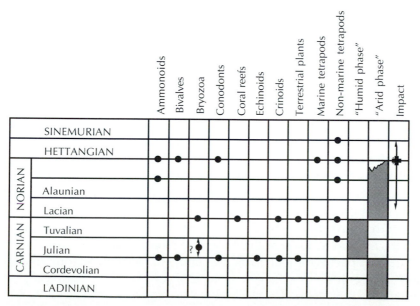

clude the Montagnais Crater on the continental shelf off Nova Scotia, 45 km in diameter and dated at 51 million years, and the Tookoonooka Crater in Queensland,
Australia, 55 km in diameter and dated at 128 million years (Raup 1991).

Summary of extinction

The only period in which worldwide mass extinction certainly had a profound effect on the evolution of vertebrates was at the end of the Cretaceous, when dinosaurs, pterosaurs, marine reptiles, and a large percentage of marsupials all became
extinct within a time span that was probably less than a million years. Nowhere is
there a complete sequence of terrestrial sedimentation across the Permo–Triassic
boundary, precluding determination of actual numbers of taxa lost over that interval; however, it seems unlikely that vertebrates suffered a mass extinction, in view
of the subsequent radiation of most lineages that were present before. The great
number of extinctions that have been tabulated at other times can largely be attributed either to gaps in the fossil record, or to periods of rapid evolution during
which there was extensive turnover in species as a result of replacement of ancestral lineages by their own descendants.

On the other hand, detailed studies of individual faunas at local and continental
scales show many episodes of significant reduction in taxonomic diversity that are
attributable to changes in climate or the immigration of large numbers of taxa from
other geographical regions (e.g., Stehli and Webb 1985a; Benton 1991; Janis 1993;
Prothero 1994; McCune 1996). Extinctions of this scale, however, are of the nature
that could be extrapolated from knowledge of modern phenomena and do not require forces beyond those envisioned by Darwin.

Summary

Processes that can be studied at the level of modern species, including population
growth, genetic change, varying behavior, speciation, and natural selection, constitute a force that has driven evolutionary change throughout the history of vertebrates. However, the patterns of change evident over longer time scales are also
influenced by other phenomena that cannot be extrapolated from studies of living
populations. Inherent factors influencing long-term trends include the anatomy,
physiology, and way of life of the ancestral forms and the relative potential for self-
perpetuating changes in integrated functional complexes (correlated progression).
External factors that are especially important in influencing large-scale evolutionary patterns include physical constraints on body form and physiology. The continually changing configuration of the continents and seas and the associated modifications in the world's climate have strongly influenced the patterns of adaptations
and distribution of both aquatic and terrestrial vertebrates throughout their history. The extinction event at the end of the Cretaceous significantly modified the patterns of evolution of a large number of major vertebrate groups. Other periods of
extinction that have been observed in the fossil record of vascular plants and invertebrates may have affected the evolution of vertebrates as well, but the magnitude of their impact is difficult to assess on the basis of the incomplete stratigraphic and fossil records.

15 Conclusions and comparisons

General features of vertebrate evolution

Evaluation of vertebrate evolution at all time scales, from that of individual populations to the origin and radiation of major clades, reveals these general features:

1. Evolutionary forces that can be studied in modern populations are sufficiently powerful to account for the amount and rate of morphological change throughout the entire course of vertebrate history.
    A.   Genetic variation and selection coefficients recorded in living populations can produce changes at a far faster rate than those observed in the fossil record, or that are required to explain any large-scale evolutionary changes.
    B.   Variation in behavior is a very powerful force of evolution, enabling populations to explore and adapt to varied environments and diets even without anatomical change.
    C.   Speciation allows populations to respond differently to selection pressures and give rise to separate clades that can diverge morphologically and occupy a broader range of adaptive space.

2. Rates of evolution are much more varied than Darwin hypothesized.
    A.   Stasis is much more common that he assumed, but is much less pervasive than argued by Eldredge and Gould on the basis of data from the fossil record of invertebrates and protists.
    B.   Individuals among species respond to selection on a generation-by-generation basis, allowing populations to change genetically over very short time scales.
    C.   Throughout the duration of species, the direction and forces of selection change measurably over short time scales, resulting in much slower net rates of change over millions of years.
    D.   Physiological plasticity and close tracking of favorable environments enable some organisms, such as modern amphibians and reptiles, to adjust to external change with a minimum of genetic and structural modification.

3. Evolutionary constraints impose limits to the extent, rate, and patterns of evolutionary change.
    A.   The nature and extent of change in all organisms are governed by the nature of their immediate ancestors.

389

B.   The properties of the chemical and material constituents of the body are limited in the extent to which they can vary and respond to selection.

C.   Long-term stasis of body form in aquatic and flying vertebrates can be explained by physical constraints of the environment. These apply in a less inclusive manner to terrestrial vertebrates.

D.   Developmental constraints may maintain basic *Bauplans* and details of particular structures for tens of millions of years, but extensive change is possible as a result of strong directional selection.

4.   Transitions between environments governed by major differences in physical constraints do not necessarily require special evolutionary processes.

A.   Periods of transition are not necessarily any less well represented in the fossil record than are other periods of evolution.

B.   The critical period of most transitions ranges from 5 to 15 million years.

C.   The rate of evolution of key characters is not significantly more rapid than that within the ancestral and descendant lineages, but it is essentially unidirectional and more nearly continuous.

D.   Basic changes in processes of development are significant factors in some major transitions.

5.   Rapid radiations may result from mass extinction of other taxa previously inhabiting the same adaptive zone or from the achievement of newly evolved capacities to invade previously unoccupied environments.

6.   Large-scale patterns of evolution cannot be fully explained by processes that are directly observable at the level of modern populations and species.

A.   Changes in the configuration of the continents and their positions relative to one another determine the nature of the climate and oceanic circulation, which in turn influence the course of evolution and patterns of dispersal of both terrestrial and aquatic vertebrates.

B.   Mass extinctions occurred at several intervals in the history of life, but only the end-Cretaceous event has been clearly demonstrated as having a major influence on the course of vertebrate evolution.

7.   The fossil record remains very incompletely known. Prior to the Cenozoic, many species and even genera are known from individual specimens. In only a few cases can the patterns or rates of radiations be established. During most postulated periods of mass extinction, the fossil record is too incomplete to determine whether or not vertebrates were significantly affected.

The most general conclusions that can be derived from these observations are that the patterns, rates, and controlling forces of evolution are much more varied than had been conceived by either Darwin or Simpson.

Is a distinct theory of macroevolution necessary?

Many authors have used the terms *microevolution* and *macroevolution* to differentiate factors that are common to the level of species from those that apply to the evolution of higher taxa. Goldschmidt (who coined the terms in 1940) and Gould (1995b) clearly felt that different processes of evolution were active at these levels. Simpson (1944) added a further category, *megaevolution,* but later stated: "At present I am inclined to think that all three of these somewhat monstrous terminological innovations have served whatever purpose they may have had and that clarity might now be improved by abandoning them" (1953, p. 339). Nevertheless, Simpson wrote of higher categories as characterized by distinct processes of origination, and concluded: "It is my opinion that tachytely [evolution at a very rapid rate] is a usual element in the origin of higher categories and that it helps to explain systematic deficiencies of the paleontological record" (p. 335).

Gould (1995b) argued that Simpson failed to develop a theory of macroevolution because he was influenced by the early-twentieth-century reductionist philosophy of science that required that all phenomena be explained by a small number of universal processes. Since Mendelian and population genetics together with Darwinian selection theory provided a satisfactory explanation of evolution at the level of populations and species, it was awkward to propose a second theory specific to macroevolution. On the other hand, one might argue that Simpson's "failure" was caused by the dearth of data documenting the patterns and rates of evolution during transitions, radiations, and other macroevolutionary events. Simpson (1944) did propose a theory and tempo of evolution to explain the lack of fossil data – quantum evolution at tachytelic rates – but in the absence of data there could be no specific theory to explain its nature.

Is macroevolution conceptionally different than microevolution? The main driving forces are the same as at the species level: population growth, genetic variation, and behavioral plasticity. At both time scales, external factors of the biological and physical environment control the rate, scope, and direction of change.

One of the outstanding problems in large-scale evolution has been the origin of major taxa, such as tetrapods, birds, and whales, that had appeared to arise suddenly, without any obvious ancestors, over a comparatively short period of time. Increased knowledge of the fossil record has greatly increased our understanding of these and other transitions, and shown that they do not necessarily require processes that differ from those known to occur at much lower taxonomic levels. To Simpson and others of his generation, higher categories were recognized by a combination of factors: morphological and adaptive distinction, a significant number of included taxa, and appreciable longevity. From examples considered in this text, it can be seen that adaptive change, morphological change, and radiation can be decoupled in that each may occur at a different time. We now see that the overall rate of evolution is not greatly faster during the origin of a group than it is within the ancestral or the descendant lineages, and with the discovery of intermediate forms, we see that they are not necessarily any more poorly represented in the fossil record than single lineages might be at other stages of evolution. The strongest link between micro- and macroevolution is speciation, which is the same whether a small clade (with one species) or a large clade (with tens of thousands of species) is the result.

Nearly all the factors that have been used to distinguish the origin of higher categories can be attributed to the same processes of speciation, behavioral adaptation, and the gradual accumulation of morphological differences that characterize evolution at the levels of populations, species, and genera. There are no fundamental differences between the early stages in the radiation of placental mammals in the earliest Cenozoic and what is known to have occurred in the origin of the species flocks in the East African Great Lakes.

On the other hand, macroevolutionary patterns cannot be directly extrapolated from microevolution because there are additional external factors that are undetectable over short time scales and whose force, frequency, and duration are unpredictable. These include plate tectonics, with corresponding changes in climate and routes of dispersal, and mass extinction, whether of earthbound or extraterrestrial causes.

A separate macroevolutionary "theory" would have to be limited to factors that are operative only over time scales longer than hundreds of thousands of years – none of which has occurred a sufficient number of times or in sufficiently similar ways to establish a general pattern. Moreover, these factors act in the same general way, although at a larger scale, as do the external forces that control evolution over short time periods. The explanation of macroevolutionary patterns can thus be incorporated into the modern evolutionary synthesis without changing its fundamental hypotheses.

Although formulation of a distinct theory of macroevolution does not appear to be justified, it may be convenient to retain the terms *microevolution* and *macroevolution* to describe the different *patterns* of evolution that are observed at the level of populations and species versus higher taxonomic levels and time spans exceeding 5–10 million years.

Macroevolution, in common with human history, is a historical phenomenon, exciting not because it can be fitted into a particular mold, but because each major event is unique and worthy of detailed study in its own right. Every major transition is an intriguing problem to be solved, and each origin and radiation a mystery of its own.

Agenda for the future

The many advances in knowledge during the past twenty-five years, from a diversity of disciplines bearing on the patterns and processes of evolution, provide a much different outlook on problems than that available to Simpson in his groundbreaking works of the 1940s–60s. The significance of recent theories and new discoveries has only just begun to influence the way in which we think of large-scale evolutionary phenomena and the manner in which they are studied. However, it is now possible to suggest an agenda that takes advantage of the potential for uniting the still diverse elements of research into a program for further investigation of large-scale patterns and processes of evolution.

1. *Genetics*. More use should be made of the growing knowledge of quantita-

tive traits, especially those that regulate developmental processes, to investigate both short- and long-term evolutionary patterns. Less emphasis should be placed on the relative frequency of readily apparent alternative alleles as a basis for understanding how evolutionary change occurs at the level of species and above.

Evidence for the diverse rates of evolution at different time scales, and the extremely rapid changes in the direction and strength of selection in living populations, demonstrate that models of population genetics based on constant selection coefficients over long periods of time are unrealistic. Models should be developed incorporating varying selection coefficients, which more accurately portray the irregular patterns of evolution produced by agents of selection acting from the level of single generations to the duration of species. Such models should be based on actual observations rather than on simulated data.

2. *Developmental biology.* Developmental biology now has the capacity to investigate genetic control over the formation and evolution of major structures. The nature of this control should become generally known among paleontologists. Paleontologists, in turn, may cooperate with developmental biologists by providing information regarding the nature and rates of major anatomical changes that can be determined from the fossil record. From the standpoint of vertebrate evolution, there are many important problems that might be addressed by a more complete understanding of developmental processes. For example: How are the specific patterns of the primitive amniote hand and foot established, and how are the exact number of digits and phalanges retained in many lineages for hundreds of millions of years? What specific developmental factors are involved in the increase or reduction in the number of phalanges? Why is the number of presacral vertebrae extremely stable in crocodiles but extremely variable in lizards? Is this correlated with different patterns or processes of development, or is it controlled primarily by selection?

Evolutionary biologists are eager to learn more of the nature and number of *Hox* genes in a wider range of vertebrates such as the hagfish, reputed to be phylogenetically very divergent from other craniates. Details of the pattern in the lamprey are still not known, and nothing has been published on the *Hox* gene complement in cartilaginous fish or in primitive bony fish such as the sturgeon, gar, and *Amia*.

3. *Paleontology.* Much still needs to be known about major transitions, radiations, and putative times of mass extinction. More attention needs to be given to specific times of origination and termination of lineages, and especially to differentiating pseudoextinctions from true extinctions. Counting taxa is never sufficient to solve the important biological aspects of these events. Another significant challenge is that of developing new approaches to evaluating the rates of evolutionary change. This is an especially serious problem when investigating changes in complex structures and functional complexes during the origin of major adaptive groups. At present, it is not possible to provide an objective measure of the rates of evolution involved in large-scale transitions and radiations, or to compare them in a biologically meaningful manner with the tiny, fluctuating variability that occurs in modern populations.

4. *Behavior.* Relatively little consideration of the importance of behavior in the early stages in the establishment of major clades is evident in the current literature. This would seem to be a fertile field for more detailed investigation.

I expect that numerous conclusions reached in the present study will be challenged, and hope that this will lead to increased concern with major problems regarding large-scale evolutionary phenomena among vertebrates.

Comparisons

This book has concentrated on a single assemblage, the vertebrates, in order to compare aspects of evolution at various time scales, taxonomic levels, and in differing habitats, among a diversity of lineages that shared a single common ancestor. The patterns and processes at the population and species level are very much as hypothesized by Darwin, but larger-scale phenomena differ significantly in both rates and patterns. The question then arises: Do vertebrates provide a good model for other organisms, or are there important differences in the patterns and processes of evolution in prokaryotes, protists, vascular plants, and nonvertebrate metazoans?

Prokaryotes

Schopf (1995) and Knoll (1995) have provided up-to-date reviews of patterns of evolution among prokaryotes and protists, with emphasis on the Precambrian fossil record. The best known fossil prokaryotes are the cyanobacteria or blue-green algae. They are totally asexual photoautotrophs and include species that are anaerobic, aerobic, or facultatively aerobic. Expanding on Simpson's (1944) terminology, Schopf referred to the rate of evolution of cyanobacteria as being "hypobradytelic," or evolving at an astonishingly slow pace.

More than forty genera have been described from the Precambrian. Most can be assigned to the modern families Oscillatoriaceae (filamentous algae) and Chroococcaceae (spheroidal), but five other living families are also represented. Almost all of the Precambrian fossils exhibit detailed similarities of morphology, development, and mode of degradation with modern genera.

Nearly three hundred species have been informally recognized from the Precambrian, compared with approximately two thousand in the living fauna, indicating that the early fossil record of cyanobacteria is surprisingly informative and that the group had achieved most of its modern diversity prior to the Phanerozoic. Seventy-five percent of the 143 Precambrian species of filamentous cyanobacteria and 25 percent of the spheroid species may be compared with particular living species. Where the fossils are found in natural assemblages, the Precambrian species also show similarities with the community organization of living forms.

Both the oscillatoriaceans and the chroococcaceans are known as early as 3.5 billion years ago. Individual genera commonly show little or no morphological change for 1–2 billion years. Many of the Precambrian genera had a worldwide distribution and very large populations, as do their living counterparts. Schopf attributes the great longevity of these taxa to being ecologically unspecialized, with the versatility to survive under a great range of environmental conditions. The simplicity of their genome and the total absence of sexual reproduction would also have greatly reduced the potential for morphological or physiological change.

Protists

Unicellular eukaryotes appear in the fossil record about 2 billion years ago. The most common and diverse Precambrian remains are organic-walled microfossils known as *acritarchs*. Most were the vegetative and reproductive walls of unicellular protists, although the reproductive cysts of multicellular algae and even egg cases of early animals may be included. Few of these remains from the Precambrian can be associated with specific phyla recognized in later horizons. Acritarchs become conspicuous elements in the fossil record as early as 1.7 billion years ago, but initially their diversity was very low. It increased from six to thirteen species during the following 700 million years and then rapidly doubled with a tenfold increase in species turnover. The numbers thereafter remained fairly constant until the Early Cambrian, when there was another doubling in species diversity, and turnover rates increased by an order of magnitude.

Knoll (1995) suggested that the first burst of diversification may have been associated with the appearance of nuclear introns and life cycles in which classical meiosis played a prominent role. Multicellular red, green, and chromophytic algae also evolved at about this time. He attributes the early Cambrian radiation to the enormous increase in the diversity of multicellular organisms, which would have provided a greater variety of adaptive niches for unicellular organisms as well.

Radiolarians, diatoms, coccoliths, dinoflagellates, and foraminifera have been common fossils for much of the Phanerozoic. They have been extensively studied as models for evolutionary change because of the possibility of making very large collections from extremely long sequences of marine sediments (Berggren and Casey 1983). They exhibit a variety of patterns and rates of evolution, from stasis to phyletic gradualism, punctuated equilibrium, and punctuated gradualism. Nevertheless, as Gould and Eldredge (1993) pointed out, even in the lineages that definitely exemplify phyletic evolution, the amount of change is so small and the rate so slow as to question the efficacy of such change in producing major new structures or morphological patterns indicative of major differences in the way of life.

Neither prokaryotes nor protists figured prominently in the development of the evolutionary synthesis, and their Precambrian fossil record has only recently become adequately known; however, they provide an informative contrast in both patterns and rates to those of multicellular organisms.

Nonvertebrate metazoans

Much of the current excitement regarding the patterns and processes of evolution has been generated by Gould and Eldredge's hypotheses based on their extensive knowledge of fossil invertebrates. The many phyla of nonvertebrate metazoans most clearly illustrate the great differences between Darwin's postulated pattern of long-term evolution and that which is documented from the fossil record. The appearance of long-term stasis in a great number of invertebrate species suggests that even the processes of evolution at the species level postulated by Darwin may be fundamentally incorrect.

Appraisal of all vertebrate groups demonstrates that their long-term evolution followed a pattern very different from that which Darwin had hypothesized, but that the basic processes that he postulated among species provide an accurate and sufficient explanation of evolution at this level throughout the history of the group.

Do invertebrates actually exhibit different processes of evolution at the species level than vertebrates, or could the processes hypothesized by Darwin be applicable to invertebrates despite the dominance of species-level stasis in this assemblage?

Variable rates of evolutionary change. One of the most important generalizations gained from this study of vertebrates was the extreme variability in the pattern of their evolution at the species level. A great number of mammalian lineages show progressive change over much of their fossil record, exactly as hypothesized by Darwin. This is particularly well documented in late Cenozoic mammals, for which the fossil record provides samples over relatively short intervals of time throughout their duration. Measurements of large numbers of morphological features demonstrate many changes in the rates and even the direction of change over intervals of a few thousand years, but more consistent patterns are evident over hundreds of thousands of years. In many cases, continuous trends are evident over long successions of species that were differentiated on the basis of morphological criteria. Mammals clearly demonstrate the accuracy of Darwin's analysis of the processes of species-level evolution among vertebrates.

On the other hand, Cenozoic amphibians and reptiles show very little skeletal change within known fossil lineages. Although studies of modern amphibians and reptiles demonstrate that they do show significant change in their soft anatomy over the period of species duration, it is obvious that their skeletal anatomy has hardly changed at all over the past 5 million years. Their stasis resembles the pattern of many fossil invertebrates. Could it be that invertebrates simply show a different balance between the prevalence of stasis and clearly defined phyletic evolution than that seen in many vertebrate groups?

Stasis, as reflected in the longevity of species, seems to be particularly conspicuous among invertebrates with simple external skeletons, such as bivalves and ostracods. Stanley (1985) estimated species durations (in millions of years) of 11–14 for that marine bivalves, 10–13 for marine gastropods, 25 for diatoms, 20–30 for benthic foraminifera, and >20 for planktonic foraminifera, compared to 2–3 for graptolites, >1 for trilobites, and ~5 for ammonites (but with a mode of 1–2). He listed freshwater fish, snakes, and mammals as having respective durations of 3, >2, and 1–2. Clearly, much less morphological information for judging the amount of change is available in a bivalve than a trilobite. There are also likely to be significant physical constraints on the nature of an external covering that would limit the degree and nature of change possible, especially in forms such as foraminifera, diatoms, and ostracods. Vertebrate skeletons, in contrast, reflect a great many aspects of the soft anatomy and behavior that is approached only in arthropods among the invertebrate phyla. It is hardly surprising that simple shells and carapaces rarely show long-term, significant modifications in shape. However, even typical arthropods appear to provide little evidence of extensive phyletic evolution.

Phyletic gradualism at the level of species. Relatively few cases of phyletic change have been adequately documented among invertebrate species. An excellent se-

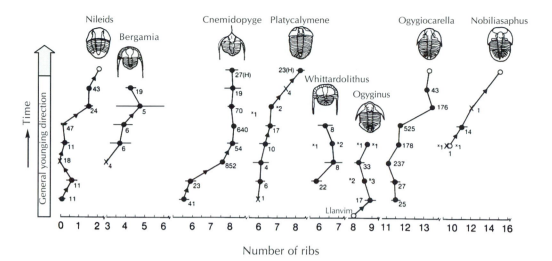

Figure 15.1. Patterns of evolution in eight lineages of trilobites from a 3-million-year sequence of nearly uniformly deposited beds in the Welsh Ordovician. In all lineages the number of ridges or "ribs" in the caudal portion of the exoskeleton increased. The rate and even the direction of change differs during the duration of each lineage and among the various lineages. This pattern clearly demonstrates that evolutionary change cannot be attributed to environmental factors that would be expected to affect all lineages in a comparable manner. Reprinted (modified) from *Nature* (Sheldon, vol. 333). Copyright © 1987, Macmillan Magazines Limited.

ries of examples was provided by Sheldon (1987, 1993); these involve Ordovician trilobites and so provide an informative contrast with Eldredge's demonstration of stasis in the number of eye facets in the same group. Sheldon counted "rib" numbers in the caudal portion of the carapace in eight lineages of trilobites from a sequence of sediments covering a time span of 3 million years. He found a variety of patterns, all of which resulted in net increase but showed considerable variation in the direction and rate of evolution (Figs. 15.1–15.3). In the nileids, the differences were sufficiently pronounced that the early and late members had been assigned to different genera, but knowledge of specimens from intervening horizons indicates that any subdivision of the lineage would be arbitrary. Specimens of *Cnemidopyge* from a succession of horizons showed progressive change in the modes of distribution of "rib" numbers, as would be postulated for phyletic evolution.

A subsequent study of the *Cnemidopyge* lineage, in which a varying number of samples were taken over different stratigraphic intervals, demonstrates that the rate and direction of evolution changed greatly over short intervals but appeared to progress more slowly and directly if only a few samples were taken (Sheldon 1993). This corresponds exactly with the observations of Gingerich that evolutionary rates are inverse to the time interval over which they are measured. These and other studies by Geary (1995) indicate that fossil invertebrates were capable of phyletic evolution, although it may not be as conspicuously expressed or as commonly recognized as among vertebrates.

Sheldon felt that the great importance of invertebrates as stratigraphic indicators may, to a degree, minimize the attention that has been placed on variation and

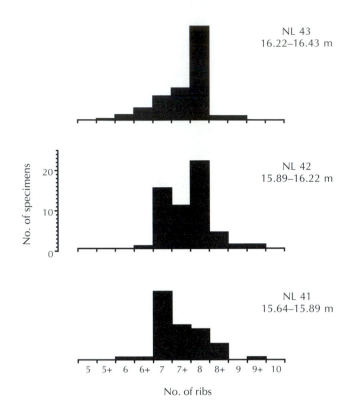

Figure 15.2. Rib count distribution for *Cnemidopyge* from the Welsh Ordovician. The modes in localities NL 41 and NL 43 are clearly 7 and 8, respectively, whereas in NL 42 the distribution is somewhat bimodal, suggesting that the mode shifted from 7 to 8 within the time span represented by this assemblage. The shifting pattern in the number of ribs from one horizon to the next suggests phyletic evolution, as hypothesized by Darwin. Reprinted (modified) from *Nature* (Sheldon, vol. 333). Copyright © 1987, Macmillan Magazines Limited.

change within lineages. All biostratigraphic charts are characterized by straight lines indicating the range of invertebrate taxa, even when references to change and possible interrelationships between lineages are mentioned in the text. Sheldon (1987, p. 563) stated: "[T]he perception of many other gradualistic patterns may have been hindered by conventional descriptive procedures, particularly the requirement to apply binomial taxonomy to fossils and the practice of lumping together specimens collected from different horizons in order to amass enough material for full 'species' description." On the other hand, conspicuously changing lineages, such as have been described among vertebrates, would be expected to have aroused more attention if they were common.

Sheldon (1993) argued that stasis among marine invertebrates may be especially common among the many forms that live in continually changing environments, as in shallow marginal seas, where selection cannot be sufficiently persistent to establish long-term directional change. He suggested that the trilobites he studied may be exceptional because they lived in a deep, isolated basin that was free from cyclical changes in sea level.

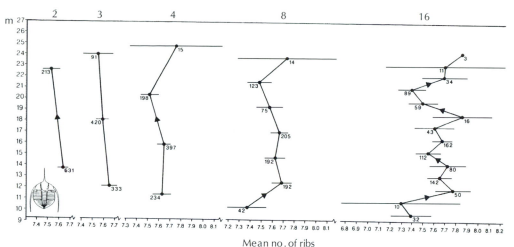

Figure 15.3. Data for part of the *Cnemidopyge* trilobite lineage from a single strati-graphic section, showing the effect of sampling strategy, time-averaging, and reversals on the perception of evolutionary patterns. The number above each column is the number of subdivisions into which the section has been partitioned. Each point repre-sents the mean number of pygidial ribs within that particular subdivision. Beside the mean is its 95-percent confidence interval and the number of specimens with ribs counted. An arrow indicates a significant difference between two successive means at the 95-percent confidence level. From Sheldon (1993).

Stasis at the community level. The importance of changes in the physical environ-ment to the evolution of invertebrates was considered on a more global scale by Brett and Baird (1995) in their analysis of successive communities from the Siluri-an to Middle Devonian. They described a series of faunas that were very stable throughout their duration but separated by episodic perturbations and the collapse of one stable ecosystem and the establishment of another. The faunal replacements were always correlated with significant modifications in the physical environ-ment. All the units are separated by evidence of marked changes in sea level that were associated with extensive migration. Some were characterized by deposition of black shales, indicative of anoxic conditions. The periods of geological and eco-logical change between each of the faunal units lasted about 0.5 million years.

The Hamilton fauna provides a particularly clear example. It lasted about 9 mil-lion years and comprised approximately 330 species spanning the range of the in-vertebrate phyla. Eighty percent of the species, none of which changed signifi-cantly, occurred throughout the sequence. Brett and Baird referred to this as **coordinated stasis**, in which not just single species but the entire fauna exhibit-ed stasis. Here and in other parts of the sequence, only 10–30 percent of the spe-cies continued from one faunal unit to the next. This phenomenon is not unique to the Silurian and Devonian, but has been identified from the Cambrian into the Cenozoic.

During the Quaternary, the marine environment was subject to even more fre-quent changes in sea level and temperature resulting from the periodic advances and retreats of the continental ice sheets. Marine organisms faced sea-level changes of up to 100 m as well as warming or cooling of the water that occurred at 100,000-year intervals, each change taking place over periods as short as 10,000 years. Surprisingly, these changes resulted not in mass extinctions but rather in periodic and rapid dispersals so that the organisms could remain in the environments to which they were best adapted (Bennett 1990). Because of different rates of dispersal by different types of organisms, all species had to adapt continuously to different community structures. This was apparently achieved primarily by behavioral means, for there is little if any evidence of consistent morphological change in any of the affected lineages.

Brett and Baird argued that periodic changes in the physical nature of the marine environment, rather than anagenetic change in the organisms themselves, were the primary force in the evolution of marine invertebrates. This appears to be in strong contrast with what has been observed in marine vertebrates. The only time when an entire community of marine fish is known to have become extinct over a relatively short period was in the Late Devonian, when all the remaining placoderm lineages became extinct. This was also accompanied by a change in environment, indicated by the end of deposition of black shales, in which these fish were frequently preserved. Other fish groups underwent a period of major radiation across the Devonian–Carboniferous boundary. There is no other period when there is an obvious break in the evolution of marine fish, although their early fossil record is very incompletely known.

Terrestrial invertebrates – specifically, insects – like their marine counterparts, show a very strong capacity for migration, which is clearly a more widely used strategy than genetic change in coping with major environmental modification over intervals of tens to hundreds of thousands of years. Coope (1995) pointed out that insects that dealt successfully with the repeated advances and retreats of the ice sheets during the Quaternary were specifically adapted to following the appropriate climatic zones as they shifted north and south, so that they remained in the same environment throughout the late Cenozoic. They show a remarkable degree of morphological stasis during this period.

The late Cenozoic climatic changes brought about comparable range changes among terrestrial vertebrates, with rapid dispersal into and out of areas subject to glaciation. The rate of advance and retreat of individual episodes of continental glaciation appear to have been too rapid for significant skeletal evolution, even among mammals. Earlier Cenozoic mammals can be recognized as belonging to a series of land mammal ages, whose boundaries are marked by periods of extensive migration and faunal turnover, but nearly all the family level units show progressive evolutionary change throughout this time, as do the marine mammals. At no time has the history of vertebrates, either terrestrial or aquatic, been subject to periodicity between stasis and rapid change involving entire communities, as was apparently common to most invertebrate phyla.

The rarity of fossil evidence of long-term evolutionary trends. Unfortunately, few major transitions or long-term evolutionary trends are recognized in the fossil record of invertebrate phyla. Many of the classes diverged within a few million years after the emergence of the phyla in the Early Cambrian. There are few if any equivalents of the protracted sequences of evolutionary change that are documented by the origin of mammals and teleost fish; nor is there adequate fossil evidence of transitions between distinct physical environments such as are exemplified by the origin of amphibians, birds, and whales.

In terms of subsequent species diversity, the origin of insect flight is one of the most important events in the history of arthropods, yet it must be studied mainly on the basis of living genera (Marden and Kramer 1994). Developmental biology now appears to provide a more informative insight into the evolution of insect wings and arthropod limbs than does the study of fossils (S. Carroll, Weatherbee, and Langland 1995; Panganiban et al. 1995). However, this may reflect primarily gaps in the fossil record, and perhaps a lesser likelihood of preservation of this and other important transitions, rather than being indicative of different processes of evolution.

Gould and Eldredge have argued that inherent genetic and/or developmental factors limit the amount of change that species can undergo, while any minor changes that do accumulate during the duration of species may be truncated by major shifts in the environment or mass extinction. Bennett (1990) argued that the short-term periodicity of the Milankovitch cycles would be expected to have a significant effect on the pattern of evolution in all environments that would reverse, dilute, or undo whatever changes had accumulated over shorter time scales within species. He proposed that they might fit into Gould's (1985) three tiers of evolutionary phenomena between ecological time and geological time. They would contribute to evolutionary stasis, since no evolutionary trends could persist for more than 23,000, 41,000, 100,000, or 400,000 years.

With the exception of the periodic drying up of lakes and the resultant extinction of their faunas (McCune 1996) and the dispersal response to glacial advances and retreats, the patterns of vertebrate evolution do not appear to be nearly as susceptible to either Milankovitch or longer-term cyclicity, in either the aquatic or the terrestrial environment, as are those of invertebrates. The degree to which this may reflect an inherently different capacity of invertebrates to respond to natural selection through change in allele frequencies remains to be determined. It is possible that the habitats and ways of life of most invertebrates make it more effective for them to track environments to which they originally adapted rather than to evolve structurally so as to adapt to changing conditions. References to occasional rapid and significant change in species from horizon to horizon have yet to be analyzed in sufficient detail to establish whether or not the rates and nature of change accord with Darwinian processes.

Labandeira and Sepkoski (1993) provided a very direct comparison of patterns of evolution of land vertebrates and insects through measures of the proportion of living families that have appeared at different times in the past (Fig. 15.4). Few if any new insect families have appeared during the Cenozoic, and even 100 mya, 84 percent of the insect fauna consisted of modern families. In contrast, only about 25 percent of early Cenozoic tetrapods belonged to modern families, and 100 mya the number was under 20 percent. Bivalves have shown even less change, with more than 50 percent of the modern families being present in the Lower Triassic.

Insects have been significantly more diverse than tetrapods since the Carboniferous. Labandeira and Sepkoski argued that this diversity was the result of a much lower rate of extinction at the family level rather than a higher origination rate.

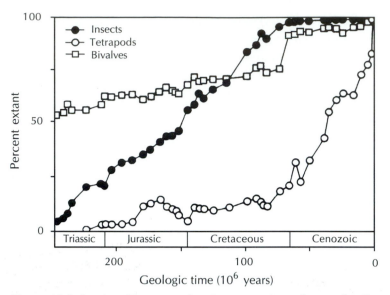

Figure 15.4. Survivorship curves showing proportions of extant families as a function of time for insects, terrestrial vertebrates, and marine bivalves. Steeper curves indicate a higher extinction rate among families. The low extinction rates of insects, especially relative to tetrapods, mean that even a relatively discontinuous fossil record provides a reasonably continuous record of insect diversity. Reprinted with permission from *Science*, vol. 261, Labandeira and Sepkoski, Jr. Copyright © 1993, American Association for the Advancement of Science.

They attributed this to ". . . the intrinsic evolutionary behavior of their constituent species" (p. 312).

Vascular plants

Vascular plants have an informative fossil record beginning in the Silurian. They have undergone a series of successive radiations during the Phanerozoic that broadly resemble those of terrestrial vertebrates. Their history began with a diversification of spore-bearing forms in the Late Silurian and Early Devonian that are represented today by horsetails, ferns, and lycopods; it continued with primitive seed-bearing plants in the Upper Devonian, exemplified today by the conifers and cycads. Since the Lower Cretaceous, the world has been dominated by the *angiosperms,* or flowering plants (Crane et al. 1995).

Vascular plants differ considerably from multicellular animals in their evolutionary potential. They have great capacity for vegetative as well as sexual reproduction, and are much more capable of forming fertile hybrids than are most metazoans, raising the potential for reticulate evolution. With or without hybridization, polyploidy has been an extremely important factor in the instantaneous appearance of new species. Plants have compensated for the absence of mobility of the adult stage by extremely effective means of dispersing spores, pollen, and seeds, so that range shifts may be faster and more extensive than those of the most mobile animals.

Dormancy of both the standing crop and the seeds enables plants to survive large-scale seasonal variations as well as catastrophic changes in the environment of as great a magnitude as the end-Cretaceous extinction (Spicer 1989). These factors suggest that vascular plants would be capable of very long-term stasis, as does the average longevity of modern species of 8–20 million years (Stanley 1985). On the other hand, the enormous and continuing radiation of angiosperms demonstrates the capacity for major and relatively rapid changes in anatomy.

Vascular plants have not played a conspicuous role in the controversy between punctuated equilibrium and phyletic gradualism – or at least they have not commonly been used as examples in the papers discussing these patterns among vertebrates and other metazoans.

Final conclusions

Throughout their history, vertebrates provide the most consistently informative fossil record of any major group of organisms. In addition, our familiarity both with their current diversity and with major extinct groups – from dinosaurs to fossil hominids – makes it simple to appreciate the pattern of their evolution and the nature of the major transitions and radiations. However, one must bear in mind that vertebrates are but one of a number of major types of organisms and may emphasize but a segment of the total spectrum of evolutionary patterns and processes.

Prokaryotes and protists have a much longer history and are distinguished by some aspects of evolution that may be more general to life than those that typify vertebrates. Evolution is commonly thought of in terms of change, as indicated by the definition in *Webster's New World Dictionary* (2d college ed., 1970): "process of development, as from a simple to a complex form, or of gradual, progressive change. . . ." Stebbins (1988, p. 4) reviewed other definitions, including those emphasizing the genetic component ("evolution consists of alteration in genetic frequency . . .") or population-environmental interactions (". . . the essence of evolution is the production of adaptive diversity)." He settled on that of Futuyma (1986, p. 7): "Biological evolution is change in the properties of populations of organisms that transcends the lifetime of a single individual." However, prokaryotes lack most of the properties on which these definitions were based, and in neither prokaryotes nor protists are individuals separated from populations in the sense used by Futuyma.

Using prokaryotes as the clearest example, we might consider the most obvious aspect of evolution to be survival, accompanied by continuing reproduction. As emphasized by Schopf (1995), cyanobacteria have maintained the capacity to remain in the same habitat, with identical external morphology, for 3.5 billion years.

Eldredge and Gould have argued that "stasis is data," but one may also say that stasis has always been a major strategy of evolution. Stasis is not specifically a character of species – a concept that has very different meanings for vertebrates and prokaryotes or other asexual organisms – but may apply at any taxonomic level. It is most conspicuously expressed in prokaryotes and single-celled eukaryotes, but progressively less so in multicellular organisms of greater levels of genetic and structural complexity. It is least commonly expressed among mammals.

Cyanobacteria are able to maintain stasis by great physiological flexibility and tolerance to changing environments. Relatively bradytelic invertebrates and vertebrates are also frequently characterized by physiological and dietary flexibility (Wake, Roth, and Wake 1983; Eldredge and Stanley 1984). Even in much more complex organisms, physiological and behavioral flexibility may be more effective than structural and physiological change in coping with rapid and radical changes in the environment, such as large shifts in water level, continental glaciation, or the impact of asteroids. Hibernation, estivation, and other forms of dormancy also enable species in a wide spectrum of taxonomic groups to survive environmental conditions with which they could not cope as active organisms. It is hardly surprising that these strategies have contributed to the range of variability in evolutionary rates and patterns that are expressed.

However, within the total span of life on Earth, there has been a shift in the balance between evolutionary history being expressed as the maintenance of stasis in a changing environment and an increasing capacity of species for genetic and structural change to meet the challenges of an ever-increasing range of environmental possibilities. The simplest organisms, with only a single copy of the genetic material and no possibility of systematic recombination, maintain a conservative genetic makeup from generation to generation and have little capability for ongoing change. Single-cell eukaryotes have achieved nearly all the metabolic processes of multicellular organisms, but without cellular differentiation they are also limited in their potential for structural change. With the evolution of complex systems for control of developmental mechanisms, the current diversity of metazoans became possible. At the same time, increased structural and physiological complexity limited the capacity for the flexibility that enabled simpler forms to adapt to such broad and long-lasting environments that structural evolution was not necessary. The vertebrate level of genetic, developmental, and structural complexity pushed the balance further with the capacity for extensive change in nearly all aspects of the skeleton so that adaptation to the widest range of environments and ways of life is possible. Physiological plasticity, behavioral flexibility, and dormancy remain as viable strategies, but structural and physiological change dominate the most actively expanding groups.

As the genetically, developmentally, and physiologically most complex organisms, vertebrates have sacrificed the stability and great species longevity common to other forms of life. On the other hand, they have emphasized other aspects of evolution in their manifest capacity to adapt to a diversity of habitats based on distinctive morphological patterns, which make them the most informative group for the study of evolutionary change.

Glossary

actinotrichia
Collagenous fibrils supporting the fins of actinopterygian fish.

adaptive topography
Wright's use of a topographic analogy to illustrate the changes in allele frequencies during adaptation.

additive effect
The influence on the genetic variance of the phenotype resulting from alleles that are additive, rather than showing dominance relative to one another.

allopatric
Used in reference to populations or species that are physically isolated from one another.

allopatric speciation
Speciation that results from physical isolation of populations.

anagenesis
Evolutionary change over time within a single lineage.

apical ectodermal fold
Ridge of tissue that forms at the distal margin of the fin in bony fish.

apical ectodermal ridge (AER)
Line of tissue that develops at the tip of the limb bud; it is essential to limb growth and differentiation.

apomorphy
Derived character.

apomorphy-based definition
Crown group that includes all the descendants of a species that had achieved a particular apomorphy, such as the feathers of birds.

aspect ratio
Ratio of a tail's span to its chord.

autapomorphy
Derived character that is unique to a particular taxon.

autopod
Distal part of the limbs including the carpals or tarsals, the metacarpals or metatarsals, and the digits.

biological species concept
Recognition of species on the basis of the potential of individual members to reproduce with one another, but not with other species.

bone morphogenetic proteins (BMPs)
Genetically specific bone- and cartilage-inducing molecules.

cartilage replacement bone
See endochondral bone.

character displacement
Evolutionary change of structure and behavior in closely related species that allows them to exist in sympatry without competing for the same resources.

chord
Average anteroposterior extent of the tail.

chronomorph
One of a series of populations originally designated as separate species within an evolving lineage that is not subdivided by speciation events.

clade
Group that can trace its origin to a single ancestral species.

cladistics
See phylogenetic systematics.

cladogenesis
Division of single species into two reproductively isolated lineages.

cline
Gradient of phenotypic or genotypic change in a population or species correlated with the orientation of features of the environment such as altitude, latitude, temperature, and dryness.

continental drift   Movement of continents relative to one another and the Earth's axis of rotation (*see* plate tectonics).

coordinated stasis   Phenomenon illustrated in many communities of marine invertebrates in which nearly all species exhibit stasis for tens of millions of years, followed by major environmental turnover and the establishment of a new community.

correlated progression   Term coined by Kemp in reference to the origin of mammals, in which the complex integration of a number of structural/functional systems acted as a force propelling subsequent evolution.

crown group   All the species that share a common ancestry with the living members of a monophyletic group.

darwin   Measure of evolutionary rate; the amount of change per million years expressed as the natural log of linear dimensions.

dermal bone   Bone that forms superficially and without a cartilaginous precursor; also called *intramembranous bone.*

digital arch   Extension of the axis of development within the limb bud into the autopod of tetrapod limbs.

dominance effect   The influence of the genetic variance of the phenotype resulting from the degree of dominance of one allele relative to another. The dominance effect does not contribute to evolutionary change.

endochondral bone   Bone that forms deep in the body, and is preformed in cartilage; also called *cartilage replacement bone.*

epistasis   Circumstance in which two or more genes acting together produce phenotypes that differ from those expected when either locus is considered individually.

evolutionary constraints   Limits to the direction, nature, rate, or amount of change that is possible, exclusive of biological factors such as competition, predation, and other features stressed by Darwin.

founder effect   Abrupt changes in allele frequencies that may result from small colonizing populations.

friction drag   Drag resulting from the resistance between the boundary layer and the surrounding fluid.

gel electrophoresis   Method of determining the presence of different alleles by the differential rate of migration of their protein products in an electric field.

gene groups   Different *Hox* genes that make up the homeobox cluster on a single chromosome.

genetic drift   Change in allele frequency by chance alone.

ghost lineage   Early portion of a lineage that is not known from the fossil record but is predicted to have existed on the basis of the earlier appearance of a sister-group.

greenhouse condition   Period in the history of Earth when the temperature was consistently above average.

growth/differentiation factors (GDF)   Genetically specific proteins influencing precartilaginous mesenchymal condensations and formation of perichondrium.

haldane   Measure of evolutionary rate; one haldane is defined as change by a factor of one standard deviation per generation.

heritability   Degree to which traits are genetically controlled.

heterochrony   The dissociation of the relative timing of events in development between ancestral and descendent ontogenies.

historical contingency   Term coined by Gould in reference to the importance of historical factors in establishing the direction of evolution.

homeobox	Highly conserved region of a gene that codes for a particular sequence of amino acids termed the *homeodomain*.
homeobox genes	Genes containing a homeobox sequence.
homeodomain	Sequence of amino acids in a protein that serves as a sequence-specific DNA-binding site for genes that control aspects of development.
homeotic genes	Genes that control the position in the embryo where a structure develops.
homeotic mutants	Mutants observed in arthropods that result in the appearance of structures in inappropriate segments of the body (e.g., legs rather than antennae on the head).
hopeful monsters	Term coined by Goldschmidt for organisms expressing systemic mutations, a very small percentage of which might give rise to important new structures.
Hox **cluster**	Arrangement of *Hox* genes in a continuous linear sequence on a single chromosome.
Hox **genes**	Genes that make up the homeobox cluster.
icehouse condition	Period in the history of Earth when the temperature was consistently below average. Icehouse conditions were characterized by extensive, long-term continental glaciation.
induced drag	Drag produced by air that flows around the wing tip and inward over the wing.
intramembranous bone	*See* dermal bone.
intrinsic evolutionary rate	Average change in a character over a single generation, measured in standard deviations.
iterative evolution	*See* racemic evolution.
large igneous province	Enormous areas of lava flows, either under the sea or on land, resulting from long-lasting mantle plumes.
lepidotrichia	Scales making up the dermal fin rays.
limb field	Area of the embryo where a limb bud will form.
living fossils	Plants or animals living today that closely resemble fossils of species tens to hundreds of millions of years old.
macroevolution	Evolution above the level of species.
mantle plume	Flow of magma from the base of the mantle to the Earth's surface without respect for the distribution of the crustal plates.
mass extinction	Extinction of large numbers of species, belonging to a number of major taxonomic groups, because of sudden changes in the physical environment (and presumably without regard to the relative fitness of the species in their normal environment).
mean	Expression (in quantitative genetics) of a quantitative trait, averaged over all members of a population.
meristic change	Change in the number of particular skeletal elements, such as vertebrae, fin rays, digits, or phalanges.
mesopodial	Carpal or tarsal.
metapodial	Metacarpal or metatarsal.
microevolution	Evolution at the level of populations and species.
monophyletic group	Originally coined in reference to a group with a single common ancestor; redefined by Hennig as a single species and all of its descendants.
morphogens	Molecules that diffuse through the embryo, controlling particular aspects of its development.

mosaic evolution   Evolution of characters at various rates both within and between species.

multifactorial traits   *See* quantitative traits.

neoteny   Particular pattern of heterochrony in which the reproductive system matures while other aspects of the body retain a level of development comparable to that of the larval stage of other individuals or taxa.

net rate of evolution   The net amount of change, in units of standard deviation, measured over many generations, divided by the number of generations. This is always a substantially lower number than the intrinsic evolutionary rate.

node-based definition   Definition of a crown group to include all the descendants of the most recent common ancestor of the living clades.

paralogous genes   Genes occupying comparable linear positions on different chromosomes in the homeobox cluster of vertebrates, and presumed to have evolved from a single ancestral gene.

paraphyletic group   Species and some, but not all, of its descendants.

peripatric speciation   Speciation as a result of reproductive isolation of populations on the periphery of the range of a species.

phyletic evolution   Evolutionary change within a lineage without speciation.

phyletic gradualism   Term coined by Eldredge and Gould in reference to slow, continuous evolutionary change within a species.

phylogenetic systematics   Methodology established by Hennig for determining relationships and classifying organisms that is based on evolutionary affinities; also called *cladistics,* reflecting the importance of recognizing clades.

phylotypic stage   Developmental stage during which embryos within each phylum are strikingly similar to one another.

plate tectonics   Movement, separation, and coalescence of lithosphere plates as a result of convection currents in the Earth's mantle (*see also* continental drift).

pleiotrophy   Single gene affecting a number of different traits.

plesiomorphy   Primitive character, as such, having no value in establishing the relationship of a clade.

polygenic traits   *See* quantitative traits.

pressure drag   Drag resulting from displacement of the water in front and behind an animal and around any irregularities of its surface.

profile drag   Drag resulting from the friction of the air against the body, the displacement of air, the formation of pressure gradients in the air, and the creation of eddies.

progress zone   Region of cell division within the limb bud that determines the proximal–distal limb axis.

punctuated equilibria   Theory of evolution proposed by Eldredge and Gould in which there is very little change within the duration of species but significant change at the time of speciation.

qualitative traits   Traits that are controlled by alternative alleles at a single locus.

quantitative traits   Traits affected by more than one pair of alleles, commonly with additive results. They are also referred to as *multifactorial traits,* since their expression is commonly affected by environmental as well as genetic factors.

quantum evolution   Term coined by Simpson in reference to periods of rapid evolution with little fossil evidence that were thought common to the origin of major taxonomic groups.

racemic evolution   Pattern of evolution in which a number of similar lineages diverge from an ancestral stock over time; by analogy with the pattern of flower structure exemplified by the delphinium. Also called *iterative evolution*.

regulatory genes   Genes controlling the timing, place, and degree of expression of other genes.

response to selection   Difference between the mean phenotypic value of the offspring population and the mean of the entire parental population before selection.

selection coefficient   Force of selection calculated as a ratio of the number of surviving progeny resulting from individuals with different alleles, which were originally present in the same percentage.

selection differential   Difference between the mean of the parents that give rise to the next generation and the mean for the entire parental population.

shifting balance theory   Wright's model to explain how species might shift from one adaptive peak to another (*see* adaptive topography).

sibling species   Closely related species that are very difficult to differentiate on morphological criteria but are nonetheless reproductively isolated.

sister-group   One of a pair of taxa that share an immediate common ancestry.

span   Maximum dorsoventral extent of the tail.

speciation   Division of a single species into two reproductively isolated lineages.

species flock   Group of species, such as the cichlids in Lake Victoria, that share an immediate common ancestry and are restricted to a single geographical area.

species selection   Concept of Stanley that large-scale evolutionary changes may proceed by selection acting at the level of species rather than individuals, as argued by Darwin.

standard deviation   The square root of the variance; a fixed fraction of the area under the curve of a normal distribution, so that 68 percent of the observations fall within one unit on either side of the mean.

stem-based definition   Crown group based on the more immediate common ancestry of all its species with one rather than another of two living groups.

stem group   Members of a monophyletic group that do not share an immediate common ancestry with any of the living clades within that group.

stylopod   Proximal element of the limb; humerus or femur.

sympatric   Used in reference to a single restricted geographical area.

sympatric speciation   Speciation occurring in a restricted geographical area, without a physical barrier to reproduction.

symplesiomorphy   Primitive character common to two or more clades; cannot be used to establish specific interrelationship.

synapomorphy   Shared derived character that unites sister-groups.

variance   Measure of the deviation of the individual members of a population from the population mean; calculated as the sum of the squares of the deviation of each member from the mean, divided by the number of individuals.

wing loading   Ratio of the weight of the body divided by the area of the wings.

zeugopod   Ulna and radius or tibia and fibula.

zone of polarizing activity (ZPA)   Group of cells on the posterior margin of the limb bud that determines the anterior–posterior axis of the limb.

References

Ahlberg, P. E. (1991). Tetrapod or near-tetrapod fossils from the Upper Devonian of Scotland. *Nature* 354:298–301.

Ahlberg, P. E. (1995). *Elginerpeton pancheni* and the earliest tetrapod clade. *Nature* 373:420–5.

Ahlberg, P. E., Clack, J. A., and Lukševičs, E. (1996). Rapid braincase evolution between *Panderichthys* and the earliest tetrapods. *Nature* 381:61–4.

Ahlberg, P. E., Lukševičs, E., and Lebedev, O. A. (1994). The first tetrapod finds from the Devonian (upper Famennian) of Latvia. *Philosophical Transactions Royal Society London B* 343: 303–28.

Alberch, P. (1985). Problems in the interpretation of developmental sequences. *Systematic Zoology* 24:46–58.

Alberch, P., and Blanco, M. J. (1994). Process and outcome: Evolutionary analysis of morphological patterns. In *Experimental and Theoretical Advances in Biological Pattern Formation*, ed. H. G. Othmer, P. K. Maini, and J. D. Murry, pp. 11–20. New York: Plenum Press.

Alberch, P., and Gale, E. (1985). A developmental analysis of an evolutionary trend. Digital reduction in amphibians. *Evolution* 39:8–23.

Alexander, R. McN. (1974). *Functional Design in Fishes*. London: Hutchinson & Co.

Alexander, R. McN. (1996). *Tyrannosaurus* on the run. *Nature* 379:121.

Allen, J. A. (1877). The geographical distribution of mammals. *U.S. Geological Survey* 4:313–76.

Alroy, J. (in press-a). Constant extinction, constrained diversification, and uncoordinated stasis in North American mammals. *Palaeogeography, Palaeoclimatology, and Palaeoecology*.

Alroy, J. (in press-b). Long-term equilibrium in North American mammalian diversity. In *Biodiversity Dynamics: A Turnover of Populations, Taxa, and Communities*, ed. M. L. McKinney. New York: Columbia University Press.

Alvarenga, H. M. F., and Bonaparte, J. F. (1992). A new flightless landbird from the Cretaceous of Patagonia. *Science Series. Natural History Museum Los Angeles* 36:51–64.

Alvarez, L. W., Alvarez, W., Asaro, F., and Michel, H. V. (1980). Extraterrestrial cause for the Cretaceous–Tertiary extinction. *Science* 208:1095–108.

Anders, M. H., Krueger, S. W., and Sadler, P. M. (1987). A new look at sedimentation rates and the completeness of the stratigraphic record. *Journal of Geology* 95:1–14.

Anderson, D. K. (1993). A method for recognizing morphological stasis. In *Morphological Change in Quaternary Mammals of North America*, ed. R. A. Martin and A. D. Barnosky, pp. 13–23. Cambridge: Cambridge University Press.

Andrews, S. M., and Westoll, T. S. (1970). The postcranial skeleton of *Eusthenopteron foordi* Whiteaves. *Transactions of the Royal Society of Edinburgh* 68:207–329.

Archibald, J. D. (1982). A study of Mammalia and geology across the Cretaceous–Tertiary boundary in Garfield County, Montana. *University of California Publications Geological Sciences* 122:xvi, 286.

Archibald, J. D. (1983). Structure of the K–T mammal radiation in North America: Speculations on turnover rates and trophic structure. *Acta Paleontologica Polonica* 28:7–17.

Archibald, J. D. (1993). The importance of phylogenetic analysis for the assessment of species turnover: A case history of Paleocene mammals in North America. *Paleobiology* 19:1–27.

Archibald, J. D. (1994). Metataxon concepts and assessing possible ancestry using phylogenetic systematics. *Systematic Biology* 43:27–40.

Archibald, J. D. (1996). Fossil evidence for a Late Cretaceous origin of "hoofed" mammals. *Science* 272:1150–3.

Archibald, J. D. (in press). Archaic ungulates ("Condylarthra"). In *Evolution of Tertiary Mammals of North America*, ed. C. M. Janis, K. M. Scott, and L. Jacobs. Cambridge: Cambridge University Press.

Archibald, J. D., Clemens, W. A., Gingerich, P. D., Krause, D. W., Lindsay, E. H., and Rose, K. D. (1987). First North American land mammal ages of the Cenozoic Era. In *Cenozoic Biochronology of North America*, ed. M. O. Woodburne, pp. 24–76. Berkeley: University of California Press.

Archibald, J. D., and Lofgren, D. L. (1990). Mammalian zonation near the Cretaceous–Tertiary boundary. In *Dawn of the Age of Mammals in the Northern Part of the Rocky Mountain Interior, North America*, ed. T. M. Bown and K. D. Rose. *Geological Society of America Special Paper* 243:31–50.

Atchley, W. R. (1993). Genetic developmental aspects of variability in the mammalian mandible. In *The Skull*, vol. 1, *Development*, ed. J. Hanken and B. K. Hall, pp. 207–47. Chicago: Chicago University Press.

Auffenberg, W. (1958). Fossil turtles of the genus *Terrapene* in Florida. *Bulletin of the Florida State Museum, Biological Series* 3:53–92.

Avise, J. C. (1994). *Molecular Markers, Natural History, and Evolution*. New York: Chapman & Hall.

Avise, J. C., Nelson, W. S., and Sugita, H. (1994). A speciational history of "living fossils": Molecular evolutionary patterns in horseshoe crabs. *Evolution* 48:1986–2001.

Ax, P. (1987). *The Phylogenetic System*. New York: John Wiley & Sons.

Ayala, J. F., and Valentine, J. W. (1979). *Evolving: The Theory and Processes of Organic Evolution*. Menlo Park, Calif.: Benjamin/Cummings.

Bailey, D. W. (1985). Genes that affect the shape of the murine mandible. *Journal of Heredity* 76:107–14.

Bailey, D. W. (1986). Genes that affect morphogenesis of the murine mandible. *Journal of Heredity* 77:17–25.

Barel, C. D. N. (1984). Form-relations in the context of constructional morphology: The eye and suspensorium of lacustrine Cichlidae (Pisces, Teleostei), with a discussion on the implications for phylogenetic and allometric form-interpretations. *Netherlands Journal of Zoology* 34:439–502.

Barel, C. D. N., Anker, G. C., Witte, F., Hoogerhoud, R. J. C., and Goldschmidt, T. (1989). Constructional constraint and its ecomorphological implications. *Acta Morphologica Neerlando-Scandinavica* 27:83–109.

Barel, C. D. N., Ligtvoet, W., Goldschmidt, T., Witte, F., and Goudswaard, P. C. (1991). The haplochromine cichlids in Lake Victoria: An assessment of biological and fisheries interests. In *Cichlid Fishes: Behaviour, Ecology, and Evolution*, ed. M. H. A. Keenleyside, pp. 258–79. London: Chapman & Hall.

Barnosky, A. D. (1987). Punctuated equilibria and phyletic gradualism: Some facts in the Quaternary mammalian record. *Current Mammalogy* 1:107–47.

Barnosky, A. D. (1990). Evolution of dental traits since latest Pleistocene in meadow voles (*Microtus pennsylvanicus*) from Virginia. *Paleobiology* 16:370–83.

Barnosky, A. D. (1993). Mosaic evolution at the population level in *Microtus pennsylvanicus*. In *Morphological Change in Quaternary Mammals of North America*, ed. R. A. Martin and A. D. Barnosky, pp. 24–59. Cambridge: Cambridge University Press.

Barton, N. H., and Charlesworth, B. (1984). Genetic revolutions, founder effect, and speciation. *Annual Review of Ecology and Systematics* 15:133–64.

Barton, N. H., and Turelli, M. (1989). Evolutionary quantitative genetics: How little do we know? *Annual Review of Genetics* 23:337–70.

Bateson, W. (1894). *Materials for the Study of Variation*. London: Macmillan.

Behrensmeyer, A. K., and Hill, A. P. (eds.) (1980). *Fossils in the Making: Vertebrate Taphonomy and Paleontology*. Chicago: University of Chicago Press.

Behrensmeyer, A. K., and Hook, R. W. (1992). Paleoenvironmental contexts and taphonomic modes. In *Terrestrial Ecosystems through Time*, ed. A. K. Behrensmeyer, J. D. Damuth, W. A. Di-

Michele, R. Potts, H.-D. Sues, and S. L. Wing, pp. 15–136. Chicago: University of Chicago Press.

Bell, G. A. C. (1996). *Selection: The Mechanism of Evolution*. New York: Chapman & Hall.

Bell, G. L., Jr. (1996). The first direct evidence of live birth in Mosasauridae (Squamata): Exceptional preservation in the Cretaceous Pierre Shale of South Dakota. *Journal of Vertebrate Paleontology, Abstracts of Papers* 16(Supplement):21A–22A.

Bell, M. A. (1988). Stickleback fishes: Bridging the gap between population biology and paleobiology. *Trends in Ecology and Evolution* 3:320–5.

Bell, M. A. (1994). Palaeobiology and evolution of the threespine stickleback. In *The Evolutionary Biology of the Threespine Stickleback*, ed. M. A. Bell and S. A. Foster, pp. 438–71. Oxford: Oxford University Press.

Bell, M. A., Baumgartner, J. V., and Olson, E. C. (1985). Patterns of temporal change in single morphological characters of a Miocene stickleback fish. *Paleobiology* 11:258–71.

Bell, M. A., and Foster, S. A. (eds.) (1994a). *The Evolutionary Biology of the Threespine Stickleback*. Oxford: Oxford University Press.

Bell, M. A., and Foster, S. A. (1994b). Introduction to the evolutionary biology of the threespine stickleback. In *The Evolutionary Biology of the Threespine Stickleback*, ed. M. A. Bell and S. A. Foster, pp. 1–27. Oxford: Oxford Univesity Press.

Bemis, W. E., and Lauder, G. V. (1986). Morphology and function of the feeding apparatus of the lungfish, *Lepidosiren paradoxa* (Dipnoi). *Journal of Morphology* 187:81–108.

Bennett, K. D. (1990). Milankovitch cycles and their effects on species in ecological and evolutionary time. *Paleobiology* 16:11–21.

Benton, M. J. (1989). Mass extinctions among tetrapods and the quality of the fossil record. *Philosophical Transactions of the Royal Society London B* 325:369–86.

Benton, M. J. (1990). Origin and interrelationships of dinosaurs. In *The Dinosauria*, ed. D. B. Weishample, P. Dodson, and H. Osmólska, pp. 11–30. Berkeley: University of California Press.

Benton, M. J. (1991). What really happened in the Late Triassic? *Historical Biology* 5:263–78.

Benton, M. J. (1993). *The Fossil Record 2*. London: Chapman & Hall.

Benton, M. J. (1995). Diversification and extinction in the history of life. *Science* 268:52–8.

Berggren, W. A., and Casey, R. E. (1983). Introduction to symposium on tempo and mode of evolution from micropaleontological data. *Paleobiology* 9:326.

Bergmann, C. (1847). Über die Verhaltnisse der Warmekonomie der Thiere zu ihrer Grosse. *Gottingen Studien* 3:595–708.

Bigelow, H. B., and Schroeder, W. C. (1948). Fishes of the Western North Atlantic. *Memoirs of the Sears Foundation for Marine Research* 1: xvii, 576.

Bigelow, H. B., and Schroeder, W. C. (1953). Fishes of the Gulf of Maine. *Fishery Bulletin of the Fish and Wildlife Service* 53: viii, 577.

Blair, W. F. (1965). Amphibian speciation. In *The Quaternary of the United States*, ed. H. E. Wright and D. G. Frey, pp. 543–56. Princeton: Princeton University Press.

Boles, W. E. (1995). The world's oldest songbird. *Nature* 374:21–2.

Bolt, J. R. (1991). Lissamphibian origins. In *Origins of the Higher Groups of Tetrapods: Controversy and Consensus*, ed. H.-P. Schultze and L. Trueb., pp. 194–222. Ithaca and London: Cornell University Press.

Bonner, J. T. (1988). *The Evolution of Complexity*. Princeton: Princeton University Press.

Bookstein, F. L. (1987). Random walk and the existence of evolutionary rates. *Paleobiology* 13:446–64.

Bookstein, F. L. (1988). Random walk and the biometrics of morphological characters. *Evolutionary Biology* 23:369–98.

Bookstein, F. [L.], Chernoff, B., Elder, R., Humphries, J., Smith, G., and Strauss, R.(1985). *Morphometrics in Evolution. Special Publication, Academy of Natural Sciences, Philadelphia* 15.

Boss, K. J. (1978). On the evolution of gastropods in ancient lakes. In *Pulmonates,* vol. 2A, ed. V. Fretter and J. Peake, pp. 385–428.

Bowman, R. I. (1961). Morphological differentiation and adaptation in the Galápagos finches. *University of California Publications in Zoology* 58:1–326.

Bowman, R. I., Berson, M., and Leviton, A. E. (1983). *Patterns of Evolution in Galápagos Organisms* San Francisco: American Association for the Advancement of Science.

Bowring, S. A., Grotzinger, J. P., Isachsen, C. E., Knoll, A. H., Pelechaty, S. M., and Kolosov, P. (1993). Calibrating rates of Early Cambrian Evolution. *Science* 261:1293–8.

Brett, C. E., and Baird, G. C. (1995). Coordinated stasis and evolutionary ecology of Silurian to Middle Devonian faunas in the Appalachian Basin. In *New Approaches to Speciation in the Fossil Record*, ed. D. H. Erwin and R. L. Anstey, pp. 285–315. New York: Columbia University Press.

Briggs, D. E. G., Fortey, R. A., and Wills, M. A. (1992). Morphological disparity in the Cambrian. *Science* 256:1293–8.

Briggs, D. E. G., Fortey, R. A., and Wills, M. A. (1993). How big was the Cambrian evolutionary explosion? A taxonomic and morphological comparison of Cambrian and Recent arthropods. In *Evolutionary Patterns and Processes,* ed. D. R. Lees and D. Edwards, pp. 33–44. Linnean Society Symposium Series no. 14. London: Linnean Society.

Broom, R. (1913). On the South-African pseudosuchian *Euparkeria* and allied genera. *Proceedings Zoological Society London* 1913:619–33.

Brown, R. P., and Thorpe, R. S. (1991). Within-island microgeographic variation in body dimensions and scalation of the skink *Chalcides sexlineatus,* with testing of causal hypotheses. *Biological Journal of the Linnean Society* 44:47–64.

Brown, R. P., Thorpe, R. S., and Baez, M. (1991). Parallel within-island microevolution of lizards on neighbouring islands. *Nature* 352:60–2.

Browne, M. (1996). Feathery fossil hints dinosaur–bird link. *New York Times,* October 19, pp. 1, 11.

Burke, A. C. (1989). Development of the turtle carapace: Implications for the evolution of a novel bauplan. *Journal of Morphology* 199:363–78.

Burke, A. C., Nelson, C. E., Morgan, B. A., and Tabin, C. (1995). *Hox* genes and the evolution of vertebrate axial morphology. *Development* 121:333–46.

Cain, A. J., and Sheppard, P. M. (1954). Natural selection in *Cepaea. Genetics* 39:89–116.

Caldwell, M. W. (1994). Developmental constraints and limb evolution in Permian and extant lepidosauromorph diapsids. *Journal of Vertebrate Paleontology* 14:459–71.

Caldwell, M. W. (1995). *Limb Ontogeny, Evolution, and Aquatic Adaptation in the Neodiapsida (Reptilia: Diapsida).* Unpublished PhD thesis. Department of Biology, McGill University; Montréal, Canada.

Caldwell, M. W. (1996). Ontogeny and phylogeny of the mesopodial skeleton in mosasauroid reptiles. *Zoological Journal of the Linnean Society* 116:407–36.

Caldwell, M. W. (in press-a). Development and evolution: Modified perichondral ossification and the evolution of paddle-like limbs in ichthyosaurs and plesiosaurs. *Journal of Vertebrate Paleontology.*

Caldwell, M. W. (in press-b). Limb ossification patterns of the ichthyosaur *Stenopterygius*, and a discussion of the proximal tarsal row of ichthyosaurs and other neodiapsid reptiles. *Zoological Journal of the Linnean Society.*

Caldwell, M. W. (in press-c). Ossification patterns of the limb skeleton in the plesiosaur *Cryptoclidus* (Upper Jurassic, Callovian). *Journal of Vertebrate Paleontology.*

Caldwell, M. W., Carroll, R. L., and Kaiser, H. (1995). The pectoral girdle and forelimb of *Carsosaurus marchesetti* (Aigialosauridae), with a preliminary phylogenetic analysis of mosasauroids and varanoids. *Journal of Vertebrate Paleontology* 15:516–31.

Cameron, R. A .D. (1992). Change and stability in *Cepaea* populations over 25 years: A case of climatic selection. *Proceedings of the Royal Society of London B* 248:181–7.

Campbell, K. S. W. (1977). Trilobites of the Haragan, Bois d'Arc, and Frisco Formations (Early Devonian) Arbuckle Mountains region. *Bulletin Oklahoma Geological Survey* 123:1–227.

Campbell, K. S. W., and Day, M. F. (eds.) (1987). *Rates of Evolution.* London: Allen & Unwin.

Cano, R. J., Poiner, H. N., Pieniazek, N. J., Acra, A., and Poinar, G. O. (1993). Amplification and sequencing of DNA from a 120–135 million-year-old weevil. *Nature* 363:536–8.

Carpenter, K., Hirsch, J. F., and Horner, J. R. (1994). *Dinosaur Eggs and Babies.* Cambridge: Cambridge University Press.

Carroll, R. L. (1964). Early evolution of the dissorophid amphibians. *Bulletin of the Museum of Comparative Zoology* 131:161–250.

Carroll, R. L. (1970). Quantitative aspects of the amphibian–reptilian transition. *forma et functio* 3: 165–78.

Carroll, R. L. (1977). Patterns of amphibian evolution: An extended example of the incompleteness of the fossil record. In *Patterns of Evolution as Illustrated by the Fossil Record*, ed. A. Hallam, pp. 405–37. Amsterdam: Elsevier.

Carroll, R. L. (1984). Problems in the use of terrestrial vertebrates for zoning the Carboniferous. *Compte Rendu of the 9th International Congress of Carboniferous Stratigraphy and Geology* 2:1–33.

Carroll, R. L. (1985). Evolutionary constraints in aquatic diapsid reptiles. In *Evolutionary Case Histories from the Fossil Record*, ed. J. C. W. Cope and P. W. Skelton. *Special Papers in Palaeontology* 33:145–55.

Carroll, R. L. (1986). The skeletal anatomy and some aspects of the physiology of primitive reptiles. In *The Ecology and Biology of Mammal-like Reptiles*, ed. N. Hotton III, P. MacLean, J. J. Roth, and E. C. Roth, pp. 25–45. Washington D.C.: Smithsonian Institution Press.

Carroll, R. L. (1987). *Vertebrate Paleontology and Evolution*. New York: W. H. Freeman.

Carroll, R. L. (1990). A tiny microsaur from the Lower Permian of Texas: Size constraints in Palaeozoic tetrapods. *Palaeontology* 33:893–909.

Carroll, R. L. (1991). The origin of reptiles. In *Origins of the Higher Groups of Tetrapods: Controversies and Consensus,* ed. H. P. Schultze and L. Trueb, pp. 331–53. New York: Cornell University Press.

Carroll, R. L. (1992). The primary radiation of terrestrial vertebrates. *Annual Review of Earth and Planetary Science* 20:45–84.

Carroll, R. L. (1995). Problems of the phylogenetic analysis of Paleozoic choanates. *Bulletin du Muséum National d'Histoire Naturelle* 17:389–445.

Carroll, R. L. (1996). Revealing the patterns of macroevolution. *Nature* 381:19–20.

Carroll, R. L. (in press). Mesozoic marine reptiles as models of long term, large scale evolutionary phenomena. In *Mesozoic Marine Reptiles*, ed. E. Nicholls and J. Callaway. London and New York: Academic Press.

Carroll, R. L., and Currie, P. J. (1975). Microsaurs as possible apodan ancestors. *Zoological Journal of the Linnean Society.* 57:229–47.

Carroll, R. L., and Gaskill, P. (1985). The nothosaur *Pachypleurosaurus* and the origin of plesiosaurs. *Philosophical Transactions of the Royal Society of London B* 309:343–93.

Carroll, R. L., and Holmes, R. (1980). *Zoological Journal of the Linnean Society* 68:1–40.

Carroll, S. B. (1995). Homeotic genes and the evolution of arthropods and chordates. *Nature* 376: 479–85.

Carroll, S. B., Weatherbee, S. D., and Langland, J. A. (1995). Homeotic genes and the regulation and evolution of insect wing number. *Nature* 375:58–62.

Castellano, S., Malhotra, A., and Thorpe, R. S. (1994). Within-island geographic variation of the dangerous Taiwanese snake, *Trimeresurus stejnegeri,* in relation to ecology. *Biological Journal of the Linnean Society* 52:365–75.

Castle, W. E. (1921). An improved method of estimating the number of genetic factors concerned in cases of blending inheritance. *Science* 54:223.

Chaline, J. (1987). Arvicolid data (Arvicolidae, Rodentia) and evolutionary concepts. *Evolutionary Biology* 21:237–310.

Chaline, J., and Laurin, B. (1986). Phyletic gradualism in a European Plio-Pleistocene *Mimomys* lineage (Arvicolidae, Rodentia). *Paleobiology* 12:203–16.

Chaloner, W. G., and Hallam, A. (eds.) (1989). Evolution and extinction. *Philosophical Transactions of the Royal Society London B* 325:239–488.

Charité, J., Graaff, W., Shen, S., and Deschamps, J. (1994). Ectopic expression of *Hoxb-8* in the forelimb and homeotic transformation of axial structures. *Cell* 78:589–601.

Chatterjee, S. (1991). Cranial anatomy and relationships of a new Triassic bird from Texas. *Philosophical Transactions Royal Society of London B* 322:277–346.

Chiappe, L. M. (1992). Enantiornithine tarsometatarsi and the avian affinity of the Late Cretaceous Avisauridae. *Journal of Vertebrate Paleontology* 12:344–50.

Chiappe, L. M. (1995a). The first 85 million years of avian evolution. *Nature* 378:349–55.

Chiappe, L. M. (1995b). The phylogenetic position of the Cretaceous birds of Argentina: Enanti-ornithes and *Patagopteryx deferrariisi. Courier Forschungsinstitut Senckenberg* 181: 55–63.

Chiappe, L. M. (1996). Late Cretaceous birds of southern South America: Anatomy and systematics of Enantiornithes and *Patagopteryx deferrariisi. Münchener Geowissenschaftliche Abhand-lungen Reihe A Geologie und Paläontologie* 30:203–44.

Chiappe, L. M., and Calvo, J. O. (1994). *Neuquenornis volans,* a new Late Cretaceous bird (Enanti-ornithes: Avisauridae) from Patagonia, Argentina. *Journal of Vertebrate Paleontology* 14: 230–46.

Chinsamy, A., Chiappe, L. M., and Dodson, P. (1995). Mesozoic avian bone microstructure: Physio-logical implications. *Paleobiology* 21:561–74.

Christie, D. M., Duncan, R. A., McBirney, A. R., Richards, M. A., White, W. M., Harppe, K. S., and Fox, C. G. (1992). Drowned islands downstream from the Galápagos hotspot imply ex-tended speciation times. *Nature* 355:246–8.

Clack, J. A., and Coates, M. I. (1995). *Acanthostega gunnari,* a primitive, aquatic tetrapod? *Bulletin du Muséum National d'Histoire Naturelle* 17:359–72.

Clarkson, E. N. K., Milner, A. R., and Coates, M. I. (1993). Palaeoecology of the Viséan of East Kirk-ton, West Lothian, Scotland. *Transactions of the Royal Society of Edinburgh: Earth Sci-ences* 85:417–25.

Clyde, W. C., and Gingerich, P. D. (1994). Rates of evolution in the dentition of early Eocene *Can-tius:* Comparison of size and shape. *Paleobiology* 20:506–22.

Coates, M. I. (1993). *Hox* genes, fin folds and symmetry. *Nature* 364:195–6.

Coates, M. I. (1994). The origin of vertebrate limbs. *Development 1994 Supplement:*169–80.

Coates, M. I. (in press). The postcranial skeleton of the Upper Devonian amphibian *Acanthostega. Transactions of the Royal Society of Edinburgh* 88.

Coates, M. I., and Clack, J. A. (1995). Romer's gap: Tetrapod origins and terrestriality. *Bulletin du Mu-séum National d'Historie Naturelle* 17:373–88.

Coffin, M. F., and Eldholm, O. (1993). Large igneous provinces. *Scientific American* 269(4):42–9.

Cohen, A. S., Soreghan, M. J., and Scholz, C. A. (1993). Estimating the age of formation of lakes: An example from Lake Tanganyika, East African Rift system. *Geology* 21:511–14.

Conway Morris, S. (1994). Why molecular biology needs paleontology. *Development 1994 Supple-ment:* 1–13.

Cope, E. D. (1881). On some Mammalia from the lowest Eocene beds of New Mexico. *Proceedings of the American Philosophical Society* 19:484–95.

Cope, E. D. (1896). *The Primary Factors of Organic Evolution.* Chicago: Open Court.

Coope, G. R. (1995). Insect faunas in ice age environments: Why so little extinction? In *Extinction Rates,* ed. J. H. Lawton and R. M. May, pp. 55–74. Oxford: Oxford University Press.

Coulter, G. W. (ed.) (1991). *Lake Tanganyika and Its Life.* London, Oxford, and New York: Oxford University Press.

Cracraft, J. (1986). The origin and early diversification of birds. *Paleobiology* 12:383–99.

Cracraft, J. (1989). Speciation and its ontology: The empirical consequences of alternative species concepts for understanding patterns and processes of differentiation. In *Speciation and Its Consequences,* ed. D. Otte and J. A. Endler, pp. 28–59. Sunderland, Mass.: Sinauer Associates.

Crane, P. R., Friis, E. M., and Pedersen, J. R. (1995). The origin and early diversification of angio-sperms. *Nature* 374:27–30.

Crow, J. F., and Kimura, M. (1970). *An Introduction to Population Genetics.* New York: Harper & Row.

Crowley, T. J., and Kim, K.-Y. (1994). Milankovitch forcing of the last interglacial sea level. *Science* 265:1566–8.

Cummings, M. R. (1988). *Human Heredity: Principles and Issues* (1st ed.). St. Paul: West Publishing.

Cummings, M. R. (1994). *Human Heredity* (3d ed.). St. Paul: West Publishing.

Daeschler, E. B., Shubin, N. H., Thomson, K. S., and Amaral, W. W. (1994). A Devonian tetrapod from North America. *Science* 265:639–42.

Darwin, C. (1859). *On the Origin of Species by Means of Natural Selection.* London: Murray.

Davis, A. P., Witte, D. P., Hsieh-Li, M. M., Potter, S. S., and Capecchi, M. R. (1995). Absence of radius and ulna in mice lacking *hoxa-11* and *hoxd-11. Nature* 375:791–5.

Davis, L. C. (1987). Late Pleistocene/Holocene environmental changes in the Central Plains of the United States: The mammalian record. *Illinois State Museum Scientific Papers* 22:88–143.

Dawley, M. R., and Bogart, J. P. (eds.) (1989). *Evolution and Ecology of Unisexual Vertebrates*. Albany, N.Y.: University of the State of New York, State Education Department.

Dawson, J. W. (1890). *Modern Ideas of Evolution as Related to Revelation and Science*. London: Religious Tract Society.

Dawson, W. R., Bartholomew, G. A., and Bennet, A. F. (1977). A reappraisal of the aquatic specializations of the Galápagos marine iguana (*Amblyrhynchus cristatus*). *Evolution* 31:891–7.

DeBraga, M., and Carroll, R. L. (1993). The origin of mosasaurs as a model of macroevolutionary patterns and processes. *Evolutionary Biology* 27:245–322.

de Queiroz, K. (1994). Replacement of an essentialistic perspective on taxonomic definitions as exemplified by the definition of "Mammalia." *Systematic Biology* 43:497–510.

de Queiroz, K., & Gauthier, J. (1990). Phylogeny as a central principle in taxonomy: Phylogenetic definitions of taxon names. *Systematic Zoology* 39:307–22.

de Queiroz, K., & Gauthier, J. (1992). Phylogenetic taxonomy. *Annual Review of Ecology and Systematics* 23:449–80.

de Queiroz, K., & Gauthier, J. (1994). Toward a phylogenetic system of biological nomenclature. *Trends in Ecology and Evolution* 9:27–31.

De Robertis, E. M., Morita, E. A., and Cho, K. (1991). Gradient fields and homeobox genes. *Development* 112:669–78.

DiMichele, W. A., and Hook, R. W. (1992). Paleozoic terrestrial ecosystems. In *Terrestrial Ecosystems through Time*, ed. A. K. Behrensmeyer, J. D. Damuth, W. A. DiMichele, R. Potts, H.-D. Sues, and S. L. Wing, pp. 205–325. Chicago: University of Chicago Press.

DeSalle, R., Gatesy, J., Wheeler, W., and Grimaldi, D. (1992). DNA sequences from a fossil termite in Oligo–Miocene amber and their phylogenetic implications. *Science* 257:1933–6.

De Vries, H. (1906). *Species and Varieties: Their Origin by Mutation* (2d ed.). Chicago: Open Court.

Diamond, J. (1992). Horrible plant species. *Science* 360:627–8.

Dickinson, W. J. (1991). The evolution of regulatory genes and patterns in *Drosophila*. *Evolutionary Biology* 25:127–73.

Diehl, S. R., and Bush, G. L. (1989). The role of habitat preference in adaptation and speciation. In *Speciation and Its Consequences*, ed. D. Otte and J. A. Endler, pp. 345–65. Sunderland, Mass.: Sinauer Associates.

Dingus, L. (1984). Effects of stratigraphic completeness on interpretations of extinction rates across the Cretaceous–Tertiary boundary. *Paleobiology* 4:420–38.

Dingus, L., and Sadler, P. (1982). The effects of stratigraphic completeness on estimates of evolutionary rates. *Systematic Zoology* 31:400–12.

Dobzhansky, T. (1937). *Genetics and the Origin of Species* (1st ed.). New York: Columbia University Press.

Dobzhansky, T. (1947). Adaptive changes induced by natural selection in wild populations of *Drosophila*. *Evolution* 1:1–16.

Dobzhansky, T. (1951). *Genetics and the Origin of Species* (3d ed.). New York: Columbia University Press.

Dobzhansky, T., Ayala, F. J., Stebbins, G. L., and Valentine, J. W. (1977). *Evolution*. San Francisco: W. H. Freeman.

Dominey, W. J. (1984). Effects of sexual selection and life history on speciation: Species flocks in African cichlids and Hawaiian *Drosophila*. In *Evolution of Fish Species Flocks*, ed. A. S. Echell and I. Kornfield, pp. 231–49. Orono: University of Maine Press.

Dorit, R. L. (1990). The correlates of high diversity in Lake Victoria haplochromine cichlids: A neotological perspective. In *Causes of Evolution: A Paleontological Perspective,* ed. R. M. Ross and W. D. Allmon, pp. 322–53. Chicago and London: University of Chicago Press.

Duboule, D. (1994). Temporal colinearity and the phylotypic progression: A basis for the stability of a vertebrate Bauplan and the evolution of morphologies through heterochrony. *Development 1994 Supplement*: 135–42.

Dudley, J. W. (1977). Seventy-six generations of selection for oil and protein percentage in maize. In *Proceedings of the International Conference on Quantitative Genetics,* ed. E. Pollak, O. Kempthorne, and T. Bailey, Jr., pp. 459–73. Ames: Iowa State University Press.

Duellman, W. E., and Trueb, L. (1986). *Biology of Amphibians.* New York: McGraw–Hill.

Eaton, C. F. (1910). Osteology of *Pteranodon. Memoirs of the Connecticut Academy of Arts and Sciences* 2:1–38.

Echelle, A. A., and Kornfield, I. (1984). *The Evolution of Fish Species Flocks.* Orono: University of Maine Press.

Eckert, R., and Randall, D. (1983). *Animal Physiology* (2d ed.). San Francisco: W. H. Freeman & Co.

Eldredge, N. (1985a). *Time Frames.* New York: Simon & Schuster.

Eldredge, N. (1985b). *Unfinished Synthesis.* New York and Oxford: Oxford University Press.

Eldredge, N., and Gould, S. J. (1972). Punctuated equilibria: An alternative to phyletic gradualism. In *Models in Paleobiology,* ed. T. J. M. Schopf, pp. 82–115. San Francisco: Freeman, Cooper.

Eldredge, N., and Stanley, S. M. (1984). *Living Fossils.* New York, Berlin, Heidelberg, Tokyo: Springer-Verlag.

Emry, R. J., and Thorington, R. W., Jr. (1984). The tree squirrel *Sciurus* (Sciuridae, Rodentia) as a living fossil. In *Living Fossils,* ed. N. Eldredge and S. M. Stanley, pp. 23–31. New York: Springer-Verlag.

Endler, J. A. (1986). *Natural Selection in the Wild.* Princeton: Princeton University Press.

Erwin, D. H. (1993). *The Great Paleozoic Crisis.* New York: Columbia University Press.

Erwin, D. H. (1994). The Permo–Triassic extinction. *Nature* 367:231–6.

Erwin, D. H. (1996). The mother of mass extinction. *Scientific American* 275 (1):72–8.

Evans, S. E. (1994). A new anguimorph lizard from the Jurassic and Lower Cretaceous of England. *Palaeontology* 37:33–49.

Evans, S. E., and Hecht, M. K. (1993). A history of an extinct reptilian clade, the Choristodera: Longevity, Lazarus-taxa, and the fossil record. *Evolutionary Biology* 27:323–38.

Falconer, D. S. (1989). *Introduction to Quantitative Genetics.* New York: Longman Scientific & Technical.

Farlow, J. O., Smith, M. B., and Robinson, J. M. (1995). Body mass, bone "strength indicator," and cursorial potential of *Tyrannosaurus rex. Journal of Vertebrate Paleontology* 15:713–25.

Fay, L. P. (1988). Late Wisconsinan Appalachian herpetolofaunas: Relative stability in the midst of change. *Annals of the Carnegie Museum* 57:189–220.

Feduccia, A. (1980). *The Age of Birds.* Cambridge, Mass.: Harvard University Press.

Feduccia, A. (1994). Tertiary bird history: Notes and comments. In *Major Features of Vertebrate Evolution,* ed. D. R. Prothero and R. M. Schoch, pp. 178–89. Knoxville: University of Tennessee Paleontological Society.

Feduccia, A. (1995). Explosive evolution in Tertiary birds and mammals. *Science* 267:637–8.

Feduccia, A. (1996). *The Origin and Evolution of Birds.* New Haven and London: Yale University Press.

Feduccia, A., and Tordoff, H. B. (1979). Feathers of *Archaeopteryx:* Asymmetric vanes indicate aerodynamic function. *Science* 203: 1021–2.

Feduccia, A., and Wild, R. (1993). Birdlike characters in the Triassic archosaur *Megalancosaurus. Naturwissenschaften* 80:564–6.

Fejfar, O., and Heinrich, W.-D. (eds.) (1990a). *International Symposium on the Evolution, Phylogeny and Biostratigraphy of Arvicolids (Rodentia, Mammalia).* Prague: Geological Survey.

Fejfar, O., and Heinrich, W.-D. (1990b). Muroid rodent biochronology of the Neogene and Quaternary in Europe. In *European Neogene Mammal Chronology,* ed. E. H. Lindsay, V. Fahlbusch, and P. Mein, pp. 91–117. *NATO ASI Series A, Life Sciences,* vol. 180. New York: Plenum Press.

Fisher, D. C. (1984). The Xiphosurida: Archetypes of bradytely? In *Living Fossils,* ed. N. Eldredge and S. M. Stanley, pp. 196–213. New York: Springer-Verlag.

Fisher, R. A. (1930). *The Genetical Theory of Natural Selection* (1st ed.). Oxford: Clarendon Press.

Fisher, R. A. (1958). *The Genetical Theory of Natural Selection* (2d ed.). New York: Dover.

Fisher, R. A., and Ford, E. B. (1947). The spread of a gene in natural conditions in a colony of the moth *Panaxia dominula. Heredity* 4:117–19.

Foote, M., and Gould, S. J. (1992). Cambrian and Recent morphological disparity. *Science* 258:1816.

Ford, E. B. (1964). *Ecological Genetics.* London: Methuen.

Fordyce, R. E., and Barnes, L. G. (1994). The evolutionary history of whales. *Annual Review of Earth and Planetary Sciences* 22:419–55.

Forey, P. L., Humphries, C. J., Kitching, I. L., Scotland, R. W., Siebert, D. J., and Williams, D. M. (1992). *Cladistics: A Practical Course in Systematics*. Systematics Association Publication no. 10. Oxford: Clarendon Press.

Forey, P. L., and Janvier, P. (1993). Agnathans and the origin of jawed vertebrates. *Nature* 361:129–34.

Fortey, R. A., Briggs, D. E. G., and Wills, M. A. (1996). The Cambrian evolutionary "explosion": Decoupling cladogenesis from morphological disparity. *Biological Journal of the Linnean Society* 57:13–33.

Fox, R. C. (1984). *Paranyctoides maleficus* (new species), an early eutherian mammal from the Cretaceous of Alberta. *Carnegie Museum of Natural History Special Publication* 9:9–20.

Fryer, G., and Iles, T. D. (1972). *The Cichlid Fishes of the Great Lakes of Africa: Their Biology and Evolution*. Edinburgh: Oliver & Boyd.

Futuyma, D. J. (1986). *Evolutionary Biology* (2d ed.). Sunderland, Mass.: Sinauer Associates.

Futuyma, D. J. (1989). Macroevolutionary consequences of speciation: Inferences from phytophagous insects. In *Speciation and Its Consequences,* ed. D. Otte and J. A. Endler, pp. 557–78. Sunderland, Mass.: Sinauer Associates.

Gagnier, P.-Y. (1993). *Sacabambaspis janvieri,* vertébré ordovicien de Bolivie: I. Analyse morphologique. *Annales de Paléontologie (Vertebrata)* 79:19–69.

Gans, C. (1989). Stages in the origins of vertebrates: Analysis by means of scenarios. *Biological Reviews of the Cambridge Philosophical Society* 64:221–68.

Gans, C. (1993). Evolutionary origin of the vertebrate skull. In *The Skull,* vol. 2, *Patterns of Structural and Systematic Diversity,* ed. J. Hanken and B. K. Hall, pp. 1–35. Chicago: University of Chicago Press.

Gans, C., and Northcutt, R. G. (1983). Neural crest and the origin of vertebrates: A new head. *Science* 220:268–74.

Gatesy, S. M., and Dial, K. P. (1996). Locomotor modules and the evolution of avian flight. *Evolution* 50:331–40.

Gauthier, J., Cannatella, D., de Queiroz, K. (1989). Tetrapod phylogeny. In *The Hierarchy of Life*, ed. B. Fernholm, K. Bremer, and H. Jörnvall, pp. 337–53. Amsterdam: Elsevier.

Geary, D. H. (1995). The importance of gradual change in species-level transitions. In *New Approaches to Speciation in the Fossil Record*, ed. D. H. Erwin and R. L. Anstey, pp. 67–86. New York: Columbia University Press.

Gehring, W. J. (1985). Homeotic genes, the homeobox and the genetic control of development. *Cold Spring Harbor Symposium on Quantitative Biology* 50:243–51.

Geyh, M. B., and Schleicher, H. (1990). *Absolute Age Determinations*. Berlin: Springer-Verlag.

Gibbs, H. L., and Grant, P. R. (1987). Oscillating selection on Darwin's finches. *Nature* 327:511–13.

Gilbert, S. F. (1988). *Developmental Biology* (2d ed.). Sunderland, Mass.: Sinauer Associates.

Gilbert, S. F. (1994). *Developmental Biology* (4th ed.). Sunderland, Mass.: Sinauer Associates.

Gilbert, S. F., Opitz, J. M., and Raff, R. A. (1996). Resynthesizing evolutionary and developmental biology. *Developmental Biology* 173:357–72.

Gingerich, P. D. (1976). Paleontology and phylogeny: Patterns of evolution at the species level in Early Tertiary mammals. *American Journal of Science* 276:1–28.

Gingerich, P. D. (1977). Patterns of evolution in the mammalian fossil record. In *Patterns of Evolution as Illustrated by the Fossil Record,* ed. A. Hallam, pp. 469–500. Amsterdam: Elsevier.

Gingerich, P. D. (1982). Time resolution in mammalian evolution: Sampling, lineages, and faunal turnover. *Third North American Paleontological Convention, Proceedings* 1:205–10.

Gingerich, P. D. (1983). Rates of evolution: Effects of time and temporal scaling. *Science* 222:159–61.

Gingerich, P. D. (1987). Evolution and the fossil record: Patterns, rates, and processes. *Canadian Journal of Zoology* 65:1053–60.

Gingerich, P. D. (1993a). Quantification and comparison of evolutionary rates. *American Journal of Science*. 293A:453–78.

Gingerich, P. D. (1993b). Rates of evolution in Plio–Pleistocene mammals: Six case studies. In *Morphological Change in Quaternary Mammals,* ed. R. A. Martin and A. D. Barnosky, pp. 84–106. Cambridge: Cambridge University Press.

Gingerich, P. D., Raza, S. M., Arif, M., Anwar, M., and Zhou, X. (1994). New whale from the Eocene of Pakistan and the origin of cetacean swimming. *Nature* 368:844–7.

Gingerich, P. D., Smith, B. H., and Simons, E. L. (1990). Hind limbs of Eocene *Basilosaurus:* Evidence of feet in whales. *Science* 249:154–7.

Gingerich, P. D., Wells, N. A., Russell, D. E., and Shah, S. M. I. (1983). Origin of whales in epicontinental remnant seas: New evidence from the Early Eocene of Pakistan. *Science* 220: 403–6.

Golding, G. B. (1987). Nonrandom patterns of mutation are reflected in evolutionary divergence and may cause some of the unusual patterns observed in sequences. In *Genetic Constraints on Adaptive Evolution,* ed. V. Loeschcke, pp. 151–72. Berlin: Springer-Verlag.

Goldschmidt, R. B. (1938). *Physiological Genetics*. New York: McGraw–Hill.

Goldschmidt, R. B. (1940). *The Material Basis of Evolution*. New Haven: Yale University Press.

Goodwin, H. T. (1993). Patterns of dental variation and evolution in prairie dogs, genus *Cynomys*. In *Morphological Change in Quaternary Mammals,* ed. R. A. Martin and A. D. Barnosky, pp. 107–33. Cambridge: Cambridge University Press.

Gould, S. J. (1977). *Ontogeny and Phylogeny*. Cambridge, Mass.: Harvard University Press.

Gould, S. J. (1982). The meaning of punctuated equilibrium and its role in validating a hierarchical approach to macroevolution. In *Perspectives on Evolution,* ed. R. Milkman, pp. 83–104. Sunderland, Mass.: Sinauer Associates.

Gould, S. J. (1985). The paradox of the first tier: An agenda for paleobiology. *Paleobiology* 11:2–12.

Gould, S. J. (1989). *Wonderful Life: The Burgess Shale and the Nature of History*. New York: W. W. Norton.

Gould, S. J. (1990). Speciation and sorting as the source of evolutionary trends, or "things are seldom what they seem." In *Evolutionary Trends,* ed. K. J. McNamara, pp. 3–27. London: Belhaven Press.

Gould, S. J. (1995a). A task for *Paleobiology* at the threshold of majority. *Paleobiology* 21:1–14.

Gould, S. J. (1995b). Tempo and mode in the macroevolutionary reconstruction of Darwinism. In *Tempo and Mode in Evolution,* ed. W. M. Fitch and F. J. Ayala, pp. 125–44. Washington, D.C.: National Academy of Sciences.

Gould, S. J., and Eldredge, N. (1977). Punctuated equilibria: The tempo and mode of evolution reconsidered. *Paleobiology* 3:115–51.

Gould, S. J., and Eldredge, N. (1993). Punctuated equilibrium comes of age. *Nature* 366:223–7.

Gould, S. J., and Vrba, E. S. (1982). Exaptation – a missing term in the science of form. *Paleobiology* 8:4–15.

Graham, R. W., Semken, H. A., Jr., and Graham, M. A. (eds.) (1987). *Late Quaternary Mammalian Biogeography and Environments of the Great Plains and Prairies. Illinois State Museum Scientific Papers* 22.

Grant, B. R., and Grant, P. R. (1989). *Evolutionary Dynamics of a Natural Population*. University of Chicago Press.

Grant, B. R., and Grant, P. R. (1993). Evolution of Darwin's finches caused by a rare climatic event. *Proceedings of the Royal Society of London B* 251:111–17.

Grant, P. R. (1986). *Ecology and Evolution of Darwin's Finches*. Princeton, N.J.: Princeton University Press.

Grant, P. R. (1994). Ecological character displacement. *Science* 266:746–7.

Grant, P. R., Abbott, I., Schluter, D., Curry, P. L., and Abbott, L. K. (1985). Variation in the size and shape of Darwin's finches. *Biological Journal of the Linnean Society* 25:1–39.

Green, D., Sharbal, T. F., Kearsley, J., and Kaiser, H. (1996). Postglacial range, genetic subdivision and speciation in the western North American spotted frog complex, *Rana pretiosa. Evolution* 50:374–90.

Green, P., Lipman, D., Hillier, L., Waterson, R., States, D., and Claverie, J. (1993). Ancient conserved regions in new gene sequences and the protein databases. *Science* 259:1711–16.

Greenberg, N. (1980). Physiological and behavioural thermoregulation in living reptiles. In *A Cold Look at the Warm-Blooded Dinosaurs*, ed. R. D. K. Thomas and E. C. Olson, pp. 141–66. Washington, D.C.: American Association for the Advancement of Science.

Greenwood, P. H. (1965). The cichlid fishes of Lake Nabugabo, Uganda. *Bulletin of the British Museum (Natural History) Zoology* 12:315–57.

Greenwood, P. H. (1974). The cichlid fishes of Lake Victoria, East Africa: The biology and evolution of a species flock. *Bulletin of the British Museum (Natural History) Supplement* 6:1–134.

Greenwood, P. H. (1981). Species flocks and explosive evolution. In *Chance, Change and Challenge – The Evolving Biosphere,* ed. P. H. Greenwood and P. L. Forey, pp. 61–74. Cambridge and London: Cambridge University Press and British Museum (Natural History).

Greenwood, P. H. (1983). On *Macropleurodus, Chilotilapia* (Teleostei, Cichlidae) and the interrelationships of African cichlid species in flocks. *Bulletin of the British Museum (Natural History) Zoology* 45:209–31.

Greenwood, P. H. (1984). African cichlids and evolutionary theories. In *Evolution of Fish Species Flocks,* ed. A. S. Echell and I. Kornfield, pp. 141–54. Orono: University of Maine Press.

Greer, A. E. (1991). Limb reduction in squamates: Identification of the lineages and discussion of the trends. *Journal of Herpetology* 25:166–73.

Gregory, W. K. (1957). *Evolution Emerging: A Survey of Changing Patterns from Primeval Life to Man.* Vol 2. New York: Macmillan.

Griffiths, A. J., Miller, J. H., Suzuki, D. T., Lewontin, R. C., and Gelbart, W. M. (1993). *An Introduction to Genetic Analysis.* New York: W. H. Freeman.

Grotzinger, J. P., Bowring, S. A., Saylor, B. Z., and Kaufman, A. J. (1995). Biostratigraphic and geochronologic constraints on early animal evolution. *Science* 270:598–604.

Grüneberg, H. (1963). *The Pathology of Development.* Oxford: Blackwell Scientific Publications.

Guilday, J. E., Hamilton, H. W., Anderson, E., and Parmalle, P. W. (1978). The Baker Bluff cave deposit, Tennessee, and the late Pleistocene faunal gradient. *Carnegie Museum of Natural History Bulletin* 11:1–67.

Haas, G. (1980). Remarks on a new ophiomorph reptile from the Lower Cenomanian of Ein Jabrud, Israel. In *Aspects of Vertebrate History,* ed. L. L. Jacobs, pp. 177–92. Flagstaff: Northern Arizona Press.

Haeckel, E. (1866). *Generelle Morphologie der Organismen: Allgemeine Grundzüge der organischen Formen-Wissenschaft, mechanisch begründet durch die von Charles Darwin reformierte Descendenz-Theorie.* Berlin: Georg Riemer.

Haldane, J. B. S. (1924). A mathematical theory of natural and artificial selection, I. *Transactions of the Cambridge Philosophical Society* 23:19–41.

Haldane, J. B. S. (1932). *The Causes of Evolution.* London: Longmans.

Haldane, J. B. S. (1949). Suggestions as to quantitative measurement of rates of evolution. *Evolution* 3:51–6.

Hall, B. K. (1984). Developmental mechanisms underlying the formation of atavisms. *Biology Review* 59:89–124.

Hall, B. K. (1992). *Evolutionary Developmental Biology.* London: Chapman & Hall.

Hall, B. K. (1996). *Baupläne,* phyotypic stages, and constraint: Why there are so few types of animals. *Evolutionary Biology* 29:215–61.

Hallam, A. (1992). *Phanerozoic Sea-level Changes.* New York: Columbia University Press.

Hanken, J. (1982). Appendicular skeletal morphology in minute salamanders, genus *Thorius* (Amphibia: Plethodontidae): Growth regulation, adult size determination, and natural variation. *Journal of Morphology* 174:57–77.

Hanken, J. (1984). Miniaturization and its effect on cranial morphology in plethodontid salamanders, genus *Thorius* (Amphibia: Plethodontidae). I: Osteological variation. *Biological Journal of the Linnean Society* 23:55–75.

Hanken, J., and Hall, B. K. (eds.) (1993). *The Skull.* 3 vols. Chicago: University of Chicago Press.

Hanken, J., and Thorogood, P. (1993). Evolution and development of the vertebrate skull: The role of pattern formation. *Trends in Ecology and Evolution* 8:9–16.

Hanken, J., and Wake, D. B. (1993). Miniaturization of body size: Organismal consequences and evolutionary significance. *Annual Review of Ecology and Systematics* 24:501–19.

Harland, W. B., Armstrong, R. L., Cox, A. V., Craig, L. E., Smith, A. G., and Smith, D. G. (1990). *A Geologic Time Scale.* Cambridge: Cambridge University Press.

Harris, J. M., and White, T. D. (1979). Evolution of the Plio–Pleistocene African Suidae. *Transactions of the American Philosophical Society* 69:1–128.

Hartl, D. L., and Clark, A. G. (1989). *Principles of Population Genetics* (2d ed.). Sunderland, Mass.: Sinauer Associates.

Hecht, M. K. (1975). The morphology and relationships of the largest known terrestrial lizard, *Megalania prisa* Owen, from the Pleistocene of Australia. *Proceedings of the Royal Society Victoria* 87:239–50.

Hecht, M. K., Ostrom, J. H., Viohl, G., and Wellnhofer, P. (eds.) (1985). *The Beginnings of Birds*. Eichstätt: Freunde des Jura-Museums Eichstätt. Proceedings of the International *Archaeopteryx* Conference.

Hedges, S. B., Parker, P. H., Sibley, C. G., and Kumar, S. (1996). Continental breakup and the ordinal diversification of birds and mammals. *Nature* 381:226–9.

Heilmann, G. (1926). *The Origin of Birds*. New York: Appleton.

Hennig, W. (1950). *Grundzüge einer Theorie der phylogenetischen Systematik*. Berlin: Deutscher Zentralverlag.

Hennig, W. (1966). *Phylogenetic Systematics*. Urbana: University of Illinois Press.

Hennig, W. (1981). *Insect Phylogeny*. Chichester: Wiley.

Hibbard, C. W. (1949). Techniques of collecting microvertebrate fossils. *Contributions of the Museum of Paleontology, University of Michigan* 8:7–19.

Higuchi, R., Dowman, B., Feidburger, M., Ryder, O.A., and Wilson, A.C. (1984). DNA sequences from the quagga, an extinct member of the horse family. *Nature* 312:282–4.

Hildebrand, A. R., Pilkington, M., Connors, M., Ortiz-Aleman, C., and Chavez, R. E. (1995). Size and structure of the Chicxulub crater revealed by horizontal gravity gradients and cenotes. *Nature* 376:415–17.

Hildebrand, M. (1995). *Analysis of Vertebrate Structure* (4th ed.). New York: John Wiley & Sons.

Hoar, W. S. (1983). *General and Comparative Physiology* (3d ed.). Englewood Cliffs, N.J.: Prentice–Hall.

Hoffman, A. (1989). *Arguments of Evolution*. Oxford: Oxford University Press.

Holland, L. Z., Pace, D. A., Blink, M. L., Kene, M., and Holland, N. (1996). Sequence and expression of amphioxus alkali myosin light chain (*AmphiMLC-alk*) throughout development: Implications for vertebrate myogenesis. *Developmental Biology* 171:666–76.

Holland, N. D., Panganiban, G., Henyey, E. L., and Holland, L. Z. (1996). Sequence and developmental expression of *AmphiDll*, an amphioxus *Distal-less* gene transcribed in the ectoderm, epidermis and nervous system: Insights into evolution of craniate forebrain and neural crest. *Development* 122:2911–20.

Holland, P. (1992). Homeobox genes in vertebrate evolution. *BioEssays* 14:267–73.

Holland, P., Garcia-Fernàndez, J., Williams, N. A., and Sidow, A. (1994). Gene duplications and the origins of vertebrate development. *Development 1994 Supplement:* 125–33.

Holland, P., Holland, L. Z., Williams, N. A., and Holland, N. D. (1992). An amphioxus homeobox gene: Sequence conservation, spatial expression during development and insights into vertebrate evolution. *Development* 116:653–61.

Holland, R. W. H., and Garcia-Fernàndez, J. (1996). *Hox* genes and chordate evolution. *Developmental Biology* 173:382–95.

Holman, J. A. (1991). North American Pleistocene herpetofaunal stability and its impact on the interpretation of modern herpetofaunas: An overview. In *Beamers, Bobwhite, and BluePoints: Tributes to the Career of Paul W. Parmalee,* ed. J. R. Purdue, W. E. Klipperl, and B. W. Styles. *Illinois State Museum Scientific Papers* 23:227–35.

Holman, J. A. (1993). Review: British Quaternary Herpetofaunas: A history of adaptations to Pleistocene disruptions. *Herpetological Journal* 3:1–7.

Holman, J. A. (1995). *Pleistocene Amphibians and Reptiles in North America*. Oxford: Oxford University Press.

Holman, J. A., and Andrews, K. D. (1994). North American Quaternary cold-tolerant turtles: Distributional adaptations and constraints. *Boreas* 23:44–52.

Holmes, A. (1913). *The Age of the Earth*. London and New York: Harper.

Holmgren, N. (1933). On the origin of the tetrapod limb. *Acta Zoologica* 14:185–295.

Holtz, T. R. (1994). The arctometatarsalian pes, an unusual structure of the metatarsus of Cretaceous Theropoda (Dinosauria; Saurischian). *Journal of Vertebrate Paleontology* 14:480–519.

Hoogerhoud, R. J. C. (1984). A taxonomic reconsideration of the haplochromine genera *Gaurochromis* Greenwood, 1980 and *Labrochromis* Regan, 1920 (Pisces, Chichlidae). *Netherlands Journal of Zoology* 34:539–65.

Hopson, J. A. (1994). Synapsid evolution and the radiation of non-eutherian mammals. In *Major Features of Vertebrate Evolution*, ed. D. R. Prothero and R. M. Schoch, pp. 190–219. Knoxville: University of Tennessee Paleontological Society.

Horner, J. R., Varricchio, D. J., and Goodwin, M. B. (1992). Marine transgressions and the evolution of Cretaceous dinosaurs. *Nature* 358:59–61.

Hou, L.[-H.], and Liu, Z. (1984). A new fossil bird from Lower Cretaceous of Gansu and early evolution of birds. *Scientia Sinica* 27:1296–302.

Hou, L.[-H.], Martin, L. D., Zhou, Z., and Feduccia, A. (1996). Earliest adaptive radiation of birds revealed by newly discovered Chinese fossils. *Science* 274:1164–7.

Hou, L.[-H.], and Zhang, J. (1993). A new fossil bird from Lower Cretaceous of Liaoning. *Vertebrata Palasiatica* 31:223–4.

Hou, L.-H., Zhou, Z., Martin, L. D., and Feduccia, A. (1995). A beaked bird from the Jurassic of China. *Nature* 377:616–18.

Houde, P. W. (1988). Paleognathous birds from the early Tertiary of the Northern Hemisphere. *Publications of the Nuttall Ornithological Club* 22:1–148.

Houde, P. W., and Olson, S. L. (1981). Paleognathous carinate birds from the Early Tertiary of North America. *Science* 214:1236–7.

House, M. R. (1989). Ammonoid extinction events. *Philosophical Transactions of the Royal Society London B* 325:307–26.

Hulbert, R. C., Jr., and MacFadden, B. J. (1991). Morphological transformation and cladogenesis at the base of the adaptive radiation of Miocene hypsodont horses. *American Museum Novitates* 3000:1–61.

Hulbert, R. C., Jr., and Morgan, G. S. (1993). Quantitative and qualitative evolution in the giant armadillo *Holmesina* (Edentata: Pampatheriidae) in Florida. In *Morphological Change in Quaternary Mammals in North America*, ed. R. A. Martin and A. D. Barnosky, pp. 134–77. Cambridge: Cambridge University Press.

Huxley, T. H. (1868). On the animals which are most nearly intermediate between the birds and reptiles. *Annals and Magazine of Natural History* 4:66–75.

Jablonski, D. (1996). Mass extinctions: Persistent problems and new directions. In: *The Cretaceous–Tertiary Event and Other Catastrophes in Earth History*, ed. G. Ryder, D. Fastovsky, and S. Gartner. *Geological Society of America Special Paper* 307:1–10.

Janis, C. M. (1993). Tertiary mammal evolution in the context of changing climates, vegetation, and tectonic events. *Annual Review of Ecology and Systematics* 24:467–500.

Janis, C. M., Archibald, J. D., Cifelli, R. L., Lucas, S. G., Schaff, C. R., Schoch, R. M., and Williamson, T. E. (in press-a). Archaic ungulates and ungulate-like mammals. In *Evolution of Tertiary Mammals of North America*, vol. 1, *Terrestrial Carnivores, Ungulates, and Ungulate-like Mammals,* ed. C. M. Janis, K. M. Scott, and L. Jacobs. Cambridge: Cambridge University Press.

Janis, C. M., and Fortelius, M. (1988). On the means whereby mammals achieve increased functional durability of their dentitions, with special reference to limiting factors. *Biological Reviews* 63:197–230.

Janis, C. M., Scott, K. M., and Jacobs, L. (in press-b). *Evolution of Tertiary Mammals of North America,* vol. 1, *Terrestrial Carnivores, Ungulates, and Ungulatelike Mammals.* Cambridge: Cambridge University Press.

Jansa, L. F., Aubry, M.-P., and Gradstein, F. M. (1990). Comets and extinctions: Cause and effect? The taxonomic structure of periodic extinction. In *Global Catastrophes in Earth History,* ed. V. L. Sharpton and P. D. Ward, *Geological Society of America Special Paper* 247:223–32.

Jarvik, E. (1996). The Devonian tetrapod *Ichthyostega. Fossils and Strata* 40:1–213.

Jarvis, J.U.M., O'Riain, M. J., Bennett, N. C., and Sherman, P. W. (1994). Mammalian eusociality: A family affair. *Trends in Ecology and Evolution* 9:47–51.

Jenkins, F. A., Jr., Dial, K. P., and Goslow, G. E., Jr. (1988). A cineradiographic analysis of bird flight: The wishbone in starlings is a spring. *Science* 241:1495–8.

Jenkins, F. A., Jr., and Goslow, G. E., Jr. (1983). The functional anatomy of the shoulder of the Savannah monitor lizard (*Varanus exanthematics*). *Journal of Morphology* 175:195–216.

Jenkins, F. A., Jr., and Walsh, D. M. (1993). An Early Jurassic caecilian with limbs. *Nature* 365:246–50.

Jepsen, G. L. (1970). *Biology of Bats,* vol. 1, *Bat Origins and Evolution.* New York: Academic Press.

Jepsen, G. L., Mayr, E., and Simpson, G. G. (1949). *Genetics, Paleontology, and Evolution.* Princeton: Princeton University Press.

Johnson, D. R., O'Higgins, P., and McAndrew, T. J. (1988). The effect of replicated selection for body weight in mice on vertebral shape. *Genetical Research* 51:129–35.

Johnson, K. R., and Hickey, L. J. (1990). Megafloral change across the Cretaceous–Tertiary boundary in the northern Great Plains and Rocky Mountains, USA. In *Global Catastrophes in Earth History*, ed. V. L. Sharpton and P. D. Ward, *Geological Society of America Special Paper* 247:433–44.

Johnson, T. C., Scholz, C. A., Talbot, M. R., Kelts, K., Ricketts, R. D., Ngobi, G., Beuning, K., Ssemmanda, I., McGill, J. W. (1996). Late Pleistocene desiccation of Lake Victoria and rapid evolution of cichlid fishes. *Science* 273:1091–3.

Johnston, R. F., and Selander, R. S. (1971). Evolution in the house sparrow. II: Adaptive differentiation in North American populations. *Evolution* 25:1–28.

Jones, C.A., Choate, J. R., and Genoways, H. H. (1984). Phylogeny and paleobiogeography of short-tailed shrews (genus *Blarina*). In *Contributions in Quaternary Vertebrate Paleontology*, ed. H. H. Genoways and M. R. Dawson, *Carnegie Museum of Natural History Special Publication* 8:56–148.

Jones, G. M., and Spells, K. E. (1963). A theoretical and comparative study of the functional dependence of the semicircular canal upon its physical dimensions. *Proceedings of the Royal Society B* 157:403–19.

Jones, S., Martin, R., and Pilbeam, D. (1992). *The Cambridge Encyclopedia of Human Evolution.* Cambridge: Cambridge University Press.

Keenleyside, M. H. A. (ed.) (1991a). *Cichlid Fishes: Behaviour, Ecology, and Evolution.* London: Chapman & Hall.

Keenleyside, M. H. A. (1991b). Parental care. In *Cichlid Fishes: Behaviour, Ecology, and Evolution*, ed. M. H. A. Keenleyside, pp. 191–208. London: Chapman & Hall.

Kellogg, D. E. (1976). Character displacement in the radiolarian genus *Eucyrtidium. Evolution* 29: 736–49.

Kellogg, D. E. (1983). Phenology of morpholgical change in radiolarian lineages from deep-sea cores: Implications for macroevolution. *Paleobiology* 9:355–62.

Kemp, T. S. (1982). *Mammal-Like Reptiles and the Origin of Mammals.* London: Academic Press.

Kettlewell, B. (1973). *The Evolution of Melanism.* Oxford: Clarendon Press.

Kielan-Jaworowska, Z., Bown, T. M., and Lillegraven, J. A. (1979). Eutheria. In *Mesozoic Mammals*, ed. J. A. Lillegraven, Z. Kielan-Jaworowska, and W. A. Clemens, pp. 221–58. Berkeley: University of California Press.

King, G. M. (1991). Terrestrial tetrapods and the end Permian event: A comparison of analyses. *Historical Geology* 5:239–55.

Klein, G. D. (ed.) (1994). *Pangea: Paleoclimate, tectonics, and sedimentation during accretion, zenith, and breakup of a supercontinent. Geological Society of America Special Paper* 288:v, 299.

Knoll, A. H. (1989). Evolution and extinction in the marine realm: Some constraints imposed by phytoplankton. *Philosophical Transactions of the Royal Society London B* 325:279–90.

Knoll, A. H. (1995). Proterozoic and Early Cambrian protists: Evidence of accelerating evolutionary tempo. In *Tempo and Mode in Evolution,* ed. W. M. Fitch and F. J. Ayala, pp. 63–83. Washington, D.C.: National Academy of Sciences.

Knoll, A. H. (1996a). Breathing room for early animals. *Science* 382:111–12.

Knoll, A. H. (1996b). Daughter of time. *Paleobiology* 22:1–7.

Knoll, A. H., and Rothwell, G. W. (1981). Paleobotany: Perspectives in 1980. *Paleobiology* 7:7–35.

Kocher, T. D., Controy, J. A., McKaye, K. R., and Stauffer, J. R. (1993). Similar morphologies of cichlid fish in Lakes Tanganyika and Malawi are due to convergence. *Molecular Phylogenetics and Evolution* 2:158–65.

Kostic, D., and Capecchi, M. R. (1994). Targeted disruptions of the murine *Hoxa-4* and *Hoxa-6* genes result in homeotic transformations of components of the vertebral column. *Mechanisms of Development* 46:231–47.

Krishtalka, L., and Stucky, R. K. (1985). Revision of the Wind River faunas. Early Eocene of central Wyoming. Part 7. Revision of *Diacodexis* (Mammalia, Artiodactyla). *Annals of the Carnegie Museum* 54:413–86.

Krumlauf, R. (1994). *Hox* genes in vertebrate development. *Cell* 78:191–201.

Kurochkin, E. N. (1985). A true carinate bird from Lower Cretaceous deposits in Mongolia and other evidence of Early Cretaceous birds in Asia. *Cretaceous Research* 6:271–8.

Kurtén, B. (1960). Rates of evolution in fossil mammals. *Cold Spring Harbor Symposium on Quantitative Biology* 24:205–15.

Kurtén, B. (1968). *Pleistocene Mammals of Europe.* London: Weidenfeld & Nicolson.

Labandeira, C. C., and Sepkoski, J., Jr. (1993). Insect diversity in the fossil record. *Science* 261:310–15.

Lacalli, T. C. (1995). Antecedents of vertebrate eye and brain: Evidence from serial EM studies of amphioxus larvae. *Journal of Vertebrate Paleontology, Abstracts of Papers* 15(Supplement):40.

Lacasa Ruiz, A. (1989). Nuevo genero de ave fosil del yacimiento Neocomiense del Montsec (Provincia de Lerida, España). *Estudios Geologia* 45:417–25.

Lack, E. (1947). *Darwin's Finches.* Cambridge: Cambridge University Press.

Lancaster, W. C. (1990). The middle ear of the Archaeoceti. *Journal of Vertebrate Paleontology* 10:117–27.

Lande, R. (1976). Natural selection and random genetic drift in phenotypic evolution. *Evolution* 30:314–34.

Lande, R. (1981). The minimum number of genes contributing to quantitative variation between and within populations. *Genetics* 99:541–53.

Lande, R. (1986). The dynamics of peak shifts and the pattern of morphological evolution. *Paleobiology* 12:343–54.

Lande, R. (1988). Quantitative genetics and evolutionary theory. In *Proceedings of the Second International Conference on Quantitative Genetics,* ed. B. S. Weir, E. J. Eisen, M. M. Goodman, and Gene Namkoong, pp. 71–84. Sunderland, Mass.: Sinauer Associates.

Langille, R. M., and Hall, B. K. (1993). Pattern formation and the neural crest. In *The Skull,* vol. 1, *Development,* ed. J. Hanken and B. K. Hall, pp. 77–111. Chicago: University of Chicago Press.

Langston, W., Jr. (1981). Pterosaurs. *Scientific American* 244:92–102.

Larson, A. (1989). The relationship between speciation and morphological evolution. In *Speciation and Its Consequences,* ed. D. Otte and J. A. Endler, pp. 579–98. Sunderland, Mass.: Sinauer Associates.

Laurin, M., and Reisz, R. R. (1996). The osteology and relationships of *Tetraceratops insignis,* the oldest known therapsid. *Journal of Vertebrate Paleontology* 16:95–102.

Lebedev, O. A., and Coates, M. I. (1995). The postcranial skeleton of the Devonian tetrapod *Tulerpeton curtum* Lebedev. *Zoological Journal of the Linnean Society* 114:307–48.

Lee, M. S. Y. (1993). The origin of the turtle body plan: Bridging the morphological gap. *Science* 261:1716–20.

Lee, M. S. Y. (1996). Correlated progression and the origin of turtles. *Nature* 379:812–15.

Levinton, J. (1988). *Genetics, Paleontology, and Macroevolution.* Cambridge: Cambridge University Press.

Lich, D. K. (1990). *Cosomys primus:* A case for stasis. *Paleobiology* 16:384–95.

Liem, K. F. (1991). Functional morphology. In *Cichlid Fishes: Behaviour, Ecology, and Evolution,* ed. M. H. A. Keenleyside, pp. 129–50. London: Chapman & Hall.

Lillegraven, J. A. (1979). Reproduction in Mesozoic mammals. In *Mesozoic Mammals,* ed. J. A. Lillegraven, Z. Kielan-Jaworowska, and W. A. Clemens, pp. 259–76. Berkeley: University of Chicago Press.

Lindsay, E. H., and Tedford, R. H. (1990). Development and application of land mammal ages in North America and Europe, a comparison. In *European Neogene Mammal Chronology,* ed. E. H. Lindsay, V. Fahlbusch, and P. Mein, pp. 601–24. New York: Plenum Press.

Lindsley, D. L., and Zimm, G. G. (1992). *The Genome of Drosophila melanogaster.* San Diego: Academic Press.

Lipps, J. H., and Signor, P. W. (eds.) (1992). *Origin and Early Evolution of the Metazoa.* New York and London: Plenum.

Lister, A. M. (1993). Evolution of mammoths and moose: The Holarctic perspective. In *Morphological Change in Quaternary Mammals in North America.* ed. R. A. Martin and A. D. Barnosky, pp. 178–204. Cambridge: Cambridge University Press.

Loeschcke, V. (1987). Introduction: Genetic constraints on adaptive evolution and the evolution of genetic constraints. In *Genetic Constraints on Adaptive Evolution,* ed. V. Loeschcke, pp. 1–2. Berlin: Springer-Verlag.

Lofsvold, D. (1986). Quantitative genetics of morphological differentiation in *Peromyscus:* I. Tests of the homogeneity of genetic covariance structure among species and subspecies. *Evolution* 40:559–73.

Lofsvold, D. (1988). Quantitative genetics of morphological differentiation in *Peromyscus:* II. Analysis of selection and drift. *Evolution* 42:54–67.

Lombard, R. E., and Bolt, J. R. (1988). Evolution of the stapes in Paleozoic tetrapods. In *The Evolution of the Amphibian Auditory System,* ed. B. Frische, pp. 37–67. New York: J. Wiley & Sons.

Longacre, S. A. (1970). Trilobites of the Upper Cambrian ptychaspid biomere Wilberns Formation, central Texas. *Journal of Paleontology Memoir* 4:vi, 70.

Losos, J. B. (1994). Integrative approaches to evolutionary ecology: *Anolis* lizards as model systems. *Annual Review of Ecology and Systematics* 25:467–93.

Lowe-McConnell, R. H. (1991). Ecology of cichlids in South American and African waters, excluding the African Great Lakes. In *Cichlid Fishes: Behaviour, Ecology, and Evolution,* ed. M. H. A. Keenleyside, pp. 60–85. London: Chapman & Hall.

Lucas, A. M., and Stettenheim, P. R. (1972). *Avian Anatomy: Integument.* Agricultural Handbook no. 362. Washington, D.C.: U.S. Government Printing Office.

Lyell, C. (1832–3). *Principles of Geology.* 3 vols. Reprint ed. 1970. Forestburgh, N.Y.: Lubrecht & Cramer.

Lynch, M. (1988). The rate of polygenic mutation. *Genetical Research* 51:137–48.

Lynch, M. (1990). The rate of morphological evolution in mammals from the standpoint of the neutral expectation. *American Naturalist* 136:727–41.

Lyon, M. F., and Searle, A. G. (1989). *Genetic Variants and Strains of the Laboratory Mouse* (2d ed.). Oxford: Oxford University Press.

McCune, A. R. (1987). Toward the phylogeny of a fossil species flock: Semionotid fishes from a lake deposit in the Early Jurassic Towaco Formation, Newark Basin. *Bulletin of the Peabody Museum of Natural History* 43:1–108.

McCune, A. R. (1990). Evolutionary novelty and atavisms in the Semionotus complex: Relaxed selection during colonization of an expanding lake. *Evolution* 44:71–85.

McCune, A. R. (1996). Biogeographic and stratigraphic evidence for rapid speciation in semionotid fishes. *Paleobiology* 22:34–48.

McCune, A. R., Thomson, K. S., and Olsen, P. E. (1984). Semionotid fishes from the Mesozoic great lakes of North America. In *Evolution of Fish Species Flocks,* ed. A. S. Echelle and I. Kornfield, pp. 27–44. Orono: University of Maine Press.

MacFadden, B. J. (1992). *Fossil Horses.* Cambridge: Cambridge University Press.

McGinnis, W., Garber, R. L., Wirz, J., Kuroiwa, A., and Gehring, W. (1984). A homologous protein-coding sequence in *Drosophila* homeotic genes and its conservation in other metazoans. *Cell* 37:403–8.

McGowan, C. (1983). *The Successful Dragons: A Natural History of Extinct Reptiles.* Toronto: Samuel Stevens.

McKaye, K. R. (1991). Sexual selection and the evolution of the cichlid fishes of Lake Malawi. In *Cichlid Fishes: Behaviour, Ecology, and Evolution,* ed. M. H. A. Keenleyside, pp. 241–57. London: Chapman & Hall.

McKusick, V. A. (1994). *Mendelian Inheritance in Man: Catalogs of Autosomal Dominant, Autosomal Recessive, and X-Linked Phenotypes* (11th ed.). Baltimore: Johns Hopkins University Press.

MacLeod, N. (1991). Punctuated anagenesis and the importance of stratigraphy to paleobiology. *Paleobiology* 17:167–88.

MacLeod, N. (1995). Biogeography of Cretaceous/Tertiary (K/T) planktic foraminifera. *Historical Biology* 10:49–101.

McNamara, K. J. (1986). The role of heterochrony in the evolution of Cambrian trilobites. *Biological Review* 61:121–56.

McNamara, K. J. (1995) *Evolutionary Change and Heterochrony*. New York: John Wiley & Sons.

Maderson, P. F. A. (1972). When? Why? and How?: Some speculations on the evolution of the vertebrate integument. *American Zoologist* 12:159–71.

Maglio, V. J. (1973). Origin and evolution of the Elephantidae. *Transactions American Philosophical Society* 63:1–149.

Malhotra, A., and Thorpe, R. S. (1991). Microgeographic variation in *Anolis oculatus,* on the island of Dominica, West Indies. *Journal of Evolutionary Biology* 4:321–35.

Malmgren, B. A., Berggren, W. A., and Lohmann, G. P. (1983). Evidence for punctuated gradualism in the Late Neogene *Globorotalia tumida* lineage of planktonic foraminifera. *Paleobiology* 9:377–89.

Marden, J. H., and Kramer, M. G. (1994). Surface-skimming stoneflies: A possible intermediate stage in insect flight evolution. *Science* 266:427–30.

Marsh, O. C. (1880). *Odontornithes: A Monograph on the Extinct Toothed Birds of North America.* Washington, D.C.: U.S. Government Printing Office.

Martin, L. D. (1984). Phyletic trends and evolutionary rates. In *Festschrift for J. Guilday,* ed. M. Dawson and H. Genoways, *Carnegie Museum of Natural History Special Publication* 8: 526–38.

Martin, L. D. (1993). Evolution of hypsodonty and enamel structure in Plio–Pleistocene rodents. In *Morphological Change in Quaternary Mammals in North America,* ed. R. A. Martin and A. D. Barnosky, pp. 205–25. Cambridge: Cambridge University Press.

Martin, L. D. (1995). The Enantiornithes: Terrestrial birds of the Cretaceous in avian evolution. *Courier Forschungsinstitut Senckenberg* 181:23–36.

Martin, P. S., and Klein, R. G. (1984). *Quaternary Extinctions: A Prehistoric Evolution.* Tucson: University of Arizona Press.

Martin, R. A. (1992). Generic species richness and body mass in North American mammals: Support for the inverse relationship of body size and speciation rate. *Historical Biology* 6:73–90.

Martin, R. A. (1993). Variation and speciation in rodents. In *Morphological Change in Quaternary Mammals in North America,* ed. R. A. Martin and A. D. Barnosky, pp. 226–80. Cambridge: Cambridge University Press.

Martin, R. A. (1996). Dental evolution and size change in the North American muskrat: Classification and tempo of a presumed phyletic sequence. In *Palaeoecology and Palaeoenvironments of Late Cenozoic Mammals,* ed. K. M. Stewart and K. L. Seymour, pp. 431–57. Toronto: University of Toronto Press.

Martin, R. A., and Barnosky, A. D. (eds.) (1993). *Morphological Change in Quaternary Mammals in North America.* Cambridge: Cambridge University Press.

Mastick, G., McKay, R., Oligino, T., Donovan, K., and Lopez, A. (1995). Identification of target genes regulated by homeotic proteins in *Drosophila melanogaster* through genetic selection of *Ultrabithorax* protein-binding sites in yeast. *Genetics* 139:349–63.

Maxwell, W. D. (1992). Permian and Early Triassic extinctions of non-marine tetrapods. *Palaeontology* 35:571–83.

Maynard Smith, J. (1989). *Evolutionary Genetics.* Oxford: Oxford University Press.

Maynard Smith, J., Burian, R., Kauffman, S., Alberch, P., Campbell, J., Goodwin, B., Lande, R., Raup, D., and Wolbert, L. (1985). Developmental constraints and evolution. *Quarterly Review of Biology* 60:265–87.

Mayr, E. (1942). *Systematics and the Origin of Species.* New York: Columbia University Press.

Mayr, E. (1954). Geographic speciation in tropical echinoids. *Evolution* 8:1–18.

Mayr, E. (1963). *Animal Species and Evolution.* Cambridge, Mass.: Harvard University Press.

Mayr, E. (1969). *Principles of Systematic Zoology.* New York: McGraw–Hill.

Mayr, E. (1982). *The Growth of Biological Thought.* Cambridge, Mass.: Harvard University Press.

Mayr, E., and Provine, W. (1980). *The Evolutionary Synthesis: Perspectives on the Unification of Biology.* Cambridge, Mass.: Harvard University Press.

Mead, J. I., Thompson, R. S., and Van Devender, T. R. (1982). Late Wisconsinan and Holocene Fauna from Smith Creek Canyon, Snake Range, Nevada. *Transactions of the San Diego Society of Natural History* 20:1–16.

Mendel, J. (1866). Versuche über Pflanzen hybriden. *Verhandlungen der Naturforschenden Vereins Brünn* (1865):3–57.

Merola, M. (1994). A reassessment of homozygosity and the case for inbreeding depression in the cheetah, *Acinonxy jubatus:* implications for conservation. *Conservation Biology* 8:961–71.

Meyer, A. (1993). Phylogenetic relationships and evolutionary processes in East African cichlid fishes. *Trends in Ecology and Evolution* 8:279–84.

Milner, A. R. (1987). The Westphalian tetrapod fauna: Some aspects of its geography and ecology. *Journal of the Geological Society (London)* 144:459–506.

Milner, A. R. (1989). Late extinctions of amphibians. *Nature* 338:117.

Milner, A. R. (1993). The Palaeozoic relatives of lissamphibians. *Herpetological Monographs* 7:8–27.

Milner, A. R., and Sequeira, S. E. K. (1993). The temnospondyl amphibians from the Viséan of East Kirkton, West Lothian, Scotland. *Transactions of the Royal Society of Edinburgh: Earth Sciences* 84:331–61.

Minelli, A. (1993). *Biological Systematics.* New York: Chapman & Hall.

Molven, A., Wright, C. V. E., Bremiller, R., De Robertis, E. M., and Kimmerl, C. B. (1990). Expression of a homeobox gene product in normal and mutant zebrafish embryos: Evolution of the tetrapod body plan. *Development* 109:279–88.

Moore, R. C., Lalicker, C. G., and Fischer, A. G. (1952). *Invertebrate Fossils.* New York: McGraw–Hill.

Morgan, B. A., Izpisua-Belmonte, J., Duboule, D., and Tabin, C. (1992). Targeted mis-expression of *Hox-4.6* in the avian limb bud causes apparent homeotic transformations. *Nature* 358:236–9.

Morgan, B. A., and Tabin, C. (1994). Hox genes and growth: Early and late roles in limb bud morphogenesis. *Development 1994 Supplement:* 181–6.

Motani, R., You, H., and McGowan, C. (1996). Eel-like swimming in the earliest ichthyosaurs. *Nature* 382:347–8.

Nelson, C. E., Morgan, B. A., Burke, A. C., Laufer, E., DiMammbro, E., Murtaugh, L. C., Gonzales, E., Tessarolle, L., Parada, L. F., & Tabin, C. (1996). Analysis of *Hos* gene expression in the chick limb bud. *Development* 122:1449–66.

Nelson, C. E., and Tabin, C. (1995). Footnote on limb evolution. *Nature* 375:630–1.

Nelson, G. (1989). Species and taxa: Systematics and evolution. In *Speciation and Its Consequences,* ed. D. Otte and J. A. Endler, pp. 60–81. Sunderland, Mass.: Sinauer Associates.

Nelson, J. S. (1984). *Fishes of the World.* New York: Wiley–Interscience.

Niswander, L., Jeffrey, S., Martin, G. R., and Tickle, C. (1994). A positive feedback loop coordinates growth and patterning in the vertebrate limb. *Nature* 371:609–12.

Norberg, U. M. (1990). *Vertebrate Flight.* Berlin: Springer-Verlag.

Norell, M. A. (1993). Tree-based approaches to understanding history: Comments on ranks, rules, and the quality of the fossil record. *American Journal of Science* 293A:407–17.

Norell, M. A., and Novacek, M. J. (1992a). The fossil record and evolution: Comparing cladistic and paleontological evidence for vertebrate history. *Science* 255:1690–3.

Norell, M. A., and Novacek, M. J. (1992b). Congruence between superpositional and phylogenetic patterns: Comparing cladistic patterns with fossil records. *Cladistics* 8:319–37.

Northcutt, R. G., and Gans, C. (1983). The genesis of neural crest and epidermal placodes: A reinterpretation of vertebrate origins. *Quarterly Review of Biology* 58:1–28.

Novacek, M. J. (1994). The radiation of placental mammals. In *Major Features of Vertebrate Evolution,* ed. D. R. Prothero and R. M. Schoch, pp. 220–37. Knoxville: University of Tennessee Paleontological Society.

Novacek, M. J., and Norell, M. A. (1982). Fossils, phylogeny, and taxonomic rates of evolution. *Systematic Zoology* 31:366–75.

Nussbaum, R. A. (1983). The evolution of a unique dual jaw-closing mechanism in caecilians (Amphibia: Gymnophiona) and its bearing on caecilian ancestry. *Journal of the Zoological Society London* 199:545–54.

Odin, G. S. (ed.) (1982). *Numerical Dating in Stratigraphy*. 2 vols. Chichester: Wiley–Interscience.

O'Leary, M. A., and Rose, K. D. (1995). Postcranial skeleton of the early Eocene mesonychid *Pachyaena* (Mammalia: Mesonychia). *Journal of Vertebrate Paleontology* 15:401–30.

Oliver, G., Wright, C. V. E., Hardwicke, J., and De Robertis, E. M. (1988). A gradient of homeodomain protein in developing forelimbs of *Xenopus* and mouse embryos. *Cell* 55:1017–24.

Olsen, P. E. (1986). A 40-million-year record of Early Mesozoic orbital climatic forcing. *Science* 234: 842–8.

Olsen, P. E., Shubin, H. H., and Anders, M. H. (1987). New Early Jurassic tetrapod assemblages constrain Triassic–Jurassic tetrapod extinction event. *Science* 237:1025–9.

Olson, E. C. (1947). The Family Diadectidae and its bearing on the classification of reptiles. *Fieldiana Geology* 7:2–53.

Olson, S. L. (1977). A Lower Eocene frigatebird from the Green River Formation of Wyoming (Pelecaniformes: Fregatidae). *Smithsonian Contributions Paleontology* 35:1–33.

Olson, S. L. (1985). The fossil record of birds. In *Avian Biology*, vol. 8, ed. D. S. Farner, J. R. King, and K. C. Parks, pp. 79–238. London: Academic Press.

Orth, C. J., Attrep, Jr., M., and Quintana, L. R. (1990). Iridium abundance patterns across bio-event horizons in the fossil record. In *Global Catastrophes in Earth History*, ed. V. L. Sharpton and P. D. Ward, *Geological Society of America Special Paper* 247:45–59.

Osborn, H. F. (1934). Aristogenesis, the creative principle in the origin of species. *American Naturalist* 68:193–235.

Ostrom, J. H. (1969). Osteology of *Deinonychus antirrhopus*, an unusual theropod from the Lower Cretaceous of Montana. *Peabody Museum Natural History Bulletin* 30:1–165.

Ostrom, J. H. (1974). *Archaeopteryx* and the origin of flight. *Quarterly Review of Biology* 49:27–47.

Ostrom, J. H. (1975). The origin of birds. *Annual Review of Earth and Planetary Sciences* 3:55–77.

Ostrom, J. H. (1976a). *Archaeopteryx* and the origin of birds. *Biological Journal of the Linnean Society* 8:91–182.

Ostrom, J. H. (1976b). Some hypothetical anatomical stages in the evolution of avian flight. *Smithsonian Contributions to Paleobiology* 27:1–21.

Ostrom, J. H. (1985). Introduction to *Archaeopteryx*. In *The Beginnings of Birds*, ed. M. K. Hecht, J. H. Ostrom, G. Viohl, and P. Wellnhofer, pp. 9–20. Eichstätt: Freunde des Jura-Museums Eichstätt. Proceedings of the International *Archaeopteryx* Conference.

Ostrom, J. H. (1990). Dromaeosauridae. In *The Dinosauria*, ed. D. B. Weishampel, P. Dodson, and H. Osmólska, pp. 269–79. Berkeley: University of California Press.

Ostrom, J. H. (1994). On the origin of birds and of avian flight. In *Major Features of Vertebrate Evolution*, ed. D. R. Prothero and R. M. Schoch, pp. 160–77. Knoxville: University of Tennessee Paleontological Society.

Otte, D. (1989). Speciation in Hawaiian crickets. In *Speciation and Its Consequences*, ed. D. Otte and J. A. Endler, pp. 482–526. Sunderland, Mass.: Sinauer Associates.

Otte, D., and Endler, J. A. (1989). *Speciation and Its Consequences*. Sunderland, Mass.: Sinauer Associates.

Owen, R. B., Crossley, R., Johnson, T. C., Tweddle, D., Kornfield, I., Davison, S., Eccles, D. H., and Engstrom, D. E. (1990). Major low levels of Lake Malawi and their implications for speciation rates in cichlid fishes. *Proceedings of the Royal Society of London B* 240:519–53.

Padian, K. (1983). A functional analysis of flying and walking in pterosaurs. *Paleobiology* 9:218–39.

Padian, K. (1996). Terrestrial locomotion in pterosaurs. *Journal of Vertebrate Paleontology, Abstracts of Papers* 16(Supplement):57A.

Panganiban, G., Sebring, A., Nagy, L., and Carroll, S. (1995). The development of crustacean limbs and the evolution of arthropods. *Science* 270:1363–6.

Patterson, C. (1994). Bony fishes. In *Major Features of Vertebrate Evolution*, ed. D. R. Prothero and R. M. Schoch, pp. 57–84. Knoxville: University of Tennessee Paleontological Society.

Patterson, C., and Smith, A.B. (1987). Is periodicity of mass extinctions a taxonomic artefact? *Nature* 330:248–51.

Pendleton, J. W., Nagai, B. K., Murtha, M. T., and Ruddle, F. H. (1993). Expansion of the *Hox* gene family and the evolution of chordates. *Proceedings of the National Academy of Science* 90:6300–4.

Peterson, K. J. (1994). The origin and early evolution of the Craniata. In *Major Features of Vertebrate Evolution,* ed. D. R. Prothero and R. M. Schoch, pp. 14–37. Knoxville: University of Tennessee Paleontological Society.

Philippe, H., Chenuil, A., and Adoutte, A. (1994). Can the Cambrian explosion be inferred through molecular phylogeny? *Development 1994 Supplement:* 15–25.

Polans, N. (1983). Enzyme polymorphism in Galápagos finches. In *Patterns of Evolution in Galápagos Organisms,* ed. R. I. Bowman, M. Berson, and A. E. Leviton. San Francisco: American Association for the Advancement of Science.

Pollak, E., Kempthorne, O., and Bailey, T. B., Jr. (eds.) (1977). *Proceedings of the International Conference on Quantitative Genetics.* Ames: Iowa State University Press.

Pregill, G. (1986). Body size of insular lizards: A pattern of Holocene dwarfism. *Evolution* 86:997–1008.

Prothero, D. R. (1994). Mammalian evolution. In *Major Features of Vertebrate Evolution,* ed. D. R. Prothero and R. M. Schoch, pp. 238–70. Knoxville: University of Tennessee Paleontological Society.

Prothero, D. R., and Berggren, W. A. (eds.) (1992). *Eocene–Oligocene Climatic and Biotic Evolution.* Princeton: Princeton University Press.

Prothero, D. R., Manning, E., and Fischer, M. S. (1988). The phylogeny of the ungulates. In *The Phylogeny and Classification of the Tetrapods,* ed. M. J. Benton, vol. 2, pp. 201–34. Oxford: Clarendon Press.

Prothero, D. R., and Schoch, R. M. (1994). *Major Features of Vertebrate Evolution.* Knoxville: University of Tennessee Paleontological Society.

Provine, W. B. (1986). *Sewall Wright and Evolutionary Biology.* Chicago: University of Chicago Press.

Raff, R. A. (1996). *The Shape of Life.* Chicago: University of Chicago Press.

Rage, J.-C., and Roček, Z. (1989). Redescription of *Triadobatrachus massinoti* (Piveteau, 1936), an anuran amphibian from the Early Triassic. *Palaeontographica Abteilung A: Palaeozoologie-Stratigraphie* 206:1–16.

Raikow, R. J. (1977). The origin and evolution of the Hawaiian Honeycreepers (Drepanididae). *Living Bird* 15:95–117.

Rancourt, D. E., Teruhisa, T., and Capecchi, M. (1995). Genetic interaction between *hoxb-5* and *hoxb-6* is revealed by nonallelic noncomplementation. *Genes and Development* 9:108–22.

Raup, D.M. (1987). Major features of the fossil record and their implications for evolutionary rate studies. In *Rates of Evolution,* ed. K. S. W. Cambell and M. F. Day, pp. 1–14. London: Allen & Unwin.

Raup, D.M. (1991). *Extinction: Bad Genes or Bad Luck?* New York: W. W. Norton.

Raup, D.M. (1995). The role of extinction in evolution. In *Tempo and Mode in Evolution,* ed. W. M. Fitch and F. J. Ayala, pp. 109–24. Washington, D.C. National Academy of Sciences.

Raup, D. M., and Sepkoski, J. J., Jr. (1982). Mass extinctions in the marine fossil record. *Science* 215: 1501–3.

Raup, D. M., and Sepkoski, J. J., Jr. (1984). Periodicity of extinctions in the geological past. *Proceedings of the National Academy of Science USA* 81:801–5.

Raup, D. M., and Sepkoski, J. J., Jr. (1986). Periodic extinction of families and genera. *Science* 231: 833–6.

Rensberger, J. M., Barnosky, A. D., and Spencer, P. K. (1984). Geology and paleontology of a Pleistocene-to-Holocene loess succession, Benton County, Washington. *Archaeological and Historical Services Eastern Washington University Reports in Archaeology and History:* 100–39.

Repenning, C. A. (1987). Biochronology of the arvicolid rodents of the United States. In *Cenozoic Mammals of North America: Geochronology and Biostratigraphy,* ed. M. O. Woodburne, pp. 236–68. Berkeley: University of California Press.

Repenning, C. A., Fejfar, O., and Heinrich W.-D. (1990). Arvicolid rodent biostratigraphy of the Northern Hemisphere. In *International Symposium on the Evolution, Phylogeny, and Biostratigraphy of Arvicolids (Rodentia, Mammalia),* ed. O. Fejfar and W.-D. Heinrich, pp. 385–417. Prague: Geological Survey.

Retallack, G. J., Veevers, J. J., and Morante, R. (1996). Global coal gap between Permian–Triassic extinction and Middle Triassic recovery of peat-forming plants. *Geological Society of America Bulletin* 108:195–207.

Ribbink, A. J., Marsh, B. A., Marsh, A. C., Ribbink, A. C., and Sharp, B. J. (1983). A preliminary survey of the cichlid fishes of rocky habitats in Lake Malawi. *South African Journal of Science* 18:149–310.

Rich, S. S., Bell, A. E., and Wilson, S. P. (1979). Genetic drift in small populations of *Tribolium. Evolution* 33:579–84.

Ridley, M. (1996). *Evolution,* 2d ed. Oxford: Blackwell Scientific Publications.

Rieppel, O. (1984). Miniaturization of the lizard skull: Its functional and evolutionary implications. *Symposia of the Zoological Society of London* 52:503–20.

Rieppel, O. (1992). Studies on skeleton formation in reptiles. I. The postembryonic development of the skeleton in *Cyrtodactylus pubisulcus* (Reptilia: Gekkonidae). *Journal of Zoology (London)* 227:87–100.

Rieppel, O. (1993). Studies on skeleton formation in reptiles: Patterns of ossification in the skeleton of *Chelydra serpentina* (Reptilia, Testudines). *Journal of Zoology (London)* 231:487–509.

Rightmire, G. P. (1990). *The Evolution of* Homo erectus: *Comparative Anatomical Studies of an Extinct Human Species.* Cambridge: Cambridge University Press.

Robinson, P. L. (1973). A problematic reptile from the British Upper Trias. *Journal of the Geological Society (London)* 129:457–79.

Rose, K. D., and Bown, T. M. (1986). Gradual evolution and species discrimination in the fossil record. In *Vertebrates, Phylogeny, and Philosophy,* ed. K. M. Flanagan and J. A. Lillegraven, *Contributions in Geology Special Papers* (University of Wyoming) 3:119–30.

Ruddle, F. H., Bartels, J. L., Bentley, K. L., Kappen, C., Murtha, M. T., and Pendleton, J. W. (1994a). Evolution of *Hox* genes. *Annual Review of Genetics* 28:423–42.

Ruddle, F. H., Bentley, K. L., Murtha, M. T., and Risch, N. (1994b). Gene loss and gain in the evolution of the vertebrates. *Development 1994 Supplement:* 155–61.

Ryder, G., Fastovsky, D., and Gartner, S. (eds.) (1996). *The Cretaceous–Tertiary Event and Other Catastrophes in Earth History. Geological Society of America Special Paper* 307.

Sanderson, M. J., and Donoghue, M. J. (1994). Shifts in diversification rate with the origin of angiosperms. *Science* 264:1590–3.

Sansom, I. J., Smith, M. M., and Smith, M. P. (1996). Scales of thelodont and shark-like fishes from the Ordovician of Colorado. *Nature* 379:628–30.

Sanz, J. L., and Bonaparte, J. F. (1992). A new order of birds (Class Aves) from the Lower Cretaceous of Spain. *Natural History Museum of Los Angeles County Science Series* 36:39–49.

Sanz, J. L., and Buscalioni, A. D. (1992). A new bird from the Early Cretaceous of Las Hoyas, Spain, and the early radiation of birds. *Palaeontology* 35:829–45.

Sanz, J. L., Chiappe, L. M., and Buscalioni, A. D. (1995). The osteology of *Concornis lacustris* (Aves: Enantiornithes) from the Lower Cretaceous of Spain and a reexamination of its phylogenetic relationships. *American Museum Novitates* 3133:1–23.

Sanz, J. L., Chiappe, L. M., Pérez-Moreno, B. P., Buscalioni, A. D., Moratalla, J. J., Ortega, F., and Poyato-Ariza, F. J. (1996). An Early Cretaceous bird from Spain and its implications for the evolution of avian flight. *Nature* 382:442–5.

Sarich, V. M. (1977). Rates, sample sizes and the neutrality hypothesis for electrophoresis in evolutionary studies. *Nature* 265:24–5.

Saunders, A. D., Storey, M., Kent, R. W., and Norry, M. J. (1992). Consequences of plume–lithosphere interactions. In *Magmatism and the Causes of Continental Break-up,* ed. B. C. Storey, A. Alabaster, and R. J. Pankhurst, *Geological Society of London Special Paper* 68:41–60.

Savage, D. E., and Russell, D. E. (1983). *Mammalian Paleofaunas of the World.* New York: Addison–Wesley.

Savard, P., and Tremblay, M. (1995). Differential regulation of *Hox C6* in the appendages of adult urodeles and anurans. *Journal of Molecular Biology* 249:879–89.

Schaeffer, B., and Thomson, K. S. (1980). Reflections on agnathan–gnathostome relationships. In *Aspects of Vertebrate History: Essays in Honor of Edwin Harris Colbert,* pp. 19–33. Flagstaff: Museum of Northern Arizona Press.

Schaeffer, B., and Williams, M. (1977). Relationships of fossil and living elasmobranchs. *American Zoologist* 17:293–302.

Schindel, D. E. (1982). Resolution analysis: A new approach to the gaps in the fossil record. *Paleobiology* 8:340–53.

Schindewolf, O. H. (1950). *Grundfragen der Paläontology*. Stuttgart: Schweizerbart.

Schleger, G., and Dickie, M. M. (1971). Natural mutation rates in the house mouse. Estimates for five specific loci and dominant mutations. *Mutation Research* 11:89–96.

Schluter, D. (1994). Experimental evidence that competition promotes divergence in adaptive radiation. *Science* 266:798–801.

Schmidt-Nielsen, K. (1975). *Animal Physiology* (2d ed.). Cambridge: Cambridge University Press.

Schmidt-Nielsen, K. (1984). *Scaling: Why Is Animal Size So Important?* Cambridge: Cambridge University Press.

Schmidt-Nielsen, K. (1990). *Animal Physiology* (4th ed.). Cambridge: Cambridge Univerity Press.

Schopf, J. W. (1995). Disparate rates, differing fates: Tempo and mode of evolution changed from the Precambrian to the Phanerozoic. In *Tempo and Mode in Evolution,* ed. W. M. Fitch and F. J. Ayala, pp. 41–61. Washington, D.C.: National Academy of Sciences.

Scott, M. P. (1992). Vertebrate *Hox* gene nomenclature. *Cell* 71:551–3.

Scott, M. P., and Weiner, A. J. (1984). Structural relationships among genes that control development: sequence homology between the *Antennapedia, Ultrabithorax* and *fushi tarazu* loci of *Drosophila. Proceedings of the National Academy of Science* 81:4115–19.

Sepkoski, J. J., Jr. (1990). The taxonomic structure of periodic extinction. In *Global Catastrophes in Earth History,* ed. V. L. Sharpton and P. D. Ward, *Geological Society of America Special Paper* 247:33–44.

Sereno, P. C., and Chenggang, R. (1992). Early evolution of avian flight and perching: New evidence from the Lower Cretaceous of China. *Science* 255:845–8.

Shackleton, N. J., and Opdyke, N. D. (1977). Oxygen isotope and palaeomagnetic evidence for early Northern Hemisphere glaciation. *Nature* 270:216–19.

Sharpton, V. L., Dalrymple, G. B., Marín, L. E., Ryder, G., Schuraytz, B. C., and Urrutia-Fucugauchi, J. (1992). New links between the Chicxulub impact structure and the Cretaceous–Tertiary boundary. *Nature* 359:819–21.

Sharpton, V. L., and Ward P. D. (eds.) (1990). *Global Catastrophes in Earth History. Geological Society of America Special Paper* 247: xi, 631.

Sheldon, P. R. (1987). Parallel gradualistic evolution of Ordovician trilobites. *Nature* 330:561–3.

Sheldon, P. R. (1993). Making sense of microevolutionary patterns. In *Evolutionary Patterns and Processes,* ed. D. R. Lees and D. Edwards, pp. 19–31. London: Linnean Society.

Shrock, R., and Twenhofel, W. (1953). *The Principles of Invertebrate Paleontology*. New York: McGraw–Hill.

Shu, D., Conway Morris, S., and Zhang, X. (1996). *Pikaia*-like chordate from the Lower Cambrian of China. *Nature* 384: 157–8.

Shu, D., Zhang, X., and Chen, L. (1996). Reinterpretation of *Yunnanozoon* as the earliest known hemichordate. *Nature* 380:428–30.

Shubin, N. H. (1995). The evolution of paired fins and the origin of tetrapod limbs: Phylogenetic and transformational approaches. *Evolutionary Biology* 28:39–86.

Shubin, N. H., and Alberch, P. (1986). A morphogenetic approach to the origin and basic organization of the tetrapod limb. *Evolutionary Biology* 20:319–87.

Shubin, N. H., and Jenkins, F. A., Jr. (1995). An Early Jurassic jumping frog. *Nature* 377:49–52.

Simpson, G. G. (1928). A catalogue of the Mesozoic Mammalia in the Geological Department of the British Museum. London: British Museum (Natural History), London.

Simpson, G. G. (1929). American Mesozoic Mammalia. *Peabody Museum Memoirs* 3:1–171.

Simpson, G. G. (1944). *Tempo and Mode in Evolution*. New York: Columbia University Press.

Simpson, G. G. (1952). How many species? *Evolution* 6:342.

Simpson, G. G. (1953). *The Major Features of Evolution*. New York: Columbia University Press.

Simpson, G. G. (1959). Mesozoic mammals and the polyphyletic origin of mammals. *Evolution* 13:405–14.

Simpson, G. G. (1961). *Principles of Animal Taxonomy*. New York: Columbia University Press.

Skelton, P. (ed). (1993). *Evolution: A Biological and Palaeontological Approach*. Wokingham, England: Addison–Wesley and Longman.

Slack, J. M. W., Holland, P. W. H., and Graham, C. F. (1993). The zootype and the phylotypic stage. *Nature* 361:490–1.

Smith, A. B. (1994). *Systematics and the fossil record*. Oxford: Blackwell Scientific Publications.

Smith, A. B., and Littlewood, D. T. (1994). Paleontological data and molecular phylogenetic analysis. *Paleobiology* 20:259–73.

Smith, A. B., and Patterson, C. (1989). The influence of taxonomic method on the perception of patterns of evolution. *Evolutionary Biology* 23:127–216.

Smith, G. R., and Todd, T. N. (1984). Evolution of species flocks of fishes in North Temperate lakes. In *Evolution of Fish Species Flocks*, ed. A. S. Echell and I. Kornfield, pp. 45–68. Orono: University of Maine Press.

Smith, M. M., and Hall, B. K. (1990). Development and evolutionary origins of vertebrate skeletogenic and odontogenetic tissue. *Biological Reviews* 65:277–374.

Smith, R. M. H. (1995). Changing fluvial environments across the Permian–Triassic boundary in the Karroo Basin, South Africa, and possible causes of tetrapod extinction. *Palaeogeography, Palaeoclimatology, and Palaeoecology* 117:81–104.

Smith, S. C., Graveson, A. C., and Hall, B. K. (1994). Evidence for a developmental and evolutionary link between placodal ectoderm and neural crest. *Journal of Experimental Zoology* 270: 292–301.

Smithson, T. R. (1985). Scottish Carboniferous amphibian localities. *Scottish Journal of Geology* 21: 123–42.

Smithson, T. R., Carroll, R. L., Panchen, A., and Andrews, S. (1994). *Westlothiana lizziae*, a stem amniote from the Lower Carboniferous of Scotland. *Transactions of the Royal Society of Edinburgh: Earth Sciences* 84:383–412.

Soltis, P. S., Soltis, D. E., and Smiley, C. J. (1992). An *rbc*L sequence from Miocene *Taxodium* [bald cypress]. *Proceedings of the National Academy of Science USA* 89:449–51.

Sondaar, P. Y. (1977). Insularity and its effect on mammal evolution. In *Major Patterns of Vertebrate Evolution*, ed. M. K. Hecht, P. C. Goody, and B. M. Hecht, pp. 671–707. New York: Plenum.

Sordino, P., van der Hoeven, F., and Duboule, D. (1995). *Hox* gene expression in teleost fins and the origin of vertebrate digits. *Nature* 375:678–81.

Spicer, R. A. (1989). Plants at the Cretaceous–Tertiary boundary. *Philosophical Transactions of the Royal Society B* 325:291–305.

Spotila, J. R. (1980). Constraints of body size and environment on the temperature regulation of dinosaurs. In *A Cold Look at the Warm-Blooded Dinosaurs*, ed. R. D. K. Thomas and E. C. Olson, pp. 233–52. Washington, D.C.: American Association for the Advancement of Science.

Stanley, S. M. (1975). A theory of evolution above the species level. *Proceedings of the National Academy of Science USA* 72:646–50.

Stanley, S. M. (1979). *Macroevolution*. San Francisco: W. H. Freeman.

Stanley, S. M. (1984). Simpson's inverse: Bradytely and the phenomenon of living fossils. In *Living Fossils*, ed. N. Eldredge and S. M. Stanley, pp. 272–7. New York: Springer-Verlag.

Stanley, S. M. (1985). Rates of evolution. *Paleobiology* 11:13–26.

Steadman, D. W., Stafford, T. W., Jr., Donahue, D. J., and Jull, A. J. T. (1991). Chronology of Holocene vertebrate extinction in the Galápagos islands. *Quaternary Research* 36:126–33.

Steadman, D. W., and Zousmer, S. (1988). *Galápagos*. Washington, D.C.: Smithsonian Institution Press.

Stearn, C., and Carroll, R. L. (1989). *Paleontology: The Record of Life*. New York: John Wiley & Sons.

Stebbins, G. L. (1988). Essays in comparative evolution: The need for evolutionary comparisons. In *Plant Evolutionary Biology*, ed. L. D. Gottlief and S. K. Jain, pp. 3–20. London and New York: Chapman & Hall.

Stebbins, G. L., and Ayala, F. J. (1981). Is a new evolutionary synthesis necessary? *Science* 213:967–71.

Stebbins, R. C. (1951). *Amphibians of Western North America*. Berkeley: University of California Press.

Stehli, F. G., and Webb, S. D. (eds.) (1985a). *The Great American Biotic Interchange*. New York: Plenum.

Stehli, F. G., and Webb, S. D. (1985b). A kaleidoscope of plates, faunal and floral dispersals, and sea level changes. In *The Great American Biotic Interchange,* ed. F. G. Stehli and S. D. Webb, pp. 3–16. New York and London: Plenum Press.

Stern, D. L., and Grant, P. R. (1996). A phylogenetic reanalysis of allozyme: Variation among populations of Galápagos finches. *Zoological Journal of the Linnean Society* 118: 119–34.

Stiassny, M. L. J. (1991). Phylogenetic intrarelationships of the family Cichlidae: An overview. In *Cichlid Fishes: Behaviour, Ecology, and Evolution* ed. M. H. A. Keenleyside, pp. 1–35. London: Chapman & Hall.

Storm, E. E., Huynh, T. V., Copeland, N. G., Jenkins, N. A., Kingsley, D. M., and Lee, S.-J. (1994). Limb alterations in *brachypodism* mice due to mutations in a member of the TGFβ-superfamily. *Nature* 368:639–43.

Storrs, G. W. (1993). The quality of the Triassic sauropterygian fossil record. *Revue de Paléobiologie* 7:217–28.

Stuart, A. J. (1982). *Pleistocene vertebrates in the British Isles*. New York: Longman.

Sues, D.-H. (1987). Postcranial skeleton of *Pistosaurus* and interrelationships of the Sauropterygia (Diapsida). *Zoological Journal of the Linnean Society* 90:109–31.

Szalay, R. W., Novacek, M. J., and McKenna, M. C. (eds.) (1993). *Mammal Phylogeny*. New York: Springer-Verlag.

Tabin, C. J. (1992). Why we have (only) five fingers per hand: *Hox* genes and the evolution of paired limbs. *Development* 116:289–96.

Tabin, C. J., and Laufer, E. (1993). *Hox* genes and serial homology. *Nature* 361:692–3.

Taliev, D. N. (1955). *Sculpins of Baikal (Cottoidei)*. Moscow: Academy of Science USSR.

Tax, S. (1960). *Evolution after Darwin: The University of Chicago Centennial*. 3 vols. Chicago: University of Chicago Press.

Templeton, A. R. (1989). The meaning of species and speciation: A genetic perspective. In *Speciation and its Consequences,* ed. D. Otte and J. A. Endler, pp. 3–27. Sunderland, Mass.: Sinauer Associates.

Thewissen, J. G. M. (1994). Phylogenetic aspects of cetacean origins: A morphological perspective. *Journal of Mammalian Evolution* 2:157–84.

Thewissen, J. G. M., and Hussain, S. T. (1993). Origin of underwater hearing in whales. *Nature* 361:444–5.

Thewissen, J. G. M., Hussain, S. T., and Arif, M. (1994). Fossil evidence for the origin of aquatic locomotion in archaeocete whales. *Science* 263:210–12.

Thomson, K. S. (1967). Mechanisms of intracranial kinetics in fossil rhipidistian fishes (Crossopterygii) and their relatives. *Journal of the Linnean Society of London (Zoology)* 178:223–53.

Thomson, K. S. (1976). On the heterocercal tail in sharks. *Paleobiology* 2:19–38.

Thomson, K. S. (1988). *Morphogenesis and Evolution*. Oxford: Oxford University Press.

Thomson, K. S., and Simanek, D. E. (1977). Body form and locomotion in sharks. *American Zoologist* 17:343–54.

Thorogood, P. (1991). The development of the teleost fin and implications for our understanding of tetrapod limb evolution. In *Development Patterning of the Vertebrate Limb,* ed. J. R. Hinchliffe, J. M. Hurle, and D. Summerbell, pp. 347–54. New York: Plenum.

Thorogood, P. (1993). Differentiation and morphogenesis of cranial skeletal tissues. In *The Skull,* vol. 1, *Development,* ed. J. Hanken and B. K. Hall, pp. 112–52. Chicago: University of Chicago Press.

Thorpe, R. S. (1987). Geographic variation within an island: univariate and multivariate contouring of scalation, size and shape of the lizard *Gallotia galloti*. *Evolution* 41:256–68.

Thorpe, R. S. (1991). Clines and cause: microgeographic variation in the Tenerife Gecko (*Tarentola delalandii*). *Systematic Zoology* 40:172–87.

Thorpe, R. S., and Baez, M. (1993). Geographic variation in scalation of the lizard *Gallatia stehlini* with the island of Gran Canaria. *Biological Journal of the Linnean Society* 48:75–87.

Thorpe, R. S., and Brown, R. P. (1989). Microgeographic variation in the colour pattern of the lizard *Gallotia galloti* within the island of Tenerife: Distribution, pattern and hypothesis testing. *Biological Journal of the Linnean Society* 38:303–22.

Thorpe, R. S., and Brown, R. P. (1991). Microgeographic clines in the size of mature male *Gallotia galloti* (Squamata: Lacertidae) on Tenerife: Causal hypothesis. *Herpetologica* 47:28–37.

Thorpe, R. S., McGregor, D. P., and Cumming, A. M. (1993). Molecular phylogeny of the Canary Island lacertids (*Gallotia*): Mitochondrial DNA restriction fragment divergence in relation to sequence divergence and geological time. *Journal of Evolutionary Biology* 6:47–88.

Thulborn, T. (1990). *Dinosaur Tracks*. London and New York: Chapman & Hall.

Trinkaus, E., and Shipman, P. (1993). *The Neanderthals*. New York: Alfred A. Knopf.

Trueb, L., and Cloutier, R. (1991). A phylogenetic investigation of the inter- and intrarelationships of the Lissamphibia (Amphibia: Temnospondyli). In *Origins of the Higher Groups of Tetrapods: Controversy and Consensus,* ed. H.-P. Schultze and L. Trueb., pp. 223–313. Ithaca and London: Cornell University Press.

Turelli, M., Gillespie, J. H., and Lande, R. (1988). Rate tests for selection on quantitative characters during macroevolution and microevolution. *Evolution* 42:1085–9.

Valentine, J. W. (1995). Late Precambrian bilaterians: Grades and clades. In *Tempo and Mode in Evolution,* ed. W. M. Fitch and F. J. Ayala, pp. 87–108. Washington, D.C.: National Academy Press.

Valentine, J. W., Erwin, D. H., and Jablonski, D. (1996). Developmental evolution of metazoan bodyplans: The fossil evidence. *Developmental Biology* 173:373–81.

van Andel, T. H. (1994). *New Views on an Old Planet* (2d ed.). Cambridge: Cambridge University Press.

Van Damme, D. (1984). *The Freshwater Mollusca of North Africa*. Dordrecht: Junk.

Van Devender, T. R., and Mead, J. I. (1978). Early Holocene and late Pleistocene amphibians and reptiles in Sonoran Desert packrat middens. *Copeia* 1978:464–75.

Van Tyne, J., and Berger, A. L. (1976). *Fundamentals of Ornithology* (2d ed.). New York: John Wiley & Sons.

Vanzolini, P. E., and Heyer, W. R. (1985). The American herpetofauna and the interchange. In *The Great American Biotic Interchange,* ed. F. G. Stehli and S. D. Webb, pp. 475–87. New York and London: Plenum Press.

Vartanyan, S. L., Garutt, V. E., and Sher, A. V. (1993). Holocene dwarf mammoths from Wrangel Island in the Siberian Arctic. *Nature* 362:337–40.

Vazquez, R. J. (1992). Functional osteology of the avian wrist and the evolution of flapping flight. *Journal of Morphology* 211:259–68.

von Baer, K. E. (1828). *Entwicklungsgeschichte der Tiere: Beobachtung und Reflexion.* Konigsberg: Borntraeger.

Vorobyeva, E., and Schultze, H.-P. (1991). Description and systematics of Panderichthyid fishes with comments on their relationship to tetrapods. In *Origins of the Higher Groups of Tetrapods: Controversy and Consensus,* ed. H.-P. Schultze and L. Trueb., pp. 68–109. Ithaca and London: Cornell University Press.

Vrba, E. S. (1980). Evolution, species, and fossils: How does life evolve? *South African Journal of Science* 76:61–84.

Vrba, E. S. (1984). Evolutionary pattern and process in the sister-group Alcelaphini–Aepycerotini (Mammalia: Bovidae). In *Living Fossils,* ed. N. Eldredge and S. M. Stanley, pp. 62–79. New York: Springer-Verlag.

Wainwright, S. A., Biggs, W. D., Currey, J. D., and Gosline, J. M. (1982). *Mechanical Design in Organisms*. Princeton: Princeton University Press.

Wake, D. B., Brame, A. H., Jr., and Thomas, R. (1982). A remarkable new species of salamander allied to *Bolitoglossa altamazonica* (Plethodontidae) from southern Peru. *Occasional Papers of the Museum of Zoology, Louisiana State University* 58:1–21.

Wake, D. B., and Elias, P. (1983). New genera and a new species of Central American salamanders, with a review of the tropical genera (Amphibia, Caudata, Plethodontidae). *Natural History Museum of Los Angeles County Contributions in Science* 345:1–19.

Wake, D. B., and Lynch, J. F. (1976). The distribution, ecology, and evolutionary history of plethodontid salamanders in tropical America. *Natural History Museum of Los Angeles County Science Bulletin* 25:1–65.

Wake D. B., Roth, G., and Wake, M. H. (1983). On the problem of stasis in organismal evolution. *Journal of Theoretical Biology* 101:211–24.

Wake, D. B., Yaney, K. R., and Frelow, M. M. (1989). Sympatry and hybridization in a "Ring Species": The plethodontid salamander *Ensatina eschscholtzii*. In *Speciation and Its Consequences,* ed. D. Otte and J. A. Endler, pp. 134–57. Sunderland, Mass.: Sinauer Associates.

Wallace, B. (1985). Reflections on the still-"hopeful monster." *Quarterly Review of Biology* 60:31–42.

Ward, R. D., Skibinski, D. O., and Woodwark, M. (1992). *Evolutionary Biology* 26:73–159.

Webb, P. W. (1994). The biology of fish swimming. In *Mechanics and Physiology of Animal Swimming,* ed. L. Maddock, Q. Bone, and J. M. V. Rayner, pp. 45–62. Cambridge: Cambridge University Press.

Webb, S. D. (1978). A history of savanna vertebrates in the New World. Part II. South America and the Great Interchange. *Annual Review of Ecology and Systematics* 9:393–426.

Webb, S. D., and Barnosky, A. D. (1989). Faunal dynamics of Pleistocene mammals. *Annual Review of Earth and Planetary Science* 17:413–38.

Weir, B. S., Eisen, E.J., Goodman, M. M., and Kamkoong, G. (eds.) (1988). *Proceedings of the Second International Conference on Quantitative Genetics.* Sunderland, Mass.: Sinauer Associates.

Wellnhofer, P. (1978). Pterosauria, Teil 19. In *Handbuch der Paläoherpetologie,* ed. O. Kuhn and P. Wellnhofer. Stuttgart: Gustav Fischer Verlag.

Wellnhofer, P. (1991). *The Illustrated Encyclopedia of Pterosaurs.* New York: Crescent Books.

Wellnhofer, P. (1993). Das siebte Exemplar von *Archaeopteryx* aus den Solnhofener Schichten. *Archaeopteryx* 11:1–48.

Wellnhofer, P. (1994). New data on the origin and early evolution of birds. *Comptes Rendus de l'Académie Science (Paris)* 319:299–308.

Welman, J. (1995). *Euparkeria* and the origin of birds. *South African Journal of Science* 91:533–7.

Werdelin, L. (1989). Constraint and adaptation in the bone-cracking canid *Osteoborus* (Mammalia: Canidae). *Paleobiology* 15:387–401.

Werner, T. K., and Sherry, T. W. (1987). Behavioral feeding specialization in *Pinarolaxias inornata,* the "Darwin's finch" of Cocos Island, Costa Rica. *Proceedings of the National Academy of Science USA* 84:5506–10.

West-Eberhard, M. J. (1989). Phenotypic plasticity and the origins of diversity. *Annual Review of Ecology and Systematics* 20:249–78.

Whiteside, D. L. (1986). The head skeleton of the Rhaetian sphenodontid *Diphydontosaurus avonis* gen. et sp. nov., and the modernizing of a living fossil. *Philosophical Transactions of the Royal Society of London B* 312:379–430.

Wild, R. (1978). Die Flugsaurier (Reptilia, Pterosauria) aus der oberen Trias von Cene bei Bergamo, Italien. *Bollettino della Società Paleontologica Italiana* 17:176–256.

Wiles, J. S., and Sarich, V. M. (1983). Are the Galápagos iguanas older than the Galápagos? Molecular evolution and colonization models for the archipelago. In *Patterns of Evolution in Galápagos Organisms,* ed. R. I Bowman, M. Berson, and A. E. Leviton, pp. 177–86. San Francisco: American Association for the Advancement of Science, Pacific Division, California Academy of Sciences.

Wiley, E. O. (1981). *Phylogenetics: The Theory and Practice of Phylogenetic Systematics.* New York: John Wiley & Sons.

Williams, G. C. (1992). *Natural Selection: Domains, Levels, and Applications.* Oxford: Oxford University Press.

Williams, M. A. J., Dunkerley, D. L., De Decker, P., Kershaw, A. P., and Stokes, T. (1993). *Quaternary Environments.* London, New York, Melbourne, and Auckland: Edward Arnold.

Wilson, A. C., Carlson, W. S., and White, T. J. (1977). Biochemical evolution. *Annual Review of Biochemistry* 46:573–639.

Wilson, E. O. (1992). *The Diversity of Life.* Cambridge, Mass.: Belknap.

Windley, B. F. (1995). *The Evolving Continents* (3d ed.). New York: John Wiley & Sons.

Witte, F., Barel, C. D. N., and Hoogerhoud, R. J. C. (1990). Phenotypic plasticity of anatomical structures and its ecomorphological significance. *Netherlands Journal of Zoology* 40:278–98.

Wolbach, W. S., Gilmour, I., and Anders, E. (1990). Major wildfires at the Cretaceous–Tertiary boundary. In *Global Catastrophes in Earth History,* ed. V. L. Sharpton and P. D. Ward, *Geological Society of America Special Paper* 247:391–400.

Woodburne, M. O. (1987). *Cenozoic Mammals of North America: Geochronology and Biostratigraphy.* Berkeley: University of California Press.

Woodburne, M. O. (1996). Precision and resolution in mammalian chronostratigraphy: Principles, practices, examples. *Journal of Vertebrate Paleontology* 16:531–55.

Woodburne, M. O., & Case, J. A. (1996). Dispersal, vicariance, and the Late Cretaceous to Early Tertiary land mammal biogeography from South America to Australia. *Journal of Mammalian Evolution* 3:121–61.

Woodburne, M. O., and Swisher, C. C. (1995). Land mammal high-resolution geochronology, intercontinental overland dispersals, sea level, climate, and vicariance. In *Geochronology, Time Scales and Global Stratigraphic Correlation,* ed. W. A. Berggren, D. V. Kent, and J. A. Hardenbol, *Society of Economic Paleontologists and Mineralogists Special Publication* 54:335–65.

Woodward, S. P. (1859). On some new freshwater shells from central Africa. *Proceedings of the Zoological Society of London* 348–9.

Wright, S. (1931). Evolution in Mendelian populations. *Genetics* 16:97–159.

Wright, S. (1932). The roles of mutation, inbreeding, crossbreeding and selection in evolution. *Proceedings of the Sixth International Congress of Genetics* 1:356–66.

Wright, S. (1934). An analysis of variability in number of digits in an inbred strain of guinea pigs. *Genetics* 19:506–36.

Wright, S. (1968). *Evolution and the Genetics of Populations,* vol. 1, *Genetic and Biometric Foundations.* Chicago: University of Chicago Press.

Wright, S. (1977). *Evolution and the Genetics of Populations,* vol. 3, *The Results and Evolutionary Deductions.* Chicago: University of Chicago Press.

Wright, S. (1982). The shifting balance theory and macroevolution. *Annual Review of Genetics* 16:1–19.

Wu, X.-C. (1994). Late Triassic–Early Jurassic sphenodontians from China and the phylogeny of the Sphenodontia. In *In the Shadow of the Dinosaurs,* ed. N. C. Fraser and H.-D. Sues, pp. 38–69. Cambridge: Cambridge University Press.

Yang, S.Y., and Patton, J. L. (1981). Genetic variability and differentiation in the Galápagos finches. *Auk* 98:230–42.

Zakrzewski, R. J. (1969). The rodents from Hagerman local fauna, upper Pliocene of Idaho. *Contributions of the Museum of Paleontology, University of Michigan* 23:1–36.

Zangerl, R. (1981). Chondrichthyes I. In *Handbook of Paleoichthyology,* ed. H.-P. Schultze. Stuttgart: Gustav Fischer Verlag.

Zhou, X., Zhai, R., Gingerich, P. D., and Chen, L. (1995). Skull of a new mesonychid (Mammalia, Mesonychia) from the late Paleocene of China. *Journal of Vertebrate Paleontology* 15:387–400.

Zhou, Z.-H., Jin, F., and Zhang, J.-Y. (1992). Preliminary report on a Mesozoic bird from Liaoning, China. *Chinese Science Bulletin* 37:1365–8.

Index

Note: For organisms, check under both the scientific and common names.
Abbreviations: f, figure; t, table.

Abdominal-B, 217f, 218, 229
abortion, spontaneous, 213–14
acid rain, 382
acritarchs, 395
actinopterygians, 273–4
  primitive, transition to teleosts, 366–7
  radiation, 350, 361
  species number, 176
Actinopterygii, 159, 380
  Amia, 273
  atherinids, 142
  beryciforms, 350
  chondrosteans, 168, 273
  cichlids, 123–39; *see also* cichlid fish;
    Cichlidae
  cyprinids, 142
  Danio (zebra fish), 229, 232, 234f
  Istiophoridae, 268f, 273
  killifish, 142
  Lates niloticus, 130
  Perciformes, 350
  percomorphs, 350
  Scombridae, 273
  sculpins, 142–3
  Semionotidae, 142
  sticklebacks, 119–23; *see also Gasterosteus*
  Synodontis, 140
  teleosts, 273, 342f, 366–7, 382
  Xiphidae, 273
  Xiphius, 273
actinotrichia, 232, 405
adaptation, rate of, 118
adaptive topography, 190–1, 405
additive effect, 195, 405
Agassiz, L., 58
Agnatha, 158, 229–30
  Astraspis, 362
  hagfish, 158
  lamprey, 158
agnathans, 341, 349–50, 349f
air sacs, 320
airfoil, 276, 280
Albert, Lake, 131
alkali myosin light chain molecules, 224
Allen's rule, 36
allopatry, 22, 405
alula (bastard wing), 318
Amniota, 160, 161f, 314, 342f; *see also individual
  taxa*
amniotes, 244
  primitive, transition to mammals, 367–8
  reproduction, 288–9, 288f

Amphibia, 159
  extant taxa
    Ambystoma spp., 214, 242f, 294f
    Amphiumidae, 214
    Andrias, 287
    anurans (frogs), 243, 284, 285f, 287–8, 291–3,
      294f, 303–4
    Ascaphus, 294f
    Batrachoseps, 289f
    Bolitoglossa, 117
    caecilians, 288, 291–3, 294f, 295
    Chiromantis, 173
    Cryptobranchidae, 214
    Cryptobranchus, 287
    Ensatina eschscholtzii, 53
    Grandisonia, 294f
    Ichthyopis, 294f
    lissamphibians, 159, 350–1
    Notophthalmus, 230
    Phyllomedusa, 173
    Plethodontidae, 55, 117, 287–9, 289f
    Proteidae, 214
    Pseudoeurycae, 293f
    Sirenidae, 214
    Thorius, 293f
    urodeles (salamanders), 243, 287, 291–3, 294f
    Xenopus, 230
  extinct taxa
    Acanthostega, 231f, 233f, 300f, 301f, 300–2,
      303f, 304, 338
    Balanerpeton, 304
    Eocaecilia, 295
    Elginerpeton, 301f, 304–5
    Hynerpeton, 300, 300f, 301f
    Ichthyostega, 231, 233f, 300, 301f, 302
    labyrinthodonts, 79, 306, 350–1
    lepospondyls, 350–1
    nectrideans, 67f
    Obruchevichthys, 304
    Prosalirus, 284, 285f
    Pseudocaecilia, 292f, 303
    seymouriamorphs, 304
    stereospondyls, 350–1
    temnospondyls, 293, 303–4
    Tersomius, 292f, 294f
    Triadobatrachus, 284, 293, 294f
    Tulerpeton, 300, 301f
    Ventastega, 300, 301f
amphibians, 115–18, 244, 383, 396
  origin of, 300–6
  radiation, 350–1
  reproduction, 287–9, 304

AmphiHox 3, 223
anagenesis, 21, 53, 355, 400, 405
anagenetic origin of species, 102–3
anapsid, 163
angiosperms, 8, 16, 380, 382, 402–3
annelids, 343
Antennapedia, 215–16
apical ectodermal fold, 232–3, 235f, 405
apical ectodermal ridge (AER), 232, 235f, 238, 263, 405
apomorphy, 151, 405
aquatic adaptation, secondary, 170, 274, 324–35, 365
Archaeopteryx, 24, 163, 277–81, 306–8, 309f, 310f, 312–19, 319f, 322–3, 336–9
    bavarica, 315, 339
    lithographica, 307f, 315, 339
Archaeornithes, 315
Archosauria, 163; *see also individual taxa*
armadillo, *Holmesia*, 91, 93f
arthropods, 343–4, 346–8, 378
    copepods, 140
    ostracods, 140, 282, 396
    see also individual taxa
artificial selection, 198–9, 199f, 202
Artiodactyla, 149–54, 329, 332, 356–7, 359
    alcelaphine bovids, 105–6, 105f
    Alces, 94, 96, 96f
    Mesochoerus limnetes, 91
    Metridiochoerus andrewsi, 91
aspect ratio, 267–8, 276, 405
autapomorphy, 151, 405
autopod, 236, 405
Aves, 118–19, 160, 161f, 168, 245, 277, 283
    origin of, 306–23
    radiation during Cenozoic, 359, 361
    taxonomic definitions of, 163–4, 164f, 315
    see also individual taxa and structural elements

Baikal, Lake, 124, 140, 142
bats, *see* Chiroptera
behavior, 12, 17, 44, 48, 77, 124, 127–8, 138–9, 192, 257, 314, 326, 337–9, 348, 360–5, 389, 391–3, 400, 404
Bergmann's rule, 36, 53
biological species concept, 21–2, 84, 148, 405
birds, *see* Aves
Biston betularia, 35, 182
blending inheritance, 193
bolide, 380–2, 385, 404
bone, 173–6
    composition, 174
    dermal (intramembranous), 174, 176, 269, 406
    endochondral (cartilage replacement), 174, 176, 269, 305, 406
    as a factor of preservation, 16, 349–50
    physical constraints on shape, 174–5
    physiological response to stress, 175–6
    resistance to stress, 174
    specific gravity, 174
    trabecular, 174
    see also individual bones; ossification
bone morphogenetic proteins, 238–9, 405
boundary layer, 267
brachiopods, 344–5, 378
Brachypodism, 239, 261
brain size
    hominid, 214
    mammalian, 360, 366

bryozoans, 344, 378
buoyancy control, 176, 269, 273, 366
Burgess Shale, 4, 259, 343f, 348

Canary Islands, 38, 116–17
carnassial teeth, 170–2, 171f
Carnivora, 170–2, 356, 359
    borophagines, 170–2
    Borophagus diversidens, 171
    Canidae, 170–2
    Canis lupus, 171f
    Crocuta, 171f
    Hyaenidae, 170–2
    Osteoborus cyonoides, 171, 171f
    see also Pinnipedia
carpometacarpus, 318, 319f, 322–3, 337–8
cartilage, 173–6, 269, 367
caudofemoralis, 309, 312f
cell differentiation, 212–13, 346, 404
cellulase, 209
Cepaea, 35, 110–11
cephalochordates, 268
    amphioxus, 179
    Branchiostoma, 222–4, 229
    Cathaymyris, 344
    transition to craniates, 223–4, 363
Cetacea, 150, 172, 274, 331f, 339, 356–7, 359, 363
    Ambulocetus, 333–4, 335f
    Archaeocetes, 329–35
    basilosaurids, 334, 359
    Basilosaurus isis, 334, 335f
    Indocetus, 334
    Pakicetus, 330–2, 330f, 333f
    Protocetus, 332, 334
    Rodhocetus, 332, 334, 335f
cetaceans, origin of, 329–36
Cete, 329–36, 359
character displacement, 22, 29, 48, 405
Chelonia (turtles), 161f, 163, 165, 256, 263–4, 382, 386
Chengjiang fauna, 259, 343f, 344, 348
Chiroptera, 172, 277–9, 279f, 308
Choanata, 159, 304, 350
chondrichthyans
    radiation, 350, 361
    species number, 176
Chondrichthyes, 158, 229, 268–73, 380, 382
    Batomorphii, 168, 269
    Chimaeriformes, 168
    Cladoselache, 269, 270f
    elasmobranchs, 268–73, 270f, 271f, 272f
    neoselachians, 168, 350
    Palaeospinax, 270f, 272
chondrification patterns, 240–4
chondroblasts, 174
chondrogenetic condensations, 240–4, 243f
chord, 268, 268f, 405
Chordata, 347
Choristodira, 297, 299f
chromosomes, 23, 210
chronomorphs, 103, 405
cichlid fish, 123–39, 363
    behavioral plasticity, 127–8, 134, 138–9
    parental care, 136–7
    pharyngeal jaws, 124, 128, 132, 134, 134f
    phylogeny, 127f
    radiation, 123–44, 147f, 357, 361, 392
    reproduction, 132, 136

sexual selection, 136–7
speciation, 55, 125, 137
trophic diversification, 124, 128, 143
Cichlidae, 123–4
  Astatotilapia, 126
  Grammatoria, 135f
  haplochromines, 128, 129f, 147f
  Haplochromis, 128
  Lamprologus, 135f
  Macropleurodus, 138
  Mbuna, 137–8
  Psammochromis, 138
  Rhamphochromis, 126
  Tilapia, 128
  Tropheini, 126
  Tylochromis, 135f
clades, 106, 405
  naming and defining, 160–5
    apomorphy-based, 163, 164f, 405
    node-based, 162, 164f, 408
    stem-based, 163, 164f, 409
cladistic gradualism, 138
cladistics, 151
cladogenesis, 21, 165, 345, 355, 405
classification, 145–66
  Hennigian, 150–5; *see also* phylogenetic systematics
  Linnean, 146–50, 160
Claudiosaurus, 245–7, 247f, 265
clavicles, 307–8, 315
climate, 369
  change, 146, 372–4, 372f, 383, 390, 392, 399–400
  instability, 382–3
clines, 52–4, 118, 405
Cnidaria, 343, 346–8
coccoliths, 395
Cocos finch, *see* Darwin's finches, *Pinaroloxias inornata*
Cocos Island, 40, 48–9
coelacanths, *see* Sarcopterygii
competition, 47–8, 108, 138, 169, 340, 348, 364
Condylarthra, 150, 152, 153f, 154, 356–7
constraints, 131–2; *see also* evolutionary constraints
continental drift, 11, 80, 146, 346, 361, 368–74, 370f, 371f, 372f, 373f, 390, 406; *see also* plate tectonics
coordinated stasis, 399–400, 406
Cope, E. D., 59
coracoid, 278f, 312, 318, 319f, 320, 322, 337–8
correlated progression, 367, 406
countercurrent heat exchange, 173
cranial kinesis, 305
craniates, origin of, 222–4, 363
Cretaceous–Tertiary boudary, *see* K–T boundary
Crocodylia, 161f, 245, 283, 290, 380, 382, 386
  radiation of, 351, 352f
crown group, 162, 163f, 406
Cuvier, G., 58
Cyanobacteria (blue–green algae), 394, 403–4
  Chroococcaceae, 394
  Oscillatoriaceae, 394

darwin (measure of evolutionary rate), 72–7, 100, 205, 356, 406
Darwin, C., 1, 2, 19–21, 24–6, 29, 34, 58, 86, 110, 112, 145, 168–9, 193, 212, 286, 359, 361, 364, 368–9, 376, 389–90, 394–6
Darwin's finches, 38–52, 42f, 43t

Cactospiza spp., 41
Camarhynchus, 46
Certhidea, 41
Geospiza, 46, 47f
  conirostris, 42, 45t, 46, 47
  difficilis, 41, 47–9
  fortis, 45, 196–7, 196f, 197f, 205
  fuliginosa, 47–8
  magnirostris, 47, 49
  scandens, 47
phylogeny of, 41, 41f
Pinaroloxias inornata, 48–9
dating, 68–72
  magnetostratigraphy, 18, 70–2
  radiometric, 18, 69–70
    ^{40}Ar/^{39}Ar, 70, 71, 84
    ^{14}C, 70, 84
    ^{40}K–^{40}Ar, 70, 84
  by rate of molecular change, 79–80
  by rate of sedimentation, 70
Dawson, J. W., 58
dentine tracts, 98, 99f
dentition, 144
  ancestral mammals, 368
  Carnivora, 170–2
  early placentals, 357–8
  horses, 204t
  mesonychids, 329
  proboscidians, 52, 95, 97f, 98
  rodents, 98–102, 99f, 101f, 102f
deuterostomes, 344–7
development, 18, 212–65, 339, 361, 393, 404
  in Cambrian explosion, 346–8
  in cephalochordate–craniate transition, 222–4
  contribution to evolutionary synthesis, 212, 258–62
  of head, 222–5
  of limbs, 227–49, 305, 338, 347–8
  of vertebrae, 225–7, 265
developmental plasticity, 135
Diapsida, 164f, 342f; *see also individual taxa*
diapsids, 163, 245, 246f
diatoms, 395–6
digital arch, 242–3, 343f, 406
dinoflagellates, 395
dinosaurs, 85, 249, 283, 290, 306–8, 314, 342f, 352, 356, 360, 380, 382
  Allosaurus, 308
  coelurosaurs, 308–9
  Compsognathus, 307–8, 309f, 310f, 313
  Deinonychus, 309, 309f, 310f, 322
  Dromaeosauridae, 308, 312, 314–15, 322–3
  ornithischian, 308
  radiation, 351, 353f, 361
  saurischian, 308
  Syntarsus, 312f
  theropods, 308–9, 309f, 314, 338
  transition to birds, 306–15
  Tyrannosaurus, 249f, 308
dispersal, 112–13, 116, 355, 362, 399–401
diversity of life, 378f
DNA, 15, 18, 79
  mitochondrial, 126, 130, 138, 210
  nuclear, 180, 183–4, 209, 215–16
  sequencing, 79–80
Dobzhansky, T., 20–4, 54, 60
dominance effect, 195, 406
dormancy, 116, 380, 403–4

drag, 267, 275–6
  friction, 267, 406
  induced, 276, 407
  pressure, 267, 408
  profile, 276, 408
Drosophila, 79, 186, 201, 215–16, 217f, 218–19, 222, 229, 260–1
dwarfing, 52, 114, 116

ear ossicles, 332, 333f; *see also* stapes
East African Great Lakes, *see* rift valley lakes, East Africa
echinoderms, 343–5, 378
ectoderm, 213, 264, 314
Ediacaran fauna, 343, 343f, 346, 348, 384
El Niño, 44, 197f
Eldredge, N., 14, 27–34, 51–4, 56, 60, 86–7, 90, 104, 106, 145–6, 168, 191, 208, 389, 395, 401
electrophoresis, 55, 186, 201, 406
embryonic tissues, 213
endoderm, 213
enamel structure, 99f
Enantiornithes, 316–19, 359
  Cathayornis, 318
  Concornis, 318–22
  Confuciusornis, 316, 317f
  Enantiornis, 319, 319f
  Eoalulavis, 318
  Gobipteryx, 319
  Iberomesornis, 316–18, 317f
  Neuquenornis, 319
  Noguerornis, 316
  Sinornis, 309f, 317f, 318, 320
endemism, 140, 141t, 143
engrailed genes, 224
epistasis, 192, 406
eudiplopodia, 239
Euparkeria, 309f
evolution, *see individual processes;* macroevolution; megaevolution; microevolution; mosaic evolution; phyletic evolution; quantum evolution; racemic evolution; saltational theory of evolution; tiers of evolution
evolutionary change
  speciation and, 28, 54–6, 363–4
  standard deviation as measure of, 75–7, 86, 191, 193–200, 203, 204t
  see also patterns of evolutionary change
evolutionary constraints, 18, 167–79, 406
  chemical, 172–3, 390
  chromosomal, 210
  definition, 169
  developmental, 228, 257, 281–2, 337–8, 340, 390, 401
  genetic, 208–11, 261, 281–2, 337–8, 340, 392, 401
  historical, 169–72, 269, 274, 282, 340, 365, 389
  material, 173–9, 276–7, 389
  physical, 266–95, 340, 365
  see also size
evolutionary rates, 3, 72–80, 121, 132, 202–6, 364, 391, 396
  genetic, 79–80, 210
  morphological, 72–7, 110–12, 210, 256–8, 274, 337
  during radiations, 344, 356–7, 361
  taxonomic, 78–9
  during transitions, 299, 306, 322–4, 336–8, 393
  see also population, size, and evolutionary rate
evolutionary synthesis, 14, 60, 145, 168, 212, 258–62, 395

evolutionary trends, 56, 71–2, 365–8, 396, 400–1
exaptation, 302
extinction, 376
  rates, 379f, 381f, 386, 387f, 401
  see also mass extinctions; pseudoextinction
extraembryonic membrane, 288–9

family longevity, 167t, 167–8
feather tracts, 314
feathers, 279–81, 281f, 306–7, 312–15
  origin of, 313–14, 338
feeding, 40–2, 44–8, 124, 128, 305, 314, 326, 332, 363, 367
Fibroblast growth factor, 238
fish
  transition to amphibians, 300–6
  see also individual taxa
fitness, 182
fluctuating variability, 193
foraminifera, 206, 282, 378–9, 395–6
  Globorotalia, 206, 207f
fossil record
  environmental, geographic, and temporal sampling, 111
  inadequacy of, 25–6, 60–8, 67f, 69f, 118, 297, 298f, 340, 342, 349, 349f, 379, 383–4, 390
  preservation, 12, 17
founder effect, 190, 406
fourth trochanter, 309, 312
frogs, *see* Amphibia, extant taxa, anurans
furcula, 278f, 307, 313, 315, 319f, 320, 322, 337–8

Galápagos
  archipelago, 38–9, 39f, 328
  finches, 38–52, 116, 138, 198, 326, 361, 363; *see also* Darwin's finches
Gasterosteus, 119–23
  aculeatus, 119, 120f, 122, 123f
  doryssus, 121, 121f
  wheatlandi, 122
gastrulation, 213
gene
  activation, 213, 219f
  duplication, 218–19, 224, 258–60, 347, 363
  expression, 213, 225–40, 227f, 234f, 235f, 237f, 241f
  groups, 217f, 218–19, 406
  paralogous, 219, 408
  regulation, 213; *see also* regulatory genes
  see also individual genes; homeobox genes; homeotic genes; *Hox* genes
genetic drift, 86, 189, 189f, 190–2, 203–5, 204f, 406
genetic variability, 183–7, 339, 364, 391
  amount of, 186, 187t, 389
  source of, 183–7
  see also heterozygosity; polymorphism
Geospizidae, *see* Darwin's finches
ghost lineages, 154, 406
glaciation, 82, 84–5, 116–17, 121, 346, 370, 373–4, 399–400
Glossopteris flora, 374
Gnathostomata, 158
Goldschmidt, R. B., 215–16, 262, 391
Gondwanaland, 369, 374
Gould, S. J., 14, 27–34, 52–4, 56, 60, 81, 86–7, 90, 104, 106–7, 145–6, 191, 208, 262, 302, 337, 365–6, 369, 377, 389, 391, 395, 401

graptolites, 396
greenhouse condition, 373, 375f, 406
growth/differentiation factors, 239, 406

hair, 360, 368
haldane (measure of evolutionary rate), 75–7, 406
Haldane, J. B. S., 72, 75
Hamilton fauna, 399
hamstring, 309
Hardy–Weinberg equation, 181–2
Hawaiian Islands, 38, 361
head, development of, 222–5
hearing
    aquatic, 302–3, 328, 331–3, 333f
    terrestrial, 303–4, 368
hemichordates, 222, 259
    Saccoglossus, 222
    Yunnanozoon, 344
Hennig, W., 18, 150–5
heritability, 195–7, 196f, 406
heterochrony, 130, 214–15, 250, 256, 258, 338, 406
heterozygosity, 186, 187t
    measure of, 186
hibernation, *see* dormancy
historical contingency, 365, 406
Holmes, A., 70
homeobox, 216, 407
homeobox genes, 215–19, 258–61, 407
    diverged, 224
homeodomain, 216, 407
homeotic genes, 215–19, 262, 407
homeotic mutants, 215–18, 407
hopeful monsters, 215–16, 407
horses, 58–9, 59f, 69f, 79, 204t, 262, 342f
    Hyracotherium, 58
        aemulor, 76
        grangeri, 76–7
    Mesohippus, 58
    Parahippus, 59
    Pliohippus, 59
    see also Perissodactyla
horseshoe crab, *see Limulus*
Hovasaurus, 245–7, 247f, 248f
Hox cluster, 217f, 218–19, 224, 407
Hox genes, 217–22, 258–61, 264, 347–8, 363, 393, 407
Huxley, T. H., 58, 306–7
hydrodynamic "tongue," 132
Hylonomus, 289f
Hyopsodontidae, 152
Hyopsodus, 203
hypobradytelic, 394
hypocleidium, 316, 318
hyracoids, 154

ice ages, *see* glaciation
icehouse condition, 407
ichthyosaurs, 239, 250, 252, 254–8, 253f–5f, 262, 274, 275f
    Mixosaurus, 253f–4f, 274, 275f
Insectivora, 358
    Blarina
        brevicauda, 88
        carolinensis, 105
insects, 79, 348, 400–1, 402f
insulation, 314
interhyoideus posterior, 294f, 295
intrinsic evolutionary rate, 76, 407

introns, 395
invertebrates, 4, 5f, 26, 27f, 61, 343–8, 383, 389, 395–402; *see also individual taxa*
island isolation, 50, 52
islands, 38, 117, 190; *see also individual islands*
isotopes, in radiometric dating, *see* dating, radiometric
iterative evolution, *see* racemic evolution

jaw mechanics, 124, 134–5, 135f, 170–2, 293, 294f, 305, 361, 366–7

K–T boundary, 353–4, 379, 386
Kaiso, Lake, 140
Kelvin, Lord (William Thomson), 69–70
Kurtén, B., 13, 72, 85, 103, 108, 111

lakes, 50, 65, 119; *see also individual lakes*
Lamarck, J. B., 58
land mammal ages, 82, 83f
large igneous provinces, 374–6, 377f, 407
    Columbia River flood basalts, 375
    Deccan traps, 375, 382
    Ontong Java Plateau, 374–5
    Siberian traps, 375–6, 383
larvae, 288, 304
lateral-line canals, 302–3
Latimeriidae, 168, 297
lava flows, 374
Lepidosauria, 163; *see also individual taxa*
lepidotrichia, 231–5, 407
lift, 276, 312, 315
limb fields, 237–8, 407
limbless, 239
limbs
    development, 227–49, 305, 338, 347–8
    evolution, 227–30, 240–58
    origin, 230–5, 256, 262, 305–6
    reduction and loss, 239, 243, 334
Limulus, 168, 211
Lingula, 168
Linnaeus, 147–8
living fossils, 168, 210, 407
lizards
    terrestrial, transition to mosasaurs, 324–9
    see also Squamata
lobe-finned fish, *see* Sarcopterygii
locomotion
    aerial, 274–82, 306–15
    aquatic, 266–74, 324–5, 327f, 328–31, 333–4, 335f, 366–7
    bipedal, 308–9, 312
    hopping, 284
    terrestrial, 282–5, 305–6
lungfish, *see* Sarcopterygii
Lyell, C., 11, 369

macroevolution, 9, 262–4, 297, 407
    and microevolution, 2, 139–44, 262, 391–2
        distinction between, 8t, 10
    theory of, 146, 391–2
magnetic polarity, 70–2, 84
magnetic-reversal time scale, 71–2, 73f, 83f
magnetostratigraphy, terminology, 71f
Major Features of Evolution, The (Simpson), 60
Malawi, Lake, 124
Mammalia, 161, 162, 386; *see also individual taxa*

mammals, 77, 160, 161f, 168
  Cenozoic, 284, 396
  cursorial, 283–4
  marsupials, 284, 354, 380
    radiation, 358, 360–1
  origin of, 367–8, 374
  placentals, 4, 6f, 66, 155f, 161, 284, 360, 382
    radiation, 352–8, 354f, 355t, 357f, 360–1, 363,
      392
  Quaternary, 53, 82–114
  reproduction, 290, 291f
mammary glands, 360, 366
mantle plume, 370f, 374, 407
Mascarene Islands, 116
mass extinctions, 10, 113–14, 146, 348, 352, 361,
      372, 374, 375f, 376–88, 390, 392–3, 401, 407
  causes, 376, 380–2
  end-Cretaceous, 379–82
  end-Permian, 382–3, 385f
  Late Triassic, 385–6
  measuring magnitude, 383–6
  periodicity, 377–8, 386, 388
Mayr, E., 14, 20–24, 29, 34, 53–4, 60, 76, 191
Mdmg-1, 200
mean, as a measure of quantitative traits, 193,
      407
megaevolution, 391
membranes, transfer of substances across, 286–
      90
Mendelian genetics, 60, 265, 391
meristic changes, 77, 407
mesenchymal condensation, 305
mesenchyme, 232–3, 239–40
mesoderm, 213, 314
mesonychids, 329–30, 332–3, 359
  Pachyaena, 329, 335f
  Sinonyx, 330f, 332
  transition to whales, 329–31
  triisodontine, 152
mesopodials, 236, 250, 254–6, 407
metabolic pathways, 172–3
metabolic rates, 173, 306, 320, 360, 367, 374
metamorphosis, 215–16, 304
metapodials, 236, 407
metazoans, initial radiation; see radiations, Cambrian
      explosion
microevolution, 10, 407
  and macroevolution, 2, 139–44, 262, 391–2
    distinction between, 8t, 10
  see also species, evolution at level of
Milankovitch cycles, 65, 85, 119, 401
Minchenella, 152
mineralized skeletons, 344–6, 361
miniaturization, 290–5
missing link, 24
mitochondria, 173, 178
molecular systematics, 80
molluscs, 4, 140, 344–5
  ammonoites, 61, 378–9, 382, 396
  belemnites, 379
  gastropods, 4, 140, 378, 396
  Littorina, 76
  monophlacophorans, 4
  pelecypods, 4, 140, 378, 396, 401, 402f
  rudists, 382
  scaphopods, 4
  thalassoid forms, 140
  Thiaridae, 140

monophyly, 152–5, 407
  Hennig's definition, 152
  Simpson's definition, 151–2
morphogens, 237–8, 407
morphology, changes in
  rates of, 72–7, 110–12, 210, 256–8, 274, 337
  and speciation, 29, 55, 104–6
mosaic evolution, 91–2, 94, 95f, 408
mosasaurs, 250, 251f, 256, 258, 274, 324–9, 338, 360,
      380, 382
  Clidastes, 325f, 327f
  Halisaurus, 327f
  origin of, 324–9
  Plotosaurus, 327f
  Tylosaurus, 327f
Msx gene family, 224
multifactorial traits, 192–202
muscles, 176–9
  constraints of fibers to elongation and contraction,
      176
  development, 224
  function, 176–8
  metabolism, 178
  molecular constituents, 176, 177f
  pinnate vs. strap-shaped, 178, 178f
  plasticity of function, 179
  strength, 176
  see also individual muscles
mutation, 208
  rate, 183–7, 185t, 209–10, 339, 363
  theory, 181
  see also individual mutations
myoblasts, 224

Nabugabo, Lake, 136
nares, internal, 152, 300, 302
natural selection, 1, 14, 86, 109, 120, 122, 138–9, 145,
      169, 173, 175, 180–1, 198, 202–3, 239, 259, 261–
      2, 265, 281–2, 337–9, 363–5, 376, 389, 401
Neoceratodontidae, 168
Neornithes, 321, 359
  Lithornis, 321f
  Passer domesticus, 36–8
  passerines, 359
  penguins, 374
  Teratornis, 277
  transitional shore birds, 359
neoteny, 214–15, 408
net rate of evolution, 77, 408
neural crest cells, 223–4
normal distribution, 193, 195f
nothosaurs, 250, 251f, 258, 265, 297, 298f

On the Origin of Species (Darwin), 1–2, 19–20, 24,
      26, 58, 145, 286
Opuntia, 41–4
Ornithurae, 316, 319–21
  Ambiortus, 320
  Chaoyangia, 317f, 320
  Enaliornis, 320
  Gansus, 323
  Hesperornithiformes, 320, 359
  Ichthyornis, 321f
  Ichthyornithiformes, 320, 359
  Liaoningornis, 320
  Patagopteryx, 320
  Teratornis, 277
ornithurines, 77

orthogenesis, 59–60
Osborn, H. F., 59
ossification
    endochondral, 254–6
    genetic control of, 239
    perichondral, 239, 254–6
    sequence of, 244–9, 245t, 254–6
    see also bone
Osteichthyes, 25; see also individual taxa
osteoblasts, 174
Osteolepiformes, 235, 256, 300, 338
    Elpistostega, 304
    Eusthenopteron, 152, 231, 233, 273, 300f, 300–1,
        301f, 303f, 304–6
    Panderichthyes, 152, 231, 301f, 303f, 304–5, 338
osteolepiforms, radiation, 350
Owen, R., 58
oxygen increase, 346, 348, 361

paleomagnetic dating, see dating, magnetostratig-
    raphy
Pangaea, 369, 371
Panthalassa, 369
Pantodonta, 356
paraphyly, 152–5, 408
patterns of evolutionary change, at the species level,
    391
    directional, 208, 265, 306, 322–3, 337, 357, 364–5,
        389–90, 396
    disruptive, 208
    iterative, 208
    mosaic, 208, 322–3
    oscillatory, 208, 364
    stabilizing, 167–75, 190, 265
    see also stasis
pelycosaurs, 283, 384
periodontal ligament, 98, 99f
Periptychidae, 152, 357f
Perissodactyla, 149, 154, 329, 356–7, 359; see also
    horses
pharyngeal jaws, see cichlid fish, pharyngeal jaws
phyletic evolution, 21, 53, 62, 81, 90–102, 101f,
    102f, 112, 395, 408
phyletic gradualism, 31, 90, 138, 395–8, 408
phylogenetic definition of taxa, see clades, naming
    and defining
phylogenetic systematics, 18, 150–5, 165–6, 408
phylotypic stage, 220, 221f, 408
Pinnipedia, 172, 333–4, 359, 380
Pistosaurus, 257
Placodermi, 158, 400
plants, see individual taxa; protists; vascular plants
plate tectonics, 11, 65, 339, 370f, 392, 408; see also
    continental drift
Platyhelminthes, 344, 346–7
pleiosaurs, 250, 252f, 256–7, 265, 380, 382
pleiotrophy, 192, 209, 264–5, 408
plesiomorphy, 152, 408
pleurosaurs, 324
polyandry, 188
polydactylous, 239
polydactyly, 202, 239
polygenic inheritance, 192–202
polygyny, 188
polymorphism
    of alleles, 186–7, 190, 363–4
    of Hox genes, 261
    of proteins, 34, 41, 46, 55, 186, 261

of quantitative traits, 202
of regulatory genes, 202, 265
restriction-fragment-length polymorphism, 126
see also genetic variability; heterozygosity
polyphalangy, 250, 252, 252f, 253f, 254f, 255f
polyploidy, 23
population
    growth, 192, 348, 362, 364, 391
    size, 60, 188
        effective, 188, 203
        and evolutionary rate, 20, 188–92, 203–5, 211
    variability of, 20, 210
population genetics, 60, 181–3, 391, 393
postaxial dominance, 242
posture, erect vs. sprawling, 283–4, 285f
preadaptation, 302
premature death mutation, 223
Primates, 78f, 356
    Cantius, 73
    Homo, 113f, 214
        erectus, 112
        sapiens, 203
    Neanderthals, 24
Principles of Geology (Lyell), 369
proboscidians, 52, 58, 98, 112, 154, 172, 262, 356–7
    Mammuthus, 52, 95, 97f, 112, 113f
processes of evolution, see dispersal; extinction; ge-
    netic variability; mutation; population, growth;
    selection; speciation; stochastic processes
procoracoid process, 319f, 320
progress zone, 238, 408
prokaryotes, 362, 394, 403–4
Proterosuchus, 249f
proteutherians, 356
protists, 389, 395, 403–4
Protoavis, 308
protostomes, 344–7
Protungulatum, 154
pseudoextinction, 384–5, 393
pterosaurs, 257, 277, 278–9, 278f, 280f, 308, 380,
    382, 386
    Eudimorphodon, 280f
    Pteranodon, 277, 280f
    Quetzalcoatlus, 277
pubic foot, 312, 318, 320
punctuated equilibrium, 14, 18, 27–33, 56, 62, 64,
    81, 145, 149, 395, 408
    tests of, 85–110
punctuated gradualism, 206, 395
pygostyle, 317f, 317

qualitative traits, 180–92, 194f, 408
quantitative traits, 192–202, 194f, 239, 392–3, 408
    change in, see variance
    nature of, 202
    number of genes contributing to, 200–1, 201t
    rate of accumulation, 201
    response to selection, 196–200
quantum evolution, 60, 296, 391, 408

racemic evolution, 122, 409
radiations, 10, 143–4, 340–61, 390, 393
    actinopterygians, 350, 361
    amphibians, 350–1
    Cambrian explosion, 4, 263, 341, 343–8, 343f, 345f
    capacity vs. opportunity, 360–1
    Cenozoic birds, 359, 361
    Chondrichthyes, 350, 361

radiations (*cont.*)
  cichlids, 123–44, 147f, 357, 361, 392
  crocodiles, 351, 352f
  dinosaurs, 351, 353f, 361
  marsupials, 358, 360–1
  Mesozoic marine reptiles, 351
  osteolepiforms, 350
  placental mammals, 352–8, 354f, 355t, 357f, 360–1, 363, 392
  synapsids, 351, 361
  vertebrates, pattern of, 17f
Radinskya, 152
radiolarians, 206, 282, 395
  Eucyrtidium, 206, 207f
random walk, 86, 88, 122
rate-limiting proteins, 172
rates of evolution, *see* evolutionary rates
ray-finned fish, *see* Actinopterygii
regulatory genes, 202, 215, 224, 260–2, 393, 409
reproduction, 337
  amniotes, 288–9
  amphibians, 288–9, 304
  cichlids, 132, 136
  constraints on, 288–9, 328, 337
  mosasaurs, 328
  sarcopterygians, 304
  whales, 331
reproductive isolation, 22, 28, 55, 136–7
reptiles, 115–18, 383, 396
  Mesozoic marine, 250–6, 351; *see also individual taxa*
  as a paraphyletic group, 159–60, 161f
Reptilia, 159; *see also individual taxa*
respiration, 286–9, 306, 320, 366–7
response to selection, 196–7, 364–5, 409
retinoic acid, 238
rift valley lakes
  East Africa, 55, 65, 123–39, 125f, 143, 361
  Eastern North America, 65, 142, 361
ring species, 53
RNA, 80, 210
Rodentia
  Arvicola, 100
    cantina, 90–1
    sapidus, 102f
  arvicolids, 104
  Cosomys primus, 88–9, 89f
  cricetids, 98, 104
  Cricetulus bursae, 88
  Cynomys, 104
  Lagurus curtatus, 90
  microtines, 89, 98
  Microtus, 104
    pennsylvanicus, 88, 91–2, 93f, 94, 112
  Mimomys, 100, 102f
  mouse, 76, 201–2, 230
  muskrats, 98, 100, 101f, 103
  Ondatra zibethicus, 100, 101f, 103
  Peromyscus, 104, 205, 205t
  Pliolemmus antiguus, 89
  Pliomys episcopalis, 89
  pocket gophers, 99f
  Reithrodontomy, 104
  Rhinocrietus ehiki, 88
  Sciurus, 168–9
  Sigmodon, 104, 107–8
  Spermophilus townsendii, 88
  Ungaromys spp., 89

rodents, 98, 112, 262
  dentition, 98–102, 99f, 101f, 102f

salamanders, *see* Amphibia, extant taxa, urodeles
saltational change, 28, 59, 262–3, 296
saltational theory of evolution, 59–60
Sarcopterygii, 159
  Coelocanthini, 79, 159, 168
  Dipnoi, 159, 234f, 235
  see also individual taxa
Sauriurae, 316
Sauropsida, 163
Schindewolf, O. H., 59
sea-level changes, 112, 372, 375f, 399–400, 404
  and dispersal and/or extinction, 372
  regression and transgression, 84, 380, 383
sedimentation, 63f
  gaps in/irregularity of, 61–5
  rates of, 61–5, 64f, 88
selection, 182, 187–8, 192, 208–9; *see also* artificial
  selection; natural selection; patterns of evolutionary change; sexual selection
selection coefficients, 182–3, 183f, 187–8, 188f,
  202–8, 210, 357, 363, 389, 393, 409
  variability of, 208, 393
selection differential, 196–8, 198f, 205, 409
  variability of, 208
semicircular canals, 291, 292f
sense organs, size as constraint to cranial changes,
  290–5
sensory placodes, 223–4
sexual selection, 45, 55
shifting balance theory, 190–2, 409
short ear, 239, 261
sibling species, 55, 409
Simpson, G. G., 1, 13, 15, 18, 20, 23, 29, 51, 60–1,
  72, 76, 78–9, 87, 152, 156, 191, 216, 296, 390–2
Sirenia, 150, 154, 172, 359, 380
sister-group, 151, 409
size
  as constraint
    to flight, 276–7, 314
    to respiration, 287–8
    to terrestrial locomotion, 283–5
    see also sense organs
  and prey size, 315, 326, 368
  and rate of speciation, 108–9, 109f
  and reproduction
    amniote, 288–9, 288f
    amphibian, 287–8
    mammalian, 290, 291f
  and temperature control, 289–90, 314
  see also brain size
small shelly fauna, 344, 384
Sonic hedgehog, 238
span, 267–8, 409
speciation, 12, 21–4, 54–6, 125, 136–7, 149, 165,
  326, 363–4, 389, 391–2, 409
  allopatric, 22, 28, 405
  association with morphological change, 29, 55,
  104–6
  definition, 21
  as a force of evolutionary change, 28, 54–6, 363–4
  peripatric, 23, 28, 408
  rates, 108, 136
  sympatric, 23–4, 409
species, 19–33, 149
  definition, 21, 24, 100

evolution at level of, 14, 137̄–9, 145, 149, 208, 362–5, 389, 391–2, 395–6
  patterns of, *see* patterns of evolutionary change, at the species level
  extinction, 376
  longevity, 14, 35, 112, 356, 396, 403
  number of, 1, 12, 15, 125, 176, 367
  recognition, 106
  sorting, *see* species selection
  typological concept, 149
  see also ring species; sibling species; subspecies
species flocks, 126, 140, 142–3, 409
species selection, 20, 30, 81, 106–10, 107f, 377, 409
Sphenodon, 161f, 168–9, 245, 248, 257
Sphenodontidae, 168, 297, 386
sponges, 346, 348
sports, 193
squalene, 269
Squamata, 116–18, 245, 249f, 257, 283, 290, 304, 382, 396
  Aigialosauridae, 324–6, 325f, 327f, 328–9, 338, 360
  Amblyrhynchus, 328, 363
  Anguimorpha, 324–6
  Anolis, 118
  Conolophus, 328
  geckos, 288f
  Lacerta agilis, 23
  Megalania prisca, 326
  Mosasauridae, 324; *see also* mosasaurs
  Varanus, 285f, 325f
    komodoensis, 326
standard deviation, 193–5, 409
  definition, 194–5
  as measure of evolutionary change, 75 7, 86, 191, 193–200, 203, 204t
Stanley, S. M., 30, 86, 106–7, 109, 146, 149, 168
stapes, as ear ossicle vs. structural element, 303–4
stasis, 28, 31, 51, 56, 62, 81, 86–90, 116, 119, 123, 142, 145, 168, 172, 209–11, 257, 278, 281, 284, 364, 394–6, 398–400, 403–4
  Darwin's view of, 20, 26
  definitions, 86–7
statistical tests
  ANOVA, 88, 94
  multivariate analysis of variance, 205
  nonparametric runs test, 88
  random walk test, 88
  Spearman rank correlation, 342
  Tukey's HSD test, 94
stem group, 162, 163f, 409
sternum, 278f, 285f, 313, 315, 317–20, 319f, 322–3, 337–8
stickleback fish, 119–23, 208; *see also Gasterosteus*
stochastic processes, 180, 188–92; *see also* genetic drift; random walk
stratigraphy, 25–7
stylopod, 236, 409
subspecies, 24, 56
supracoracoideus muscle and tendon, 319f, 320
survivorship curves, 402
sweat glands, 368
swim bladder, 176, 366
symbiosis, 209
sympatry, 22, 409
symplesiomorphy, 151, 409
synapomorphy, 151, 409
Synapsida, 159, 342f, 360
  Tetraceratops, 351

synapsids, radiation, 351, 361
synsacrum, 318
systemic abnormalities, 202
systemic mutations, 215

tachytely, 391
Taeniodonta, 356
tail shape, 267–9, 268f, 272
  heterocercal, 269, 271–3
  lunate, 268f, 274
Taiwan, 117
Tanganyika, Lake, 124, 139–41, 141t
tarsometatarsus, 316, 320
Teleostomi, 159, 161
temperature, climatic
  changes, 83f, 84–5, 111, 399–400
  means of measuring, 84
temperature control, 289–90
  birds, 314
  dinosaurs, 290, 314
  lizards, 290
  small mammals, 290
  snakes, 173
  tuna, 173
Tempo and Mode in Evolution (Simpson), 60
Tethys Sea, 369
Tethytheria, 154
Thadeosaurus, 245–7, 247f
thecodonts, 307, 386
therapsids, 351, 384
thyroxin, 214–15
tiers of evolution, 377, 401
tracks and burrows, 343, 345
transitions, 296–339, 391, 393, 400–1
  cephalochordate–craniate, 223–4, 363
  dinosaur–bird, 306–15
  fish–amphibian, 300–6
  mesonychid–whale, 329–31
  primitive actinopterygian–teleost, 366–7
  primitive amniote–mammal, 367–8
  terrestrial lizard–mosasaur, 324–9
transposons, 201
Tribolium, 189, 189f, 201
trilobites, 344, 396
  Cnemidopyge, 397–8, 397f, 398f, 399f
  Phacopinae, 30–1, 30f, 31f
  phyletic evolution, 397–9, 397f
  stasis, 31
  stratigraphic ranges, 27f, 31f
triosseal canal, 319f, 320
trophic differentiation, 124–5, 131, 133f
truncation mortality, 203, 205
Tubulidentata, 359
Turkana, Lake, 131
turtles, *see* Chelonia
tympanic annulus, 303

Ultrabithorax, 261
uncinate processes, 320–1
ungulates, 150, 153f, 358
  South American, 150, 356–7
  see also individual taxa
uniformitarian theory of geology, 11, 369

variability, *see* genetic variability; polymorphism
variance, as measure of change in quantitative traits, 193–6, 409
varved lake deposits, 65, 66, 119, 121

vascular plants, 4, 6, 7f, 8, 16, 79, 380, 383,
    402–3
vertebrae, development of, 225–7, 265
vertebrates
    classification, 155–60, 156f, 157f
    as model of evolutionary patterns and processes,
        14–16, 394
    pattern of radiation, 17f
    see also individual taxa
Victoria, Lake, 124
volcanic eruptions, 374, 382

West Indies, 116–17
whales, see Cetacea
wing loading, 276, 315, 409
wing span, 276, 277f
Wright, S., 190–1

Zanthos of Sardis, 58
Zea mays, 198–9, 199f
zebra fish, see Actinopterygii, Danio
zeugopod, 236, 409
zone of polarizing activity (ZPA), 238, 263, 409